Practical Medical Microbiology for Clinicians

Practical Medical Microbiology for Clinicians

Frank E. Berkowitz

Robert C. Jerris

WILEY Blackwell

Published by John Wiley & Sons, Inc., Hoboken, New Jersey

Published simultaneously in Canada

Library of Congress Cataloging-in-Publication Data:

Berkowitz, Frank E. (Frank Ellis), 1948– , author.
Practical medical microbiology for clinicians / Frank E. Berkowitz, Robert C. Jerris.
p. ; cm.
Includes bibliographical references and index.
ISBN 978-1-119-06674-3 (pbk.)
I. Jerris, Robert C., author. II. Title.
[DNLM: 1. Microbiological Phenomena. 2. Microbiological Techniques. QW 4]
QR46
616.9'041–dc23

2015028776

Cover image courtesy Robert Jerris

10 9 8 7 6 5 4 3 2 1

MIX
Paper from responsible sources
FSC® C013604
FSC www.fsc.org

This book is dedicated to our wives, Zahava and Kerry.

Contents

Preface

Microorganisms cause a large proportion of human disease. Newly recognized organisms and newly recognized diseases caused by known organisms continue to be reported. Yet less time in medical school curricula is devoted to microbiology than previously.

The goal of this book, therefore, is to give clinicians, practicing in all branches of medicine, an insight into microorganisms that infect their patients, how these organisms are related to one another, what takes place in the microbiology laboratory to isolate and identify them, and how they can best utilize the laboratory for the benefit of their patients. It is designed to give clinicians the knowledge to facilitate their communication with the microbiologist in the laboratory. The approach is systematic, with a description of taxonomy of key (but not all) human pathogens and, for the most part, consideration of organisms within taxonomic groups. The emphasis of the book is not so much on the biology of the organisms, but rather on their epidemiology, and the use of the laboratory in managing individuals infected with them. It describes microorganisms and the diseases they cause, but it is not intended as a book about infectious diseases and their management.

Acknowledgments

We want to thank the following individuals who helped to make the writing of this book possible: the staff of the microbiology laboratory, Children's Healthcare of Atlanta at Egleston, who helped gather material for many of the illustrations: Charles Ash, Theresa Stanley, Kathy Shauger, Becky De Ridder, Salome Tesfay, Mona Dillard, Scott Brown, Heather MacDonald, and Danielle Ingebrigtsen; the fellows in Pediatric Infectious Diseases, Emory University School of Medicine, for their assistance in obtaining some of the pictures, and for their stimulating questions; Carlos Abramowski MD, who provided several histologic pictures, and for his enthusiasm in teaching us the histology of infections; Satyen Tripathi, who helped with some of the illustrations; and Joni Lewis, who helped to prepare the manuscript.

SECTION I

Laboratory methods in clinical microbiology

CHAPTER 1

Introduction

Taxonomy

There are different methods for classifying or grouping microorganisms, for example based on genetic relatedness, on phenotypic features, on epidemiologic characteristics, or on clinical effects. In this book, the genetic relatedness is used for taxonomy in most circumstances. Five main categories are used: prions, viruses, bacteria, fungi, and parasites, and within each (except for prions), there are several different subcategories. The value of classifying and naming organisms is as follows.

- Names carry information about pathogenesis, epidemiology, and antimicrobial susceptibility.
- A systematic approach might assist in constructing a microbiological differential diagnosis for solving a clinical problem.

The classification of microorganisms causing disease in humans is shown in Table 28.1 of the Appendices. Classification of organisms is also shown for each chapter or section.

Largely as a result of advances in genetics and the consequent ability to better classify organisms, taxonomy and nomenclature are changing rapidly. The following websites offer the most up-to-date classifications and nomenclatures of microorganisms.

- Viruses: http://ictvonline.org; www.ncbi.nlm.nih.gov/ICTVdb/chars.htm
- Bacteria: www.bacterio.net/-alintro.html
- Fungi: Mycobank (www.mycobank.org) and Index Fungorum (www.indexfungorum.org/)
- Parasites: www.cdc.gov/dpdx

Practical Medical Microbiology for Clinicians, First Edition. Frank E. Berkowitz and Robert C. Jerris.

Purposes of the clinical microbiology laboratory

The purpose of the clinical microbiology laboratory is the detection and identification of microorganisms, susceptibility testing of isolated organisms to antimicrobial agents and, in some circumstances, the quantification of the number of organisms in body fluids.

Principles of diagnostic testing

Diagnostic testing can be used for clinical purposes (patient management), epidemiologic purposes (recognition of disease patterns, including trends and outbreaks), and for research. The following discussion applies primarily to testing for clinical purposes.

A diagnostic test should be considered when its results may help in deciding about a patient's management. A patient's clinical features may be so suggestive of the diagnosis, and the withholding of treatment may be so deleterious, that you would give therapy without any further ado (or diagnostic testing). For this patient, your belief in the probability of the diagnosis is above a threshold, which is called the **test-treat threshold** (Fig. 1.1).

Another patient's clinical features may not be highly suggestive of the diagnosis, and withholding therapy may not carry a significant penalty. In this case, your belief in the probability of the diagnosis is so low that you think that neither testing nor treatment is appropriate. The probability of the diagnosis is below a threshold called the **no test-test threshold** (see Fig. 1.1).

Therefore deciding about diagnostic testing requires an appreciation of the following probabilities.

- The probability (what you believe to be the probability) of the diagnosis before the test is performed. This is called the **pretest probability**.

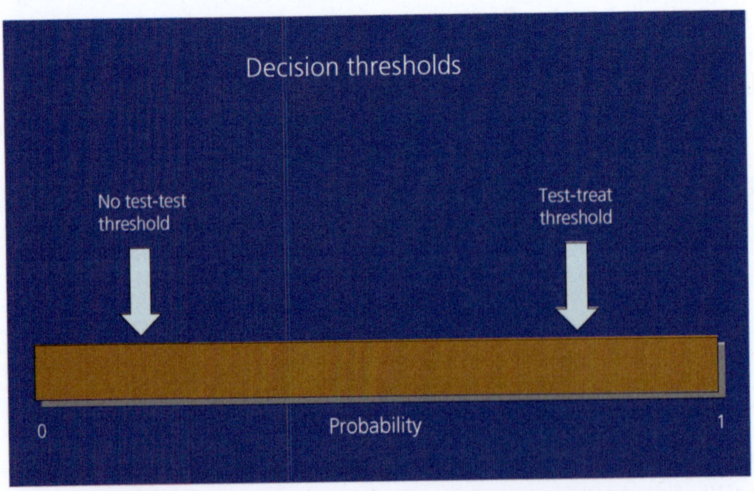

Fig. 1.1 Decision thresholds.

- The probability of the diagnosis above which you would treat the patient, irrespective of the results of a diagnostic test (test-treat threshold).
- The probability of the diagnosis below which you would not treat, irrespective of a test result (no test-test threshold).

Thus there are three zones of probability regarding treating and testing.

- Probability below the no treat-test threshold: NO ACTION.
- Probability between the two thresholds: TEST.
- Probability above the test-treat threshold: TREAT.

How do we know what the probabilities should be for these thresholds? These are determined by the benefits of treatment of patients with the disease (diagnosis), and the harm inflicted by treatment of the non-diseased as well as the diseased, and the harm inflicted by the test itself.

Each test has parameters of performance. For many blood tests, these are known, for example, as determined by the manufacturer or developer of the test. For some, these parameters are not really known, especially imaging tests.

Clinicians are interested in parameters called **sensitivity** and **specificity**. These are demonstrated in Table 1.1. This table is commonly used, and readers should become very familiar with it. The columns indicate the TRUE state of the patients (disease or no disease); the rows indicate the test results (test positive or negative).

Sensitivity means:

- the proportion of patients who really have the disease who have a positive test (a/a + c); this is also called the **true-positive rate** (TPR)
- (c/a + c) is the proportion of patients who really have the disease but who have a negative test; this is the **false-negative rate**, and is (1 – sensitivity).

Specificity means:

- the proportion of patients who really do not have the disease who have a negative test (d/b + d); this is also called the **true-negative rate** (TNR)
- (b/b + d) is the proportion of patients who really do not have the disease, but who have a positive test. This is also called the **false-positive rate**, and is (1– specificity).

In clinical medicine, the question of interest is often as follows: If the test is positive, what is the probability of the patient having disease or, if the test is negative, what is the probability that the patient does not have disease? These are the predictive values.

Table 1.1 Structure of a table used to determine the diagnostic parameters and interpretation of diagnostic tests.

	Disease	No disease	
Test positive	a	b	a + b
Test negative	c	d	c + d
	a + c	b + d	a + b + c + d

a = **true positives**; these are the cases in which the patient HAS the disease AND the test is POSITIVE.
b = **false positives**; these are the cases in which the patient DOES NOT have the disease BUT the test is POSITIVE.
c = **false negatives**; these are the cases in which the patient HAS the disease BUT the test is NEGATIVE.
d = **true negatives**; these are the cases in which the patient DOES NOT have the disease AND the test is NEGATIVE.

- The positive predictive value, i.e. the probability of disease if the test is positive, is a/a + b.
- The negative predictive value, i.e. the probability of no disease if the test is negative, is d/c + d.

These are determined not only by the sensitivity and specificity of the test, but also by the prevalence of the disease in the population from which the patient is drawn, or the pretest probability.

Sensitivity

Let us start with an example.

Example 1

You want to establish a test for screening blood donors for a viral infection. The donors are asymptomatic for the infection. You want to eliminate, to the best of your ability, any chance that infected blood could enter your donor pool, even if it means rejecting blood that actually might be fine. Therefore you want a very sensitive test. Such a test should detect everyone who has the infection (even if it means calling someone infected if they are not really infected). This means you want to minimize the number of **false negatives**. Conversely, the **true-positive** rate (sensitivity) is very high.
If the test is negative, there is no disease. A sensitive test is used to "**rule out**" a disease. **SeN**sitivity is to rule **OUT** (SNOUT) (Fig. 1.2).

Specificity

Example 2

You want a test to test for an illness, for example a cancer, for which therapy is very toxic. You do not want to be giving toxic therapy to someone who does not really have the disease. Thus you want to eliminate, to the best of your ability, any chance of making the diagnosis of this disease in someone who does not really have the disease (**false positives**). That means you want a test with a very high **true-negative** rate (specificity).
If the test is positive, there is disease. A specific test is used to "**rule in**" a disease. **Sp**ecificity is to rule **IN** (SPIN) (Fig. 1.3).
Diagnostic testing is like fishing with a net.

Example 3

Scenario: You want to catch large fish (3–5 cm across). If you use a net with small holes (2 cm across), you will catch all the large fish. However, you will also catch small fish that you do not want. This is analogous to using a sensitive test that is not specific. You will catch all the cases that you want (the large fish), but you will also catch cases that you do not want (the small fish) (Fig. 1.4).
On the other hand, if you use a net with larger holes (4 cm), you will not catch any small fish, and you will catch most large fish, but you will also miss some of the large fish that you do want. This is analogous to using a specific test that is not sensitive (Fig. 1.5).
There is always a tension between the sensitivity and the specificity of tests. As the sensitivity increases, the specificity decreases, and vice versa (Fig. 1.6).

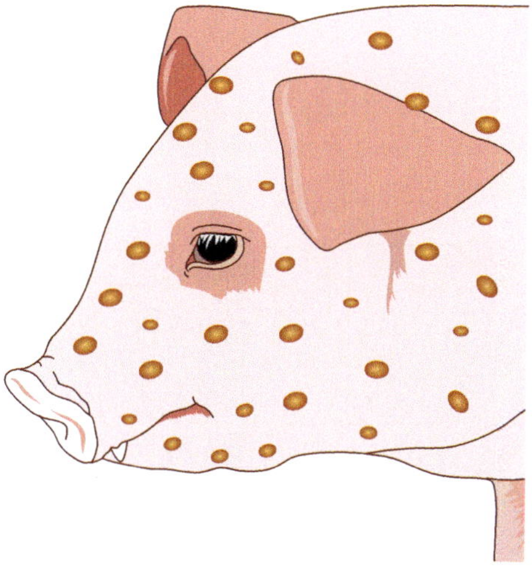

Fig. 1.2 SNout for Sensitivity.

Fig. 1.3 SPin for specificity.

How do we know the true state (disease or no disease)?

As can be seen in Table 1.1, determining the sensitivity and specificity of a test depends on knowing the patient's **true state**, that is, is disease present or not? The method by which the true state is determined is often referred to as "**the gold standard**." This is the method which is often "accepted" as the definitive way to make the diagnosis. Because the parameters of a new test are dependent on the gold standard, the dependability of the gold standard is of the utmost importance. Unfortunately, attainment of a suitable gold standard may be difficult, and there are several potential pitfalls in studies of diagnostic tests in which a suitable gold standard is not used.

In the microbiology laboratory, the gold standard has, for many years and in many circumstances, been culture of the microorganism. The disadvantages of this are the following.

- An organism may grow poorly in culture, or not at all, e.g. *Treponema pallidum*, the cause of syphilis. (There may be many organisms that are unknown because they cannot be cultured in artificial media.) This reduces the **sensitivity** of culture.

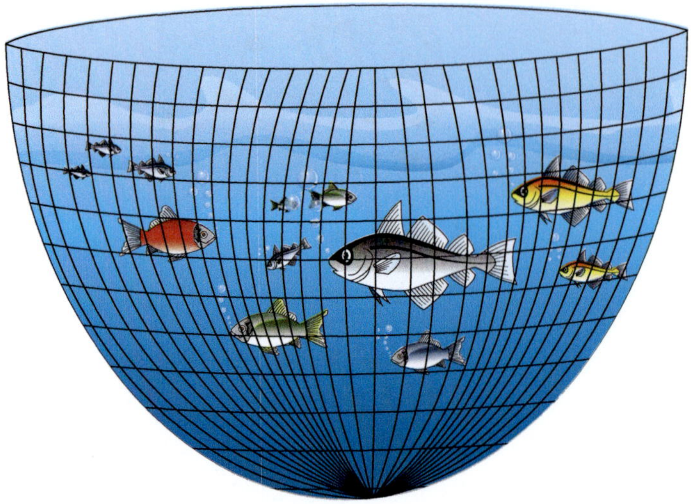

Fig. 1.4 A sensitive "net."

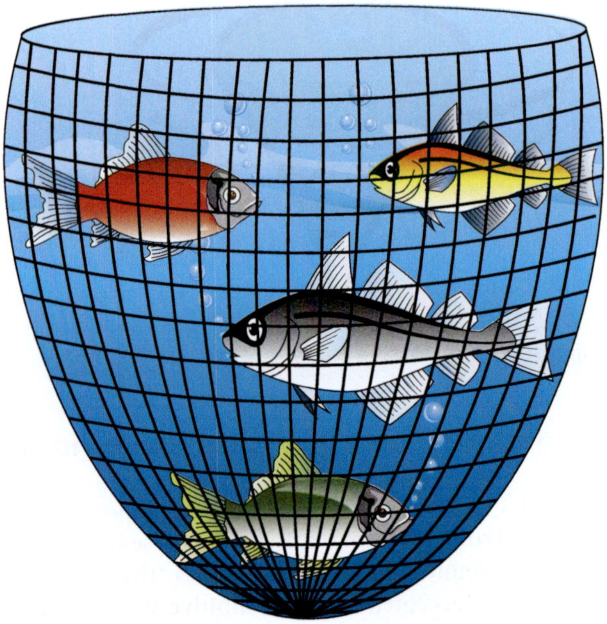

Fig. 1.5 A specific "net."

- Although an organism may be cultivable, culture may take many days or weeks, which might not be practicable for clinical medicine, e.g. *Mycobacterium tuberculosis*.
- Because specimens are taken from sites that might harbor organisms other than the pathogen of interest, culture might detect an organism that is a "contaminant." This reduces the **specificity** of culture.

Therefore, in many circumstances, molecular tests have become the gold standard (see Chapter 2).

Fig. 1.6 The tension between sensitivity and specificity.

Microbiologic tests can be narrow spectrum, i.e. specific for a single organism (e.g. polymerase chain reaction, serologic tests, antigen detection tests), broad spectrum (e.g. culture in medium supporting the growth of many different organisms), or intermediate in spectrum, i.e. able to detect a limited number of organisms (e.g. blood smear).

When considering microbiologic testing, the following should be borne in mind.

- The general medical differential diagnosis.
- The microbial differential diagnosis.
- How knowing whether there is an organism and what it is will help in patient management (for therapy, for withholding therapy, or for public health measures such as isolation of the patient).
- To what level of specificity (i.e. genus, species, serotype, strain) an organism's identification should be made.

IMPORTANT: to make a microbiologic diagnosis, you need specimens appropriate for microbiologic testing.

Antimicrobial resistance

The ability of pathogenic microorganisms to resist the effects of antimicrobial agents, antimicrobial resistance, is a very important and challenging problem in clinical medicine. Although the molecular mechanisms vary according to the different categories of organism (discussed separately within each category), the basic principles are the same. The measure of susceptibility of an organism is determined, generally, by allowing the organism to grow, in culture, in a medium containing varying concentrations of the antimicrobial agent. The lower the concentration that inhibits the growth of the organism, the more susceptible the organism is to that agent. For bacteria and fungi, the measure used is the minimal inhibitory concentration (MIC), while in viruses and parasites the measure usually used is the inhibitory concentration $_{50}$ (ID_{50}), the concentration that causes 50% inhibition of growth (see Chapter 2 on laboratory methods). There are conceptually two types of resistance: microbiologic resistance, meaning that the organism is more resistant than other members of its species; and clinical resistance, meaning that the organism is resistant to concentrations of the drug that can be safely achieved in the infected tissue.

Resistance is, ultimately, determined by the genetic attributes of the organism. Some organisms are inherently resistant and, to our knowledge, have always been resistant to certain agents. This is sometimes called "native resistance." Other organisms have acquired resistance over time since the antimicrobial agent has been in existence (prior to its existence, one could not have demonstrated susceptibility or resistance). The ability to acquire resistance depends on the organism undergoing a genetic change. This can occur by mutation or acquisition of new genetic material (discussed in the section on antibacterial resistance in Chapter 9). The frequency of mutations varies among different organisms. However, because microorganisms generally have very short generation times compared with that of their hosts, mutations can occur relatively frequently.

Once an organism has become resistant to an antimicrobial agent, it can become prevalent within a population of organisms by two processes.

- Darwinian selection: in circumstances in which the relevant antimicrobial agent is present in the organism's environment, the susceptible organisms are inhibited or killed, while the resistant ones multiply and thrive, and eventually become the predominant or only population (Fig. 1.7).

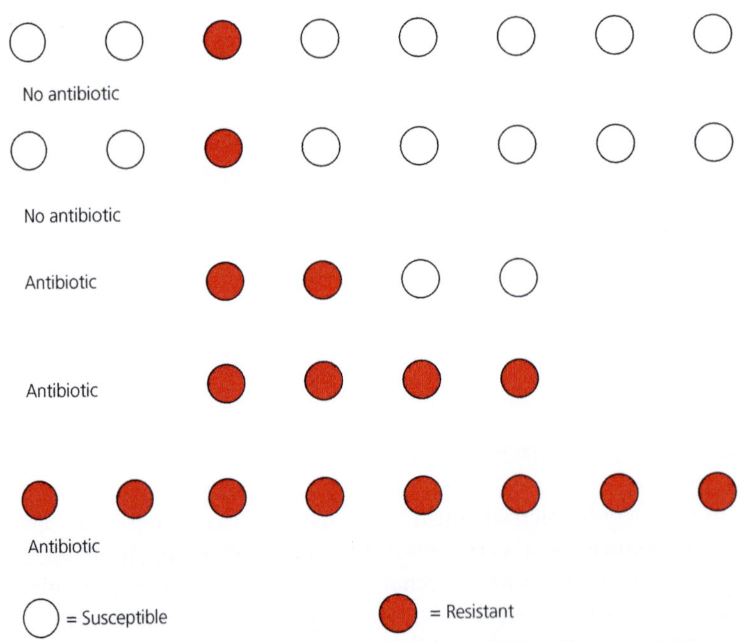

Fig. 1.7 How exposure to an antibiotic results in the resistant organisms becoming the dominant organisms and then the only organisms.

- Resistant organisms spread to new areas: this occurs via the same routes by which susceptible organisms spread, e.g. by personal contact, by droplets, by the airborne route, or by arthropod vectors. In hospitals, where there is a high prevalence of resistant organisms, the hands of healthcare workers are an important mode of spread.

Further reading

Baron EJ, Miller JMM, Weinstein M, et al. (2013) A guide to utilization of the microbiology laboratory for diagnosis of infectious diseases: 2013 recommendations by the Infectious Diseases Society of America (IDSA) and the American Society for Microbiology (ASM). Clin Infect Dis 57: e22–e121.

Fletcher RH, Fletcher SW, Wagner EH (1988) *Clinical Epidemiology. The Essentials*, 2nd edn. Baltimore: Williams and Wilkins.

Hunink M, Glasziou P, Siegel J. et al. (2001) *Decision Making in Health and Medicine. Integrating Evidence and Values*. Cambridge: Cambridge University Press.

Miller JMM (1996) *A Guide to Specimen Management in Clinical Microbiology*. Washington DC: American Society for Microbiology Press.

Sox HC, Blatt MA, Higgins MC, Marton KI (1988) *Medical Decision Making*. Boston: Butterworth-Heinemann.

CHAPTER 2

Microbiology laboratory methods

Reasons for making a microbial diagnosis

In medical microbiology laboratories, a lot of time (and money) is spent on detection and identification of microorganisms. Why is this important?

In short, **NAMES CONTAIN INFORMATION**. The identification contains the following very important information.

- Epidemiologic: where in the world the organism might have come from. Is there an outbreak caused by this organism? Potential for spread to other individuals.
- Clinical: the anatomic source of the organism, and the possible underlying disease of the host.
- Antimicrobial susceptibility information and optimal therapy.

For example, an isolate of *Escherichia coli* (*E. coli*) and *Enterobacter cloacae* might have very similar susceptibility patterns, e.g. resistance to ampicillin and first- and second-generation cephalosporins, and susceptibility to third-generation cephalosporins. However, *Enterobacter cloacae* is known to produce inducible broad-spectrum β-lactamases, which should make one wary of using cephalosporins for treating patients with infections caused by this organism. This is not the case with *E. coli*.

The clinical microbiology laboratory is a dynamic, ever-changing entity. It constantly adapts to change as procedures are balanced between outcomes for patient care, test complexity, and cost. Conventional microbiologic techniques of culture and subsequent organism identification are slowly being replaced by non-culture methodologies. This chapter will briefly describe methodologies currently extant in the laboratory.

Basic methods used in microbiology

To make a microbiologic diagnosis, you need specimens appropriate for microbiologic testing.

Practical Medical Microbiology for Clinicians, First Edition. Frank E. Berkowitz and Robert C. Jerris.

Detection and phenotypic identification

These methods are based on observational studies of an organism's physiologic and/or metabolic characteristics. They include microscopic staining morphology, macroscopic growth (colony morphology), environmental growth requirements, nutritional requirements, metabolic capacities, and, in some cases, resistance/susceptibility to antimicrobial agents.

Some organisms require very few tests for identification (e.g. catalase and coagulase to identify *Staphylococcus aureus*), while others require a full battery of tests. The number and type of tests depend on the class of organism to be identified. For most organism groups, identification is achieved using commercial "kit" systems that may detect preformed enzymes (results in a matter of hours) or metabolic use of substrates (generating colorimetric or turbidimetric endpoints detected after overnight incubation).

Direct visualization

The naked eye is adequate to visualize large organisms such as worms and the colonies produced by millions of bacteria or fungi. However, to visualize individual bacteria, fungi, or protozoa, one must use a microscope, with a magnification of at least 400×. For adequate visualization of stained bacteria, a magnification of 1000× is necessary. This requires use of an oil immersion lens. For visualizing viruses, an electron microscope is necessary.

Wet preparation

A drop of the specimen of fluid to be tested is placed on a microscope slide and a cover slip placed on top. Some specimens, e.g. stool or vaginal fluid, should be mixed with a drop of saline, and then placed on the slide. The following can be seen.

- Leukocytes
- Erythrocytes
- Bacteria (sometimes their shape and motility can be determined)
- Fungi
- Protozoa, e.g. *Trichomonas vaginalis* (motile), *Giardia intestinalis*
- Parasite ova

Parasites and ova can be stained in the wet preparation, e.g. with iodine, which enhances one's ability to see them. Lowering the condenser of the microscope can also facilitate this.

Stained preparations

For the detection of many bacteria, fungi, and protozoa, wet preparations are neither adequately sensitive nor discriminating. Detection and discrimination are vastly improved by the use of stains. The most useful stain, by far, is the Gram stain, developed by the Danish microbiologist Hans Gram in 1882.

Gram stain

The microscope slide, on to which the specimen has been smeared, is heated briefly for fixation. Then the staining solutions are dripped on sequentially, with a water wash between each step.

- Crystal violet 10–60 seconds
- Iodine 10–60 seconds (mordant step)

- Alcohol (10 seconds) or acetone alcohol (2 seconds) (decolorizing step)
- Safranin 60 seconds (counterstain step)

Organisms staining blue or purple with this stain have retained the crystal violet after the decolorizing step: they are called Gram positive (Fig. 2.1). Those staining red or pink are called Gram negative. Gram-negative bacteria, which contain more lipid in their cell walls, do not retain the crystal violet, and are stained by the red counterstain (Fig. 2.2).

The division of bacteria according to their Gram-staining properties is widely used in the taxonomy of bacteria.

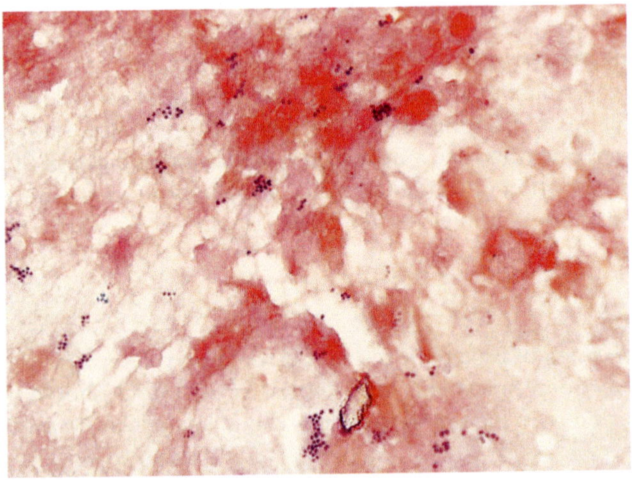

Fig. 2.1 Gram-positive cocci in pus.

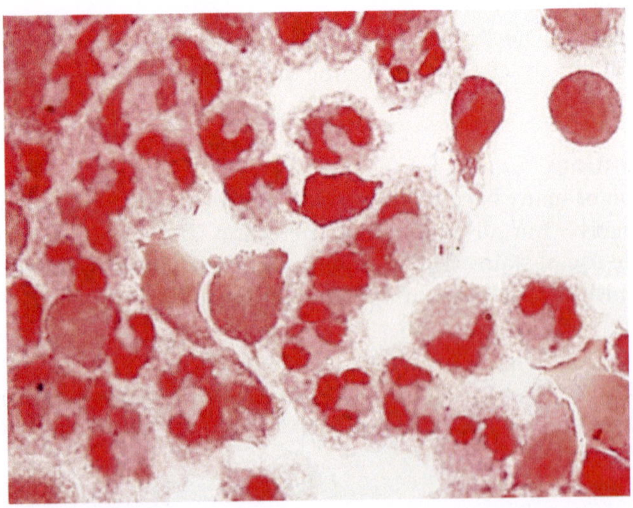

Fig. 2.2 Gram-negative rods in CSF.

The values of the Gram stain are as follows.

- It provides some degree of identification of organisms, not merely their detection.
- It is semi-quantitative (it requires a concentration of about 10^5 bacteria per mL of fluid to see bacteria in a Gram-stained preparation under 100× objective (magnification of 1000×).
- It is the **ultimate** rapid diagnostic microbiologic test.

Pitfalls in reading a Gram stain

Color If the slide is underdecolorized, Gram-negative bacteria will appear Gram positive, and the converse will occur if the slide is overdecolorized.

- Gram-positive bacteria that are "sick" due to antibiotic effects, or from old cultures, may appear Gram negative.
- Some bacteria may appear "Gram variable," e.g. *Clostridium* spp., *Acinetobacter* spp.
- Some bacteria do not stain with the Gram stain, e.g. mycobacteria, mycoplasmas.

Shape Some bacteria are small Gram-negative rods, and may have the appearance of slightly elongated cocci. They are sometimes referred to as cocco-bacilli, e.g. *Haemophilus influenzae, Acinetobacter baumanii.*

- Streptococci, especially *Streptococcus pneumoniae,* which may be elongated and occur as diplococci joined end-to-end, may be misidentified as bacilli.

Conformation Staphylococci, which classically form tetrads or clusters, may appear singly or as pairs. This is because they have not undergone enough divisions to form tetrads or clusters.

Other stains

Methylene blue

This is useful for determination of bacterial shape, but it does not convey as much information as the Gram stain. Because it is incorporated into the Romanowsky stains, used for blood smears, bacteria can be visualized in specimens stained with these stains (Figs 2.3 & 2.4). It can also stain molds, which are not stained by the Gram stain.

Acridine orange

This fluorescent stain detects the presence of DNA and RNA. It is useful for distinguishing between bacteria and bacteria-like objects seen in a Gram stain, especially in blood cultures and body fluids, when the Gram stain is difficult to interpret or when bacteria are suspected but the Gram stain is negative (Fig. 2.5).

Ziehl–Neelsen stain ("acid-fast stain")

This stain utilizes heated carbol fuchsin to detect the presence of mycobacteria (see Chapter 17). Modifications of this stain are used for detecting *Nocardia* spp. (Kinyoun stain) and *Mycobacterium leprae* (Fite stain).

Fluorochrome stain

This has largely replaced the Ziehl–Neelsen stain for detecting mycobacteria.

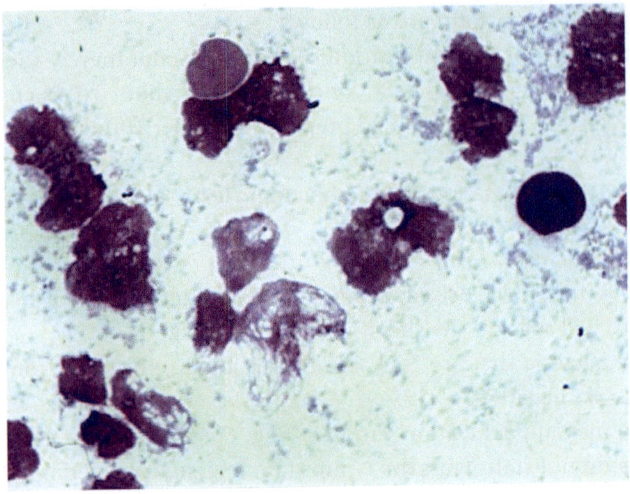

Fig. 2.3 A blood smear stained with Wright's stain showing diplococci. This was from a fatal case of *Streptococcus pneumoniae* sepsis. Copyright ©2007 Frank E. Berkowitz. Reprinted with permission of Cambridge University Press.

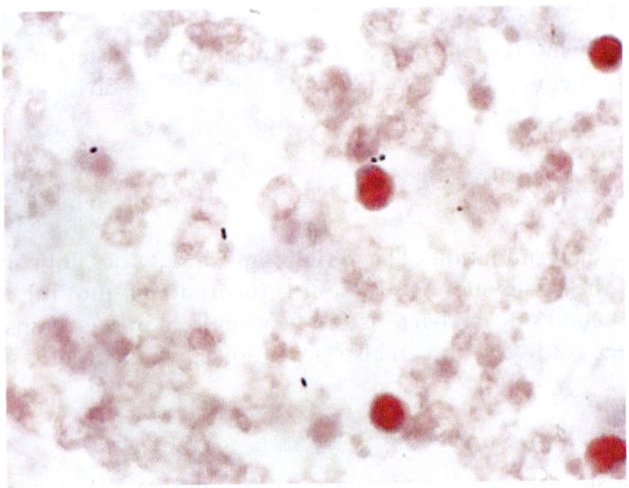

Fig. 2.4 Gram stain of a smear of the same blood as in the previous figure, showing Gram-positive diplococci. Copyright ©2007 Frank E. Berkowitz. Reprinted with permission of Cambridge University Press.

Calcofluor white

This is a fluorescent stain that binds to chitin in fungal cell walls.

Fluorescent antibody stains

These are fluorescein-labeled antibodies directed against specific microorganisms. They are used in direct fluorescent-antibody tests (see Virologic methods).

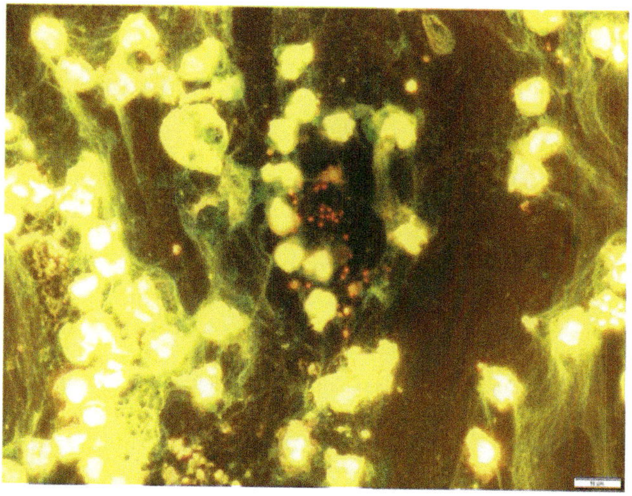

Fig. 2.5 Acridine orange preparation showing orange-staining bacteria, which are staphylococci. Courtesy of Scott Brown, Children's Healthcare of Atlanta.

Romanowsky-derived stains (Wright, Giemsa, Leishman stains)

These are used for staining blood films for examination of blood cells. In microbiology, they are used for detection of protozoa (*Plasmodium* spp., *Babesia* spp., *Trypanosoma* spp., *Leishmania* spp.), *Borrelia* spp., and *Bartonella bacilliformis*.

Iodine

This is used mainly for staining intestinal protozoa, but also for staining chlamydial inclusions in tissue culture.

India ink

Used to detect capsules, primarily of *Cryptococcus* spp.

Culture

This means the propagation of microorganisms in the laboratory, almost always done in an *in vitro* system. (Culture by animal or egg inoculation is performed in rare circumstances.)

Bacteria and fungi can be grown on solid media (generally meaning agar in a petri dish) (Fig. 2.6) or liquid media (in a tube or bottle). Viruses must be cultured in cells (tissue culture). Figure 2.7 shows an agar plate being inoculated with a specimen.

Bacteriologic methods

Bacterial isolation and identification

Culture requires that the medium supports the growth (i.e. multiplication) of the organism(s) being sought (i.e. it is nutritious and in an appropriate environment – appropriate concentration of oxygen, carbon dioxide, and temperature). Because many specimens are obtained from areas of the body where normal flora is mixed with the potential pathogen (e.g. stool, respiratory tract), selective media might be

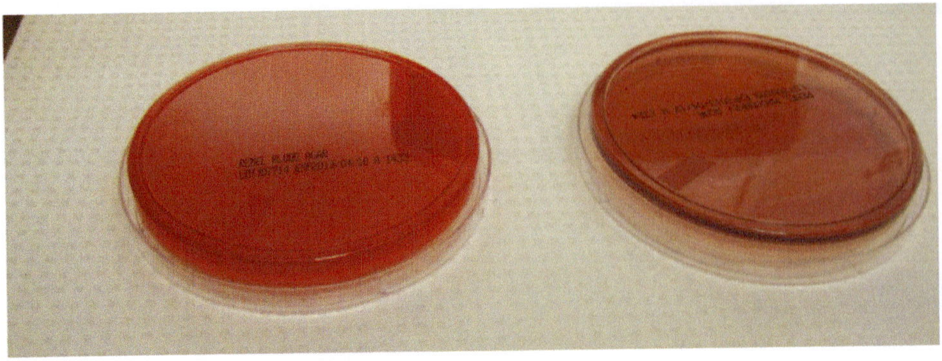

Fig. 2.6 Petri dishes with blood agar (5% sheep blood) on the left and MacConkey agar on the right.

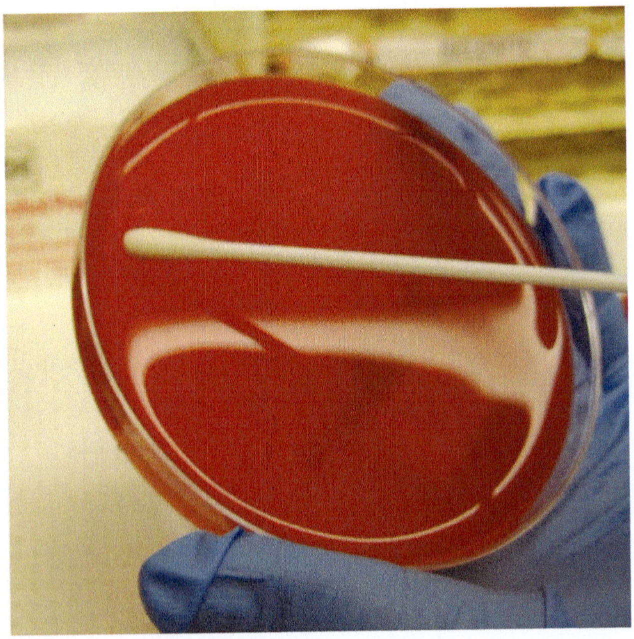

Fig. 2.7 A specimen being inoculated onto a blood agar plate.

required (i.e. media containing chemicals or antimicrobial agents that inhibit the growth of the "contaminating" organisms). Examples of selective media for bacterial culture are MacConkey agar used for selecting enteric rods, and Thayer–Martin agar used for selecting *Neisseria meningitidis* and *N. gonorrhoeae*.

As the bacteria multiply, they form colonies, which contain billions of organisms. The appearance of the colonies themselves provides information useful for identification of the organism. They provide the material that can be used for biochemical and other tests, as well as for antimicrobial susceptibility testing and advanced methods for epidemiologic typing.

One benefit of solid media is that colonial morphology, important in bacterial and fungal identification, can be readily visualized. The benefit of liquid media is that they

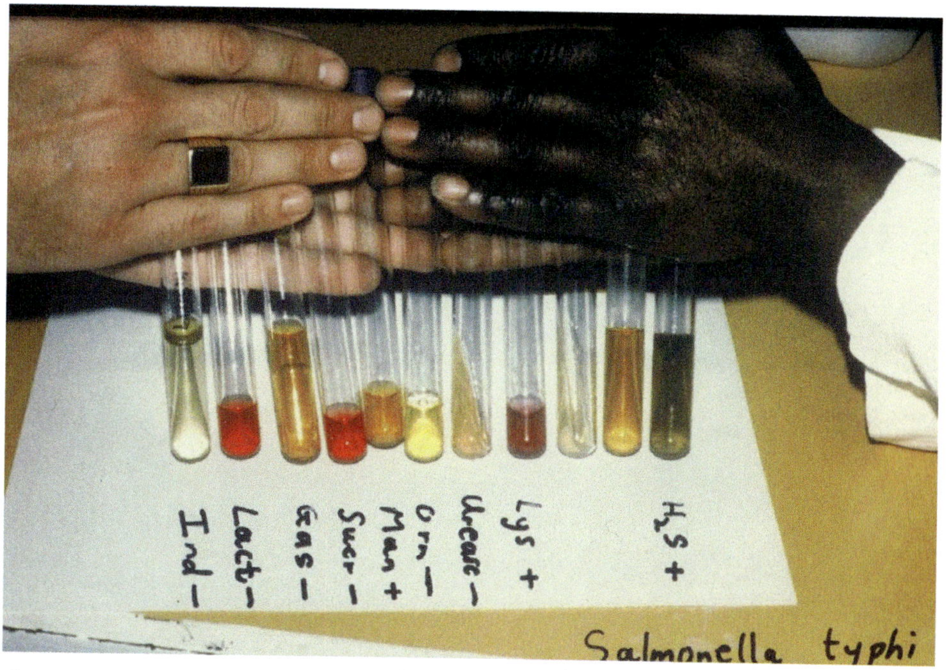

Fig. 2.8 Various biochemical tests used to identify bacteria, in this case *Salmonella typhi.*

can be used for culturing fluids containing very few organisms. Unfortunately, inherent in the use of liquid media is the amplification of contaminants as well as the target organism.

The features that are widely used for identification of bacteria are based on their physiology, as follows.

- Their biochemical reactions (or lack thereof) with various chemical compounds; most of these tests are performed in test tubes or, nowadays, the microtiter well equivalent (Figs 2.8 & 2.9).
- Their ability (or lack thereof) to grow in the presence of particular compounds or under particular environmental circumstances, e.g. anaerobically.
- Their nutritional requirements for growth.
- Their susceptibility or resistance to certain antimicrobial agents.

Additional characterization based on innate unique characteristics of organisms include their ribosomal RNA profile (pulse field gel electrophoresis), protein profile MALDI-TOF MS (see p. 24), and gene sequence.

The usual course of identification of bacteria is shown in Figure 2.10.

For deep wound specimens, also use anaerobic media and culture conditions. Anaerobic cultures are usually examined after 48 hours.

Blood culture: because, in bacteremia, the concentration of bacteria in the blood is usually low, culturing a few drops of blood on an agar plate might not detect the bacteria. Therefore blood for culture is performed in a liquid medium, in which the number of bacteria contained in a few mL of blood can be amplified. The course is shown in Figure 2.11.

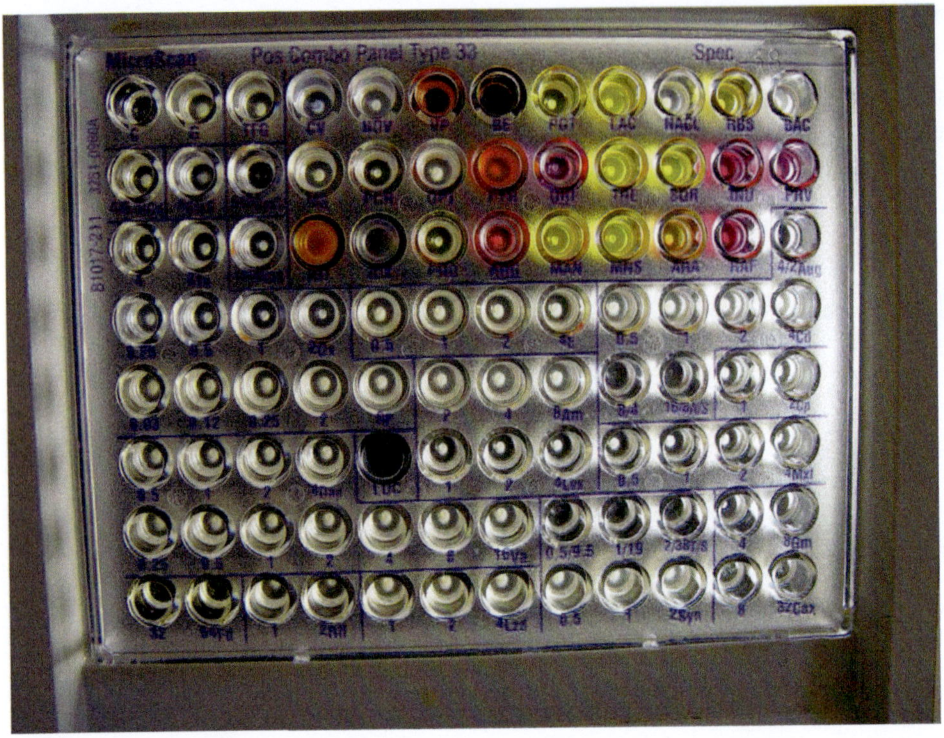

Fig. 2.9 A microtiter well plate with biochemical tests used to identify Gram-positive organisms, and to test for their antimicrobial susceptibilities.

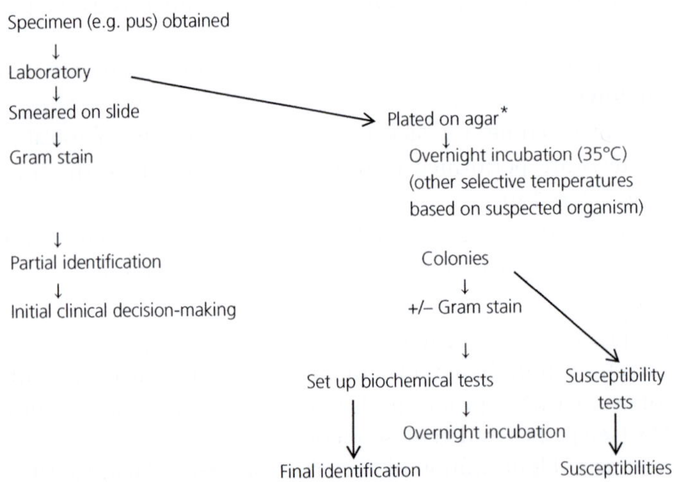

Fig. 2.10 The flow of bacterial isolation and identification.

Figure 2.12 shows blood being removed from the blood culture bottle to be smeared on a slide for Gram stain and inoculated on to agar plates.

Figure 2.13 shows a simplified algorithm of bacterial identification using Gram stain and a few culture characteristics.

Blood obtained and 5–10 mL inoculated into bottle containing broth, at the bedside.
↓
Laboratory
↓ Incubation (at least several hours) in a continuously monitored blood culture system
Detection system alerts that there is growth
↓
Aliquot removed → (see Fig. 2.12) → Inoculated on agar plates
↓
Smeared on slide
↓
Gram stain
↓
Partial identification
↓
Reported
↓
Clinical decision-making

Inoculated on agar plates
↓ Overnight incubation (35°C)
Colonies → Susceptibility tests
↓
+/– Gram stain
↓
Set up biochemical tests

Set up biochemical tests ↓ Overnight incubation → Final identification
Susceptibility tests ↓ Overnight incubation → Susceptibilities

Fig. 2.11 The flow of bacterial isolation and identification using blood culture.

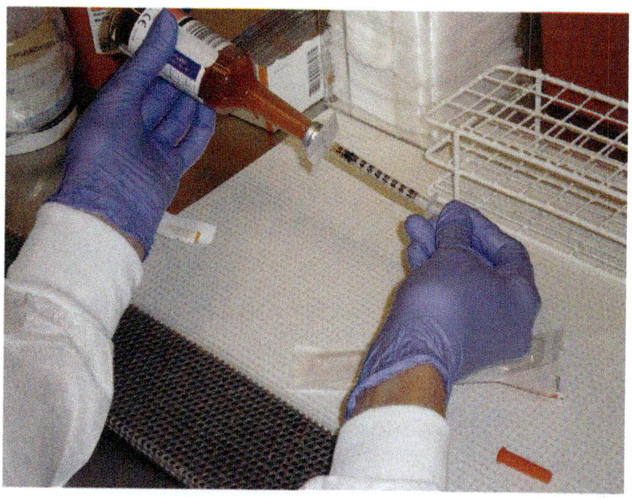

Fig. 2.12 Broth being removed from a blood culture bottle, to be smeared on a slide for Gram stain, and to be inoculated onto agar plates.

Molecular (nucleic acid) identification

The field of molecular biology has seen an unprecedented increase in the numbers and types of tests available for use in the clinical laboratory. Methods that detect RNA or DNA are based on the simple principle that unique and consistent gene sequences are found in all organisms. These characteristics can be exploited by laboratory methods to hybridize, amplify, or sequence the nucleic acids for identification (Fig. 2.14).

Hybridization uses a tagged probe complementary to a gene sequence. When admixed with a sample containing the target gene sequences, a duplex is formed and the tagged probe signal is generated. This assay is most commonly used to detect

Gram stain morphology (aerobic growth)

Gram-positive cocci→**catalase**→ Positive *Staphylococcus* spp.

→Negative *Streptococcus/Enterococcus* spp.

Gram-positive bacilli→**spores** →Positive *Bacillus* spp.

→Negative → catalase →Positive *Corynebacterium* spp., *Listeria* spp., others

→Negative *Lactobacillus* spp., *Actinomyces* spp., others

Gram-negative cocci→**Growth on Thayer–Martin Media**→ Positive *Neisseria gonorrheae, N. meningitidis*

→Negative saprophytic *Neisseria* spp., *Moraxella* spp.

Gram-negative bacilli→**Growth on MacConkey agar** → Positive Enterobacteriaceae, *Pseudomonas* spp.

→Negative *Haemophilus* spp.

Fig. 2.13 A simplified algorithm for bacterial identification.

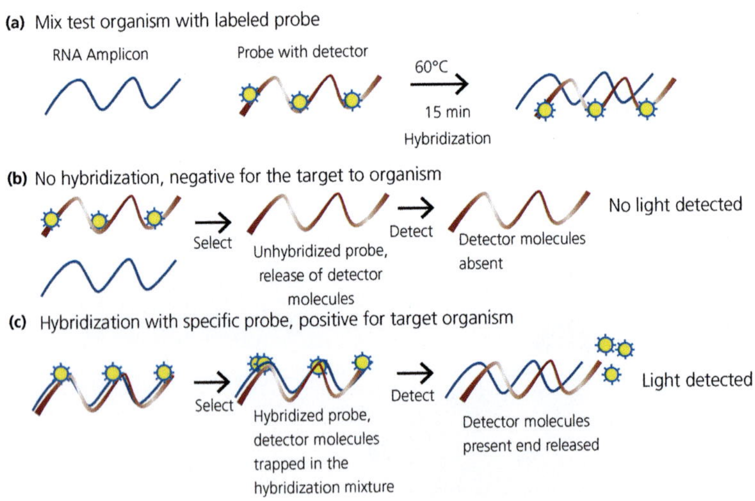

Fig. 2.14 The essence of hybridization of a nucleic acid target with a specific complementary probe. When target molecules are present, hybridization occurs and in this assay light is emitted and detected.

ribosomal RNA where the target is present in abundant numbers, unlike the single strand of DNA which requires amplification of target genes for detection.

Amplification tests, commonly called nucleic acid amplification tests (NAATs), have become a major method for detection of organisms. The most widely employed NAAT is the polymerase chain reaction (PCR) which utilizes the basic principles of hybridization and nucleic acid replication to amplify specific gene sequences unique to an organism. In principle, the assay involves repetitive cycling of three sequential reactions: denaturation of target DNA by heating (e.g. 94°C), complementary target annealing (joining with the specific sequence at a lower temperature [50–58°C]), and extension of the duplex via DNA polymerase at its optimal operation temperature of

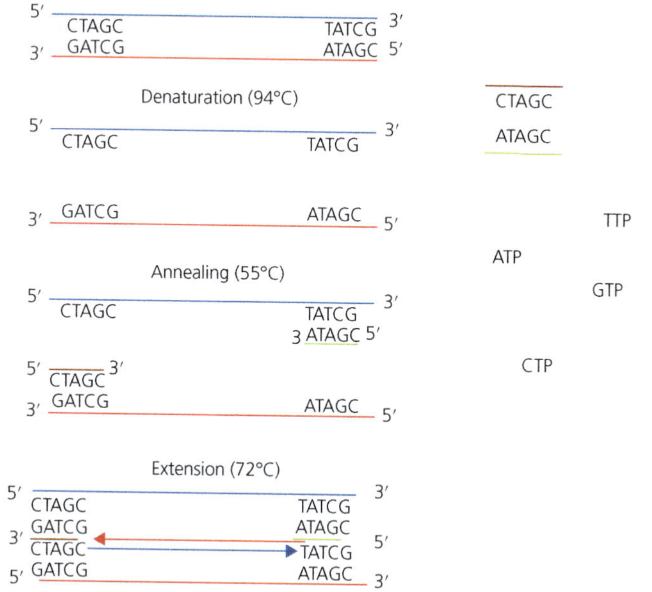

Fig. 2.15 The polymerase chain reaction (PCR).

72°C. Repetitive cycling at these temperatures (that may vary by organism) allows amplification of a unique gene to millions of copies. This is shown in Figure 2.15.

Detection of the target is accomplished by many different methods, including gel electrophoresis, in real time by use of labeled probes detecting amplification products as they cycle (hence a cycle time [CT]), or by endproduct analysis like melt curve analysis of the amplified product.

While PCR requires cycling at specific, different temperatures, there are numerous NAATs that now use alternative amplification techniques that are isothermal and require less sophisticated and costly instrumentation for amplification.

The advantages of molecular methods are:

- rapid turnaround time for detection of slow-growing organisms
- detection of organisms that cannot be cultured, or for which culture is insensitive (they have thus become the "gold standard" for the diagnosis of many infections)
- specificity can be adjusted by the choice of primers; they can be automated, so that large numbers of specimens can be processed
- they can detect dead organisms, although this can also be a disadvantage because they cannot differentiate between an active and a treated infection
- they can detect the genes coding for virulence factors that cause particular diseases or syndromes, e.g. toxins, and genes coding for antimicrobial resistance
- they can be used for quantitation of organisms. This is used for management of viral infections such as HIV, hepatitis B, hepatitis C, Epstein–Barr virus and cytomegalovirus infections.

Currently, many of the NAATs have been designed into multiplex assays which allow detection of multiple targets (for example, as many as 20 targets for a respiratory disease profile – BioFire Diagnostics, Inc., several others).

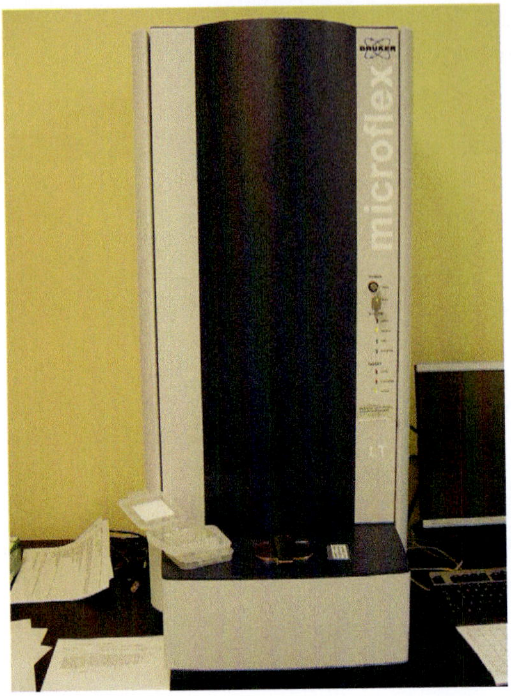

Fig. 2.16 A MALDI-TOF instrument.

Matrix-assisted laser desorption ionization time of flight (MALDI-TOF MS)

The most recent addition to identification of organisms from culture is MALDI-TOF. This technique utilizes mass spectrometry to measure particles (primarily proteins) based on their mass-to-charge ratio. In practice, a sample is embedded in a matrix on a target template and placed into the instrument where it is exposed to a laser burst. Its particles are ionized and separated based on their mass-to-charge ratio. The generated mass spectrum is detected and compared with a library of mass spectra. To date, spectra are available for bacteria (routine Gram positive, routine Gram negative, non-fermenting Gram negative, and anaerobic bacteria, *Mycobacterium* spp., *Nocardia* spp., yeasts, and filamentous fungi). This technique holds the potential to replace many of the previously mentioned methodologies with a simple, standardized, reproducible, cost-effective system for rapid, accurate identifications. One of the major instruments used to perform MALDI-TOF MS and the principle behind its use are shown in Figs 2.16 and 2.17.

Antibody detection (serologic tests)

Serodiagnosis depends on the host's response to the infection by the production of antibodies. This depends on the host's ability to make antibodies, which could be curtailed in some immunodeficiency states or abrogated by the use of antimicrobials early in the course of the infection. It also takes time from the onset of the illness for antibodies to appear in the blood (days to weeks), unless the illness has a long incubation period, e.g. syphilis. Therefore measurement of antibodies may be negative early in the infection,

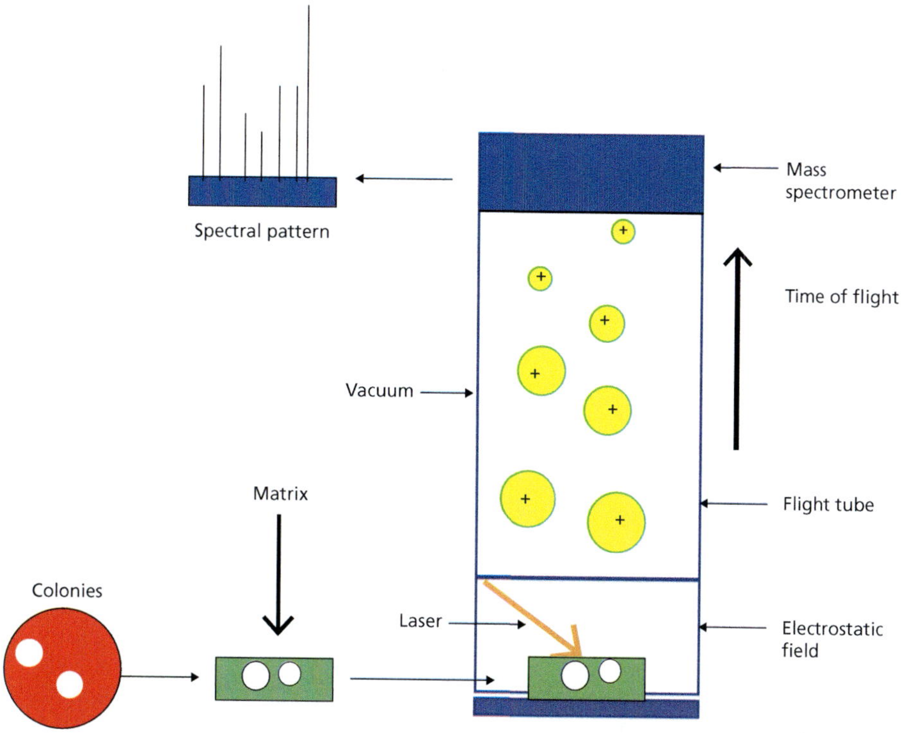

Fig. 2.17 The MALDI-TOF process. (1) Colonies are mixed with the matrix chemical on a template. (2) The template is placed in the instrument, where it is subject to (3) a laser burst, resulting in "soft" ionization of proteins; (4) these ionized particles are drawn by a vacuum, flying at a speed determined by their size and charge (m/hz), towards the mass spectrometer, where (5) a spectral pattern is produced. (6) The pattern is compared with those in a database to provide an identification.

and **they are not useful for management decisions for diseases that are acute or rapidly progressive**, e.g. Rocky Mountain spotted fever, ehrlichiosis, and leptospirosis.

Their value lies in the diagnosis of infections in which the causative organism cannot readily be cultured (e.g. *Treponema pallidum*) or detected by any other manner (see above), especially if the diagnosis and management are not emergencies. They are also useful for epidemiologic studies, to demonstrate how many individuals in a population have been infected by a particular agent in the past (e.g. serosurveys).

Many methods are used for the detection of antibodies. Traditionally, the concentration of antibodies in serum has been expressed as a titer, using doubling dilutions, i.e. 1 : 1, 1 : 2, 1 : 4, 1 : 8, etc. The higher the dilution at which antibodies can be detected, the higher the concentration of antibodies present. Such tests often describe the methods used or the type of antibody detected, and include the following.

- Neutralizing antibody tests (the serum is mixed with the infective agent, and the viability of the agent is then tested *in vitro* or in an animal).
- Complement fixation tests.
- Hemagglutination antibody tests.
- Immunofluorescent antibody tests.
- Opsonophagocytic antibody test, which tests the ability of the serum to enable phagocytes to opsonize the organism.

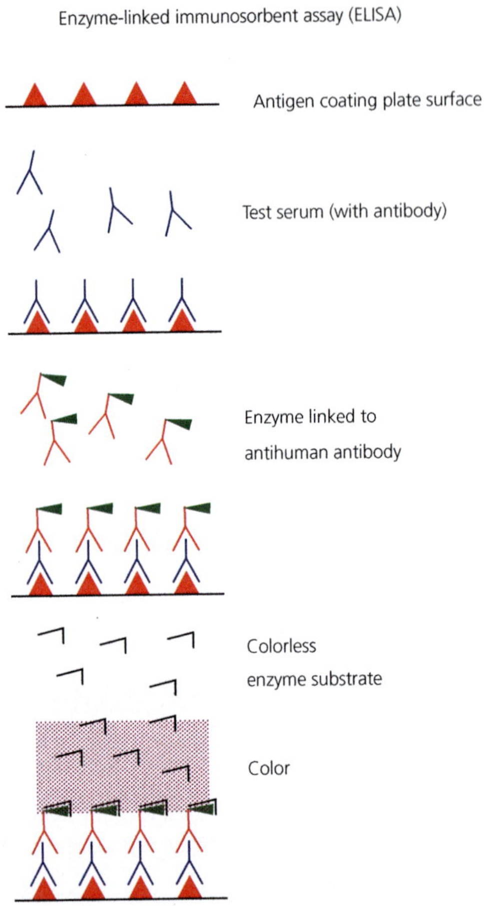

Fig. 2.18 Enzyme-linked immunosorbent assay (ELISA). A known antigen is attached to the surface of the microtiter well; the test serum (possibly containing antibodies) is added. During an incubation step, the antibody, if present, attaches to the antigen. After a wash, an antibody to human immunoglobulin, raised in a different animal, e.g. a goat, and to which an enzyme has been linked, is added, with a second incubation step and wash. Then a substrate to the enzyme is added. When acted upon by the enzyme, this produces a color, the intensity of which is measured by a spectrophotometer. The result is expressed as an optical density (OD).

Many serological tests nowadays use the enzyme-linked immunosorbent assay (ELISA). This utilizes the principle that a particular enzyme, linked to an antibody, reacts with its substrate to produce a color, the intensity of which is proportional to the concentration of antibody present, and which can be measured with a spectrophotometer. The result is expressed as an optical density (OD), not as a titer. The main enzymes used are alkaline phosphatase and horseradish peroxidase (Fig. 2.18). The same principle can be used to detect antigens in fluids and in tissue, by reversing the reagents, i.e. where the unknown is the antigen and the known is the antibody.

How precise should a microbiologic diagnosis be?

An important question to ask is: to what degree of precision is a microbiologic identification necessary? (See section on diagnostic testing in Chapter 1.)

Although names of organisms contain important information, such as epidemiologic, potential clinical problems, and antimicrobial resistance patterns, this information is not always necessary. Therefore the degree to which the precision in microbiologic diagnosis is necessary depends on the purpose of the testing and, in clinical situations, on the specific clinical scenario. In most clinical situations, viral diagnostic testing (by culture methods) is not of value for patient management, and the most important information in bacterial diagnosis is the genus identification and the antimicrobial susceptibilities (usually to only a limited number of drugs). In the situation of an outbreak, identification to the strain level is important.

Additional methods used in epidemiology for strain identification include: pulse field gel electrophoresis (PFGE) typing, and molecular methods such as restriction endonuclease typing, multilocus sequence analysis, and complete genome sequencing.

Antimicrobial susceptibility testing of bacteria

Antimicrobial susceptibility testing is a very important function of a microbiology laboratory. This is used as a guide for clinicians to determine optimal antimicrobial therapy for patients, and which antimicrobial agents will likely be ineffective. There are several different methods and standards established for interpretation of the results.

The standard measurement for describing the susceptibility or resistance of a bacterial organism to a particular antimicrobial agent is the **minimal inhibitory concentration (MIC)** (see Chapter 9). This is the lowest concentration of drug that inhibits growth of the organism. It is a property of the organism with respect to a particular drug, e.g. the MIC of a particular isolate of *Staphylococcus aureus* to oxacillin is 1 μg/mL (this is the same if the units used are mg/L) (Fig. 2.19). The lower the MIC, the more susceptible is the organism.

What is antimicrobial resistance?

- Clinical resistance: this means that the concentration of drug SAFELY attainable in the infected tissue is inadequate to inhibit growth of or kill the organism.

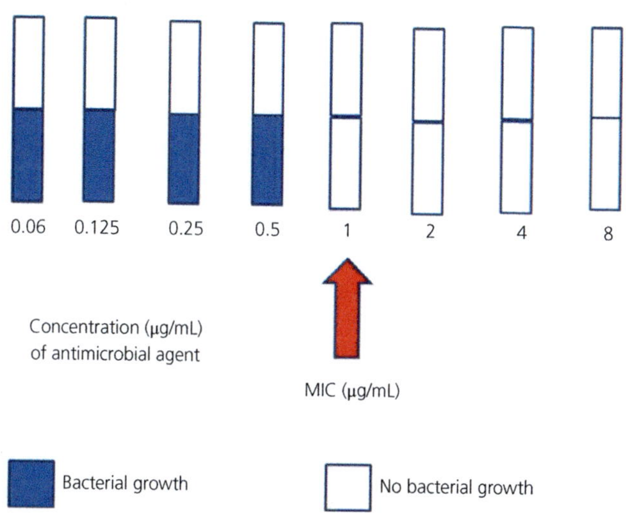

Fig. 2.19 Tube dilution measurement of minimal inhibitory concentration (MIC). Bacterial growth is indicated in blue. In this case the MIC is 1 μg/mL.

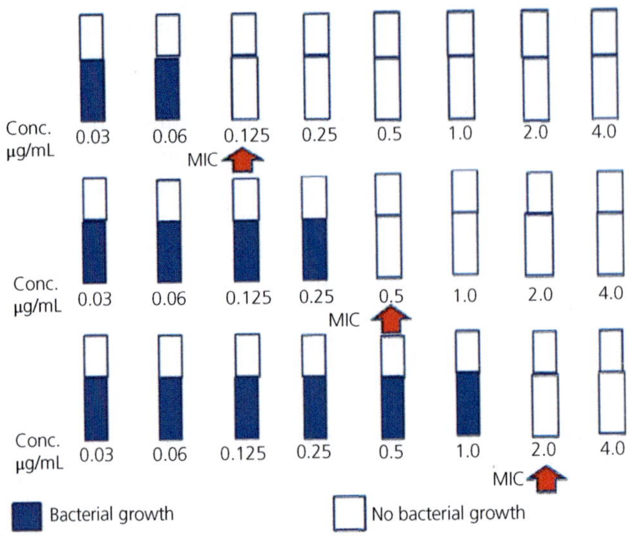

Fig. 2.20 Demonstrating why "microbiologic resistance" (rising MICs) does not necessarily constitute "clinical resistance." Bacterial growth is indicated in blue. If clinical resistance were defined by a minimal inhibitory concentration (MIC) greater than 1 µg/mL, then only the isolate in the third row would be considered resistant (MIC of 2 µg/mL), although the isolate in the second row (MIC of 0.5 µg/mL) is less susceptible than that in the first row (MIC of 0.125 µg/mL).

- Microbiologic resistance is a concept describing the susceptibility of a strain of a species of an organism in relation to that of other strains of the same species. For example, the MIC of *Neisseria meningitidis* to penicillin has increased over time, although it has not crossed the threshold rendering it CLINICALLY resistant (Fig. 2.20).

Several different methods are used for determining antimicrobial susceptibilities of bacteria.

- Tube dilution method: doubling dilutions of the drug are made in serial tubes. Each tube is inoculated with the same number of bacteria ($1–5 \times 10^5$/mL), and the tubes are incubated at 35°C overnight. The tubes in which there is bacterial growth (the lower concentrations of drug) are turbid, and those in which there is no growth (the higher concentrations of drug) are clear. The concentration of drug in the tube with the lowest concentration of drug in which there is no growth is the MIC (see Fig. 2.19). This type of test can be done more easily in plates with 96 microtiter wells. Several different concentrations of several different drugs can be tested simultaneously (Figs 2.21 & 2.22).
- Disc diffusion: a suspension of a standard concentration of bacteria is made in a tube, and plated on agar suitable for growth of the organism. Disks of filter paper impregnated with a fixed amount of the drug are placed on the agar, and the plate is incubated overnight. The drug diffuses into the agar, creating a zone of inhibition around the disk. The larger the zone of inhibition, the more susceptible is the organism to the drug. Several different drugs can be tested against a single organism on one plate. Standards have been established for zone size cut-offs to differentiate between susceptible and resistant organisms. These standards are specific for drug and organism. The Kirby–Bauer test is one disk diffusion test (Fig. 2.23).

Fig. 2.21 A tube dilution MIC test on *Staphylococcus aureus* performed in a microtiter well plate. The top row is for daptomycin in which there is growth in the 0.5µg/mL well but not in the 1 µg/mL well. The MIC is therefore 1 µg/mL. The lower row is for vancomycin. There is growth in 0.25, 0.5, and 1 µg/mL wells, but not in the 2 µg/mL well. Therefore the MIC is 2 µg/mL.

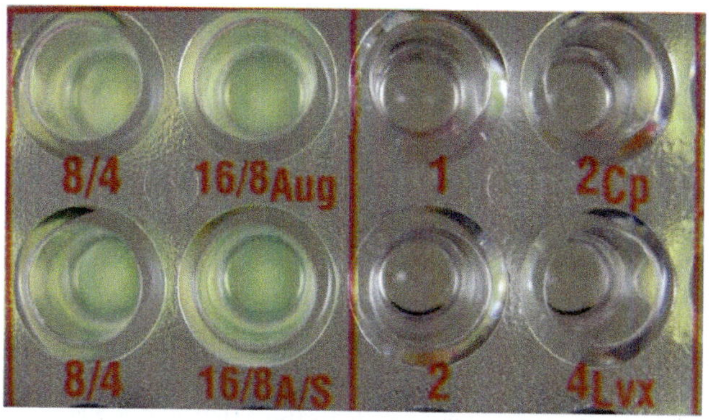

Fig. 2.22 A tube dilution MIC test on *Pseudomonas aeruginosa* performed in a microtiter well plate. There is growth (green due to the production of pyocyanin) in all the wells containing Aug, which is Augmentin® (amoxicillin/clavulanic acid) and A/S, which is ampicillin/sulbactam, as expected for this organism, and no growth in the wells containing Cp (ciprofloxacin) and Lvx (levofloxacin). The MIC for ciprofloxacin is <= 1 µg/mL, and to levofloxacin is <= 2 µg/mL.

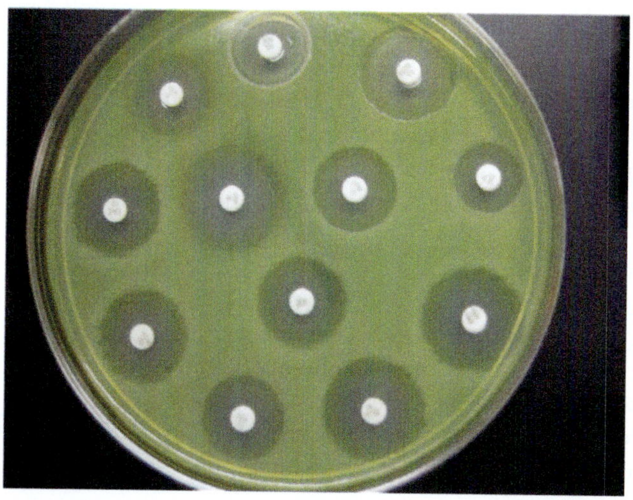

Fig. 2.23 Kirby–Bauer test using *Pseudomonas aeruginosa* as the test organism. Note the zones of inhibition around the disks.

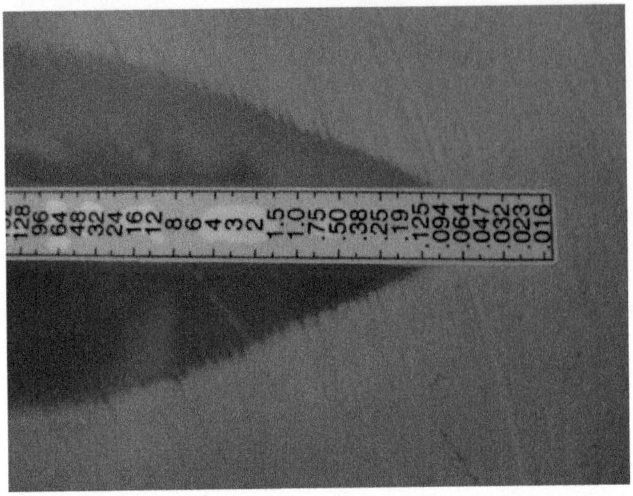

Fig. 2.24 An E-test.

- Antimicrobial concentration gradient method (Epsilometer or E-test): plastic strips are impregnated on the undersurface with a logarithmic$_2$ concentration of antimicrobial agent, and there are corresponding markings of the concentration on the upper surface of the strip. The strip is placed on the surface of an agar plate that has been inoculated as in the disk diffusion method. Growth is inhibited according to the concentration of the drug that has diffused into the agar. The line on the strip where growth meets the strip corresponds to the MIC (Fig. 2.24).
- Rate of growth of the organism in the presence of an antimicrobial agent.
- Detection of a gene conferring resistance, e.g. mec A gene conferring methicillin resistance in *Staphylococcus aureus*.
- Detection of an enzyme causing resistance. There is a rapid colorimetric test for β-lactamase, using a cephalosporin, nitrocefin, which turns from a pale yellow to a deep orange when acted upon by β-lactamase (Fig. 2.25).

Bacteriology culture reports, including susceptibility testing reports, generally contain the following information, in addition to patient identifiers.

- date on which the specimen was submitted
- date of the final report
- type of specimen and its source (e.g. pus/exudate)
- Gram stain findings
- identification of the organism.

This is followed by a table showing a list of antimicrobials tested against the isolate, an interpretation (susceptible [S], intermediately susceptible [I], or resistant [R]), and the MIC (in μg/mL or mg/L, numerically the same) on which the interpretation is based, and which is expressed in one of the following three ways:

- a number preceded by the < sign, which means that the test does not test antimicrobial concentrations lower than that number, and implies susceptibility of the organism

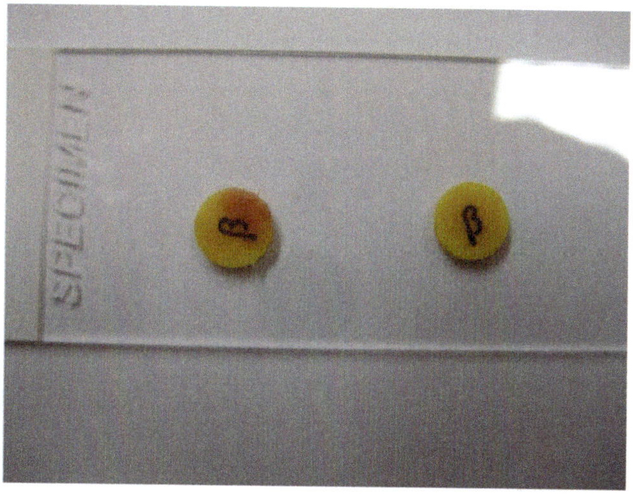

Fig. 2.25 The color change caused by β-lactamase acting on nitrocefin. The right disk is a negative control, and the left disk is a positive.

- a number preceded by the > sign, which means that the tests does not test antimicrobial concentrations higher than that number, and implies resistance of the organism
- a number alone, which could imply susceptibility or resistance, depending on its value and the clinical circumstance.

The lower the MIC, the more susceptible the organism. The "breakpoints," i.e. the MIC below which the isolate is susceptible and above which it is resistant, are based on standards determined by the Clinical Laboratory Standards Institute (CLSI). There may be additional comments if the susceptibility pattern suggests a specific pattern of resistance, e.g. methicillin-resistant *Staphylococcus aureus* (MRSA), extended-spectrum β-lactamase-producing Gram-negative rod, or the fact that an organism is known to produce inducible broad-spectrum β-lactamases, e.g. *Enterobacter cloacae* or *Serratia marcescens*.

Examples of such reports are shown in Tables 2.1–2.8.

Virologic methods

Diagnosis of viral infections

The virology laboratory utilizes many different diagnostic modalities in order to optimize detection of viral agents. Analytical procedures include culture, requiring specific cell lines receptive to individual virus types, and taking days to weeks for identification; antigen detection, by enzyme immunoassay or direct fluorescent antibody testing, taking hours for detection; histopathology, taking hours to a day; serology, taking days for single titer or weeks for acute and convalescent titers; and nucleic acid amplification assays (NAAT) taking hours for detection. The analysis of choice depends on many variables, including the suspected virus, the stage of the disease, the test methods

Table 2.1 Specimen description: abscess.
Site: flank; Gram stain: white blood cells; rare Gram-positive cocci in clusters
Culture: ***Staphylococcus aureus***

Antibiotic	Interpretation	MIC (μg/mL)
Ceftriaxone	S	<=8
Ciprofloxacin	S	<=1
Clindamycin	R*	<=0.5*
Erythromycin	R	>4
Gentamicin	S	<=4
Linezolid	S	2
Oxacillin	S	<0.25
Rifampin	S	<=1
Trimethoprim/sulfamethoxazole	S	<=0.5/9.5
Tetracycline	S	<=4
Vancomycin	S	1

* Although the minimal inhibitory concentration (MIC) is less than the breakpoint, this isolate was shown to have inducible clindamycin resistance (see D-test in Chapter 5), and is therefore considered resistant.
This organism is susceptible to oxacillin, and is therefore a methicillin-susceptible *Staphylococcus aureus* (MSSA).
R, resistant; S, susceptible.

Table 2.2 Specimen description: abscess.
Site: thigh
Gram stain: white blood cells; rare Gram-positive cocci in clusters
Culture: ***Staphylococcus aureus***
PBP 2a detected; presumptive methicillin-resistant *Staphylococcus aureus*

Antibiotic	Interpretation	MIC (μg/mL)
Ceftriaxone	R*	<=8
Ciprofloxacin	I	2
Clindamycin	S	<=0.5
Erythromycin	R	>4
Gentamicin	S	<=4
Linezolid	S	2
Oxacillin	R	>2
Rifampin	S	<=1
Trimethoprim/sulfamethoxazole	S	<=0.5/9.5
Tetracycline	S	<=4
Vancomycin	S	1

* This organism is resistant to oxacillin, and is therefore a methicillin-resistant *Staphylococcus aureus* (MRSA), confirming the rapid test for penicillin-binding protein 2a, a rapid test which detects the protein that confers methicillin resistance. It is considered resistant to all β-lactam antibiotics, which explains why the ceftriaxone is resistant, despite the minimal inhibitory concentration (MIC) being below the cut-off.
I, intermediate; R, resistant; S, susceptible.

available and the intent of the assay (e.g. diagnostic versus immune response). Specimen collection (in unique viral transport media), transport (on ice in most cases for viral isolation and NAAT), and storage are key components for successful detection of these agents. Communication directly with the laboratory personnel for assistance with these preanalytical parameters will ensure optimal processing.

Table 2.3 Specimen description: blood.
Culture: ***Streptococcus pneumoniae***

Antibiotic	Interpretation	MIC (μg/mL)
Amoxicillin/clavulanate	S	<=0.5/0.25
Cefotaxime meningitis	S	<=0.25
Cefotaxime non-meningitis	S	<=0.25
Ceftriaxone meningitis	S	<=0.25
Ceftriaxone non-meningitis	S	<=0.25
Clindamycin	R	>0.5
Erythromycin	R	>0.5
Levofloxacin	S	0.5
Vancomycin	S	<=0.12

This isolate of *Streptococcus pneumoniae* is susceptible to penicillin (amoxicillin/clavulanic acid), third-generation cephalosporins (cefotaxime and ceftriaxone), levofloxacin, and vancomycin, but resistant to clindamycin and erythromycin. Two additional comments should be made.

- Because the mechanism of penicillin resistance in *Strep. pneumoniae* is not due to β-lactamase, clavulanic acid does not add any benefit to amoxicillin therapy.
- Because the concentrations of antibiotics attained in CSF are lower than those in blood, the minimal inhibitory concentration (MIC) breakpoints used for meningitis are lower than those used for non-meningeal infections.

R, resistant; S, susceptible.

Table 2.4 Specimen description: cerebrospinal fluid.
Gram stain: white blood cells; Gram-positive cocci in pairs
Culture: ***Streptococcus pneumoniae***

Antibiotic	Interpretation	MIC (μg/mL)
Cefotaxime meningitis	S	<=0.25
Ceftriaxone meningitis	S	<=0.25
Meropenem	S	0.12
Vancomycin	S	0.5

Note that only antibiotics that are considered suitable for treating patients with meningitis are reported.
MIC, minimal inhibitory concentration; S, susceptible.

Table 2.5 Specimen description: cerebrospinal fluid.
Site: cerebrospinal fluid
Gram stain: many white blood cells, moderate Gram-positive cocci in pairs
Culture: ***Streptococcus pneumoniae***

Antibiotic	Interpretation	MIC (μg/mL)
Penicillin (meningitis)*	R	2.0
Cefotaxime (meningitis)*	I	1.0
Ceftriaxone (meningitis)*	I	1.0
Levofloxacin	S	≤2.0
Linezolid	S	≤2.0
Vancomycin	S	≤1.0
Meropenem	S	≤0.5

* Note that *Strep. pneumoniae* isolated from cerebrospinal fluid has lower minimal inhibitory concentration (MIC) breakpoints for S, I, and R for certain drugs compared to respiratory isolates due to lower penetration of antimicrobial into the site of infection.
Note: drugs reported are limited to those that may be used to treat meningitis.
I, intermediate; R, resistant; S, susceptible.

Table 2.6 Specimen: blood.
Culture: ***Escherichia coli***

Antibiotic	Interpretation	MIC (μg/mL)
Amikacin	S	<=4
Ampicillin	S	<=8
Aztreonam	S	<=8
Cefazolin	S	<=4
Ceftazidime	S	<=1
Ceftriaxone	S	<=8
Cefuroxime	S	<4
Ciprofloxacin	S	<=1
Gentamicin	S	4
Meropenem	S	<=4
Piperacillin/tazobactam	S	<=8
Trimethoprim/sulfamethoxazole	R	>2/38
Ticarcillin/clavulanic acid	S	<=16
Tobramycin	I	8

This isolate is susceptible to all β-lactams tested.
I, intermediate; R, resistant; S, susceptible.

Table 2.7 Specimen: blood.
Culture: ***Escherichia coli***

Antibiotic	Interpretation	MIC (μg/mL)
Amikacin	S	16
Ampicillin	R	>16
Aztreonam	R	>16
Cefazolin	R	>16
Ceftazidime	R	16
Ceftriaxone	R	>32
Cefuroxime	R	>16
Ciprofloxacin	R	>2
Gentamicin	R	>8
Meropenem	S	<=4
Piperacillin/tazobactam	R	>64
Trimethoprim/sulfamethoxazole	S	<2=/38
Ticarcillin/clavulanic acid	I	64
Tobramycin	R	>8

This organism harbors an extended-spectrum β-lactamase. It is resistant to all β-lactam antibiotics except carbapenems (e.g. meropenem); use of a β-lactam/β-lactamase inhibitor is controversial.
I, intermediate; R, resistant; S, susceptible.

Challenges

- Because viruses require living cells for their replication, they cannot be cultured in non-living media such as agar. For culture, they require living tissue (live animal, organ culture, or tissue culture).
- Viruses (with the exception of poxviruses) are too small to be seen with a light microscope. Visualizing them requires the use of an electron microscope. This is not practicable for a diagnostic microbiology laboratory. Therefore, they must be detected indirectly by some other method, or by the host's immune response.

Table 2.8 Specimen: blood.
Culture: ***Enterobacter cloacae***

Antibiotic	Interpretation	MIC (μg/mL)
Amikacin	S	<=4
Ampicillin	R	>16
Aztreonam	Comment	<=8
Cefazolin	R	>16
Ceftazidime	Comment	<=1
Ceftriaxone	Comment	<=8
Cefuroxime	Comment	16
Ciprofloxacin	S	<=1
Gentamicin	S	<=1
Meropenem	S	<=4
Piperacillin/tazobactam	Comment	<=8
Trimethoprim/sulfamethoxazole	R	>2/38
Ticarcillin/clavulanic acid	Comment	<=16
Tobramycin	S	<=1

Comment: This organism is known to possess inducible broad-spectrum β-lactamases, so that it might become resistant during therapy, despite appearing susceptible on initial testing.
R, resistant; S, susceptible.

Tissue culture

Tissue culture, the main method used for virus culture, entails growing cells, such as certain types of epithelial cells or fibroblasts, as a monolayer on a glass or plastic surface. This is then infected with material suspected to contain the virus, such as blood, urine, respiratory secretions, or cerebrospinal fluid. Different viruses require different cell types for their growth, although some cell lines support the growth of several different viruses. Not all viruses can be grown in tissue culture. Because viruses cannot be seen, their growth in tissue culture requires a detection system. The following methods are used to detect viral growth in tissue or tissue culture.

Cytopathic effect (CPE)

Many viruses cause changes in the infected cells that can be seen under magnification. These changes, sometimes initially characterized by the cell "rounding up" and becoming refractile, are called cytopathic effects (Fig. 2.26). They can be fairly specific for the virus. Some viruses, which spread from cell to cell, cause changes in enough adjacent cells to be recognized by the naked eye or under low magnification. These are called plaques. The major disadvantage of using CPE for virus detection is that it can take several days, or even weeks, for CPE to become apparent. The time to detect positivity can be markedly reduced by the detection of expressed viral antigens before CPE is apparent (see later in this chapter). Cytopathic effects can be seen in stained cytologic and histopathologic specimens. These include intracellular inclusions, which can be specific for the virus (e.g. herpes simplex or varicella virus), syncytia, seen in herpes virus and paramyxovirus infections, and large cells with inclusions, as in cytomegalovirus infection.

Antigen detection

Viral antigens, expressed inside the infected cell or on its surface, can be detected by immunofluorescence. This can be performed on tissue culture, even when no cytopathic effect is apparent, as is done in the "shell vial" technique. It can also be

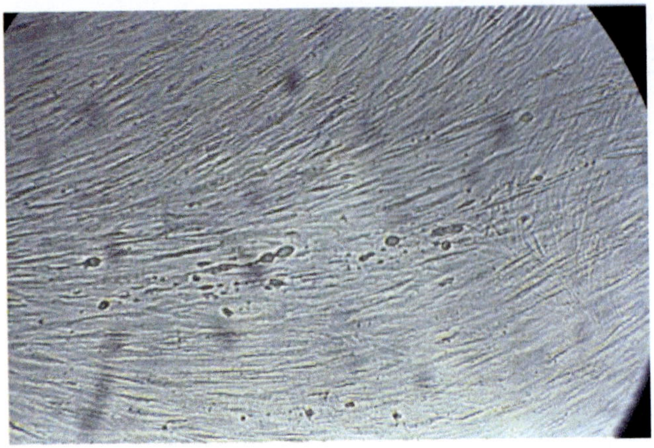

Fig. 2.26 A fibroblast monolayer with cytopathic changes ("rounding up" of cells) caused by herpes simplex virus. Courtesy of PHIL, CDC.

performed directly on clinical specimens. Because immunofluorescence is performed with specific antisera for a particular virus, the clinician must specify which virus is being sought. The specimen is placed on a glass or plastic surface (e.g. microscope slide); an antibody against the specific virus is placed on the specimen and the slide is incubated. For direct immunofluorescent tests, the antibody used is labeled with fluorescein. For indirect immunofluorescent tests, a second antibody, made against the first antibody and labeled with fluorescein, is placed on the slide and the slide is incubated. For example, if the first antibody is raised in a rabbit, the second antibody might be anti-rabbit antibody raised in a goat. The purpose of the second antibody stage is to amplify the signal. The slide is then examined under a microscope equipped to detect fluorescence. Each test uses antibodies to a specific virus. The principle of these tests is shown in Figure 2.27, and a specific example is shown in Figure 2.28.

Viral antigens can also be detected in histologic preparations, using the same principle as is used for ELISA testing (immunohistochemistry). The color is precipitated in the tissue section (Figs 2.29 & 2.30).

Molecular tests

These tests, which involve detection of part or all of a viral genome, usually after amplification, are becoming more widely used. The advantages of the tests are that they can be used to detect viruses that cannot readily be cultured, they are specific, and they can be performed quickly (see discussion of bacterial diagnosis, earlier in this chapter). Because PCR amplifies DNA, detection of RNA viruses requires the transcription of the RNA to DNA using reverse transcriptase.

Reasons for making a laboratory-confirmed diagnosis of a viral infection

- Therapeutic: antiviral chemotherapy is available for treatment of patients infected with a limited number of viruses. These include influenza virus, several herpes viruses (HSV, VZV, CMV), hepatitis B and hepatitis C viruses, and HIV. Therefore confirmation of these infections is useful and, in some cases, essential.
- A valuable therapeutic implication of proving a viral cause of a patient's disease is the withholding of other therapy, especially antibacterial therapy.

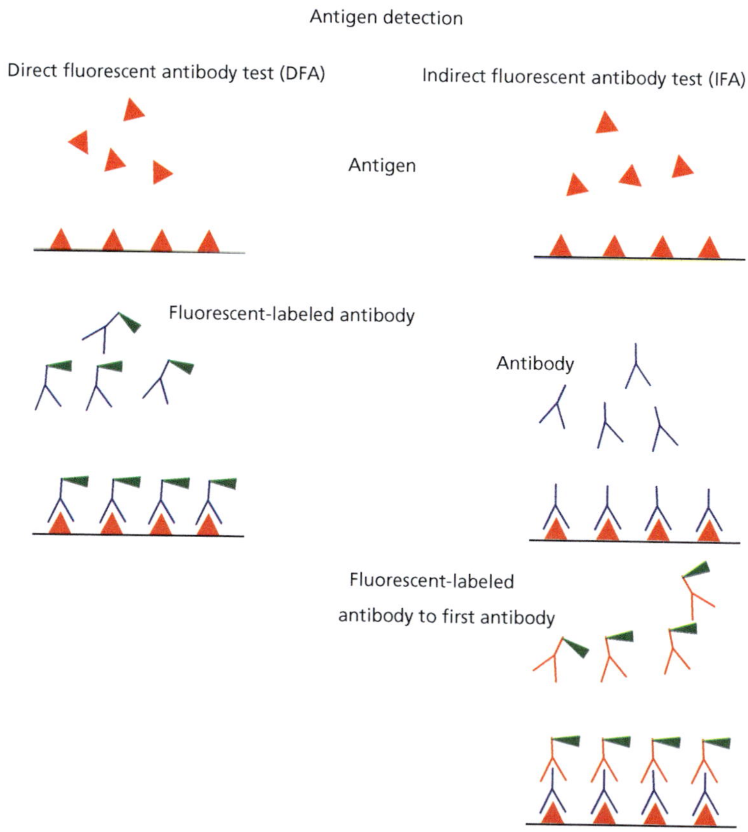

Fig. 2.27 Principle of immunofluorescence tests for antigen detection.

Fig. 2.28 Immunofluorescence for herpes simplex virus. Courtesy of Dr Jumi Yi.

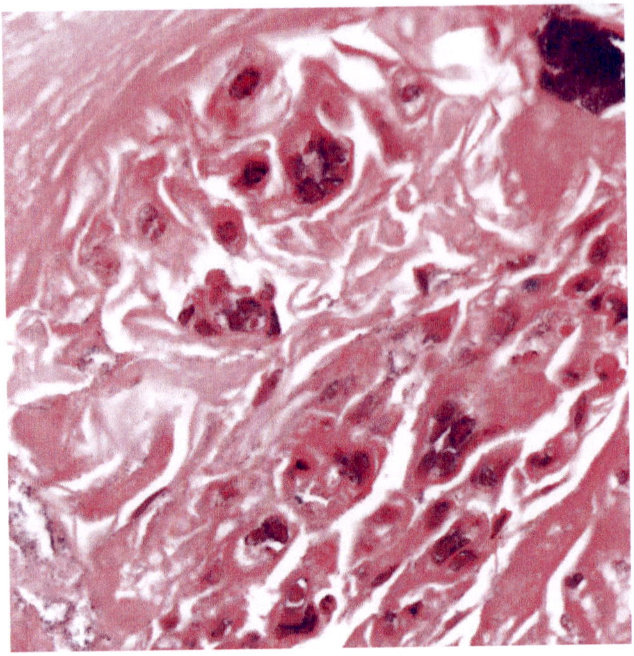

Fig. 2.29 Tissue from mouth biopsy of an immunocompromised boy with herpes zoster. Note the eosinophilic intranuclear inclusions. This is not easily appreciated. Courtesy of Dr Carlos Abramowsky.

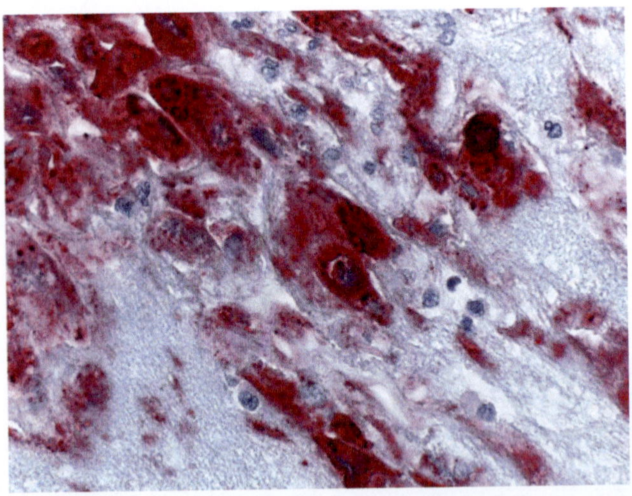

Fig. 2.30 The same tissue as shown in Figure 2.29 showing the positive immunoperoxidase staining for varicella zoster virus. Courtesy of Dr Carlos Abramowsky.

- Epidemiologic: only by determining the cause of infections can one know the epidemiology of the agent. Whether this should be done in all cases or only by sentinel clinical facilities depends on the virus in question. If the viral agent causing a patient's infection is determined, but the data are not included in an epidemiologic database, then this value of the test is lost.

- Public health: viruses which cause severe illness, and which have the potential to spread, pose a serious public health menace. Therefore, their confirmation is very important for appropriate public health interventions, such as quarantine or mass immunization, e.g. pandemic influenza and measles.
- Research: the advancement of knowledge is a good reason for performing a test, provided that it is done in an approved study, and that the patient knows that it falls under the umbrella of research.

If none of these reasons applies in a particular patient, then one should consider very carefully whether viral diagnostic tests should be performed. In our opinion, just "to know" is not a valid reason. Furthermore, one should consider who should bear the cost of performing the test.

A convenient method to evaluate viral infections is via the syndromic approach. Table 2.9 offers a guide to viral agents causing different clinical syndromes, and Table 2.10 shows the diagnostic tests used to detect these viruses. It is critical to

Table 2.9 Viral agents, common clinical syndromes, diagnostic tests, and specimen types. The tests for each virus are listed in Table 2.10.

System	Syndrome	Common agents	Less common agents	Specimen type
Respiratory	Coryza (common cold)	Rhinovirus Coronavirus Adenovirus para-influenza (esp. 3)	Influenza A,B Para-influenza 1, 2 RSV enterovirus Human metapneumovirus	Nasopharyngeal swab (NP) or washings in viral transport media (VTM) on ice; serum for serology
	Pharyngitis	Adenovirus HSV 1 Enterovirus EBV	Influenza A,B Para-influenza 1, 2 RSV rhinovirus Coronavirus CMV	Throat swab or washings in VTM on ice; serum for serology
	Croup	Para-influenza 1, 2, 3	Influenza A RSV Human metapneumovirus Rhinovirus	NP or washings in VTM on ice
	Bronchiolitis	RSV Para-influenza 3	Adenovirus Human metapneumovirus Para-influenza 1, 2 Influenza A,B Rhinovirus Enterovirus Coronavirus	NP or washings in VTM on ice; serum for serology
	Pneumonia	Influenza A RSV Para-influenza 3 Adenovirus CMV	Para-influenza 1, 2 Rhinovirus EBV Influenza B RSV VZV Coronavirus Human metapneumovirus Hantavirus HSV	Tracheal aspirate; lung biopsy; NP in VTM on ice; serum for serology

Continued

Table 2.9 Continued

Central nervous system	Aseptic meningitis	Enteroviruses (echovirus, Coxsackievirus A, B), HSV 1	LCMV mumps HSV 2 Parecho-viruses JCV arbovirus	Spinal fluid submitted on ice
	Encephalitis	HSV 1 enterovirus	Hemorrhagic fever viruses EBV Rabies CMV VZV Arboviruses HIV Measles mumps Parecho-viruses	Spinal fluid submitted on ice; brain biopsy; serum for serology
Gastrointestinal	Diarrhea	Rotavirus Norovirus Adenovirus 40, 41	Sapovirus CMV Parecho-Viruses Astrovirus	Feces
	Hepatitis	Hepatitis A Hepatitis B Hepatitis C	EBV CMV Hepatitis D Hepatitis E Adenovirus	Serum for serological and NA evaluation biopsy for histology, immune-staining
	Parotitis	Mumps	Para-influenza Adenovirus CMV Enteroviruses EBV HHV-6 HIV-1	NP or washings in VTM on ice PLUS urine; serum for serology
Cutaneous	Vesicular rash	HSV 1, 2 VZV	Enteroviruses Poxviruses	Aspirate or swab of lesion in VTM on ice; scraping of lesion for direct fluorescent antibody stain or histology
	Maculopapular	Enteroviruses HHV-6, 7	Adenovirus Parvovirus B19 Measles Rubella	NP or washings in VTM on ice PLUS feces; serum for NA or serology
Urogenital	Hemorrhagic cystitis	None frequent	Adenovirus (type 11) BK virus	Urine; serum for NA
	Urethritis Genital herpes Genital warts Molluscum contagiosum Cervicitis	HSV VZV HPV Molluscum contagiosum virus Adenovirus HSV		

Continued

Table 2.9 Continued

Cardiac	Myocarditis Pericarditis	Enterovirus (Coxsackievirus A, B; echovirus)	Adenovirus Influenza A Human metapneumovirus	NP aspirate/swab in VTM on ice PLUS feces; serum for serology
Ocular	Chorioretinitis	CMV HSV VZV		Conjunctival or corneal scraping
	Conjunctivitis Keratoconjunctivitis	Adenovirus Adenovirus HSV VZV	Enteroviruses	in VTM on ice; NP swab or aspirate in VTM on ice (adenovirus)
Fetal newborn		CMV Hepatitis B HIV Parecho (type 3) Parvovirus B19 Rubella Enterovirus	LCM	Tissue for culture; serum for serology or NA
Hematopoietic	Lymphoid disorders	EBV HIV HTLV-1		Serum for serology or NA
	Erythropoietic	Parvovirus B19		

Cult, culture; CMV, cytomegalovirus; EBV, Epstein–Barr virus; HSV, herpes simplex virus; HPV, human papillomavirus; LCMV, lymphocytic choriomeningitis virus; NA, nucleic acid amplification test; RSV, respiratory syncytial virus; VTM, viral transport medium; VZV, varicella zoster virus.

Table 2.10 Tests used for detecting different viruses.

Virus	Tests
Adenovirus	NA, culture
Adenovirus 40, 41	IA, culture
Arboviruses	Serology, NA
Astroviruses	NA
BK (polyomavirus)	NA, Histo
Caliciviruses	NA
Coronaviruses	NA, serology
Cytomegalovirus (mononucleosis)	Serology
Cytomegalovirus (pneumonia)	NA, culture
Cytomegalovirus (gut, liver)	NA, Histo, culture
Cytomegalovirus (congenital)	Culture, NA
Cytomegalovirus (eye)	NA
Enteroviruses	Culture, NA
Epstein–Barr virus (mononucleosis)	Serology
Epstein–Barr virus (PTLD, CNS)	Histo, NA
Hantaviruses	NA, serology
Hemorrhagic fever viruses	NA, serology*
Hepatitis A virus	Serology
Hepatitis B virus	Serology, NA, IA
Hepatitis C virus	NA, serology

Continued

Table 2.10 Continued

Hepatitis D virus	Serology
Hepatitis E virus	Serology
Herpes simplex virus (skin, mucosa)	Culture, NA
Herpes simplex virus (CNS)	NA
Human herpes virus 6	Serology, NA
Human immunodeficiency virus	Serology, NA
Human metapneumovirus	NA
Human papillomavirus	Histo, Cyto, NA
Human T lymphotropic virus	NA, serology
JC polyomavirus	NA, Histo
Lymphocytic choriomeningitis virus	Serology
Measles virus	NA, serology, culture*
Molluscum contagiosum virus	Histo
Mumps virus	Culture, NA, serology
Parainfluenza viruses	NA, culture
Parechoviruses	NA, culture
Parvovirus B19	NA, serology
Pox (smallpox, vaccinia)	NA, EM, culture*
Rabies virus	Histo, IA, serology#
Rhinoviruses	NA, culture
Rotavirus	IA
Rubella virus	Serology, NA
Varicella zoster virus	NA, DFA

CNS, central nervous system; Cyto, cytology; Histo, histology; IA, immunoassay; NA, nucleic acid amplification test; PTLD, posttransplant lymphoproliferative disease.
* Contact public health authority immediately.
\# Contact reference laboratory, Centers for Disease Control and Prevention.

remember that the highest viral titers in clinical specimens are present early in the course of most syndromes (first 3–4 days) and dramatically decrease with time.

Antiviral susceptibility testing

As more antiviral drugs become available, testing of a virus's susceptibility to them is becoming more important. The principles of testing are similar to those used for bacteria. They include (a) the virus's ability to grow in the presence of the antiviral agent (instead of the MIC used for antibacterial and antifungal susceptibility, the concentration that inhibits viral replication by 50% (ID_{50}) is often used); and (b) detection of gene mutations conferring resistance. As examples, these methods, especially the latter, are used for detecting resistance of HIV to antiretroviral drugs, and cytomegalovirus resistance to ganciclovir and foscarnet.

Detecting and identifying fungi

These methods are discussed in more detail in Chapter 18.

- Direct visualization: wet preparation after clarification of skin squames with KOH
- Stains:
  - Gram stain (see earlier in this chapter): yeasts
  - Methylene blue: yeasts, molds

  - Calcofluor white
  - Periodic acid Schiff (PAS)
  - Methenamine silver impregnation
  - Lactophenol cotton blue
- Culture:
  - Blood agar
  - Selective media, containing various antibacterial agents (many)
  - Chromogenic agars
  - Different temperatures
  - Morphology of cultured fungus: rate of growth, colony characteristics, conidia characteristics
- Antigen detection in body fluids:
  - blood: galactomannan, 1,3-β-D-glucan
  - blood, CSF: cryptococcal antigen
- Molecular methods, e.g. PCR
- Histology, with above-mentioned stains

Detecting and identifying parasites

These methods are discussed in more detail in Section V.

Laboratory safety

All specimens for microbiological work-up are deemed infectious and capable of transmitting disease. Specimens should be handled so as to prevent contamination of the specimen and the environment, and to prevent transmission of infection to the personnel obtaining and transporting the sample, and to those processing the sample in the laboratory. Microbiology technologists are at particular risk for exposure to infectious agents through manipulating isolates while performing identification assays and susceptibility testing because they are working with large numbers of organisms.

Therefore, clinicians should *warn laboratory personnel* when they submit specimens that they think might harbor particularly hazardous organisms, e.g. *Bacillus anthracis, Francisella tularensis, Yersinia pestis, Histoplasma capsulatum*, and hemorrhagic fever viruses.

Regulations from the Occupational Health and Safety Administration (OSHA) and the Hospital Infection Control Advisory Committee have guidelines in place to help prevent exposures. In addition, the Centers for Disease Control and Prevention (CDC) and the National Institutes of Health (NIH) publish consensus documents dealing with both general and specific biosafety issues (www.cdc.gov/biosafety/). The general laboratory practices that can reduce the risk of transmission are detailed below and include the use of biological safety cabinets or hoods (BSC). The most commonly used BSC in the microbiology laboratory is designated as Class IIa, where air is drawn into the cabinet by negative pressure, passes through a HEPA filter, and is recirculated in vertical sheets. This flow serves as a barrier between the outside and inside of the cabinet, containing aerosols within (protecting the user) and protecting samples from external contamination.

The following general guidelines are required for safe work practices.

Specimen handling

Personal protective equipment (PPE) – gloves and gowns – should be worn at all times when handling and processing clinical specimens; all specimens should be in leak-proof containers and transported in sealed plastic bags; all syringes must have needle-locking devices or plastic disposable needle units; tubes should be plastic and always carried in racks; centrifugation should take place with specimens in safety cups, which should be opened inside the BSC. Hands should be thoroughly washed after glove removal.

Specimen processing

All specimens must be processed within a BSC; when used, wire loops must be sterilized (an electric incinerator is preferred) and cooled prior to streaking samples, to avoid generation of aerosols; liquids should be mixed and transferred using sterile disposable pipettes; when vortexing or blending liquids, a secure lid or a Parafilm must be used to seal the containers; one should work over an absorbent covering or disinfectant-containing pad when working with fungi or mycobacteria, and use common sense and standard precautions to avoid exposures.

General

Laboratory personnel should apply the following practices: use a hospital/laboratory-approved disinfectant (note: daily prepared 10% household bleach is an excellent all-purpose disinfectant); disinfect all work benches before and after processing; immediately cover any spills or known contaminated areas with disinfectant-soaked wipes and notify the supervisor as needed; keep all work areas neat and without clutter; forbid food, drink or application of cosmetics within the work area; remove PPE when entering designated "clean" areas (lounge, bathroom, offices, etc.); dispose of all biohazardous waste and needles (or other sharps, such as blades) in appropriately labeled and designated areas.

Biosafety levels

Biosafety levels (BSLs) are guidelines that describe appropriate containment equipment, facilities, and procedures for use by laboratory workers and range from BSL 1 to BSL 4, based on the increased risk associated with the pathogenicity of the microorganisms. Routine clinical microbiology laboratories follow BSL 2 practices; however, when working with known or possible highly infectious agents where the risk of aerosol transmission is greater (e.g. *Brucella* spp., *Francisella* spp., *Mycobacterium tuberculosis*, and systemic fungi), BSL 3 practices are required. In cases of suspected bioterrorism agents, prions and prion diseases, specific guidelines can be found at www.cdc.gov/OD/ohs/biosfty/bmbl5/sections/SectionIV-LaboratoryBiosafetyLevel, and on the American Society for Microbiology website www.asm.org/index.php/guidelines/sentinel-guidelines.

Biosafety level 1 is recommended for work with microorganisms not known to cause disease in healthy adults. This includes access only to authorized personnel; have available sinks for hand washing and eyewash stations; ensuring that appropriate PPE is available and in use; use of bench-tops that are impervious to liquids and resistant to chemicals; disinfecting routine laboratory surfaces daily and ensuring that equipment is cleaned and disinfected on a regular basis or whenever contaminated; decontaminating infectious waste by autoclaving or other acceptable means or transporting it off-site to an approved disposal environment.

Biosafety level 2 is recommended for microorganisms associated with human disease but not transmitted by aerosols (e.g. *Salmonella* spp.). These include all BSL 1 recommendations plus: displaying universal biohazard signs prior to entry; performing all specimen processing in a BSC IIa hood (plates post incubation may be examined on a bench-top); using centrifuge safety cups for centrifugation and opening them under the hood; stressing use of the appropriate PPE (e.g. gowns, gloves, and facial barriers); ensuring that sharps are disposed of in puncture-resistant containers; trained personnel must adhere to uniform standard operating procedures in clinical microbiology and must be competency tested yearly.

Biosafety level 3 is for hazardous microorganisms primarily transmitted by aerosols (e.g. *M. tuberculosis*). These recommendations include BSL 2 plus the following: restricted and controlled access to the laboratory; manipulations of all specimens and cultures in a BSC (Class IIa); maintaining and monitoring negative-pressure airflow into the laboratory; it is recommended that the laboratory design includes double doors and an anteroom; discharging HEPA-filtered exhaust air from BSCs to the outside of the facility; stringent use and monitoring of all appropriate PPE and containment devices; use of HEPA-filtered respirators or N-95 masks when aerosols are anticipated; storing serum samples for baseline serology on all personnel in order to determine immune status before and after an exposure.

Biosafety level 4 is recommended for agents causing life-threatening or untreatable diseases, transmitted by aerosols or by unknown routes (e.g. Ebola virus). These include BSL 3 practices plus the following: changing into protective clothing before entering the laboratory and use of a full-body, air-supplied, positive-pressure suit for all procedures, and showering prior to exit; decontaminating all waste immediately on exit; use of a BSC (Class III) (such cabinets are completely enclosed, ventilated and of airtight construction; see CDC website for details); such facilities with specialized ventilation and waste management systems are generally separate and unique from other laboratories.

Finally, when specimens or organisms are to be transported to other facilities or reference laboratories for any level of follow-up (identification, ancillary susceptibility testing, typing, etc.), federal regulations exist for appropriate packaging and shipping to ensure safety. A summary of requirements can be found at www.cdc.gov/biosafety/publications/bmbl5/BMBL5_appendixC.pdf. Facilities shipping specimens and cultures are required to have technologists validated in the procedures through appropriate accreditation (e.g. state health laboratories, national training network, online courses, etc.).

Further reading

Baron EJ, Miller JMM, Weinstein M, et al. (2013) A guide to utilization of the microbiology laboratory for diagnosis of infectious diseases: 2013 recommendations by the Infectious Diseases Society of America (IDSA) and the American Society for Microbiology (ASM). Clin Infect Dis 57: e22–e121.

Baskova L, Buchta V (2012) Laboratory diagnostics of invasive fungal infections: an overview with emphasis on molecular approach. Folia Microbiol 57: 421–30.

Biswas S, Rolain J-M (2013) Use of MALDI-TOF mass spectrometery for identification of bacteria that are difficult to culture. J Microbiol Methods 92: 14–24.

Clerc O, Prod'hom G, Vogne C, et al. (2013) Impact of matrix-assisted laser desorption ionization time-of-flight mass spectrometry on the clinical management of patients with gram-negative bacteremia: a prospective observational study. Clin Infect Dis 56: 1101–7.

Denys GA, Gray LD (2010) Biohazards and safety. In: Garcia LS (ed.) *Clinical Microbiology Procedures Handbook*. Washington, DC: ASM Press.

Diekema DJ, Pfaller MA (2013) Rapid detection of antibiotic-resistant organism carriage for infection prevention. Clin Infect Dis 56: 1614–20.

Emonet S, Shah HN, Cherkaoui A, et al. (2010) Application and use of various mass spectrometry methods in clinical microbiology. Clin Microbiol Infect 16: 1604–13.

Ferreira-Gonzalez A, Versalovic J, Habbebu S, Calienda AM (2007) Molecular methods in diagnosis and monitoring of infectious diseases. In: Bruns DE, Ashwood ER, Burtis CA (eds) *Fundamentals of Molecular Diagnosis*. St Louis: Saunders Elsevier, pp. 171–96.

Jorgensen JH, Ferraro MJ (2009) Antimicrobial susceptibility testing: a review of general principles and contemporary practices. Clin Infect Dis 49: 1749.

Lagier J-C, Edouard S, Pagnier I, et al. (2015) Current and past strategies for bacterial culture in clinical microbiology. Clin Microbiol Rev 28: 208–36.

McClelland R (2001) Gram stain: the key to microbiology. Med Lab Obs 33: 20–31.

Mitsuma SF, Mansour MK, Dekker JP, et al. (2013) Promising new assays and technologies for the diagnosis and management of infectious diseases. Clin Infect Dis 56: 996–1002.

Muldrew KL (2009) Molecular diagnostics of infectious diseases. Curr Opin Pediatr 21: 102–11.

Rand KH, Tillan M (2006) Errors in interpretation of Gram stains from positive blood cultures. Am J Clin Pathol 126: 686–90.

Tissari P, Zumla A, Tarkka E, et al. (2010) Accurate and rapid identification of bacterial species from positive blood cultures with a DNA-based microarray platform: an observational study. Lancet 375: 224–30.

Resource

www.idsociety.org

SECTION II
Prions and viruses

CHAPTER 3
Prions

These agents cause a group of diseases called the spongiform encephalopathies. They are named for **Pro**teinaceous **In**fectious particles. They are abnormal proteins with an abnormal folding structure (β-pleated sheets), that have an equivalent protein with the same primary structure within the host, that folds normally (2 α-helices). The function of this protein, whose gene is located on chromosome 20, is unknown. The misfolded protein molecules form aggregates ("amyloid") that interfere with cellular function. The normal protein is called PrP^{c} (for prion protein cellular), and the abnormal protein is called PrP^{sc} (for scrapie, a transmissable infectious spongiform encephalopathy of sheep and goats). The abnormal proteins transmit their information to the normal proteins, causing them to fold abnormally. The abnormal aggregates lead to tissue damage, which affects primarily the brain, leading to a slowly progressive encephalopathy and death.

The diseases they cause can be sporadic, such as Creutzveldt–Jakob disease (sCJD), genetic, such as fatal familial insomnia, or infectious. The infectious diseases caused by prions are:

- bovine spongiform encephalopathy (BSE) also known as "mad cow disease." Cows contracted it from feeds containing animal products, and humans contract it by eating infected cows, causing a disease called variant Creutzveldt–Jakob disease (vCJD). It has an incubation period of many years
- kuru: this disease, which used to occur in New Guinea, was transmitted by cannibalism. It has disappeared since the outlawing of this practice.

Several prion diseases occur in animals, such as scrapie (see above), chronic wasting disease in various species of deer, and exotic ungulate encephalopathy in several species of African antelope.

Prions are not susceptible to proteases nor to usual sterilization methods. This has led to several outbreaks of disease transmission among humans, through the use of human growth hormone derived from human pituitary glands and dura mater. A case has been reported following blood transfusion.

Practical Medical Microbiology for Clinicians, First Edition. Frank E. Berkowitz and Robert C. Jerris.

Diagnosis

The definitive diagnostic test is histopathology. Clinical features and electroencephalography can suggest the diagnosis of the sporadic and genetic varieties, and the presence of certain proteins in cerebrospinal fluid (CSF) (protein 14-3-3 and Tau protein) might help in the diagnosis. A particular magnetic resonance imaging (MRI) finding, the "pulvinar sign" (enhancement of the posterior thalamus), is helpful in the diagnosis of vCJD. Recently, two *in vitro* tests employing amplification technology have been used to detect prion protein of CJD in CSF and olfactory mucosa, and in urine respectively.

Further reading

Collins SJ, Lawson VA, Masters CL (2004) Transmissible spongiform encephalopathies. Lancet 363: 51–61.

Knight R (2008) Clinical features and diagnosis of human prion diseases. Future Neurol 3: 473–81.

Marandi Y, Farahi N, Sadeghi A, et al. (2012) Prion diseases – current theories and potential therapies. Folia Neuropathol 50: 46–9.

Moda F, Gambetti P, Notari S, et al. (2014) Prions in the urine of patients with Variant Creutzveld-Jakob Disease. N Engl J Med 371: 530–9.

Norrby E (2011) Prions and protein-folding diseases. J Intern Med 270: 1–14.

Orru CD, Bongianni M, Tonoli G, et al. (2014) A test for Creutzveld-Jakob Disease using nasal brushings. N Engl J Med 371: 519–29.

CHAPTER 4
General virology

Properties of viruses

Viruses are small microorganisms that replicate by subverting the host cell's biochemical machinery to make more of themselves. They therefore require a host cell for replication. However, many can survive in the environment outside a cell. The duration of their survival depends on viral factors, such as whether or not it has a lipid envelope (which makes it more susceptible to external influences) and environmental factors such as temperature and humidity. They spread from within the host cell to other cells, either directly from one cell to the next, by causing fusion of adjacent cell membranes, or via extracellular fluid.

Viruses consist of a nucleic acid core, either DNA or RNA (not both) and structural proteins. In most viruses, these are arranged in one of two basic structural forms:

- icosahedral, in which the nucleic acid core is surrounded by a protein capsid
- helical, in which the nucleocapsid (nucleic acid core with attached proteins) is arranged in a helix).

In addition, a few viral families are spherical, brick-shaped, or filamentous in shape. The manner by which they invade cells, replicate, and spread varies according to virus type, but they share the following basic mechanisms:

- attachment to the host cell
- entry
- uncoating
- production of proteins
- replication of nucleic acid
- assembly of new viral particles
- extrusion of new particles from the cell, by budding through the cell membrane or by cell lysis.

These stages are very important for the design of antiviral agents. Different agents can target different stages of the replication cycle. The antiretroviral agents used for the treatment of patients with HIV infection demonstrate this very well, with different agents attacking different stages of the cell cycle (see Retroviruses in Chapter 6).

Practical Medical Microbiology for Clinicians, First Edition. Frank E. Berkowitz and Robert C. Jerris.

Taxonomy of viruses

The scheme

Viruses can be classified into:

- Order: virales
- Family: viridae, e.g. paramyxoviridae
- Subfamily: e.g. paramyxovirinae
- Genus: e.g. pneumovirus
- Species: e.g. respiratory syncytial virus.

The classification system proposed by David Baltimore is based on the method by which the virus codes for messenger RNA, as follows.

i. Double-stranded DNA: herpesviridae, adenoviridae, poxviridae, papillomaviridae
ii. Single-stranded (+ve sense) DNA: parvoviridae
iii. Double-stranded RNA: reoviridae
iv. Single-stranded (+ve sense) RNA: picornaviridae, togaviridae, flaviviridae, caliciviridae, coronaviridae, astroviridae
v. Single-stranded (–ve sense) RNA: orthomyxoviridae, paramyxoviridae, rhabdoviridae, arenaviridae, bunyaviridae, filoviridae
vi. Single-stranded RNA – reverse transcripting: retroviridae
vii. Double-stranded DNA – reverse transcripting: hepadnaviridae

The names

DNA viruses

Herpesviridae (herpesviruses)
  α-Herpesviruses
    Herpes simplex 1 and 2
    Varicella zoster virus
  β-Herpesviruses
    Cytomegalovirus
    Herpesviruses 6 and 7
  γ-Herpesviruses
    Epstein–Barr virus
Adenoviridae (adenoviruses)
Poxviridae (poxviruses)
  Orthopox viruses
    Smallpox
    Vaccinia
    Cowpox
    Monkeypox
Hepadnaviridae (hepadnaviruses)
  Hepatitis B
Papillomaviridae (papillomaviruses)
  Multiple types
Polyomaviridae (polyomaviruses)
  BK
  JC
  Merkel tumor

Parvoviridae (parvoviruses)
- B19
- Boca

RNA viruses

Picornaviridae (picornaviruses)
- Enteroviruses
- Hepatitis A
- Parechoviruses

Paramyxoviridae (paramyxoviruses)
- Parainfluenzaviruses
- Respiratory syncytial virus
- Human metapneumovirus
- Mumps virus
- Measles virus
- Hendra virus
- Nipah virus

Orthomyxoviridae (orthomyxoviruses)
- Influenza viruses A, B, and C

Flaviviridae (flaviviruses)
- Yellow fever
- Dengue
- West Nile
- Japanese encephalitis virus
- Many arthropod-borne viruses
- Hepatitis C

Togaviridae (togaviruses)
- Rubiviruses
  - Rubella
- Alphaviruses
  - Equine encephalitis viruses
  - Many arthropod-borne viruses

Bunyaviridae (bunyaviruses)
- Nairoviruses
- Bunyaviruses
- Phleboviruses
- Hantaviruses

Arenaviridae (arenaviruses)
- Lassa fever
- Lymphocytic choriomeningitis virus

Rhabdoviridae (rhabdoviruses)
- Rabies
- Several bat-borne viruses

Caliciviridae (caliciviruses)
- Norovirus
- Sapovirus

Astroviridae (astroviruses)

Reoviridae (reoviruses)
  Rotavirus
  Colorado tick fever
Filoviridae (filiviruses)
  Ebola virus
  Marburg virus
Retroviridae (retroviruses)
  Human immunodeficiency viruses (HIV), 1 and 2
  Human T-cell lymphotropic virus
Coronaviridae (coronaviruses)
  SARS coronavirus
  MERS coronavirus
  Several serotypes
Hepeviridae (hepeviruses)
  Hepatitis E

Further reading

Baltimore D (1971) Expression of animal virus genomes. Bacteriol Rev 35: 235–41.
Collier L, Kellam P, Oxford J (2011) Human Virology, 4th edn. Oxford: Oxford University Press.
International Committee on Taxonomy of Viruses. www.ictvonline.org

CHAPTER 5

DNA viruses (excluding hepatitis B virus)

Herpesviruses (Herpesviridae)

Herpesviruses are large, enveloped, double-stranded DNA viruses that include several viruses that are of importance in humans. They all have in common the property of latency, i.e. once they have been acquired, whether or not they cause overt illness, they remain dormant within the host and can become reactivated. Clinical syndromes caused by the herpesviruses are shown in Table 5.1.

α-Herpesviruses (α-herpesvirinae)

Herpes simplex viruses (HSV) 1 and 2

These alpha herpesviruses are common causes of infection. They cause mainly mucosal and cutaneous infections, which are not usually severe, but they can spread to cause visceral disease, especially of the brain. HSV 1 affects mainly the oral mucosa, and is frequently acquired by young children, in whom it causes oral, tongue, and gum infection. HSV 2 causes mostly genital infection, as a result of sexual transmission. However, type 1 can cause genital infection and type 2 can cause oral infection. Reactivations cause mucocutaneous lesions. The newborn can acquire the infection during the birth process, and can suffer devastating disease in the form of encephalitis or disseminated multiorgan infection (Fig. 5.1). After initial infection, the virus becomes latent in nerve ganglia. It can reactivate recurrently, causing mucocutaneous lesions, typically at the mucocutaneous junction of the lip (herpes labialis, commonly called fever blister or cold sore) or of the genital area.

Herpes simplex virus can be detected readily by viral culture. It causes cytopathic effects, destroying the cell monolayer within a few days. Using immunofluorescent staining for early antigen expression in monolayers, it can be detected by culture within 1–2 days. Rapid testing by immunofluorescence can be performed directly on scrapings of the base of a skin or mucosal lesion (direct fluorescence assay – DFA, see Chapter 2). The virus can also be detected in cerebrospinal fluid and blood by polymerase chain reaction (PCR). This has facilitated the diagnosis of encephalitis, for which therapy is available. The standard treatment of patients with severe disease is

Practical Medical Microbiology for Clinicians, First Edition. Frank E. Berkowitz and Robert C. Jerris.

Table 5.1 Infections caused by herpesviruses, diagnostic tests, and treatment.

Virus	Clinical syndromes	Diagnosis	Treatment
HSV 1	Gingivostomatitis, keratitis, skin lesions, encephalitis, pneumonia, disseminated disease; reactivation: herpes labialis	Culture Immunofluorescence PCR-CSF, blood, lesions	Acyclovir, foscarnet; trifluorothymidine (keratitis only)
HSV 2	Genital infection, cervicitis neonatal infection (see text)	As for HSV 1	As for HSV 1
VZV	Chickenpox Zoster	Clinical Immunofluorescence, PCR – lesions	Acyclovir, foscarnet
CMV	Fever, lymphadenopathy; hepatitis OI – pneumonia, retinitis, colitis IU – encephalitis, systemic disease	Culture (usually urine), antigen detection, PCR, serology	Ganciclovir, foscarnet, cidofovir
EBV	Fever, infectious mononucleosis; PTLD, neoplasia	Serology, PCR	None
HHV-6	Fever, roseola infantum, OI	Serology, PCR	Ganciclovir, foscarnet, cidofovir
HHV-7	Fever, roseola infantum	Serology, PCR	None
HHV-8	Kaposi sarcoma	Histology	Chemotherapy

CMV, cytomegalovirus; CSF, cerebrospinal fluid; EBV, Epstein–Barr virus; HHV, human herpesvirus; HSV, herpes simplex virus; IU, intrauterine infection; OI, opportunistic infection; PCR, polymerase chain reaction; PTLD, posttransplant lymphoproliferative disorder; VZV, varicella zoster virus.

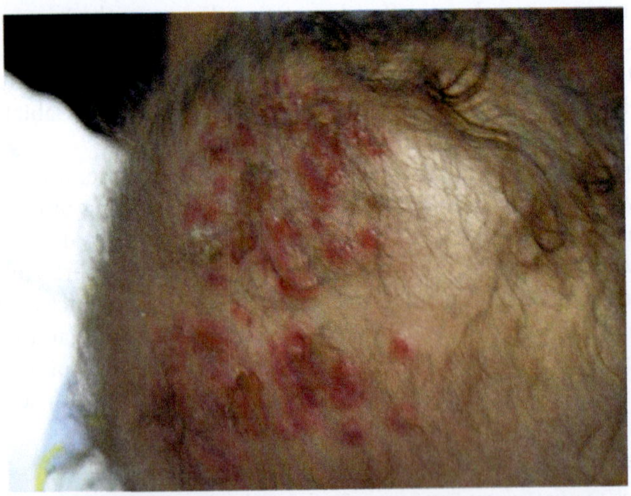

Fig. 5.1 Cutaneous herpes simplex infection in a newborn infant.

the nucleotide analog acyclovir (see later in this chapter). In cases of resistance to acyclovir, foscarnet, a drug with a different mode of action, should be used.

Varicella zoster virus (VZV) (human herpesvirus 3)

This virus causes two main clinical syndromes, varicella (chickenpox) and zoster, which is the clinical manifestation of reactivated virus. Varicella is a systemic infection, acquired by inhalation or direct contact with skin lesions. It has an incubation period of 10–21 days. It is characterized by fever and a rash that progresses from papules to vesicles to pustules, which scab over. All stages are present at one time (Fig. 5.2). Although it is usually benign, it can be complicated by secondary bacterial infection with *Streptococcus pyogenes* or *Staphylococcus aureus*, by pneumonia in immunocompromised individuals and normal adults, and by encephalitis. The virus becomes dormant in sensory ganglia, and when it becomes reactivated (an unpredictable event), it causes a painful vesicular rash along 1–3 dermatomes. This is zoster or shingles (named for the Latin word *cingulum*, meaning girdle) (Fig. 5.3).

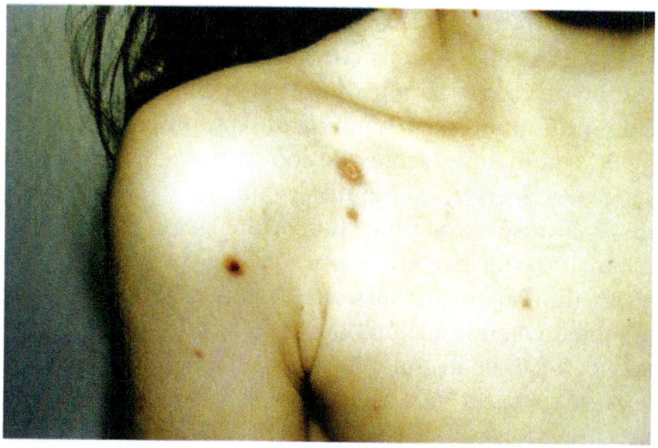

Fig. 5.2 A normal child with varicella. Note different staged lesions. Copyright © 2007 Frank E. Berkowitz. Reprinted with permission of Cambridge University Press.

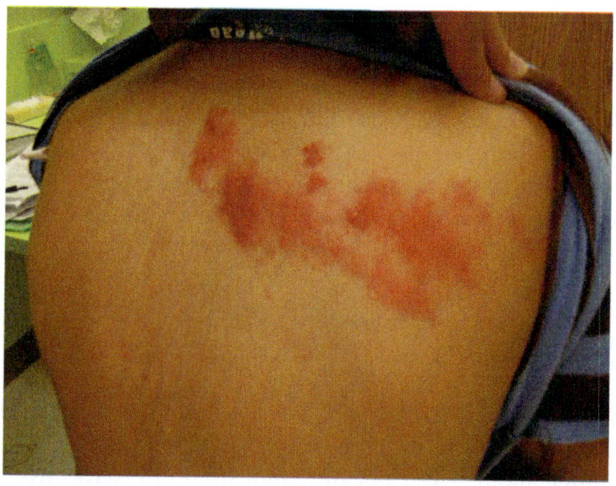

Fig. 5.3 Zoster in a teenage boy.

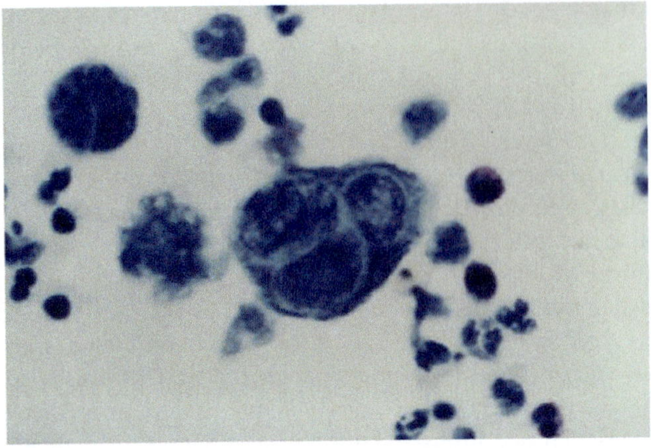

Fig. 5.4 Tzanck preparation, showing a multinucleate giant cell, which can occur with herpes simplex and varicella zoster virus infections. Courtesy of PHIL, CDC.

The diagnosis of VZV infection can usually be made clinically. The main differential diagnosis is infection with herpes simplex virus. Confirmation, seldom necessary, can be made by direct immunofluorescent staining of scrapings of the base of a lesion (see HSV earlier) or by PCR of material from a lesion. Although the virus can be cultured in tissue culture, it grows slowly, making this impractical.

Both HSV and VZV cause cytopathic changes in tissue characterized by multinucleate giant cells and intranuclear inclusions. These can be detected by staining of lesion scrapings with the Papanicolaou stain. This is called a Tzanck preparation. It does not differentiate between infections caused by the two kinds of virus. Although it is not very sensitive, it can be useful when a rapid diagnosis is required (Fig. 5.4).

Patients with severe forms of VZV infection or immunocompromised hosts should be treated with acyclovir. Resistance to this drug is due to the same mechanism as in HSV, though rare. An alternative therapy is foscarnet.

β-Herpesviruses (β-herpesvirinae)

Cytomegalovirus

This virus infects many humans during childhood. In normal individuals, it causes a range of illnesses including a non-specific febrile illness, an infectious mononucleosis-like illness (fever, pharyngitis, cervical lymphadenopathy, and, splenomegaly; see Epstein–Barr virus infection, later in this chapter), and hepatitis. It is a very important cause of opportunistic infection in individuals with impaired cell-mediated immunity, in whom it can cause pneumonia, colitis, retinitis, and encephalitis. In the fetus, it can cause encephalitis, hepatitis, and anemia. It is the most common cause of intrauterine infection in the USA.

The optimal diagnostic test depends on the clinical circumstance.

Serology

Detection of IgG antibodies should be used to determine whether an individual has ever been infected, such as early in pregnancy, and in an organ donor and recipient; repeated testing after about 2 weeks can demonstrate, by seroconversion, whether infection has occurred. Detection of IgM antibodies should be used to determine whether a normal individual has a current or recent infection (if this information is deemed necessary, e.g. pregnancy), but it is not as reliable as seroconversion. In pregnancy, serologic testing should utilize avidity assays, which measure how tightly the antibody binds to the antigen. Early in infection, the antibodies have low avidity, while later they have high affinity.

Culture

The virus grows slowly in tissue culture, taking weeks to produce cytopathic effects. However, by using the shell vial technique, in which early-expressed antigens on infected cells can be detected by immunofluorescence, the culture result can be obtained within 2 days. Culture of a newborn's urine should be used to detect intrauterine infection; this should be done within 3 weeks of life to eliminate the possibility that a positive culture is the result of postnatal infection. Cultured virus could be used for phenotypic antiviral resistance testing, although genotypic testing (detecting resistance gene mutations, e.g. in the UL97 gene) by a molecular method is the method usually used.

Polymerase chain reaction

Polymerase chain reaction on blood specimens should be used to detect and quantitate the virus in immunocompromised individuals. This method, performed on cerebrospinal fluid, should also be used to diagnose infection of the central nervous system.

Pathology

The organism produces histologic and cytologic changes characterized by large cell cells (megalo-) containing intranuclear inclusions (Fig. 5.5).

Treatment

Three drugs are available for treating patients (generally immunocompromised individuals) with CMV infection, namely ganciclovir, foscarnet, and cidofovir.

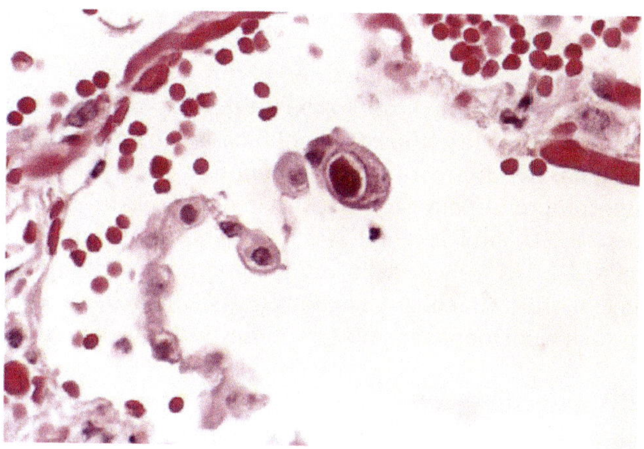

Fig. 5.5 Cytomegalovirus infection of the lung in a patient with AIDS, showing a large cell with an intranuclear inclusion. Courtesy of PHIL, CDC.

Human herpesvirus 6 (HHV-6)

This virus, of which there are types A and B, is lymphotropic. Type B is the main cause of human infection. It infects most individuals during infancy and early childhood. It becomes latent as circular DNA in various cells, and in a minority of cases it becomes integrated into the host chromosome. Clinically it can cause a non-specific febrile illness, and it is the main cause of roseola infantum (exanthem subitum). This typically manifests in infants with acute onset of high fever, sometimes associated with a febrile seizure. The fever lasts for a few days, and after it abates, a diffuse maculopapular rash appears. However, fever in the absence of rash is common. (Together with HHV-7, HHV-6 belongs to the Roseolovirus genus of β-herpesviruses.) The virus can cause severe disease, affecting various organs, such as encephalitis, in immunocompromised hosts such as bone marrow transplant patients.

Diagnosis

When necessary to make the diagnosis (only in special circumstances), serology, with IgG and IGM or seroconversion, should be used. Although PCR is widely used for demonstrating the presence of the virus in immunocompromised hosts, a positive test is difficult to interpret, because it could be the result of virus integrated into the host cell DNA, that is not playing a role in the patient's illness. The virus is susceptible to ganciclovir, foscarnet, and cidofovir.

Human herpesvirus 7 (HHV-7)

This is also a lymphotropic virus that can cause similar illness to that caused by HHV-6.

γ-Herpesviruses (γ-herpesvirinae)

Epstein–Barr virus (EBV)

This virus infects B-cells, in which it can integrate into the host cell genome. It is transmitted mainly in saliva. Like CMV, it is frequently acquired during childhood, when it manifests as a non-specific febrile illness. The classic manifestation is infectious mononucleosis, characterized by fever, pharyngotonsillitis, cervical lymphadenopathy, frequently generalized lymphadenopathy, and splenomegaly. This infection can cause inflammation of any organ, such as encephalitis and hepatitis; as a result of the production of non-specific antibodies, it can be associated with a wide variety of autoimmune complications, such as hemolytic anemia. The tonsillitis can be very severe, with pseudomembrane formation and upper airway compromise. Splenic rupture, though rare, is a potentially lethal complication of this infection.

Epstein–Barr virus plays an important role in the development of several neoplasms, the most important of which are Burkitt lymphoma in Africans (Fig. 5.6) and nasopharyngeal carcinoma, which is particularly common in eastern Asia.

In the immunocompromised host, it is associated with posttransplant lymphoproliferative disorder (PTLD) and cerebral B-cell lymphoma. In an X-linked primary immunodeficiency disease, called X-linked lymphoproliferative disease (XLP) or Duncan syndrome, there is an inability to control EBV. The males die of complications of EBV infection such as hemophagocytic lymphohistiocytosis, or lymphoma.

Diagnosis of EBV infection

The main question to be asked is the following: is a viral diagnosis necessary? In a normal host, the reasons to make a definitive diagnosis are threefold: to limit the performance of other diagnostic tests; to withhold the administration of antimicrobial

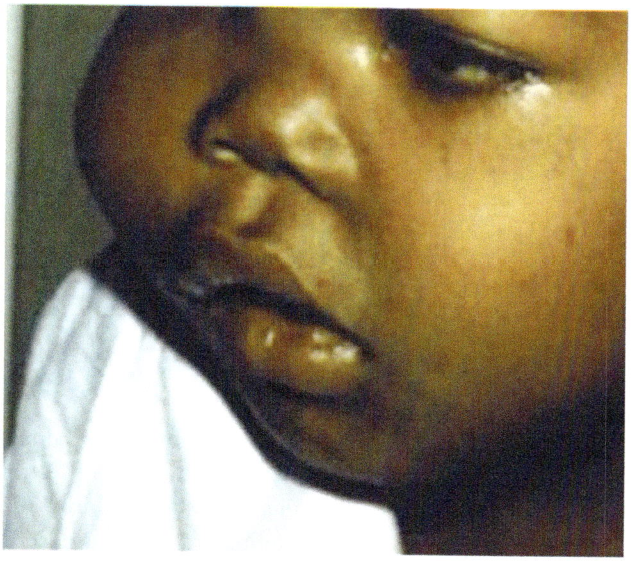

Fig. 5.6 African child with Burkitt lymphoma. Copyright © 2007 Frank E. Berkowitz. Reprinted with the permission of Cambridge University Press.

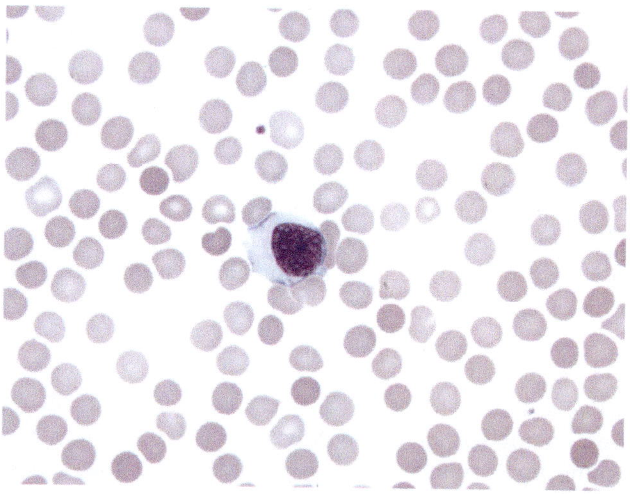

Fig. 5.7 Large "atypical lymphocyte" called Downy type II cells, in a teenage boy with infectious mononucleosis. Courtesy of Dr Sunita Park.

therapy in a febrile patient; and to advise patients against activities that could result in splenic injury, for 1 month. The following tests can be used in helping to make the diagnosis.

Blood smear

In infectious mononucleosis, a blood smear may show large lymphocytes called Downy type II cells, which are T-lymphocytes proliferating to eliminate infected B-lymphocytes (Fig. 5.7).

Table 5.2 Serological responses to different EBV antigens and the timing of their presence.

Antibody to	No infection	Acute infection	Recent infection	Past infection
EA	–	+	+	–
VCA IgM	–	+	+/–	–
VCA IgG	–	+/–	+	+
NA	–	–	–	+

EA, early antigen; NA, nuclear antigen; VCA, viral capsid antigen.

Serology

This is the main method by which acute infectious mononucleosis due to this virus is diagnosed. This can be done by the following tests:

- demonstration of antibodies to sheep red cells after absorption by beef kidney cells (Paul Bunnell test, Monospot test). This demonstrates "nonsense antibodies," is fairly specific, cheap, and fairly rapid
- specific antibody tests to viral antigens. Generally, a series of tests to different viral antigens are performed and from the pattern, the occurrence and timing of the infection can be determined to some extent. These are shown in Table 5.2.

Although the virus can be cultured, this requires the demonstration of immortalization of B-cells, which is impractical for clinical diagnostic laboratories.

Polymerase chain reaction

This is performed on serum and can be used quantitatively (i.e. viral load determination). Its value is primarily in immunocompromised hosts, such as transplant patients, in whom a rising viral load can indicate the development of posttransplant lymphoproliferative disease. It should not be used for the diagnosis of acute infectious mononucleosis in normal hosts.

Treatment

There is no effective antiviral therapy available for patients with EBV infection. For those with posttransplant lymphoproliferative disease, the mainstay of therapy is reduction of immunosuppression as much as possible. If the disease has progressed to lymphoma, chemotherapy is used.

Human herpes virus 8 (HHV-8)

This virus has a worldwide distribution, but its prevalence is very different among different regions of the world, being most common in Africa and parts of the Amazon region, least common in North America, northern Europe and Asia, and of intermediate prevalence in the Mediterranean and Caribbean areas. It is transmitted by close contact, including sexual intercourse. It becomes latent in infected cells and can become lytic, which is a necessary precursor to causing disease. It is associated with Kaposi sarcoma, of which there are several different patterns, primary effusion lymphoma, and Castleman disease. Individuals who are immunosuppressed, such as those with acquired immunodeficiency syndrome (AIDS), or who have undergone organ transplantation are at particular risk for these conditions, although in high-risk areas, normal individuals can develop Kaposi sarcoma. The initial infection can be asymptomatic or associated with mononucleosis-like manifestations.

Drugs used for the treatment of patients with herpesvirus infections

Most of these act by interfering with viral DNA synthesis.

Acyclovir

This is an acyclic deoxyguanosine analog with activity against herpes simplex viruses and varicella zoster virus. It is monophosphorylated by a virus-encoded thymidine kinase and then by cellular kinases to acyclovir triphosphate. This enters the nucleus and, after pyrophosphate is removed, it is incorporated into the DNA, resulting in chain termination. It also inhibits DNA polymerase. Resistance is due to mutations of thymidine kinase or altered viral DNA polymerase. This drug is poorly absorbed from the intestine. Therefore, a prodrug, valaciclovir, can be used for oral therapy in many situations.

Penciclovir and its prodrug, famciclovir, are acyclic guanosine analogs which also inhibit DNA polymerase, but do not cause DNA chain termination.

Ganciclovir

This is a guanosine analog, active against herpes simplex viruses, cytomegalovirus, and HHV-6. It is also monophosphorylated by virus-encoded enzymes, namely thymidine kinase in herpes simplex infected cells, and by a phosphotransferase (encoded by the UL97 gene) in CMV-infected cells. It inhibits viral DNA polymerase and prevents deoxyguanidine triphosphate incorporation into DNA. It is predictably myelotoxic. Valganciclovir is a prodrug that is well absorbed by the intestine.

Cidofovir

This is a cytidine nucleotide analog that is phosphorylated by host cell-encoded enzymes. Therefore, there is no cross-resistance with acyclovir- and ganciclovir-resistant viruses, in which the mechanism of resistance is in the monophosphorylation. The active form, a diphosphonate, causes DNA chain termination. It is used for infections resistant to acyclovir or ganciclovir. It has also been used in bone marrow transplant patients with severe adenovirus infections and in renal transplant patients with BK virus infections.

Foscarnet (trisodium phosphonoformate)

This is an analog of inorganic pyrophosphate, which inhibits herpesvirus DNA polymerase, and HIV reverse transcriptase. It is used mainly for individuals with CMV infection resistant to ganciclovir.

Idoxuridine

This is a thymidine analog, used only topically.

Trifluridine

This is a pyrimidine nucleoside analog that inhibits thymidylate synthase. It is used topically only for ocular infection with HSV.

Fomivirsen

This is a 21-base oligonucleotide that is used only for intravitreous injection in cases of CMV retinitis.

Docosanol

This is a long chain alcohol that prevents HSV attachment to the host cell. It is used topically only.

Adenoviruses (Adenoviridae)

Adenoviruses are double-stranded DNA viruses, with icosahedral symmetry, measuring 70–90 nm in diameter. They have 252 capsomeres, consisting of 240 hexagons and 12 pentagons. The pentagons, which form the points of the icosahedron, each have a long filament with a terminal knob extending outward from them, which attaches to a cellular receptor. Because adenoviruses are non-enveloped, they can survive in the environment for relatively long periods. There are eight species of adenovirus (A, B1, B2, C, D, E, F, and G) and 52 different serotypes. Some serotypes have several different genotypes. They have tropism for several different cells, and they can become latent. Reactivation of latent virus accounts for most cases of disease in severely immunocompromised hosts (see later in this chapter). They cause a wide range of clinical manifestations, according to their serotype, in normal and immunocompromised hosts. They have caused major outbreaks, especially of respiratory tract infections, in closed communities, such as the military.

In normal hosts they cause mostly upper and lower respiratory tract infections, including severe pneumonia (mostly serotypes 3, 4, 5, 7, and 21), conjunctivitis and keratitis (serotypes 1, 2, 3, 7, 8, 11, 19, and 31) and hemorrhagic cystitis (serotypes 7, 11, 34, and 35). Serotypes 40 and 41 cause acute gastroenteritis.

In immunocompromised hosts, such as transplant patients, adenoviruses can cause severe and sometimes fatal infections. These include, in addition to the above, hepatitis, encephalitis, pancreatitis, colitis, and disseminated disease.

Diagnosis

Adenoviruses can be grown in tissue culture, detected by antigen, and by PCR-based methods, for detection and quantitation. The latter have become the methods of choice. Serotypes 40 and 41 cannot be grown in the usual tissue culture cells (HeLa, A549, or Hep-2 cells), but require Chinese hamster ovary (CHO) cells for their growth *in vitro*.

When should diagnostic testing be performed?

In most situations, a clinical diagnosis of a viral respiratory tract infection or conjunctivitis is adequately precise, and a microbiological diagnosis is not necessary. However, knowing that an illness is caused by adenovirus might result in the withholding of antibacterial therapy, which is of value. When an outbreak occurs, identification of the pathogen is important for public health interventions. In this circumstance, the serotype and genotype should be determined. If attempts to detect the organism are undertaken, specimens from the site of infection should be submitted. (Pharyngeal specimens should be used in cases of respiratory disease.) In immunocompromised individuals, such as bone marrow transplant patients, in whom adenovirus can cause fatal disease, a diagnosis is important, because antiviral therapy might be attempted. In such patients, a blood specimen for PCR should also be taken. Drugs that have been used in this circumstance are ribavirin and cidofovir. Although there are anecdotal reports of success with these forms of therapy, controlled trials have not been performed.

An oral vaccine against types 4 and 7 is available for the US military.

Polyomaviruses (Polyomaviridae)

Polyomaviruses are small, double-stranded, icosahedral, non-enveloped DNA viruses. Their name is derived from their ability to cause multiple types of tumors in animals. Although they commonly infect humans, mostly in childhood, they cause disease only in immunocompromised hosts.

BK virus

This infects urinary tract epithelium and may be excreted in the urine in normal individuals. In transplant patients, almost entirely renal transplant patients, it causes a nephropathy (polyomavirus-associated nephropathy – PVAN, or BK virus-associated nephropathy, BKVAN), which is an important cause of renal allograft rejection. It causes an interstitial nephritis, which is diagnosed histologically by the presence of intranuclear viral inclusions in tubular cells, with associated inflammation. The virus can be demonstrated in the urine by the presence of viral DNA (by PCR), VP 1 mRNA, or "decoy cells." These are renal tubular epithelial cells containing basophilic intranuclear inclusions, when stained with the Papanicolaou stain. The persistent presence of the urinary virus predicts viremia, which in turn predicts nephropathy. The management entails reducing immunosuppression, if possible. Cidofovir has been used, but its value has not been demonstrated in controlled trials.

JC virus

This causes progressive multifocal leukoencephalopathy in immunocompromised hosts, such as those with AIDS. This is a slowly progressive, ultimately fatal white matter disease, characterized by demyelination due to infection of myelin-producing glial cells. The virus can be detected in cerebrospinal fluid by PCR. There is no treatment, other than improving the immune function of the patient.

SV 40 virus

This is a simian polyoma virus, that entered into humans through polio vaccine made in infected monkey kidney cells. It is not known to cause any disease in humans.

Recent developments

Three polyomaviruses have been found in humans recently:

- KIV (named for the Karolinska Institute in Finland)
- WUV (named for Washington University)
- MCV – Merkel cell virus. Merkel cell carcinoma is a rare skin cancer of neuroendocrine skin cells, occurring in the elderly and immunocompromised hosts. This virus has been found in the tumors of about 80% of cases.

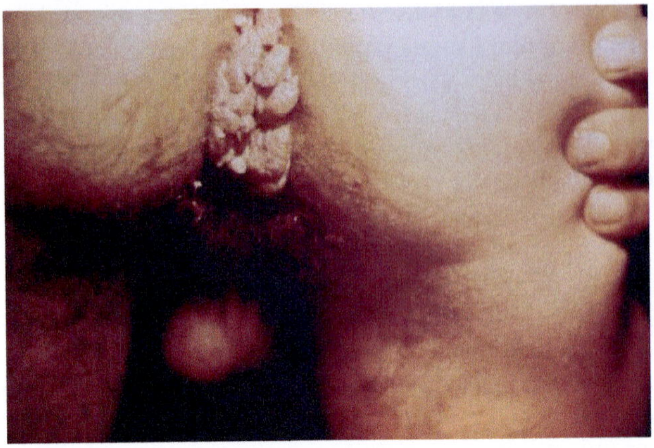

Fig. 5.8 Anal warts (condylomata acuminata). Courtesy of PHIL, CDC.

Papillomaviruses (Papillomaviridae)

These are icosahedral, non-enveloped, double-stranded DNA viruses. They infect the basal layer of stratified squamous epithelia, namely skin, oral, laryngeal, vaginal, and cervical mucosae, with their DNA being integrated into the host cell DNA. There are about 130 genotypes of human papillomavirus (HPV), different types being associated with disease in different sites. They have not been cultured *in vitro*, so their study has depended largely on molecular techniques.
Infections caused by papillomaviruses include:

- common warts, present on skin
- flat warts
- plantar and palmar warts
- genital warts (condylomata acuminata) (Fig. 5.8)
- laryngeal papillomas: in children, these are due to entry of virus from genital warts in the mother.

Cervical infection

Almost all cases of cervical cancer are due to human papillomavirus. Although only a small proportion of cases of infection develop cancer, the infection is so common that this is a very common cancer worldwide. Of the many types that cause cervical infection, the main types incriminated in cervical cancer are 16 and 18, which are included in the human papillomavirus vaccine (see later in this chapter), 31, and 45. Screening for cervical cancer has been practiced for decades, using exfoliative cytology with the Papanicolaou stain ("Pap smear"); however, screening for high-risk genotypes of papillomavirus is now being recommended.

Other cancers

Human papillomavirus causes most cases of cancer of the vulva, vagina, penis, and anus, and about half of the cases of oropharyngeal cancer.

Diagnosis

The diagnosis of papillomavirus infection is usually made clinically, but can be made histologically. The diagnosis of infection of the cervix and of the various stages of malignant transformation is very important. It can be accomplished by the use of cytopathology, using the Pap smear, molecular techniques, and, if necessary, biopsy. At present, serology is not useful.

Treatment

Various chemical agents (podophylline, salicyclic acid) and physical agents (freezing, electrocautery), and surgery have been used for the treatment of patients with various types of warts.

Discussion of the management of patients with cervical changes due to HPV is beyond the scope of this book.

Prevention

Human papillomavirus vaccines have had a very significant impact on the rate of cervical infections and cancer. Currently, in the USA, two vaccines, consisting of non-infectious capsid material, are offered to boys and girls from the age of 9 years. One vaccine targets HPV 16 and 18, and the other targets HPV 6, 11, 16, and 18. A nonavalent vaccine against types 6, 11, 16, 18, 31, 33, 45, 52, and 58 has been approved for use in the USA.

Poxviruses (Poxviridae)

The Poxviridae are very large double-stranded DNA viruses, containing more than 150 genes (Fig. 5.9). They are the only DNA viruses that replicate their DNA in the host cell cytoplasm, using their own enzymes.

There are two subfamilies, namely the Entomopoxvirinae, that infect only insects, and the Chordopoxvirinae, that infect vertebrates. Within the Chordopoxvirinae,

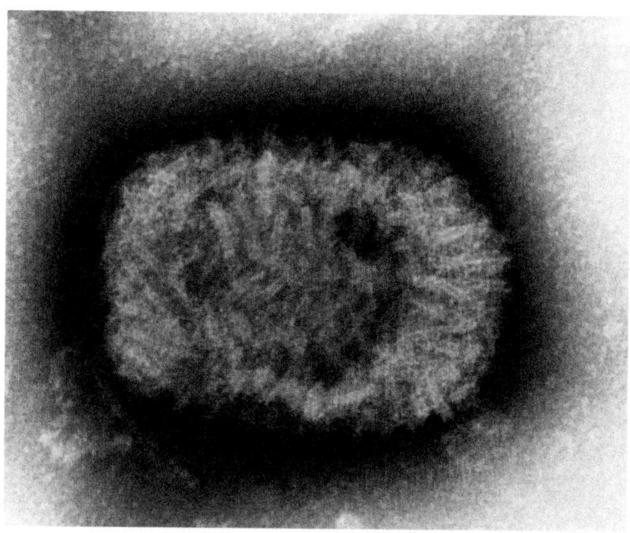

Fig. 5.9 Electron micrograph of smallpox virus. Courtesy of PHIL, CDC.

there are several genera, with cross-protection occurring to infections within the genus. From the human infection point of view, the most important is the Orthopox genus, which contains variola virus, the cause of smallpox, cowpox virus, vaccinia virus (used as vaccine against smallpox), and monkeypox virus. The other genera are Parapoxvirus, Yatapoxvirus, and Molluscipoxvirus. There are many poxviruses that infect different animals.

Smallpox (variola)

This infection has had a profound effect on human history. Due to very effective vaccination strategies, it was the first infection to be eradicated from the earth, the last case having been diagnosed in Somalia in 1977. It spread rapidly with a very high fatality rate. The infection was characterized by high fever and a pustular rash (Fig. 5.10). Involvement of the respiratory tract and other viscera occurred, as did secondary bacterial infection of skin lesions. There were two forms of the disease: variola major, which carried a fatality rate of about 30%, and variola minor (alastrim), which was a much milder disease.

Inoculation of susceptible individuals with fluid from a pustule of an infected individual (a controlled infection) was widely practiced as a form of immunization in the Middle East and Europe. However, the observation that women who milked cows seemed to be immune to smallpox infection led to the assumption that infection with

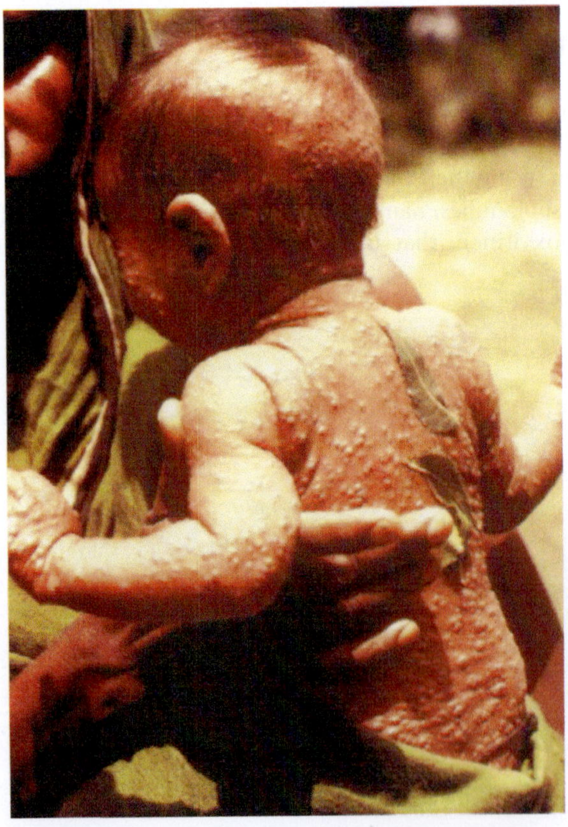

Fig. 5.10 A child with smallpox. Courtesy of James Hicks/Centers for Disease Control and Prevention.

cowpox could protect humans from smallpox. This led Edward Jenner to test the ability of inoculation with fluid from lesions of a cow with cowpox to protect a boy from a smallpox challenge. (This occurred in 1796, before the advent of institution review boards!) This experiment was very successful. This led to widespread "vaccination," the word being derived from *vacca*, the Latin word for cow. Although this term was used for immunization against smallpox, it is now used for all types of active immunization.

Diagnosis of smallpox

Because the consideration of smallpox has very important public health implications and, nowadays, implies the likelihood of bioterrorism, the diagnosis has always been of the utmost urgency. An algorithm for clinical and laboratory evaluation of a suspected case of smallpox has been published by the CDC (www.cdc.gov/smallpox). This is summarized in a review by Moore et al. (2006). Although many conditions can cause fever and a rash, a limited number cause fever and a pustular rash, the most important being varicella. Public health authorities must be involved in the diagnostic work-up. In high-risk circumstances, the personnel involved should have been vaccinated recently, and the testing should be performed in a high-containment laboratory (BSL-4) (see Chapter 2). The testing, which should be applied to specimens from the pustules, as well as blood, tonsillar swabs, and biopsy specimens, involves PCR specifically for variola virus. Electron microscopy can be used to detect poxviruses, but cannot distinguish between them. The virus can be grown in tissue culture and on the chorioallantoic membrane of chick embryos.

Cowpox virus

This is a virus affecting cows (see earlier in this chapter).

Vaccinia virus

This is the virus used for immunization (vaccination) against smallpox. Its origin is unclear, but it is different from cowpox virus. It is currently used only in specific populations, who might be exposed to smallpox used as an act of bioterrorism (Fig. 5.11).

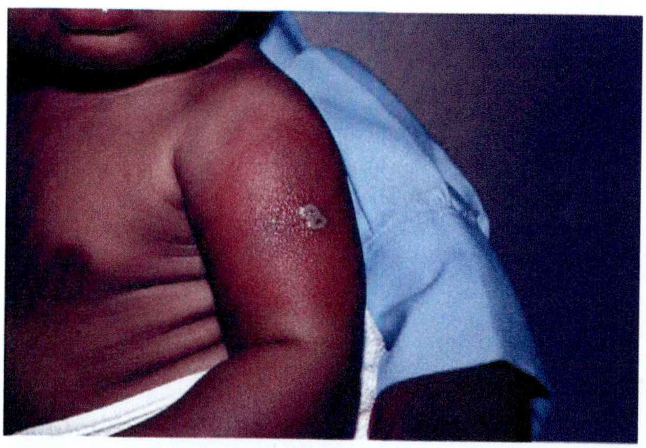

Fig. 5.11 The typical reaction to vaccination with vaccinia virus. Copyright © 2007 Frank E. Berkowitz. Reprinted with the permission of Cambridge University Press.

Monkeypox virus

This virus affects mainly rodents in Africa, and has caused large outbreaks, including a recent outbreak in the USA, traced to imported animals. It can cause fatal disease (Fig. 5.12).

Molluscum contagiosum

This is a common infection of humans, causing firm papules, which progress to become umbilicated, and which resolve over a period of a few months (Fig. 5.13). It is spread by direct contact with a lesion. Immunocompromised individuals may have extensive skin involvement.

Other poxvirus infections include orf, an infection of sheep that can occasionally cause skin lesions in farmers, and tanapox, which occurs in central and east Africa.

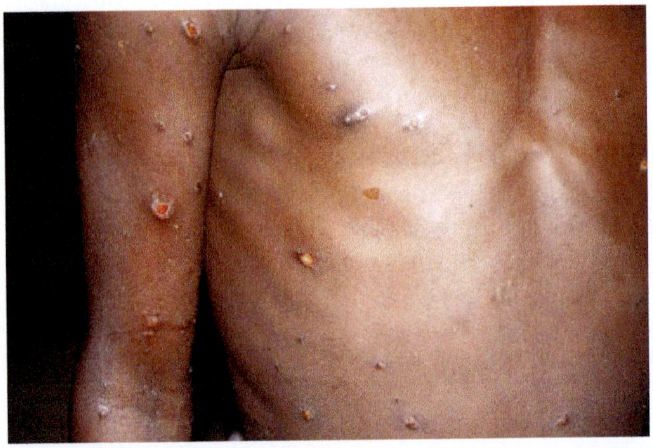

Fig. 5.12 Patient with monkeypox during an outbreak in the Democratic Republic of the Congo in 1996. Courtesy of PHIL, CDC.

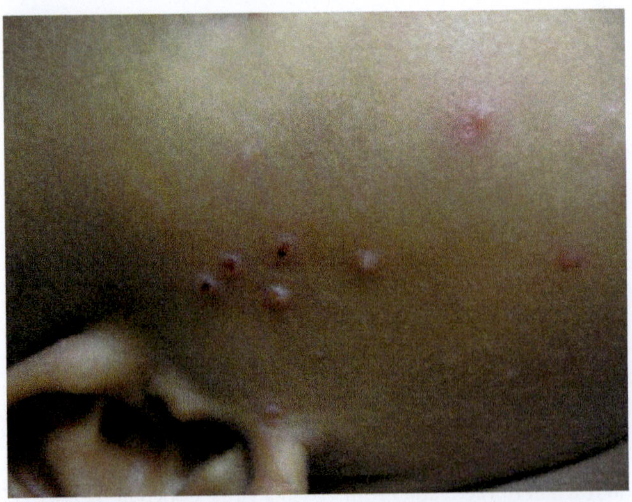

Fig. 5.13 Child with molluscum contagiosum.

Parvoviruses (Parvoviridae)

These are small, non-enveloped DNA viruses. The most important affecting humans is parvovirus B19, which belongs to the genus Erythrovirus.

Parvovirus B19

This virus was recognized in the UK in 1974, and in the 1980s was found to cause aplastic crises in patients with chronic hemolytic anemias, such as sickle cell disease, and to cause a benign exanthem called erythema infectiosum or fifth disease (Fig. 5.14). The virus is widespread and many individuals acquire the infection in childhood. It infects red cell precursors in the bone marrow, causing cessation of erythropoiesis for about 1 week. For most individuals this is of no consequence, but for those with a shortened red cell life span, such as those with chronic hemolysis, this can result in the rapid development of anemia, called "aplastic crisis." Two other types of patients can suffer severe disease from this virus: the fetus, who, when infected as a result of maternal infection, can develop severe anemia, leading to heart failure, leading, in turn, to hydrops; and immunocompromised individuals who are unable to clear the virus, which causes a chronic infection, resulting in chronic anemia.

The virus can be detected in the blood during an acute infection by PCR; it can also be diagnosed serologically. In bone marrow, intranuclear inclusions can be seen within red cell precursors (Fig. 5.15). For most individuals diagnosis of the infection is unnecessary. For individuals with the risk of severe disease, diagnosis should be confirmed. For pregnant women this is very important.

Illustrative case

A 6-year-old boy with sickle cell disease presented with abdominal and shoulder pain, and while hospitalized, developed worsening anemia. His blood count results are shown in Table 5.3.

This shows a temporary cessation of erythrocyte production, with subsequent recovery, typical of an aplastic crisis. His parvovirus B19 serology, measured by ELISA (see Chapter 2), revealed an IgG optical density of 1.8 (normal <0.9) and an IgM optical density of 11.4 (normal <0.9), confirming the diagnosis of a recent parvovirus B19 infection.

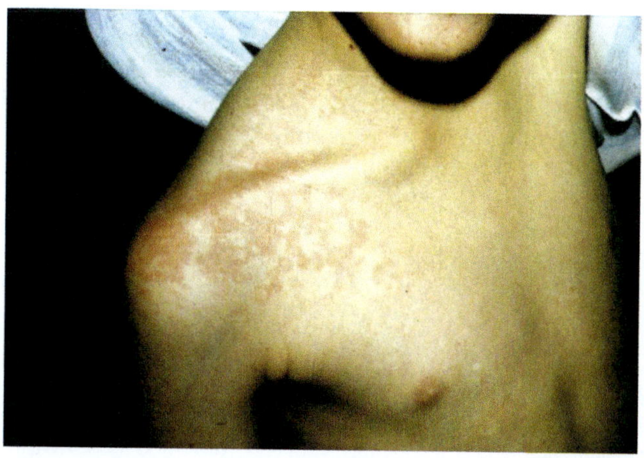

Fig. 5.14 Young girl with erythema infectiosum, showing the characteristic lacy rash. Copyright © 2007 Frank E. Berkowitz. Reprinted with the permission of Cambridge University Press.

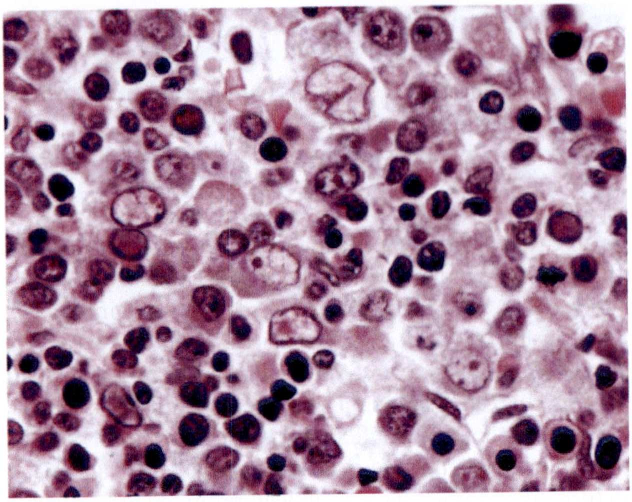

Fig. 5.15 Liver tissue of a newborn dying of hydrops fetalis caused by parvovirus B19 infection. Note the intranuclear inclusions caused by the virus. Courtesy of Dr Carlos Abramowsky.

Table 5.3 Development of anemia and then recovery in a child with sickle cell disease who developed an "aplastic crisis" caused by parvovirus B19.

Date	5/8	5/23	5/24	5/26	5/28	6/1	6/1	6/2
Hemoglobin(g/dL)	9.5	8.1	7.7	7.0	6.2	5.9	5.5	5.6
Reticulocytes (%)	3.2	1.1	0.7	0.8	0.9	1.6	1.5	11.3
Absolute reticulocytes (k/mm^3)	95	28.4	16	19	18	31	26	201
Nucleated red cells (per mm^3)	0	0	0	0	0	8	19	35
Platelets (k/mm^3)	696	178	143	123	175	526	845	629

Bocavirus

This is a recently discovered parvovirus. It has been found in respiratory secretions of children with respiratory tract infection, as well as in some controls. Whether it actually causes disease is still open to question. Human bocavirus can be detected using PCR.

Reference

Moore ZS, Seward JF, Lane JM (2006) Smallpox. Lancet 367: 425–34.

Further reading

Acosta EP, Flexner C (2011) Antiviral agents. In: Brunton LL, Lazo JS, Parker KL (eds) *Goodman and Gilman's The Pharmacological Basis of Therapeutics*, 12th edn. New York: McGraw-Hill.

Adalja AA, Toner S, Inglesby TV (2015) Clinical management of potential bioterrorism-related conditions. N Engl J Med 372: 954–62.

Antonelli G, Turriziani O (2012) Antiviral therapy: old and current issues. Int J Antimicrob Ther 40: 95–102.

Ariza-Heredia EJ, Razonable RR (2011) Human herpes virus 8 in solid organ transplantation. Transplantation 92: 837–44.

Berthet N, Nakoune E, Whist E, et al. (2011) Maculopapular lesions in the Central African Republic. Lancet 378: 1354.

Boothpur R, Brennan DC (2010) Human polyoma viruses and disease with emphasis on clinical BK and JC. J Clin Virol 47: 306–12.

Cohen JI (2013) Herpes zoster. N Engl J Med 369: 255–63.

Corey L, Wald A (2009) Maternal and neonatal herpes simplex virus infections. N Engl J Med 361: 1376–85.

Dijkmans AC, de Jong EP, Dijkmans BAC, et al. (2012) Parvovirus B19 in pregnancy: prenatal diagnosis and management of fetal complications. Curr Opin Obstet Gynecol 24: 95–101.

Dunn JJ, Miller MB (2014) Emerging respiratory viruses other than influenza. Clin Lab Med 34: 409–30.

Echavarria M (2008) Adenoviruses in immunocompromised hosts. Clin Microbiol Rev 21: 704–15.

Fishman JA (2013) Overview: cytomegalovirus and the herpesviruses in transplantation. Am J Transplant 13(s3): 1–8.

Green M, Michaels MG (2013) Epstein-Barr virus infection and posttransplant lymphoproliferative disorder. Am J Transplant 13 (s3): 41–54.

Gupta R, Warren T, Wald A (2007) Genital herpes. Lancet 370: 2127–37.

Hathaw JK (2012) HPV: diagnosis, prevention, and treatment. Clin Obstet Gynecol 55: 671–80.

Heininger U, Seward JF (2006) Varicella. Lancet 368: 1365–76.

Hoffman JA (2009) Adenovirus infections in solid organ transplant recipients. Curr Opin Organ Transplant 14: 625–33.

Hughes AL, Irausquin S, Friedman R (2010) The evolutionary biology of poxviruses. Infect Genet Evol 10: 50–9.

Huh WK, Ault KA, Chelmow D, et al. (2015) Use of primary high-risk human papillomavirus testing for cervical cancer screening: interim clinical guidance. Obstet Gynecol 125: 330–7.

Johnson EM (2010) Structural evaluation of new human polyoma viruses provides clues to pathobiology. Trends Microbiol 18: 215–23.

Kahn JA (2009) HPV vaccination for the prevention of cervical intraepithelial neoplasia. N Engl J Med 361: 271–8.

Kimberlin DW (2007) Herpes simplex virus infections of the newborn. Semin Perinatol 31: 19–25.

Kumar D (2010) Emerging viruses in transplantation. Curr Opin Infect Dis 23: 374–8.

Lautenschlager I, Razonable RR (2012) Human herpesvirus-6 infections in kidney, liver, lung, and heart transplantation: review. Transplant Int 25: 493–502.

Lion T (2014) Adenovirus infections in immunocompetent and immunocompromised patients. Clin Microbiol Rev 27: 441–62.

Luzuriaga K, Sullivan JL (2010) Infectious mononucleosis. N Engl J Med 362: 1993–2000.

Markowitz LE, Dunne EF, Saraiya M, et al. (2014) Human papillomavirus vaccination: recommendations of the Advisory Committee on Immunization Practices (ACIP). MMWR RR05; 1–30.

McCollum AM, Damon IK (2014) Human monkeypox. Clin Infect Dis 58: 260–7.

Razonable RR (2011) Antiviral drugs for viruses other than human immunodeficiency virus. Mayo Clin Proc 86: 1009–26.

Schiffman M, Castle PE, Jeronimo J, et al. (2007) Human papillomavirus and cervical cancer. Lancet 370: 890–907.

Slavov SN, Kashima S, Pinto ACS, et al. (2011) Human parvovirus B19: general considerations and impact on patients with sickle-cell disease and thalassemia and on blood transfusion. FEMS Immunol Med Microbiol 62: 247–62.

Soper DE (2013) Congenital cytomegalovirus infection: an obstetrician's point of view. Clin Infect Dis 57(S4): S171–S173.

Ward KN (2005) Human herpesviruses-6 and -7 infections. Curr Opin Infect Dis 18: 247–52.

Ward KN (2005) The natural history and laboratory diagnosis of human herpesviruses-6 and-7 infections in the immunocompetent. J Clin Virol 32: 183–93.

CHAPTER 6

RNA viruses (excluding hepatitis viruses, arthropod-borne viruses, and bat and rodent excreta viruses)

Picornaviruses (Picornaviridae)

Picornaviruses are PICO (small) RNA viruses. They have a positive RNA strand, and lack an envelope. There have been several changes in their taxonomy over the past few years. The current classification of those affecting humans is as follows.

- Genus: Enterovirus, which is divided into the species enteroviruses A, B, C, and D, and rhinoviruses A, B, and C.
- Genus: Parechovirus
- Genus: Hepatovirus (hepatitis A) (discussed in the section on hepatitis viruses)

However, there are benefits of considering them according to the clinical syndromes that they cause.

Rhinoviruses

There are more than 100 serotypes of rhinoviruses. They multiply in respiratory epithelium at relatively low temperature. Thus they infect mainly the upper respiratory tract (nose and paranasal sinuses). Although generally not invasive, they cause a large amount of morbidity, school and work absences, and visits to physicians. The complications they cause (secondary bacterial sinusitis and otitis media) can, occasionally, lead to significant further complications, as a result of spread of infection intracranially.

Rhinoviruses are not readily cultured. However, they can be detected by polymerase chain reaction (PCR). Their detection is not usually of clinical value, but can be of epidemiologic value.

Antiviral therapy is not available at this time.

Enteroviruses

There are many enteroviruses, which were formerly classified into four groups: polioviruses, serotypes 1, 2, and 3; Coxsackie A (six serotypes) and B (23 serotypes); and echoviruses (named for **E**ntero**C**ytopathic **H**uman **O**rphan viruses (orphan because initially they were viruses without a disease that they were known to cause).

Practical Medical Microbiology for Clinicians, First Edition. Frank E. Berkowitz and Robert C. Jerris.

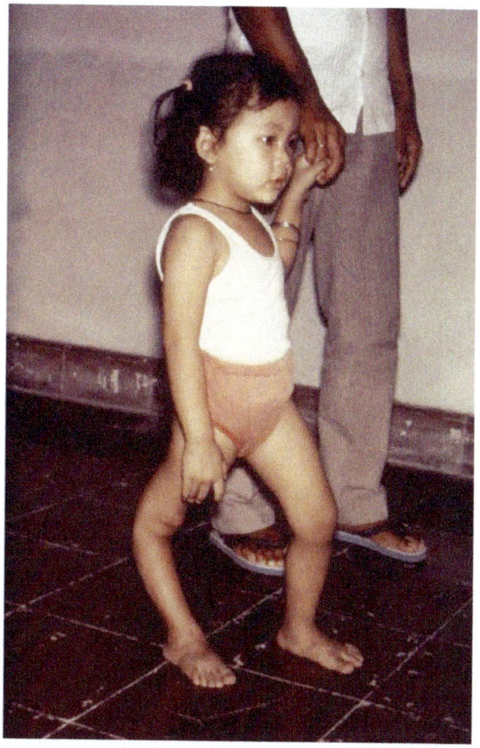

Fig. 6.1 Child with weakness due to polio. Courtesy of PHIL, CDC.

As more serotypes were discovered, they were given serotype numbers, e.g. enterovirus 70 and 71. Although these serotypes retain these names, they fall into one of the enterovirus A–D species. Two viruses formerly classified within the echoviruses (echoviruses 22 and 23) are now classified within the parechoviruses as parechovirus 1 and 2 (see later in this chapter).

Polioviruses

There are three serotypes (1, 2, and 3), which are now classified within the species enterovirus C. They cause poliomyelitis.

Polioviruses, like other enteroviruses, are spread by the fecal–oral route. Being non-enveloped, they can survive in the inanimate environment, such as water, for long periods. They enter the host via the pharynx and gut, and spread to the local lymphoid tissue and then systemically. From there, they can spread to the central nervous system. Poliomyelitis is characterized initially by a non-specific febrile illness, and sometimes myalgia, especially of the back. In a small proportion of cases, infection progresses to paralysis, due to infection of the anterior horn cell. Groups of muscles can be affected, but paralysis can be generalized, including muscles innervated by cranial nerves (bulbar paralysis). Recovery can occur, but this is not invariable. Therefore, significant long-term disability can result (Fig. 6.1). In addition, nerve disease can occur many decades after an initial infection. Although polioviruses formerly had a worldwide distribution, major efforts to eradicate them, through vaccination, have resulted in confinement of the viruses to a few countries in Asia and Africa. However, until they are totally eradicated, there is continuous risk that they could be spread to countries free of the infection. The last recognized case of wild-type poliovirus type 2 occurred in India in 1999.

The diagnosis of polio is based on clinical features (acute flaccid paralysis) and culture of the virus. The virus can be readily grown in tissue culture from stool and pharyngeal specimens. However, it is not readily grown from cerebrospinal fluid. Culture of the virus is important epidemiologically, because this allows for analysis of its genome. This makes it possible to determine whether the virus is wild type or derived from the live attenuated vaccine.

The treatment of polio is supportive (e.g. ventilation if the muscles of respiration are paralyzed), and rehabilitation for motor disabilities.

Other enteroviruses

The enteroviruses, which in temperate climates are most prevalent in warmer months, cause a wide range of illnesses, which are mostly mild but can be severe. Many cases of infection are asymptomatic. The common illnesses are fever, fever and rash, fever, rash and enanthem (hand-foot-mouth disease) (Fig. 6.2), conjunctivitis, and meningitis. The latter is the most common syndrome for which children with enterovirus infection are hospitalized. More severe illnesses include encephalitis, pericarditis, myocarditis, chest wall myositis (Bornholm disease), and polio-like myelitis.

Because the microbiological differential diagnosis of meningitis includes bacterial causes, many of these patients are treated with antibiotics, until a bacterial cause has been excluded. The ability to diagnose enteroviral meningitis rapidly by PCR performed on cerebrospinal fluid has eliminated the need for treatment of many of these cases for bacterial meningitis. In the febrile stage of infection, the virus can also be demonstrated by PCR performed on blood. Many of the enteroviruses can be grown in tissue culture, but PCR has replaced this for most situations.

There is no available antiviral therapy for use in patients with enterovirus infection.

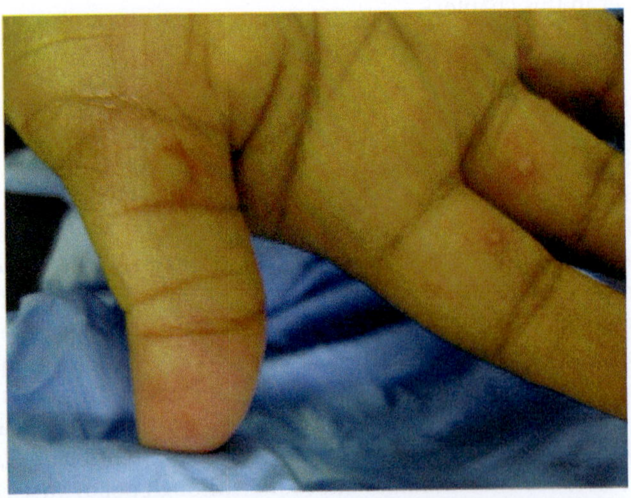

Fig. 6.2 Rash of a child with hand-foot – mouth disease, occurring during a community outbreak of Coxsackie A 6 infection.

Parechoviruses

There are 16 known genotypes of parechoviruses. Although they cause illness similar to that caused by enteroviruses, genotype 3 seems to cause more severe illness, particularly in neonates and very young infants. Of importance is the fact that they are not detected by reverse transcriptase PCR tests used to detect enteroviruses, but specific parechoviruse RT-PCR tests have been developed.

Orthomyxoviruses (Orthomyxoviridae)

The orthomyxoviruses are the influenza viruses, of which there are three species: influenza A, influenza B, and influenza C (Fig. 6.3). They are single-stranded, negative-sense RNA viruses. This family of viruses probably has had a greater influence on human history than any other virus family. They cause significant morbidity and, in high-risk individuals, significant mortality. They mutate frequently, and they are highly contagious. Influenza is also a pathogen of animals, such as birds, including poultry, and pigs (see later in this chapter).

Influenza viruses have a segmented genome – eight segments in influenza A and B, and seven in influenza C. This makes reassortment relatively easy (see later in this chapter). Influenza A and B have two major glycoprotein virulence factors that are susceptible to change, namely the surface hemagglutinin and neuraminidase (Fig. 6.4).

Influenza A

Each strain of this species has one of 16 different hemagglutinins (H1–16) and nine different neuraminidases (N1–9). Until recently, only H1, H2, H3, and H5, and N1, N2, and N9 had caused human disease. In 2013, an outbreak of human disease caused by

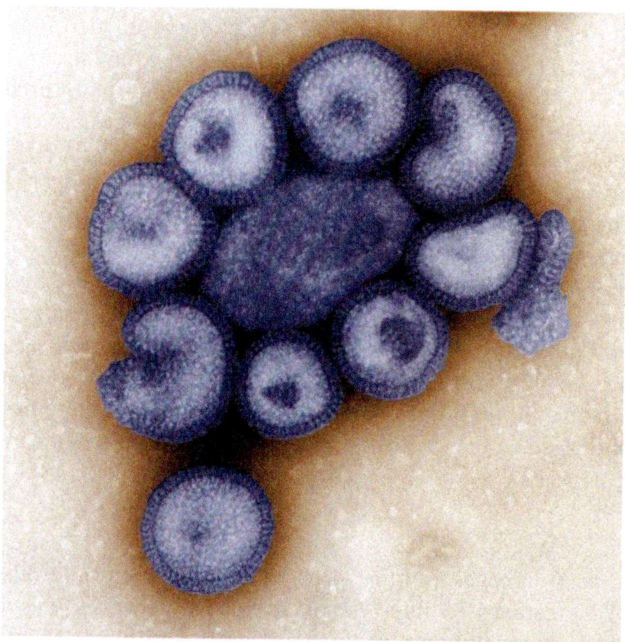

Fig. 6.3 Electron micrograph of an influenza virus. Courtesy of PHIL, CDC.

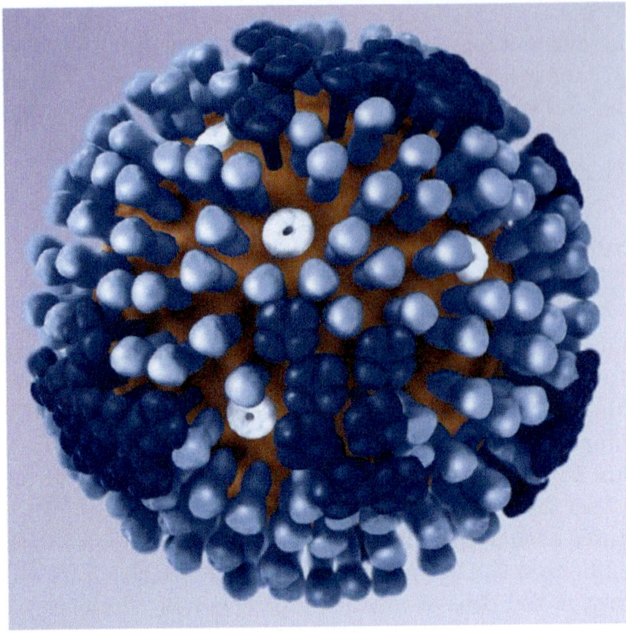

Fig. 6.4 Model of influenza virus. The dark blue structures represent neuraminidase and the light blue ones represent hemagglutinin. Courtesy of PHIL, CDC.

H7N9, originating in poultry, occurred in China. In temperate climates, this virus causes annual and predictable outbreaks during the winter. Each year, there are minor antigenic changes in the hemagglutinin and neuraminidase, such that immunity to previous strains does not necessarily confer immunity to the new strain. These changes are called "antigenic drift." At several year intervals, which are not predictable, there are major changes due to reassortment of genes. These changes are called "antigenic shifts" and are associated with pandemics. Examples of this are the "Spanish flu" in 1918, the "Asian flu" in 1957, and the "Hong Kong flu" in 1968. The most recent pandemic was the 2009 "H1N1 flu," which started in Mexico.

The gene reassortment occurs when a host cell becomes infected with two strains of influenza virus, and reassortment of gene segments occurs. If the reassortant is virulent for the host and can be spread among the host species, contagion can occur. Important animals in which such reassortment occurs are fowl and pigs. Eastern Asia is a high-risk area for this to occur because there are many of these animals living in close contact with one another. It is also a high-risk area for spread to humans because the humans are in close contact with the animals. Thus influenza A can be viewed as a zoonosis.

Influenza B

There are two main lineages of this virus, namely B/Victoria/2/87-like (Victoria lineage) and B/Yamagata/16/88-like (Yamagata lineage). Because this virus does not infect animals, it does not undergo the reassortment, and hence the antigenic shift, that occurs in influenza A.

The nomenclature of strains of influenza is as follows: species/place where first recognized/laboratory number/year of isolation/(H and N subtype), e.g. A/Perth/16/2009/(H3N2)-like or B/Brisbane/60/2008-like. Because the hemagglutinin and neuraminidase of influenza B do not change as much as in influenza A, the H and N types are not described.

Influenza viruses infect primarily the respiratory tract, from the upper to the lower tract. They can also spread systemically to other organ systems.

Clinical features

Influenza has a wide range of clinical manifestations. The typical presentation is sudden onset of fever, malaise, myalgia (all due to the elaboration of cytokines by the body's immune system), and cough. This can last for several days before recovery occurs. It can be complicated by pneumonia, caused by the virus itself, as well as by secondary invading bacteria, such as *Staphylococcus aureus* and *Streptococcus pneumoniae*. Other complications include encephalitis, encephalopathy, myocarditis, pericarditis, and myositis.

Considering that the clinical manifestations of influenza are not specific, the ability to make a specific and rapid diagnosis of the infection would be useful, for the following reasons:

- specific antiviral therapy is available, and should be used especially in individuals at risk for complicated disease, e.g. young children and the elderly
- for epidemiologic monitoring of outbreaks.

The following tests, performed on upper respiratory tract specimens, are available.

- Culture of the virus in tissue culture. This is not adequately rapid for therapeutic decisions to be made.
- Antigen detection: the sensitivities and specificities of these tests are extremely variable.
- PCR: this is considered the gold standard. Tests are now widely available with results in as little as 1–2 hours, and the results are rapid enough for immediate decisions about instituting antiviral therapy to be made.

Antiviral therapy

Several drugs are active against influenza viruses.

Adamantines

Amantadine and rimantadine, which are tricyclic amines, are active against only influenza A. They act by interfering with viral uncoating in the initial stage of infection of the cell, and with hemagglutinin function in the later stages of infection. The high rate of resistance has limited their usefulness.

Neuraminidase inhibitors

These are sialic acid analogs. Viral neuraminidase functions by cleaving the terminal residues of sialic acid, which enables progeny virus to spread to other cells. The neuraminidase inhibitors prevent this from occurring. Oseltamivir (which is administered by mouth), zanamivir (which is administered by inhalation), and peramivir (which can be administered intravenously) are active against both influenza A and B. Resistance to these drugs can develop.

Paramyxoviruses (Paramyxoviridae)

These are negative-stranded RNA viruses. They include several important human and zoonotic pathogens. Their taxonomy is shown below.

Family: Paramyxoviridae
Subfamily: Paramyxovirinae
Genus: Respirovirus
  Parainfluenza virus 1, 3
Sendai (mouse parainfluenza virus)
  Rubulavirus
  Mumps
  Parainfluenza viruses 2, 4A, 4B
  Tioman virus (bats)
  Menangle virus (zoonosis)
Morbillivirus
  Measles
  Canine distemper
  Rinderpest (bovine pathogen extinct 2011)
Henipaviruses
  Hendra virus (zoonosis)
  Nipah virus (zoonosis)
  Avulovirus
    Newcastle disease (poultry pathogen)
Subfamily: Pneumovirinae
Genus: Pneumovirus
  Respiratory syncytial virus (RSV)
Genus: Metapneumovirus
  Human metapneumovirus

These viruses have a fusion factor that causes infected cells to fuse, resulting in syncytia formation. This is seen in tissue culture and histologically as multinucleate giant cells.

The **parainfluenza viruses** (1, 2, 3, 4A, 4B), **respiratory syncytial virus** (RSV), and **human metapneumovirus** all cause respiratory tract infection of the upper tract (nose, pharynx), middle tract (larynx), and lower tract (bronchi, bronchioles, and alveoli), particularly in infants. Clinically, the illness caused by one cannot be distinguished from that caused by another of these viruses. However, in infants, respiratory syncytial virus is the most common cause of bronchiolitis and pneumonia, and parainfluenza viruses 1 and 2 are the most common causes of croup (laryngotracheobronchitis). Although in older children and adults they cause primarily upper respiratory tract infections, in the elderly and immunocompromised individuals, such as transplant patients, they can cause significant lung disease. They can be significant causes of nosocomial infection, so hospitalized patients suspected of being infected with them must be isolated. Transmission is primarily by contact with infected secretions, so contact precautions are important.

Although they can be readily detected in the laboratory by culture, by antigen detection tests such as immunofluorescence (Fig. 6.5), and by nucleic acid detection

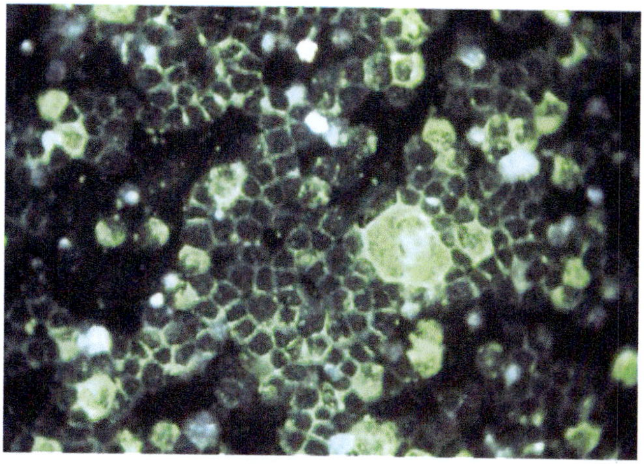

Fig. 6.5 Immunofluorescence microscopy showing respiratory syncytial virus-infected cells. Courtesy of PHIL, CDC.

methods, demonstrating their presence is not always of clinical value, because it rarely influences management. However, it can be of use for the following purposes:

- cohorting patients with respiratory tract infection in the same rooms if they have the same virus
- deciding to withhold antibiotics in cases of pneumonia
- some practitioners might use ribavirin by inhalation in cases of severe pneumonia or bronchiolitis caused by RSV, but it is of borderline value. We do not recommend its use.

Mumps virus (Fig. 6.6)

This causes a systemic infection, manifesting most prominently with parotitis. It frequently causes meningitis and pancreatitis and, in postpubertal males, orchitis. Occasionally, it causes myocarditis and glomerulonephritis. It has an incubation period of 12–25 days. The differential diagnosis of parotitis includes infections caused by other viruses such as enteroviruses and Epstein–Barr virus, and suppurative parotitis caused by blockage of Wharton's duct, or associated with severe dehydration. Chronic parotid swelling can be seen in HIV infection.

Confirmation of the diagnosis of mumps can be difficult. Serology using complement fixation or other tests can be used. However, in previously immunized individuals antibodies will be positive. The virus can be cultured in tissue culture. A mouth swab, preferably from the area of the parotid duct, is the optimal specimen.

Measles virus (also called morbilli and rubeola)

This virus (Fig. 6.7) causes a very severe infection. (The name is derived from the Latin word *misellus*, which means wretched.)

It is spread by the airborne route, enters via the pharynx, and spreads from lymphoid tissue systemically. The incubation period is usually 10 days. Initially there is fever and cough (which is a very prominent symptom of the infection). This is followed by the appearance of white spots on the oral mucosa (Koplik spots), and then

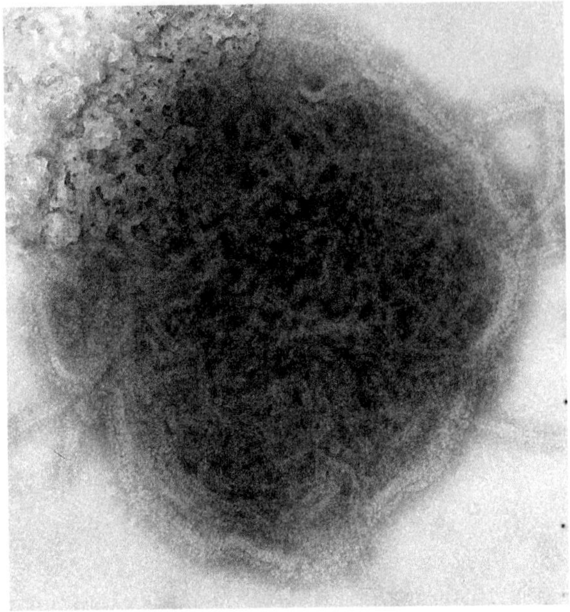

Fig. 6.6 Mumps virus: electron microscopy, negative staining. Courtesy of PHIL, CDC.

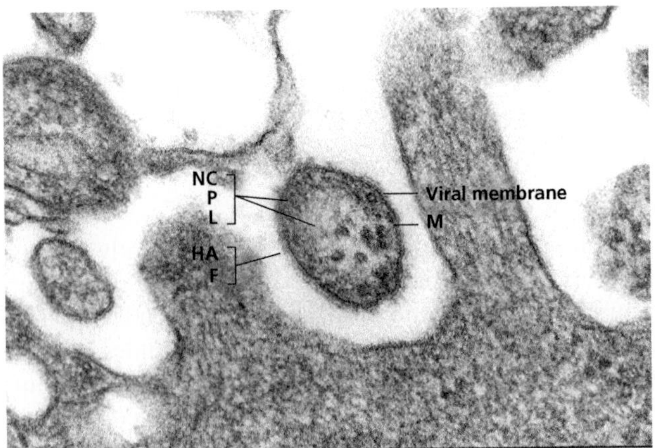

Fig. 6.7 Measles virus. Courtesy of PHIL, CDC.

by the appearance of a rash, which is macular, slightly elevated, and which becomes almost confluent (Figs 6.8 & 6.9).

The whole of the respiratory tract can be affected (otitis media, laryngotracheobronchitis, and pneumonia). Pneumonia, caused by the virus itself and by secondary viral or secondary bacterial infection, is the main cause of morbidity and mortality. Other complications include acute encephalitis, encephalitis in immunocompromised individuals, and late-onset subacute sclerosing panencephalitis, keratitis, and malnutrition.

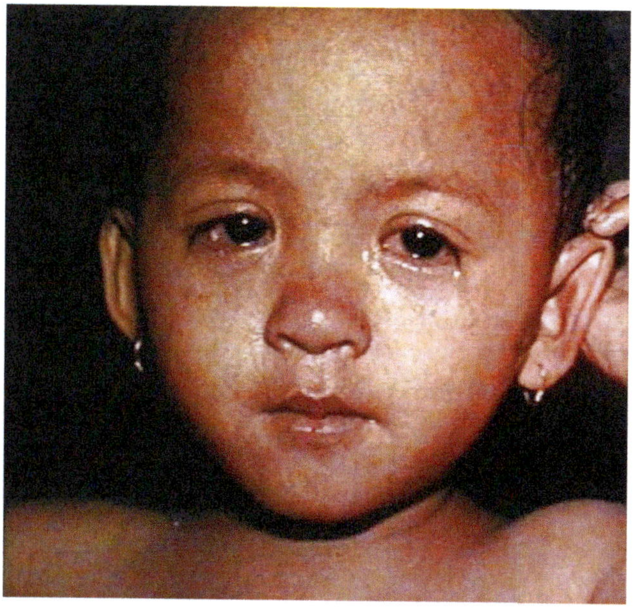

Fig. 6.8 Child with measles, showing the conjunctivitis and miserable appearance. Courtesy of PHIL, CDC.

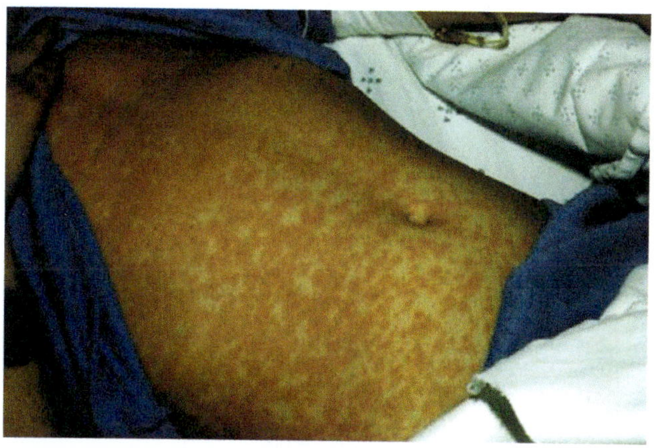

Fig. 6.9 A child with measles. Copyright © 2007 Frank E. Berkowitz. Reprinted with the permission of Cambridge University Press.

The diagnosis of measles can usually be made clinically, especially in the setting of an outbreak or endemicity. The virus can be cultured from pharyngeal secretions and urine, and detected in these specimens by reverse transcriptase PCR (RT-PCR) (available at the CDC). Rapid diagnosis can be accomplished by immunofluorescent staining of pharyngeal or mouth swabs, or urinary sediment. However, the most widely available test is the detection of IgM antibodies, which is usually performed by the local reference laboratory.

Although there is no specific antiviral therapy available, administration of vitamin A reduces the fatality rate in malnourished individuals.

Recognition of a case is of great importance because it carries huge public health implications, being a highly contagious disease with a very high morbidity and the potential for mortality.

Henipaviruses

Hendra virus was discovered in 1993 after horses in Hendra, Australia, became ill, and their caretakers developed encephalitis and respiratory disease. Most died. Nipah virus was found in 1998 in Malaysia to have caused severe infection, mainly encephalitis, in humans who worked with pigs. The fatality rate was about 38%. Subsequent outbreaks of this infection occurred in India and Bangladesh, with much higher fatality rates. These viruses are spread by the excretions (saliva and urine) of fruit bats of the family Pteropodidae (flying foxes) (Figs 6.10 & 6.11). Nipah virus causes infection of endothelial cells.

The diagnosis can be made by serologic tests and by PCR. A vaccine against Hendra virus has been licensed for use in horses.

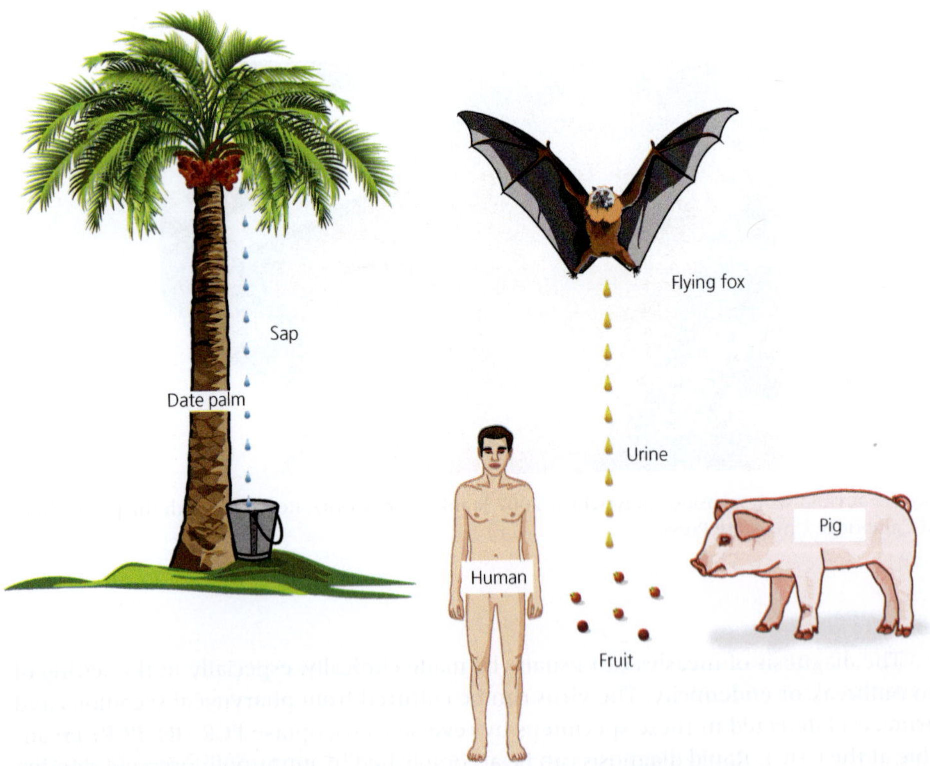

Fig. 6.10 Diagram showing how Nipah virus spreads from flying foxes to humans.

Fig. 6.11 Flying fox (Courtesy of PHIL/CDC).

Coronaviruses (Coronaviridae)

These viruses have multiple spikes on the surface, giving the appearance of a crown (corona). They have a helical nucleocapsid containing positive-sense, single-stranded RNA. Those infecting humans belong to the α-coronaviruses (CoV229E and N26L) and the β-coronaviruses (OC43, HKU1, SARS, and MERS). Except for SARS and MERS coronaviruses (see later in this chapter), they cause mainly rhinitis. In 2003, a pandemic of **S**evere **A**cute **R**espiratory **S**yndrome (ultimately called SARS) occurred, which started in China at the end of 2002. The illness was characterized initially by an upper respiratory infection, associated with systemic manifestations, and it then progressed to cause severe lung disease. It spread to Canada and other countries, and was associated with a high fatality rate. The etiology was determined very rapidly to be a coronavirus, which is now called SARS-CoV. The source was civet cats and rats, which had acquired the infection from bats (see "Batex" infections, Chapter 15).

Diagnostic testing is performed by RT- PCR.

In 2012, a cluster of cases of severe pneumonia and renal failure, associated with a very high fatality rate, occurred in Saudi Arabia, Qatar, and Jordan. They were found to be caused by a novel coronavirus, which has been named Middle East respiratory syndrome coronavirus (MERS-CoV). By June 2014, about 700 cases had been reported. Human-to-human transmission does occur, but not at the same rate as SARS. The original source of the virus has not been determined, but is suspected to be bats. Camels might be a source. Diagnosis is made by RT-PCR of respiratory specimens or feces. Suspect cases should be reported to the public health department.

Reoviruses (Reoviridae) (Respiratory Enteric Orphan viruses)

These are icosahedral, double-stranded RNA viruses. The family contains three genera, namely rotavirus, coltivirus (which is tick-borne) (see Chapter 8), and orbivirus (bluetongue virus of animals).

Rotavirus

Rotavirus (named for its wheel-like appearance under the electron-microscope; Fig. 6.12) is the most important viral cause of acute infectious diarrhea (infectious gastroenteritis) worldwide. The viral genome has 11 separate segments, which facilitates reassortment (see Influenza virus, earlier in this chapter). Serotypes are based on the G protein (10 types) and the P protein (11 types). The most common types worldwide are G1 P1A, G3 P1A, G4 P1A, and G2 P1B. Rotavirus infects small bowel epithelial cells, causing a watery diarrhea, mainly in infants and young children. It accounts for many infant deaths from dehydration worldwide. The use of vaccines, which are live attenuated viruses administered orally, has had a profoundly beneficial impact on the epidemiology of the infection in areas where it has been used, such as the USA.

The diagnosis can be confirmed by antigen testing, or immune electron microscopy on fecal specimens. Although the virus can be grown in tissue culture, this is not practicable for routine diagnosis. However, a microbiological diagnosis is seldom necessary, because there is no specific treatment, and the management of most cases of acute infectious diarrhea entails ensuring that the patient is adequately hydrated, irrespective of the cause.

Caliciviruses (Caliciviridae)

These are small, non-enveloped, icosahedral, positive-sense, single-stranded RNA viruses. The family, named for the cup-like structures seen on electron microscopy (Fig. 6.13), contains four genera, of which two, namely noroviruses and sapoviruses,

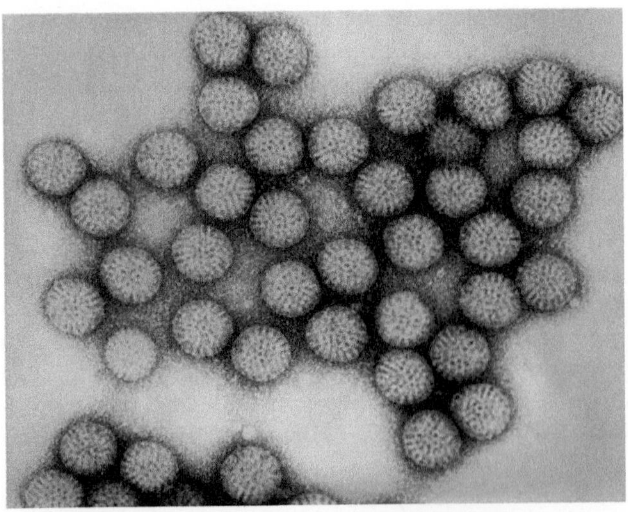

Fig. 6.12 Electron micrograph showing rotavirus. Courtesy of PHIL, CDC.

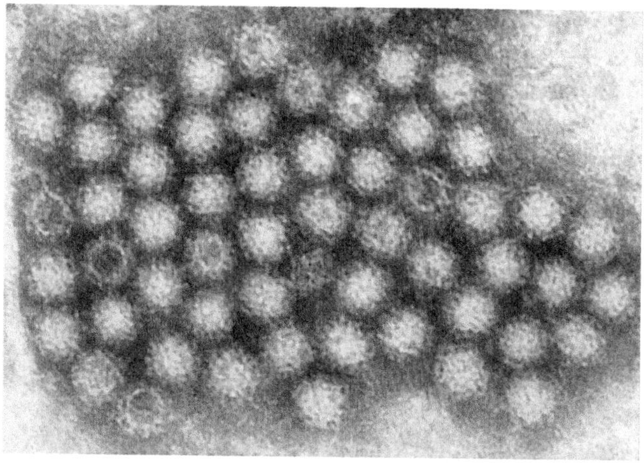

Fig. 6.13 Electron micrograph of noroviruses. Courtesy of PHIL, CDC.

are important causes of acute infectious diarrhea. Norovirus is a contraction of Norwalk virus, discovered in Norwalk, Ohio. These viruses cause outbreaks of vomiting and diarrhea, affecting all ages. They are highly contagious by close contact and via fomites, and they have caused large outbreaks on cruise ships. In normal hosts, symptoms usually last 1–2 days, but immunocompromised individuals can develop chronic infection.

They can be detected by RT-PCR, and by immune electron microscopy, which are not yet readily available in clinical laboratories.

Astroviruses (Astroviridae)

These star-shaped viruses are also causes of acute infectious diarrhea.

Rhabdoviruses (Rhabdoviridae)

These are bullet-shaped, single-stranded, negative-sense RNA viruses. This family includes the genus Lyssavirus and vesicular stomatitis virus (a horse pathogen).

The lyssaviruses (named for the Greek goddess of madness) are transmitted in the saliva of animals. These cause rabies or a rabies-like illness.

Rabies

This is primarily a disease of animals, in which it causes fatal encephalitis. It has an almost worldwide distribution. Humans are infected by being bitten by an infected animal. In most of the world this animal is a dog. In the USA, the most common source of human cases of rabies (which are very rare) is the fruit bat, although many other animals, such as wild canines, raccoons, and skunks, may harbor the virus. Human-to-human transmission has never been documented. The infection has an incubation period from 1 week to up to about 5 years. After inoculation, the virus ascends along nerve axons to the brain, where it causes encephalitis, particularly of the brainstem. Clinically, it presents with initial agitation and anxiety, sometimes

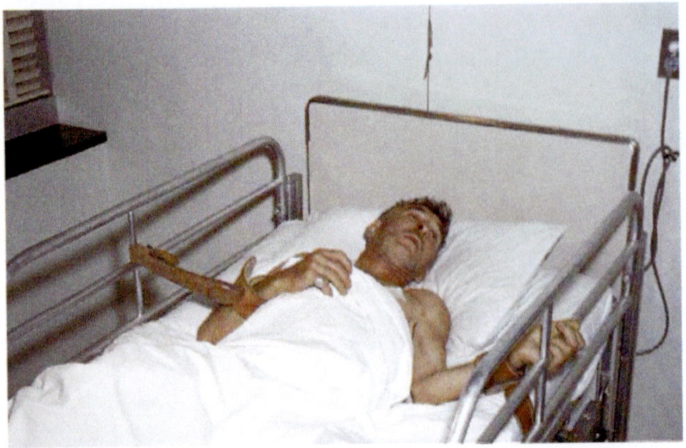

Fig. 6.14 A man with rabies. Courtesy of PHIL, CDC.

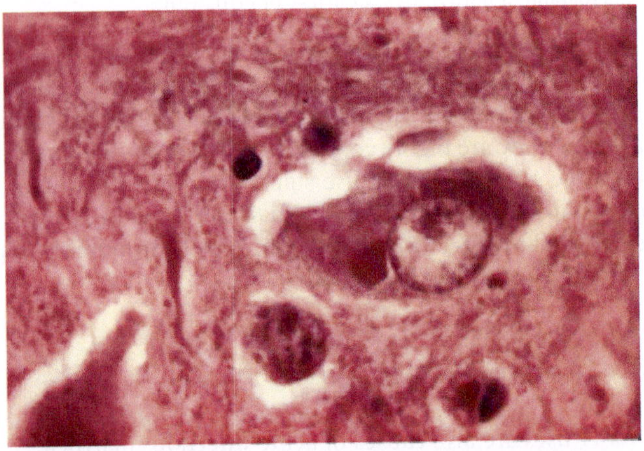

Fig. 6.15 Negri bodies (cytoplasmic inclusions) in neuron of a rabies victim. Courtesy of PHIL, CDC.

spasms of the swallowing mechanism when the patient attempts to drink ("hydrophobia" or "aerophobia"), and progresses to coma and death over a few days to weeks (Fig. 6.14). Autonomic disturbances are also prominent. There is a report of a patient surviving, with neurologic deficits, after treatment with ribavirin and interferon, and induction of pentobarbital coma.

Diagnosis (see www.cdc.gov/rabies)

This is done in reference laboratories, e.g. Centers for Disease Control and Prevention. In animals which have bitten humans, this is done optimally by examination of the brain for evidence of infected neurons. Testing of humans for rabies infection in the USA is done by several methods: by immunofluorescence and immunohistochemistry performed on a skin biopsy of the nape of the neck, which should contain 10 hair follicles (the nerves in the follicles are necessary for demonstrating evidence of the virus); antibody testing of serum (if the patient is unimmunized) and cerebrospinal fluid; and viral culture and PCR performed on saliva. Histologically, the brain shows inflammation, with characteristic Negri bodies, which are cellular inclusions (Fig. 6.15).

Prevention of rabies

The second vaccine ever used was made by Pasteur and Roux against rabies, in 1885. This had many adverse effects. The current vaccine, which is very safe and effective, consists of virus grown in human diploid cells and then inactivated. Preexposure prophylaxis (i.e. for individuals at risk of exposure, such as veterinarians) entails three doses of the vaccine. Postexposure prophylaxis entails administration of human rabies immune globulin (HRIG) around the wound and intramuscularly, and entails four doses of the vaccine, given at a remote site in the body from the HRIG. Cleaning of the wound with soap is also very important. Reducing the prevalence of rabies in animals can be accomplished by using an oral vaccine in bait.

Several other lyssaviruses include Lagos bat virus, Mokola virus, Duvenhage virus, European lyssavirus, Australian lyssavirus, West Caucasian bat virus, Aravan virus, Khajand virus, and Irkut virus. Rabies vaccine does not protect against infection with these viruses.

Togaviruses (Togaviridae)

The togaviruses contain the alphaviruses, which are arthropod borne and are discussed separately, and a single rubivirus, namely rubella virus.

Rubella virus

This causes rubella, a febrile illness associated with a diffuse rash. Although the disease is mild and seldom causes significant morbidity, if it infects a pregnant woman, the fetus can be severely affected, causing a syndrome called congenital rubella syndrome (CRS). This affects most organs, and can be manifest by the following: severe brain injury due to encephalitis; congenital heart disease such as patent ductus arteriosus and peripheral pulmonary stenosis; hepatitis; ocular disease manifested by cataracts (Fig. 6.16), glaucoma, and corneal opacities; linear bone lucencies ("celery stalk" appearance), and a skin abnormality, called the "blueberry muffin" rash, due to extramedullary erythropoiesis (Fig. 6.17).

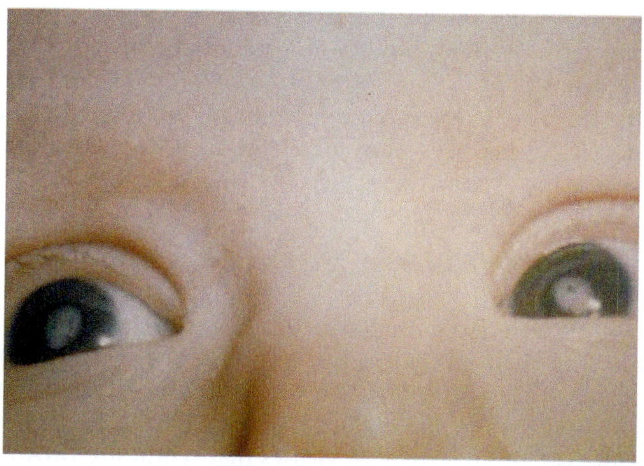

Fig. 6.16 Congenital cataracts due to the congenital rubella syndrome. Courtesy of PHIL, CDC.

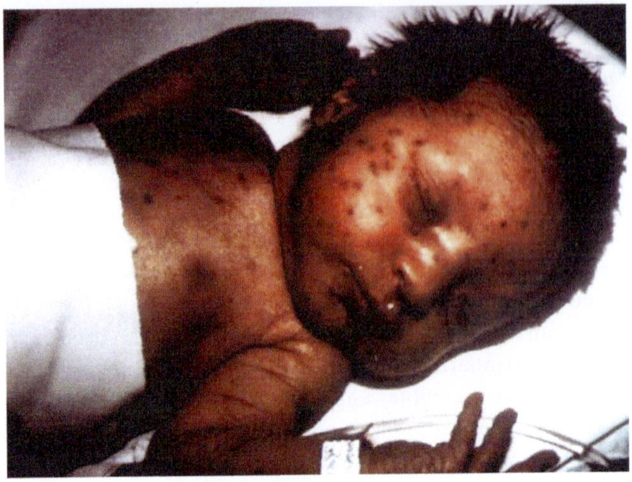

Fig. 6.17 Newborn infant with congenital rubella infection, showing the "blueberry muffin" skin lesions, due to extramedullary erythropoiesis. Courtesy of PHIL, CDC.

Therefore, the diagnosis of infection occurring during pregnancy is very important, so that consideration can be given to termination of the pregnancy. Diagnosis of congenital infection is also important in order to differentiate this infection from other intrauterine infections for which treatment can be given, such as toxoplasmosis, and so that early supportive interventions can be provided.

Diagnostic tests

Although the virus can be cultured, this is not practical in a clinical diagnostic laboratory. In cases other than pregnant women or newborns, the diagnosis is not usually important, but can be made serologically. Making a diagnosis is very important in a pregnant woman, the fetus, and the newborn, using different tests and specimens in each, as follows.

- Pregnant woman: serology, using IgG, IgM, and IgG avidity, as well as RT-PCR of a nasopharyngeal specimen.
- Fetus: RT-PCR on amniotic fluid.
- Newborn infant: serology (IgM) and RT-PCR performed on urine, oral, and nasopharyngeal specimens.

Prevention with a live attenuated vaccine (usually administered together with measles and mumps vaccine) has been very successful in rendering congenital infection extremely rare where the vaccine is used, e.g. the United States.

Retroviruses (Retroviridae)

Although retroviruses were recognized in 1910, they were not considered important in human beings until their recognition as the cause of human immunodeficiency virus (HIV) infection in the 1980s. The retroviruses known to affect humans are the lentiviruses (HIV-1 and HIV-2), and the deltaviruses (HTLV-1 and HTLV-2). The hallmark of retroviruses is that they are RNA viruses that synthesize DNA (the reverse of usual transcription).

Human immunodeficiency virus (HIV)

This virus has caused a pandemic infecting about 70 million individuals and causing about 35 million deaths, over a period of about 30 years. In the late 1970s, an acquired immunodeficiency syndrome (AIDS) was recognized. A few years later, it was shown to be caused by HIV. Its origins have been traced back to Africa, where it had entered humans from apes. There are two serotypes, namely HIV-1, which accounts for the vast majority of cases, and which is thought to have been transmitted to humans originally from chimpanzees, and HIV-2, which is limited in distribution to parts of West Africa, and which is thought to have entered humans originally from the sooty mangabey. By profoundly damaging the immune system, particularly CD4 lymphocytes, it renders its hosts susceptible to opportunistic infections and cancers (described later in this section).

The virus consists of an envelope surrounding a protein capsid, which in turn surrounds two strands of single-stranded RNA, which are linked to proteins. The proteins are structural and regulatory. After attaching to the host target cell via the CD4 receptor and a co-receptor (CXCR4 or CCR5), the virus fuses with the host cell membrane. The RNA serves as a template for the reverse transcriptase, which synthesizes complementary DNA. The RNA is destroyed by RNAse, and a second strand of DNA is synthesized. The viral DNA can now be integrated into the host cell genome. The virus can remain dormant as a provirus or it can replicate (lytic). After viral proteins have been synthesized, new viruses are assembled. A protease step is required for their maturation (Fig. 6.18). Those steps above that are underlined are targets of current antiretroviral therapy (see later in this section). HIV-1 is genetically a highly varied virus, with four lineages (M, N, O, and P), M being the most prevalent. Lineage M has 11 subtypes

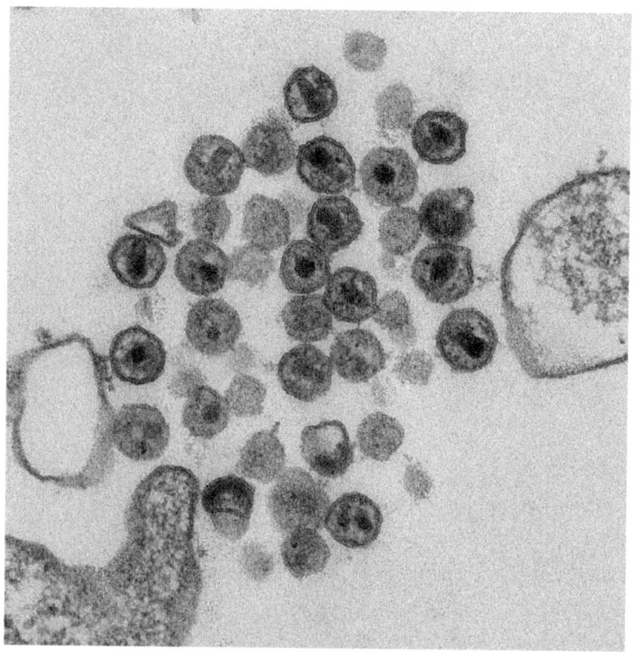

Fig. 6.18 Electron micrograph of HIV. Courtesy of PHIL, CDC.

(also called clades), which have different geographic distributions. In addition, within an infected individual, there are several different viruses (called "quasispecies"). The reasons for the variability of the virus are the following: a large number of viruses (about 10^{10}) are produced each day within the host; the reverse transcriptase is not precise, allowing for many mutations in the DNA; and the two DNA strands produced can recombine.

The main target cells are CD4 lymphocytes, macrophages, and dendritic cells in the skin.

Pathogenesis

Transmission of the infection occurs by the following routes: sexual intercourse, heterosexual or homosexual; exchange of blood, by sharing of needles used for intravenous drug abuse or transfusion of blood or blood-derived products; and mother to child, prenatally, during birth, or by breastfeeding. Premastication of food for babies has also been implicated. Rapid viral reproduction causes large numbers of immune cells to be destroyed. Although there is replenishment, over time these become severely depleted, resulting in immunodeficiency. This develops more rapidly in infants than in older individuals, because their immune systems are still developing.

Clinical manifestations

Acute HIV infection

This manifests as an acute to subacute illness, with many features of infectious mononucleosis, namely fever, malaise, sore throat, myalgias, generalized lymphadenopathy, and rash. However, this stage of infection may be asymptomatic and is often unrecognized.

Chronic illness

Weight loss, fever, lymphadenopathy, parotid swelling, anemia.

Organ disease

Encephalopathy, cardiomyopathy, nephropathy, lymphocytic interstitial pneumonia; enlargement of liver and spleen.

Opportunistic infections

Some infections are co-infections, acquired in the same manner as is HIV, and which can interact with HIV infection and its management in adverse ways. These include syphilis, HSV, hepatitis B, and hepatitis C.

Fungal (see also Chapter 18)

- Candidiasis (oral, esophageal) (Fig. 6.19)
- *Pneumocystis jiroveci* pneumonia
- *Penicillium marneffei*
- *Cryptococcus neoformans* meningitis
- Histoplasmosis and coccidioidomycosis
- Microsporidiosis

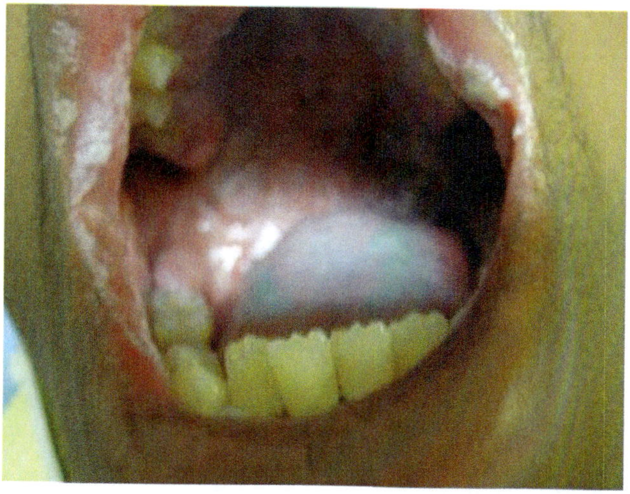

Fig. 6.19 Oral candidiasis (note white patches on inner aspects of the lips and on the palate) in a teenage boy with HIV infection.

Viral (see Chapter 7, DNA viruses)

- Cytomegalovirus infection affecting the intestine, eye, central nervous system
- HSV – mucosal and cutaneous
- VZV – severe, extensive, or persistent herpes zoster
- Polyomavirus (JC virus causing progressive multifocal leukoencephalopathy)
- Papillomaviruses: extensive verrucae

Bacterial (see Chapters 10, 12, 13, and 17)

Recurrent bacterial infections, e.g. pneumonia, bronchiectasis, otitis media.

Invasive (systemic)

- *Streptococcus pneumoniae* (especially in children)
- *Salmonella* spp.
- *Listeria monocytogenes*
- Mycobacteria, especially *M. tuberculosis* and *M. avium complex.* Worldwide, tuberculosis is the most important infection co-infecting HIV-infected individuals. The combination of HIV infection and tuberculosis has been termed the "duo from Hell".
- *Bartonella henselae* – bacillary angiomatosis and peliosis hepatis

Protozoal (see Chapters 23 and 24)

- *Toxoplasmosis gondii*, especially affecting the brain
- *Cryptosporidium* spp. and *Cystoisospora belli*, causing chronic diarrhea

Malignancies

Lymphomas, including cerebral B-cell lymphoma and Burkitt lymphoma, Kaposi sarcoma (Fig. 6.20), leiomyosarcoma.

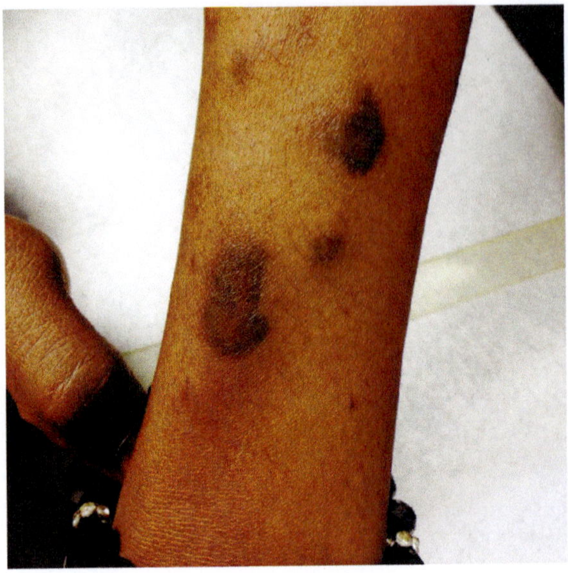

Fig. 6.20 Kaposi sarcoma. Courtesy of Dr Andres Camacho-Gonzalez.

Testing for diagnosis

Does this patient have HIV infection? This question addresses screening as well as diagnosis of symptomatic but non-acute infection. The earliest diagnostic tests available were antibody tests. These are still used for screening and for diagnosis. They are not useful in two circumstances:

- infants who are born to HIV-infected mothers have antibody from transplacental acquisition, without them necessarily being infected. These antibodies can persist for up to 18 months
- in acute HIV infection, antibodies might not be detectable by current tests at the time the patient presents. This is the so-called "window period." A combination of an antibody test and p24 antigen detection, that can be performed rapidly (see later in this chapter), has reduced the duration of this window.

The diagnosis of acute infection is very important because individuals are particularly contagious at this stage.

Antibody tests were previously performed in two stages: an initial ELISA test which, if positive, was confirmed by a Western blot test. This is no longer recommended in the USA.

The testing algorithm currently recommended by the CDC and Association of Public Health Laboratories is as follows. The initial test looks for antibodies to HIV-1 and HIV-2 and HIV-1 p24 antigen. If this is positive, differential testing for specific antibodies to HIV-1 and HIV-2 is performed. Indeterminate results should be followed by nucleic acid testing.

Point of care tests

These have the advantages that they are rapid and do not require the patient to return for the results. They are also useful for cases of needlestick injuries in healthcare workers. Several such tests are approved for use in the USA. They are ELISA based, and although they are very sensitive and specific, they do require confirmation by a non-ELISA test. Although most of the tests utilize whole blood, serum or plasma, some can be performed on saliva. Also some, but not all, can detect antibodies to HIV-2 (uncommon in the USA), as well as to HIV-1.

In infants and in patients in whom acute infection is suspected, a direct test of viral presence in the blood should be performed. The early tests for this were tests for the p24 antigen. These have largely been replaced by nucleic acid amplification tests. In infants, a viral cDNA PCR test is performed. If two tests are negative by the time the child is 4 months of age, he or she can be considered uninfected. (A final antibody test is performed at age 18 months.) In older patients viral RNA PCR is performed.

Testing for management

If the patient is infected, what is the viral load?

This is determined to gauge the degree of control of infection, and is used to monitor the effectiveness of antiviral therapy. This is accomplished by measuring the number of viral copies in the patient's blood using quantitative RNA PCR.

What are the antiviral resistance patterns of the patient's virus?

This can be determined by phenotype (determining growth of the virus in the presence of the drug, *in vitro*), by genotype (detecting mutations in the viral genome known to be associated with resistance to specific drugs), and by "virtual phenotype" (predicting the phenotype by comparing the genotype with those in a large database). This is important for optimal management, so that patients are not treated with antiviral agents to which the virus is resistant. A specific test for co-receptor usage by the virus (CCR5 or CXCR4) is used to determine whether CCR5 receptor antagonists could be used for therapy.

Does the patient have markers for severe adverse reactions to specific antiviral agents?

For example, HLA B*5701 is associated with abacavir hypersensitivity reaction. Individuals with this genetic marker should not be treated with this drug.

What is the patient's CD4 lymphocyte count?

This is used as a measure of the patient's degree of immunosuppression. The lower the count (or percentage of T-lymphocytes composed of CD4 cells), the more immunosuppressed the patient. In children, normal values vary according to age, younger children normally having higher counts.

Treatment

Detailed guidelines are provided at http://aidsinfo.nih.gov.

Antiretroviral therapy has rendered this disease, previously fatal, a chronic illness. Drugs used for therapy fall into the following classes, listed here according to the stage in the viral replication on which they exert their effect.

- Entry inhibitor
- Fusion: enfuvirtide
- CCR5 receptor inhibitor: maraviroc
- Nucleoside reverse transcriptase inhibitors: zidovudine, lamivudine, abacavir, emtricitabine, didanosine

- Nucleotide reverse transcriptase inhibitor: tenofovir
- Non-nucleoside reverse transcriptase inhibitors: nevirapine, efavirenz, etravirine, rilpivirine
- Integrase strand transfer inhibitors (often referred to as integrase inhibitors): raltegravir, dolutegravir, elvitegravir
- Protease inhibitors: fosamprenavir, lopinavir, ritonavir, darunavir, atazanavir, tipranavir, nelfinavir, saquinavir

The challenges of therapy include the following.

- Adherence, especially to regimens that have many adverse effects.
- Resistance to the antiviral agents: mutations occur frequently, and resistance can emerge readily if therapy is taken intermittently.
- Drug–drug interactions, of which there are many.

Prevention

This must address the multiple factors predisposing to infection, which are largely behavioral.

- Sexual transmission: use of condoms, monogamous relationships, male circumcision, antiretroviral therapy – by reducing the viral load in individuals, their contagiousness is reduced; preexposure antiviral therapy.
- Antiviral prophylaxis after risk events, e.g. rape, needlestick injuries in healthcare workers.
- Blood transmission: screening blood for donation; clean needle exchange programs for intravenous drug abusers.
- Maternal–child transmission interruption: screening of pregnant women, treating those who are infected, and antiviral prophylaxis of infants born to infected women have been extremely successful; formula-feeding infants of infected women in developed countries; cesarean section.
- Vaccine: this is being vigorously sought, but has, so far, proved elusive.

Human T-lymphotropic virus (HTLV)

This deltaretrovirus (Fig. 6.21) inserts itself into the host DNA and, as a provirus, remains cell associated. It is transmitted by the following methods: vertically by breast-feeding (intrauterine and intrapartum transmission is relatively uncommon); by sexual intercourse; and via blood, by transfusion or needle sharing. It has a worldwide distribution, but there is a wide variation in prevalence between different countries. The highest prevalence is in Japan.

The clinical effects can be considered in three categories.

- Adult T-cell leukemia and lymphoma.
- Inflammatory syndromes: the most important of these is HTLV-associated myelopathy/tropical spastic paraparesis. This is due to inflammation of the spinal cord, mainly the white matter, characterized by a slowly progressive spinal cord syndrome, with sensory, motor, and bladder dysfunction. Other inflammatory syndromes include arthropathy and uveitis.
- Infections: HTLV-infected individuals are at increased risk of developing strongyloidiasis, including hyperinfection (see Chapter 25), tuberculosis, leprosy, and skin infections.

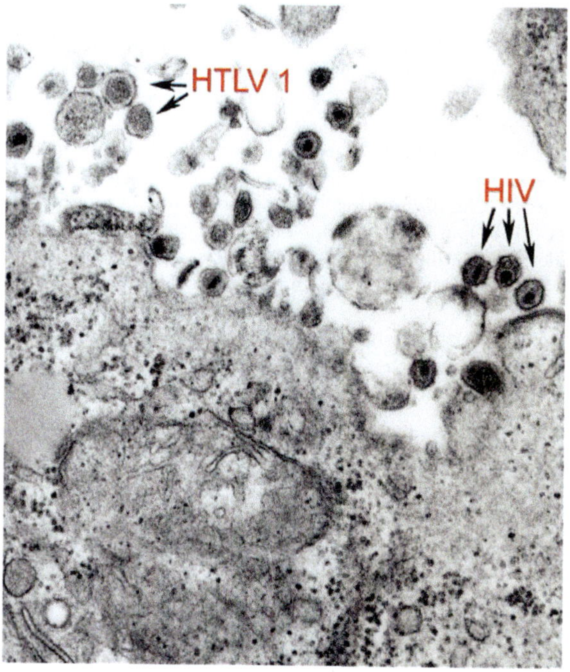

Fig. 6.21 Electron micrograph of HTLV and HIV. Courtesy of PHIL, CDC.

Diagnosis

This is made mainly serologically. Enzyme immunoassays and indirect fluorescent assays are available. These should be confirmed by Western blot. However, the latter may be associated with indeterminate results, in which case PCR should be used. Because the virus exists as a provirus in the host cell, the viral load is measured as provirus per given number of peripheral blood mononuclear cells. Currently no treatment is available.

Further reading

Abzug MJ (2014) The enteroviruses: problems in need of treatments. J Infect 68: S108–14.

Aljofan M (2013) Hendra and Nipah infection: emerging paramyxoviruses. Virus Res 177: 119–26.

Al-Tawfiq JA, Zumla A, Memish ZA (2014) Coronaviruses: severe acute respiratory syndrome coronavirus and Middle East respiratory syndrome coronavirus in travelers. Curr Opin Infect Dis 27: 411–17.

Antonelli G, Turriziani O (2012) Antiviral therapy: old and current issues. Int J Antimicrob Ther 40: 95–102.

Bok K, Green KY (2012) Norovirus gastroenteritis in immunocompromised patients. N Engl J Med 367: 2126–32.

Bouthry E, Picone O, Hamdi G, et al. (2014) Rubella and pregnancy: diagnosis, management and outcomes. Prenat Diagn 34: 1246–53.

Centers for Disease Control and Prevention and Association of Public Health Laboratories. *Laboratory Testing for the Diagnosis of HIV Infection. Updated Recommendations.* http://dx.doi.org/10.15620/cdc.23447

Dunn JJ, Miller MB (2014) Emerging respiratory viruses other than influenza. Clin Lab Med 34: 409–30.

Fischer TK, Midgley S, Dalgaard C, et al. (2014) Human parechovirus infection, Denmark. Emerg Infect Dis 20: 83–7.

Fooks AR, Banyard AC, Horton DL, et al. (2014) Current status of rabies and prospects for elimination. Lancet 384: 1389–99.
Glass RI, Parashar UD, Estes MK (2009) Norovirus gastroenteritis. N Engl J Med 361: 1776–85.
Greer D, Robbins G, Lijewski V, et al. (2013) Case 1-2013: a 63-year-old man with paresthesias and difficulty swallowing. N Engl J Med 368: 172–80.
Harvala H, Wolthers KC, Simmonds P (2010) Parechoviruses in children: understanding a new infection. Curr Opin Infect Dis 23: 224–30.
Hui DS, Chan PKS (2008) Clinical features, pathogenesis and immunobiology of severe acute respiratory syndrome. Curr Opin Pulm Med 14: 241–7.
Hviid A, Rubin S, Mühlemann K (2008) Mumps. Lancet 371: 932–44.
Joki-Korpela P, Hyypia T (2001) Parechoviruses, a novel group of human picornaviruses. Ann Med 33: 466–71.
Ksiazek TG, Rota PA, Rollin PE (2011) A review of Nipah and Hendra viruses with an historical aside. Virus Res 162: 173–83.
Lo MK, Rota PA (2008) The emergence of Nipah virus, a highly pathogenic paramyxovirus. J Clin Virol 43: 396–400.
McCutchan FE (2006) Global epidemiology of HIV. J Med Virol 78: S7–S12.
Moss WJ, Griffin DE (2012) Measles. Lancet 379: 153–64.
Owen SM (2012) Testing for acute HIV infection: implications for treatment as prevention. Curr Opin HIV AIDS 7: 125–30.
Razonable RR (2011) Antiviral drugs for viruses other than human immunodeficiency virus. Mayo Clin Proc 86: 1009–26.
Rupprecht CE, Hanlon CA, Hemachudha T (2002) Rabies re-examined. Lancet Infect Dis 2: 327–43.
Sampathkumar P (2014) Middle East respiratory syndrome: what clinicians need to know. Mayo Clin Proc 89: 1153–8.
Stockman LJ, Massoudi MS, Helfand R, et al. (2007) Severe acute respiratory syndrome in children. Pediatr Infect Dis J 26: 68–74.
Uyeki TM (2014) Preventing and controlling influenza with available interventions. N Engl J Med 370: 789–91.
Verdonck K, Gonzalez E, van Dooren S, et al. (2007) Human T-lymphotropic virus 1: recent knowledge about an ancient infection. Lancet Infect Dis 7: 266–81.
Warrell MJ, Warrell DA (2004) Rabies and other lyssavirus diseases. Lancet 363: 959–69.
Wong KT (2010) Emerging epidemic viral encephalitides with a special focus on henipaviruses. Acta Neuropathol 120: 317–25.
Yerly S, Hirschel B (2012) Diagnosing acute HIV infection. Expert Rev Anti Infect Ther 10: 31–41.
Zaki AM, van Boheemen S, Besterbroer TM, et al. (2012) Isolation of a novel coronavirus from a man with pneumonia in Saudi Arabia. N Engl J Med 367: 1814–20.

CHAPTER 7

Hepatitis viruses

Many viruses can infect and affect the liver, but the liver is not their main or usual target. This chapter discusses the viruses that are primarily hepatotropic, namely hepatitis A, hepatitis B, hepatitis C, hepatitis D, and hepatitis E viruses.

Hepatitis A virus (HAV)

This picornavirus is transmitted by the fecal–oral route, as are the enteroviruses. The organism is ingested and spreads hematogenously to the liver, affecting the hepatocyte. It is excreted into the bile and thence into the feces. It causes acute hepatitis. In children, this is often mild or asymptomatic, but can be severe, and can cause acute liver failure. The incubation period is 2–6 weeks. Acute hepatitis is characterized by anorexia, nausea, vomiting, fever, jaundice, and tender hepatomegaly. The clinical illness caused by hepatitis A is the same as that caused by the other viral causes of hepatitis, namely hepatitis B, hepatitis C, and hepatitis E. However, unlike hepatitis B and C, hepatitis A does not cause chronic liver disease. Laboratory tests reveal elevated transaminases and conjugated and unconjugated hyperbilirubinemia. Once clinical illness has developed, the amount of virus shed in feces and, consequently, the patient's contagiousness is very low.

The virus can be cultured in tissue, which is the method used for manufacture of the vaccine. However, this is not suitable for diagnosis of the infection. The diagnosis is made by the detection of IgM antibodies in serum. The presence of IgG in the blood is evidence of past infection or immunization. Confirming the diagnosis of hepatitis A is of some urgency because it is contagious by the fecal–oral route, and because preventive measures for close contacts, such as family members and co-attendees at day care centers, are available. These measures are immune globulin, by intramuscular injection, and vaccination. The vaccine is inactivated virus, and in the USA universal vaccination is recommended for children older than 1 year.

Practical Medical Microbiology for Clinicians, First Edition. Frank E. Berkowitz and Robert C. Jerris.

Hepatitis B virus (HBV)

This belongs to the family Hepadnaviridae. These are DNA viruses that replicate by transcription to RNA, and then by reverse transcription back to DNA (as do retroviruses). Their DNA becomes integrated into the host genome. They affect several different animals. Among its several structural and functional proteins are antigens that are used as markers of infection, including contagiousness. Antibodies to these antigens can be used to determine the state of infection or immunity. The replication cycle, in particular the stage of reverse transcription, provides mechanisms for the use of antiviral drugs, some of which are also used for treating patients with HIV infection.

The virus has a worldwide distribution (Fig. 7.1). It is transmitted in body fluids in three main ways: sexual intercourse, by blood inoculation (transfusion, intravenous drug abuse, or occupational needlesticks), and perinatally. The incubation period is about 6 weeks to 6 months. It can cause both acute and chronic hepatitis. Acute hepatitis manifests in the same manner as do other types of acute hepatitis. Recovery usually occurs but it can, rarely, cause liver failure and death. Chronic infection, which develops in about 1% of individuals, can lead to chronic inflammation, cirrhosis, and hepatocellular carcinoma. Perinatally acquired disease, which is associated with immune tolerance, leads to chronic infection in almost all cases.

Diagnosis

The different markers shown in Table 7.1 are used for diagnosing the presence of infection, whether it is acute or chronic, and immunity. The presence of antibody to surface antigen indicates recovery from infection, or immunization. Detection of hepatitis B e antigen and antibody to e antigen, and viral load, are used for management of patients with chronic infection.

Treatment

Management consists of supportive care for liver disease, public health interventions, e.g. management of contacts (sexual, newborns to infected mothers) with passive immunization with hepatitis B immune globulin, prevention through active immunization, and management of cases of chronic infection, which includes antiviral therapy. The drugs used are interferon α, nucleoside reverse transcriptase inhibitors (lamivudine, entecavir, and telbivudine), and nucleotide reverse transcriptase inhibitors (adefovir and tenofovir).

Hepatitis C virus (hepacivirus)

This is the only known flavivirus that is not arthropod borne. Its RNA is translated into a polyprotein that is cleaved by proteases. This provides a site for antiviral therapy. There are six genotypes, the most common in the USA being genotype 1. Disease caused by genotype 1 is the most resistant to therapy. Hepatitis C has a world-wide distribution. It is a very important cause of chronic liver disease, and is the most common reason for liver transplantation in the USA. The virus is transmitted primarily through blood, as a result of needle sharing, blood transfusion, or occupational needlesticks. It can be transmitted sexually, but not frequently, and about 5% of infected mothers transmit the virus to their infants. The incubation period is 2–12 weeks. Acute hepatitis,

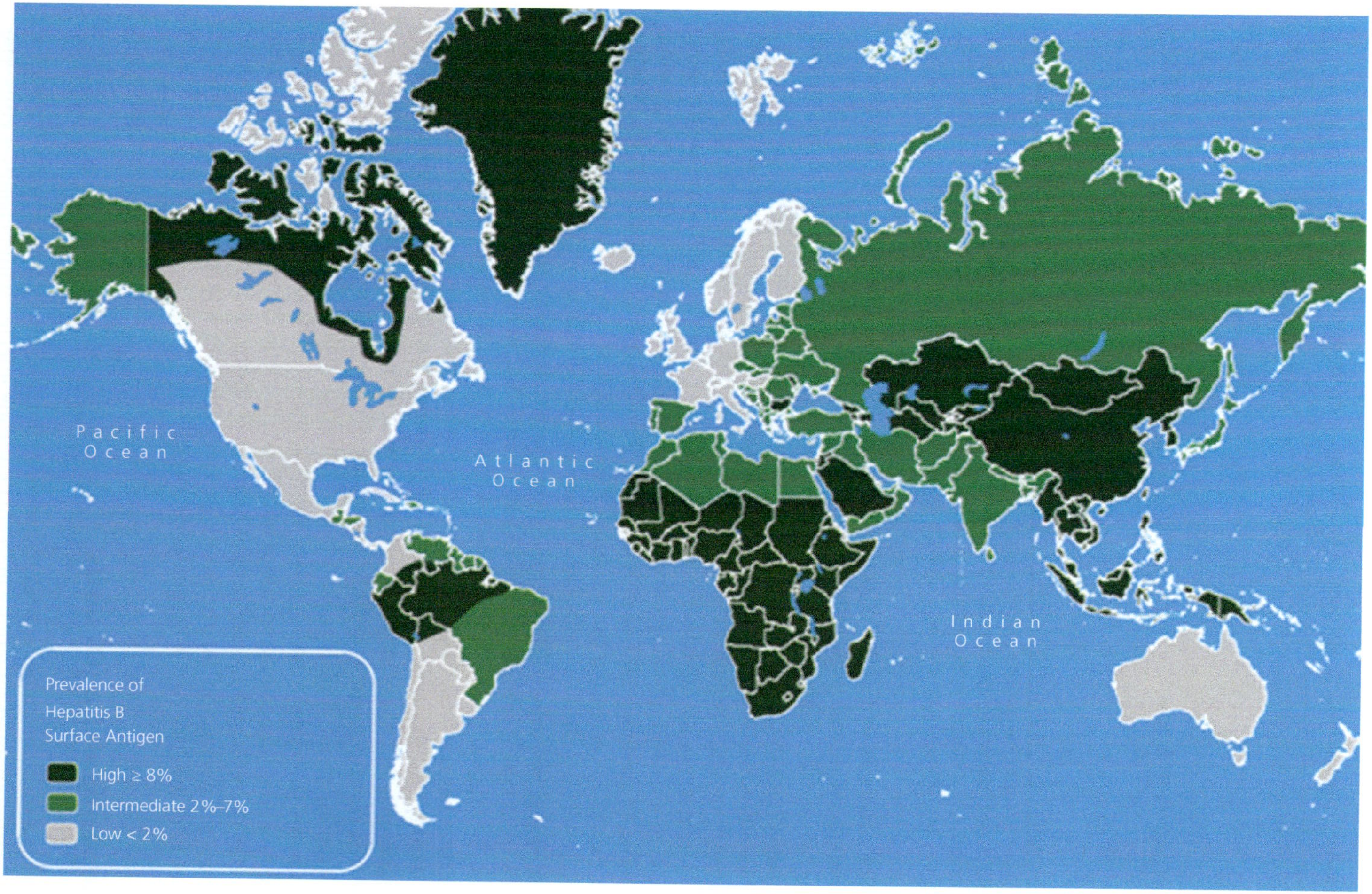

Fig. 7.1 Geographic distribution of hepatitis B (Yellow Book, 2012 CDC).

Table 7.1 Markers of hepatitis B infection.

HBsAg	Total core Ab	IgM core	Anti-HBsAg	Interpretation
–	–	–	–	Never infected
+	–	–	–	Early infection; up to 18 days after immunization
–	–	–	+	Immunized, post receipt of hepatitis B immune globulin
+	+	+	–	Acute infection
–	+	+	+/–	Acute infection, resolving
–	+	–	+	Past infection, immune
+	+	–	–	Chronic infection
–	+	–	–	False positive; past infection; infant of infected mother; low-level chronic infection

ab, antibody; HBsAg, hepatitis B surface antigen.
Source: Mast E, Margolis H, Fiore A, et al. (2005) A comprehensive immunization strategy to eliminate transmission of hepatitis B virus infection in the United States: recommendations of the Advisory Committee on Immunization Practices. Part I: Immunization of Infants, Children, and Adolescents. MMWR 54 (RR-16): 1–23.

which occurs in about 17% of cases, rarely results in fulminant disease. More symptomatic acute disease is more likely to be associated with viral clearance. Up to 80% of infected cases develop persistent infection. This leads to chronic hepatitis, cirrhosis, and hepatocellular carcinoma (see hepatitis B earlier in this chapter).

The diagnosis of hepatitis C depends on antibody tests and molecular tests. The diagnosis of acute hepatitis C can be particularly difficult to make because antibody tests are not very sensitive for this purpose. Hepatitis C nucleic acid assays are the most sensitive tests and, in acute infection, will be positive without antibody tests being positive. In screening for hepatitis C, which should be performed on individuals with risk factors for acquiring the infection, if antibody tests are positive, they should be confirmed by a nucleic acid assay. Asymptomatic individuals with evidence of infection should be evaluated for chronic liver disease, and have the virus genotype and RNA viral load determined with a view to antiviral therapy. Most individuals with acute hepatitis C infection can be followed for at least 6 months to see whether the infection has resolved, before antiviral therapy is contemplated.

In the past, the drugs that were used for patients with hepatitis C infection were interferon α and ribavirin, which interferes with viral RNA synthesis. However, this is an area of rapid progress in antiviral therapy, with several newer drugs, which act as polymerase chain inhibitors (sofosbuvir, dasbuvir, ombitasvir, ledipasvir) or protease inhibitors (simeprevir, paritaprevir), showing marked efficacy, when used in various combinations, depending on the genotype of the patient's virus. Recommendations for the diagnosis and management of hepatitis C infection have been published by the American Association for the Study of Liver Diseases and the Infectious Diseases Society of America (www.hcvguidelines.org).

Hepatitis delta (D) virus

This is a small, incomplete, RNA virus whose outer coat contains hepatitis B proteins. The virus can infect only those individuals already infected with hepatitis B virus. The viral genome resembles that of a plant virioid. It is prevalent in central Africa, the

Horn of Africa, the Amazon area of South America, the Mediterranean area, eastern Europe, and parts of Asia. Transmission is through body fluids. By infecting individuals already infected with hepatitis B, it can aggravate chronic liver disease. Infection is diagnosed by demonstrating the presence of antibodies to hepatitis D antigen. IgG antibodies are present in anyone who has ever had the infection, and IgM antibodies are present early in infection, but can remain positive in chronic infection. Hepatitis D RNA can be detected in the blood in chronic infection.

Hepatitis E virus

This gave its name to the Hepeviridae, to which it belongs. There is one serotype and four genotypes. Genotypes 1 and 2 infect humans only, and occur in epidemics, while 3 and 4 infect mainly pigs, but also infect humans. The virus is a non-enveloped, icosahedral, single-stranded, positive-sense RNA virus. It is transmitted by the fecal–oral route, but can also be transmitted by blood. It causes acute hepatitis after an incubation period of 2–10 weeks. It is spread in contaminated food and water, and is endemic in many underdeveloped regions, where large outbreaks can occur. It causes acute hepatitis, which can be particularly severe in pregnant women. Chronic infection can occur, mainly in immunocompromised individuals. The diagnosis is made by the detection of IgM antibodies in serum, but nucleic acid tests are becoming available. These have the advantage of allowing genotyping. A recombinant vaccine, which has been used in China, appears to be very effective.

Further reading

Aggarwal R, Jameel S (2011) Hepatitis E. Hepatology 54:2218–26.

Dienstag JL (2008) Hepatitis B virus infection. N Engl J Med 359: 1486–500.

Feld JJ, Kowdley KV, Coakley E, et al. (2014) Treatment of HCV with ABT-450/r – ombitasvir and dasabuvir with ribavirin. N Engl J Med 370: 1594–603.

Hoofnagle JH, Nelson KE, Purcell RH (2012) Hepatitis E. N Engl J Med 367: 1237–44.

Hughes SA, Wedemeyer H, Harrison PM (2011) Hepatitis delta virus. Lancet 378: 73–85.

Kamar N, Bendall R, Legrand-Abravanel F, et al. (2012) Hepatitis E. Lancet 379: 2477–88.

Liang TJ, Ghany MG (2013) Current and future therapies for hepatitis C virus infection. N Engl J Med 368: 1907–17.

Liang TJ, Ghany MG (2014) Therapy of hepatitis C – back to the future. N Engl J Med 370: 2043–7.

Maheshwari A, Ray S, Thulavath PJ (2008) Acute hepatitis C. Lancet 372: 321–32.

Martin A, Lemon SM (2006) Hepatitis A virus: from discovery to vaccines. Hepatology 43: S164–S172.

Trepo C, Chan HLY, Lok A (2014) Hepatitis B virus infection. Lancet 384: 2053–63.

US Public Health Service (2013) Testing for HCV infection: an update of guidance for clinicians and laboratorians. MMWR 62 (18): 362–5.

Webster DP, Klenerman P, Dusheiko GM (2015) Hepatitis C. Lancet 385: 1124–35.

Zhang Z, Zhang X-F, Huang S-J, et al. (2015) Long-term efficacy of a hepatitis E vaccine. N Engl J Med 372: 914–22.

CHAPTER 8

Arthropod-borne viruses (arboviruses), hantaviruses, arenaviruses, and filoviruses

Arboviruses are arthropod-borne viruses. This is an epidemiologic category and crosses taxonomic (virus family) lines. These viruses, which are transmitted by mosquitoes, ticks, or flies, belong to the following virus families.

- Flaviviridae (many) (all flaviviruses except hepatitis C are arboviruses)
- Alphaviruses (many)
- Bunyaviridae (many)
- Rhabdoviridae (Chandipura virus, of the vesiculoviruses)
- Reoviridae (coltiviruses of the orbiviruses)

Table 8.1 shows important arthropod-borne viruses, listed by virus family, with their geographic distributions, vectors, clinical syndromes they cause, and methods used for their diagnosis.

Flaviviruses (Flaviviridae)

These are single-stranded, positive-sense RNA viruses. This family contains the most important of the arboviruses, namely yellow fever virus, dengue viruses, and Japanese B encephalitis virus, in addition to West Nile and many other viruses.

Yellow fever virus (YFV)

This virus is the prototype flavivirus, and the family derives its name from this virus (*flavus* is the Latin for yellow). It is endemic in many countries of tropical Africa and tropical South America, where it also causes major outbreaks (Figs 8.1 & 8.2). It is transmitted by the mosquitoes *Aedes aegypti* and other species in Africa, and *Haemogogus* spp. in South America (Fig. 8.3).

There are two epidemiologic patterns: the jungle (sylvatic) pattern, in which monkeys are the hosts and humans contract the infection if they enter forests; and urban disease, in which humans are the main host.

Practical Medical Microbiology for Clinicians, First Edition. Frank E. Berkowitz and Robert C. Jerris.

Table 8.1 Arthropod-borne viruses, listed by virus family, showing their geographic distributions, vectors, clinical syndromes they cause, and diagnostic methods.

	Distribution	Vector	Clinical syndromes	Diagnosis
Flaviviruses				
Yellow fever	Tropical Africa, South America	*Aedes aegypti*	Systemic, multiorgan failure	Culture (blood), N1 Ag, RT-PCR, serology
Dengue	South and SE Asia, Africa, Caribbean, South and Central America	*Aedes aegypti*, *Aedes albopictus*	Severe musculoskeletal pain, shock, hemorrhage	Culture (blood), RT-PCR, N1 Ag, serology
Japanese encephalitis	East, South East and South Asia	*Culex* spp.	Severe encephalitis	Serology – blood, CSF
Tick-borne encephalitis	Europe and Asia	Tick (*Ixodes* and *Dermacentor* spp.)	Encephalitis	RT-PCR blood (early phase), serology
West Nile	Africa, Mediterranean, USA,	*Culex* spp.	Systemic, encephalitis	Culture blood, serology (blood, CSF)
St Louis encephalitis	USA	*Culex* spp.	Encephalitis	Serology
Powassan encephalitis	USA, Canada	Tick (*Ixodes*, *Dermacentor* spp.)	Encephalitis	Serology (blood, CSF)
Murray Valley	Australia	*Culex* spp.	Encephalitis	Culture, RT-PCR, serology
Bunyaviruses				
Rift Valley fever	Middle East, Africa	Mosquito (*Aedes* spp.)	Systemic, HF	RT-PCR, serology
Congo Crimean HF	Asia, Africa	Tick (*Hyalomma* spp.)	HF	Antigen, RT-PCR,
Sandfly fever	Middle East, Mediterranean	*Phlebotomus* spp.	Fever	Culture, RT-PCR, serology
LaCrosse	USA	*Culex* spp.	Encephalitis	Serology
Alphaviruses				
Eastern equine encephalitis	USA	Mosquito	Encephalitis	Culture, RT-PCR, serology
Western equine encephalitis	USA	Mosquito	Encephalitis	Culture, RT-PCR, serology
Chikungunya	Africa, Asia, Indian Ocean, Caribbean,	*Aedes* spp.	Fever, polyarthralgia	Culture, RT-PCR, serology
Venezuelan equine encephalitis	South America, southern USA	Mosquito (several)	Encephalitis	Culture, RT-PCR, serology
Reoviruses				
Colorado tick fever	USA	Tick	Fever	Culture, immune fluorescence of blood smear, serology

Ag, antigen; CSF, cerebrospinal fluid; HF, hemorrhagic fever; PCR, polymerase chain reaction; RT-PCR, reverse transcriptase polymerase chain reaction.

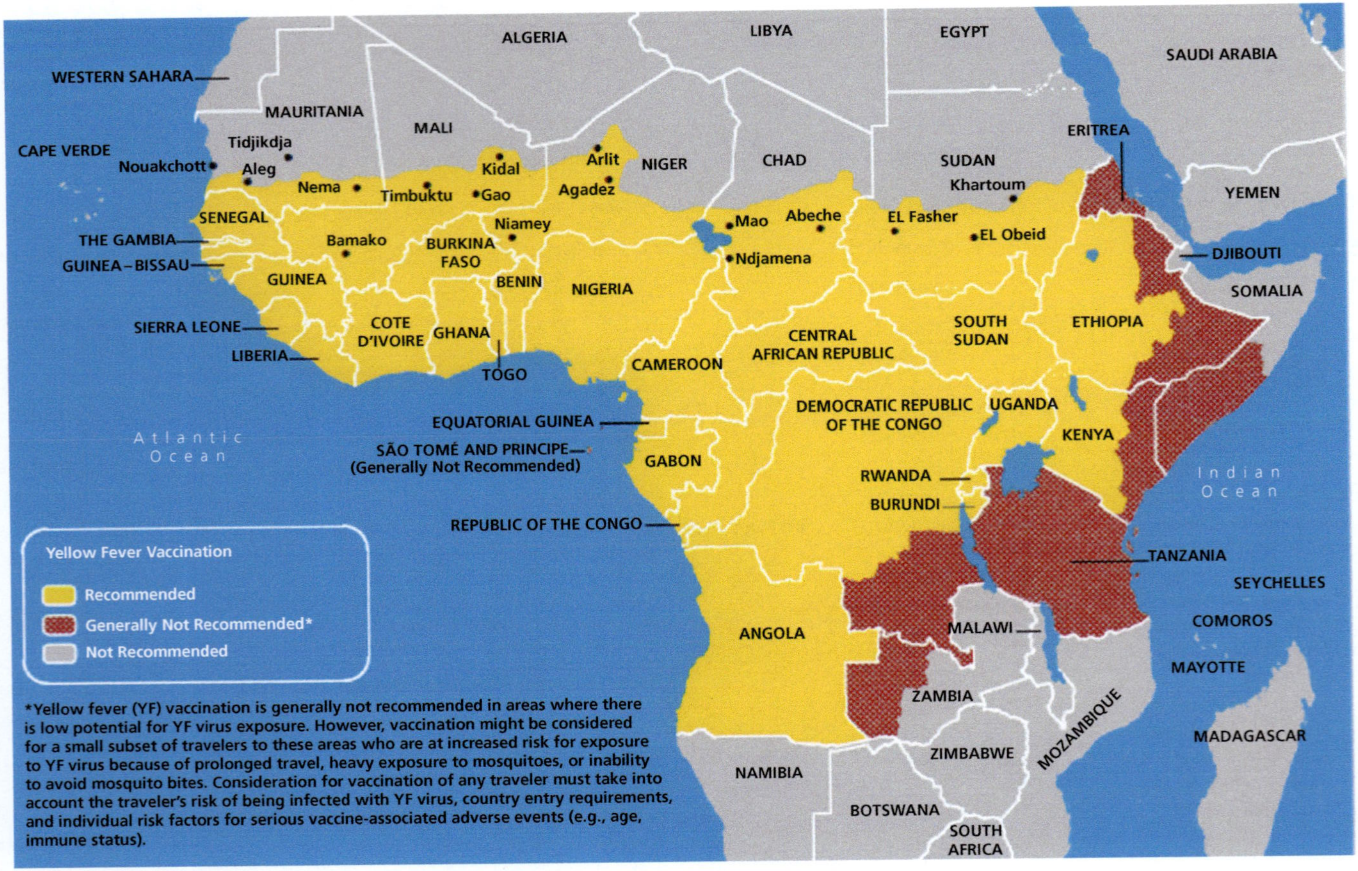

Fig. 8.1 Distribution of yellow fever in Africa. Source: Yellow Book, CDC, 2014.

Fig. 8.2 Distribution of yellow fever in South America. Source: Yellow Book, CDC 2014.

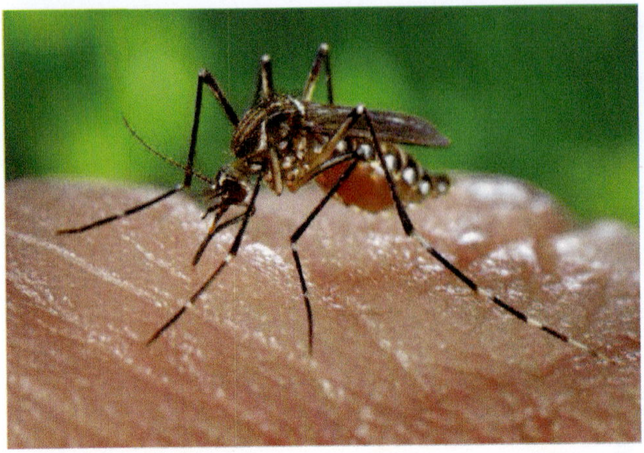

Fig. 8.3 *Aedes aegypti*. Courtesy of PHIL, CDC.

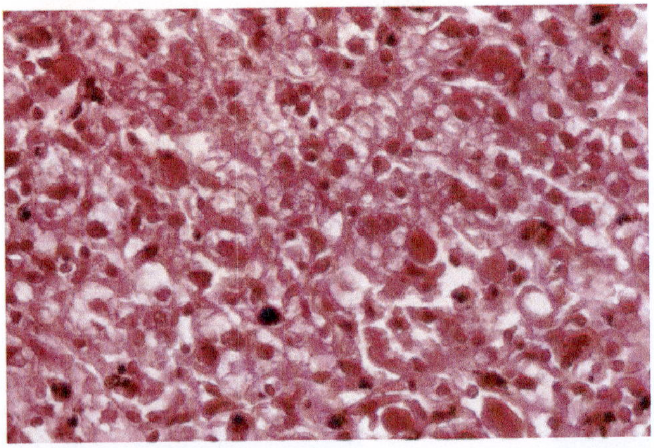

Fig. 8.4 Liver histology in yellow fever. Note the eosinophilic "Councilman" bodies. Courtesy of PHIL, CDC.

The incubation period is 3–6 days. Clinical illness is characterized by fever, vomiting, and myalgia. This is followed by evidence of hepatitis, coagulopathy, renal failure, and multiorgan failure. The fatality rate is about 30%.

The diagnosis is made by culture of the virus from blood, detection of N1 antigen in blood, by serology, and by polymerase chain reaction (PCR). There are characteristic but non-pathognomonic histopathologic findings in the liver (Fig. 8.4). Specimens should be sent to a reference laboratory.

Treatment is supportive. The infection can be prevented by a very effective live attenuated vaccine, developed by Max Theiler, who received the Nobel Prize for this work. Although rare, this vaccine can cause severe visceral disease, especially in the elderly.

Dengue virus

This is a flavivirus that has a widespread distribution, in southern and South East Asia, tropical Africa, the Indian Ocean, the Caribbean, and Central and South America. It is currently also present in Key West, Florida (Figs 8.5 & 8.6). It is transmitted by *Aedes aegypti*, the same mosquito that transmits yellow fever, and *Aedes albopictus*.

Fig. 8.5 Distribution of dengue in the Western hemisphere. Courtesy of Centers for Disease Control and Prevention. CDC Health Information for International Travel 2014. New York: Oxford University Press; 2014.

Fig. 8.6A Distribution of dengue in Asia and Oceania. Courtesy of Centers for Disease Control and Prevention. CDC Health Information for International Travel 2014. New York: Oxford University Press; 2014.

Fig. 8.6B Distribution of dengue in Africa. Courtesy of Centers for Disease Control and Prevention. CDC Health Information for International Travel 2014. New York: Oxford University Press; 2014.

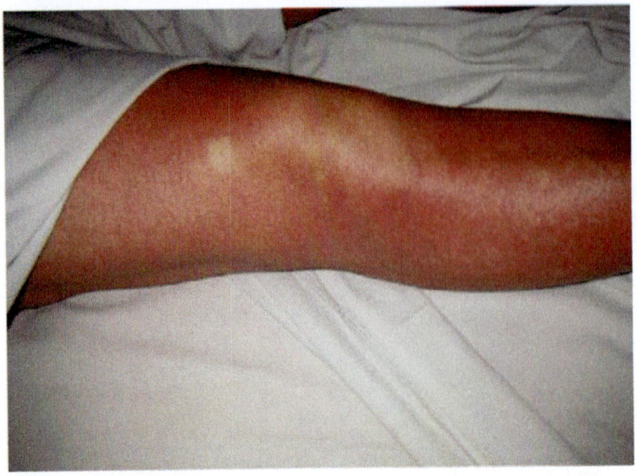

Fig. 8.7 The rash occurring in dengue.

There are four serotypes of dengue, which do not stimulate cross-protection. In fact, a previous infection with one serotype renders one more susceptible to severe disease caused by a different serotype. The disease is characterized by fever, headache and severe muscle and back pain, and rash (Fig. 8.7). Severe disease (dengue hemorrhage/shock syndrome) manifests with capillary leak, hemoconcentration, anasarca, hypotension, thrombocytopenia, and hemorrhage.

The diagnosis is made by demonstration, in blood, early in the illness of RNA (<7 days of illness), of antigen (non-structural 1 protein – NS1) (<9 days) or, if later, by the demonstration of IgM. The management is supportive. Vaccines are currently under investigation.

Japanese B encephalitis virus

This virus is prevalent in southern, south-eastern and eastern Asia (Fig. 8.8). It is transmitted by *Culex* spp. mosquitoes, the natural reservoirs being mainly pigs and aquatic birds. It causes severe encephalitis, associated with a very high fatality rate. The diagnosis can be made by the detection of IgM antibodies in blood or cerebrospinal fluid, or by the demonstration of seroconversion. An effective vaccine is available.

West Nile virus

This flavivirus was first recognized in Uganda. It was introduced into the USA from the Middle East in about 1990, and spread very quickly across the country. It causes a febrile illness but can cause encephalitis and myelitis. The elderly are most susceptible to neurologic complications, but these can occur at any age. The virus can be cultured *in vitro*, but the diagnosis is usually made serologically. Demonstration of IgM antibodies in cerebrospinal fluid confirms the diagnosis of West Nile encephalitis. Management is supportive.

Tick-borne encephalitis

This is an important cause of encephalitis in Europe and northern Asia (Fig. 8.9). It is transmitted by *Ixodes* spp. and *Dermacentor* spp. ticks.

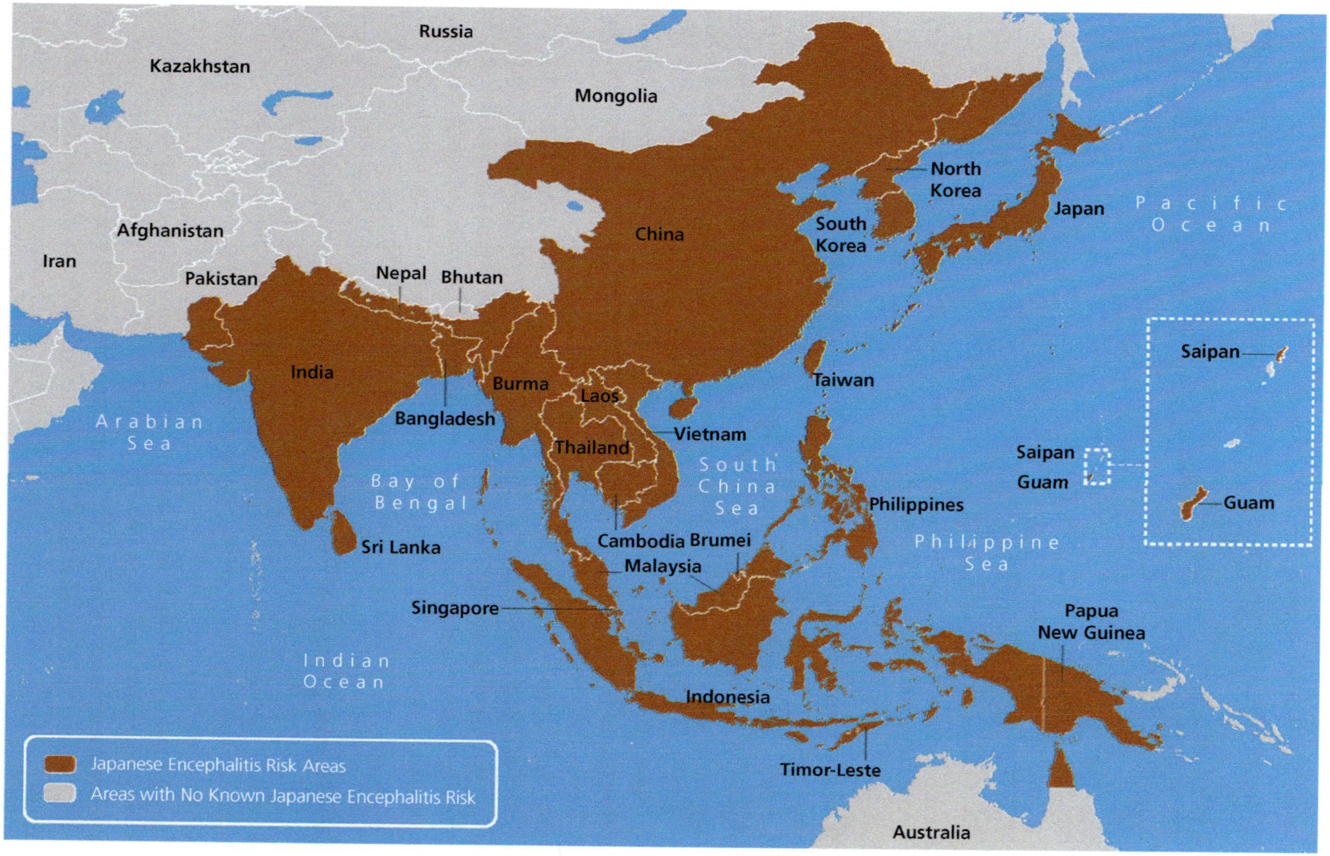

Fig. 8.8 The distribution of Japanese B encephalitis. Source: Yellow Book 2012, CDC.

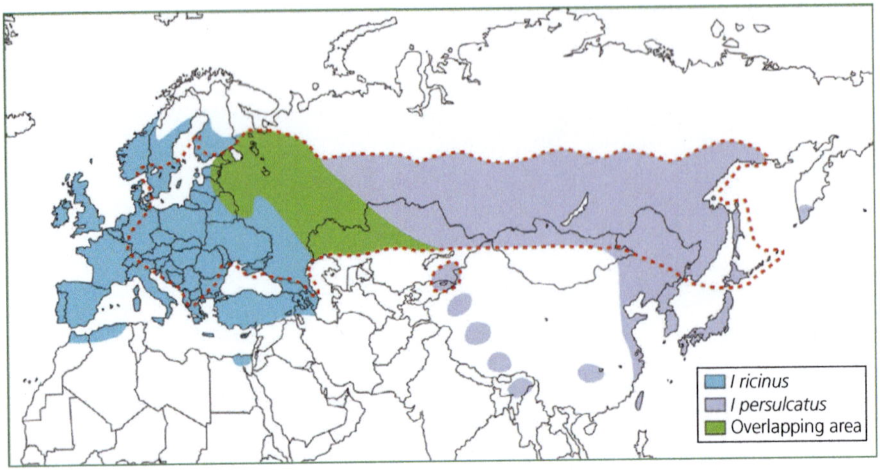

Fig. 8.9 Distribution of tick-borne encephalitis. Source: Lancet 2008; 371: 1861–71; reprinted with permission from Elsevier.

The clinical illness is biphasic: the first phase, the viremic phase, is manifested by a non-specific febrile illness with headache and myalgias. After a few asymptomatic days, the second phase occurs, manifesting with features of meningoencephalitis of varying severity. It can target the cranial motor nuclei and cervical anterior horn cells, resulting in weakness of the muscles innervated by cranial nerves and the upper limbs. Diagnosis can be made in the viremic stage by reverse transcriptase (RT)-PCR of blood, and in the neurologic phase by serology of blood or cerebrospinal fluid. A killed vaccine is available in Canada and Europe, but not in the USA.

Other arthropod-borne flavivirus infections include St Louis encephalitis, transmitted by *Culex* spp. mosquitoes in the USA, Powassan encephalitis, transmitted by *Ixodes* spp. and *Dermacentor* spp. ticks in northern USA and Canada, and Murray Valley encephalitis, transmitted by *Culex* spp. mosquitoes in Australia. In early 2014, an outbreak of Zika virus infection occurred in the Pacific island of New Caledonia.

Togaviruses (Togaviridae)

This family of positive-stranded RNA, enveloped viruses contains two genera, namely, the alphaviruses, which are arthropod borne, and one rubivirus, namely rubella virus (see Chapter 6).

The alphaviruses, which are transmitted by mosquitoes, may be considered in terms of Old World and New World viruses. The Old World alphaviruses cause mainly fever and arthralgia, which can be prolonged. These include Chikungunya, O'nyong nyong, Sindbis, and Semliki Forest virus (which were initially discovered in Africa), and Ross River and Barmah Forest virus which occur in Australia. In recent years, Chikungunya has spread across the Indian Ocean and southern Asia, causing major epidemics with significant morbidity. It has now spread to the Western hemisphere ("New World"), where its vectors, *Aedes aegypti* and *Aedes albopictus*, are present. The New World alphaviruses, namely Western, Eastern, and Venezuelan equine encephalitis viruses, are primarily pathogens of horses, in which they cause a high fatality rate. However, they can cause severe encephalitis in humans, with significant long-term morbidity.

The diagnosis can be made early in the illness by viral culture or nucleic acid amplification of serum, or by the demonstration of antibodies in serum or cerebrospinal fluid.

Bunyaviruses (Bunyaviridae)

These are negative-stranded RNA viruses, whose genome is divided into three segments. This family, named for Bunyamwera virus, contains five genera, namely Orthobunyavirus, Nairovirus, Phlebovirus, Hantavirus, and Tospovirus (members of this latter genus are plant pathogens). Many are arthropod-borne viruses, discussed here. The hantaviruses are rodent excreta-borne ("rodex") viruses that are discussed separately (see later in this chapter).

Orthobunyavirus

These include the California encephalitis virus group, to which La Crosse virus belongs, Oropouche virus, and Cache Valley virus. LaCrosse virus encephalitis can vary in severity, and does not have specific features. It can be focal, and indistinguishable from herpes simplex virus encephalitis.

Nairovirus

Congo-Crimean hemorrhagic fever virus is tick borne. It is endemic in western Asia and parts of Africa. It causes a multiorgan disease with hemorrhage, which is associated with a very high fatality rate. Nosocomial transmission can occur. The diagnosis is based on culture and PCR of blood, which should be performed only in a P4 (BSL 4) laboratory.

Phlebovirus

Rift Valley fever virus

This virus, borne by *Aedes* spp. mosquitoes, affects mainly ruminants, especially sheep. Several major outbreaks of infection have occurred in Africa, including Egypt, and in the Arabian Peninsula. In humans, it causes severe systemic disease, including hemorrhagic fever.

Toscana virus

This is transmitted by the sandfly (*Phlebotomus* spp.) in the Mediterranean area and Middle East. It is characterized by fever and myalgia which usually does not result in serious sequelae.

Severe fever with thrombocytopenia syndrome (SFTS)

This is a systemic infection recently described in eastern China. It is transmitted by the tick *Haemophysalis longicornis*. A related virus, Heartland virus (HRTV), has recently been reported from Missouri, USA. It is transmitted by the tick *Amblyomma americanum*.

Reoviruses (Reoviridae)

One of the reoviruses that is transmitted by the *Dermacentor andersoni* tick is Colorado tick fever (coltivirus), which invades red cells.

Rodent excreta ("rodex") and bat excreta ("batex") viruses

These viruses include a group that have a similar mode of transmission, in the secretions or excretions of rodents (mainly mice) and bats. They include the following families.

- Paramyxoviridae (Henipa viruses), discussed in the section on paramyxoviruses (Chapter 6)
- Bunyaviridae (hantaviruses)
- Arenaviridae
- Filiviridae

Hantavirus

In the late 1970s, a syndrome characterized by fever, hemorrhage, and renal failure was reported from Korea, and the implicated virus, Hantaan virus, acquired from rodents, was named for the Hantaan River there. Subsequently, a more benign form of the infection, called nephropathia epidemica, was reported from Europe, and the implicated virus is called Puumala virus. Although serologic evidence of infection with this virus was found in the USA, no clinical illness was reported until 1993. In that year, a newly recognized clinical syndrome occurred in the four-corners area of the USA (where Utah, New Mexico, Colorado, and Arizona meet), which is called Hantavirus pulmonary syndrome. It is characterized by non-specific features of fever, headache, backache, diarrhea, and malaise, followed by the sudden onset of respiratory disease with rapid progression to respiratory failure, with a very high fatality rate. Myocardial depression is a major factor in the deterioration. The virus does not cause cytopathic effects. The clinical illness is caused by the inflammatory cascade initiated by the virus, which has a significant effect on the endothelium, leading to endothelial failure and pulmonary edema (Figs 8.10 & 8.11).

Diagnostic tests that can help in suspecting the condition are evidence of hemoconcentration, elevated leukocyte count with a marked left shift, presence of atypical lymphocytes, and thrombocytopenia. The causative virus, acquired from the mouse *Peromyscus maniculatus* (Fig. 8.12), is called Sin Nombre virus. More than 500 cases have been reported, mostly from western states, but across most of the USA. Other hantaviruses reported from the USA and South America are Bayou virus, Black Creek virus, New York virus, Andes virus, Juquitiba virus, and Laguna virus. These viruses are excreted in the urine of mice, and humans are infected by inhalation of the virus. Diagnosis is made serologically and by PCR. The treatment is supportive.

Arenaviruses (Arenaviridae)

These are single-stranded RNA viruses with a segmented genome. They cause chronic, asymptomatic infection in rodents, and are excreted in their urine. Humans are infected by contact with the urine.

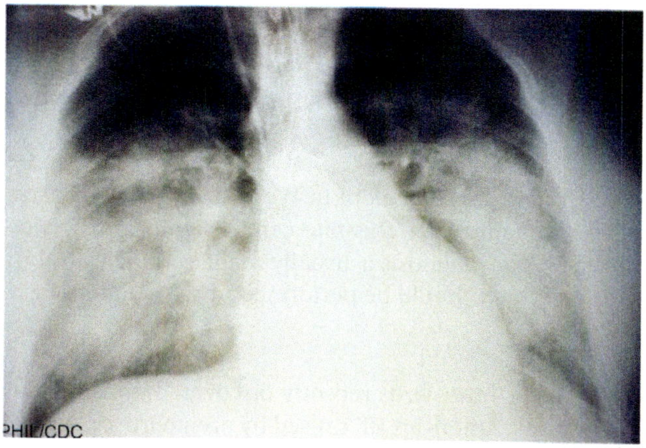

Fig. 8.10 Radiologic appearance of the lungs in a patient with hantavirus pulmonary syndrome. Courtesy of PHIL, CDC.

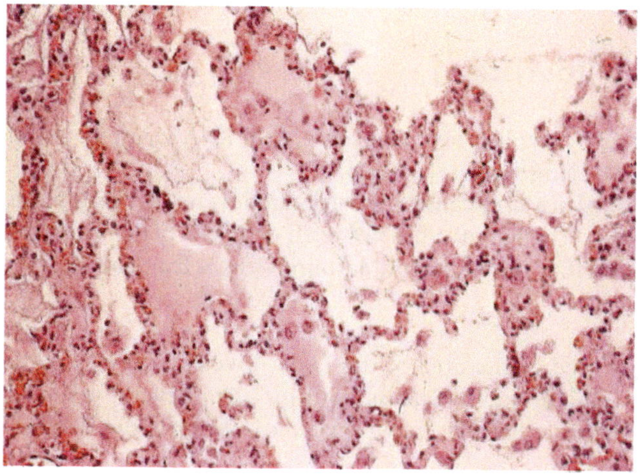

Fig. 8.11 Appearance of the lung in hantavirus pulmonary syndrome. Note the intraalveolar fluid. Courtesy of PHIL, CDC.

Fig. 8.12 *Peromyscus maniculatus*. Courtesy of PHIL, CDC.

Lymphocytic choriomeningitis virus

This causes aseptic/viral meningitis. It can cause intrauterine infection in the fetus, resulting in severe brain injury. Diagnosis is usually made serologically.

Lassa fever virus

This causes Lassa fever, which is prevalent in West Africa. It can cause a hemorrhagic fever (HF), with a high fatality rate. This rate can be significantly reduced by the use of intravenous ribavirin. The diagnosis is usually made serologically, but the virus can be detected by culture, which should be performed in a P4 laboratory.

Other arenaviruses

Lujo virus is a hemorrhagic fever virus recently discovered in Zambia.

There are several South American HF caused by arenaviruses, namely Argentinian HF caused by Junin virus, Bolivian HF caused by Machupo virus, another Bolivian HF caused by Chapare virus, a Brazilian HF caused by Sabia virus, and Venezuelan HF caused by Guanarito virus.

Filoviruses (Filoviridae)

This family of single-stranded, negative-sense, thin RNA viruses contains two species: Ebola virus and Marburg virus (Fig. 8.13).

They occur in Africa and cause hemorrhagic fevers, carrying an extremely high fatality (up to 90%). Their source has not been definitively demonstrated, but there is evidence to suggest that the source of Marburg virus is fruit bats.

Ebola virus

This was first reported in the area of the Ebola River in Zaire (now the Democratic Republic of the Congo), and in Sudan, both in 1976. Several major outbreaks have occurred since. Currently (2014), a huge epidemic is raging in the West African countries

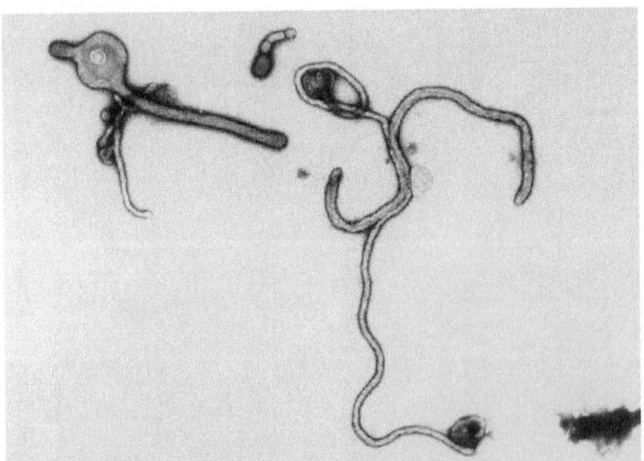

Fig. 8.13 Ebola virus. Courtesy of PHIL, CDC.

of Guinea, Liberia, and Sierra Leone, with spread to Nigeria, Senegal, Spain, and the USA. Four species have caused disease in humans: Zaire, Sudan, Bundibugyo, and Tai ebola viruses. Reston ebola virus, isolated from monkeys from the Philippines, has not infected humans.

Marburg virus

This virus was first recognized in 1967 when laboratory workers in Marburg, Germany, contracted the infection from monkeys originating in Uganda. Several outbreaks have since occurred in east, central, and southern Africa, the largest being in Angola in 2005.

The pathogenesis of both these infections involves activation of the reticuloendothelial system, with release of cytokines, and multiorgan dysfunction. Hemorrhage, which occurs in some patients, is due, at least in part, to endothelial dysfunction. Spread is through body fluids, and the infection can be spread within the healthcare setting. Many of the victims of the current Ebola epidemic have been healthcare workers. The diagnosis should be suspected in individuals who have been in tropical Africa, and present with fever, headache, conjunctival injection, and body aches. A macular rash may occur. Diarrhea, with ensuing dehydration, is a significant factor in mortality in Ebola virus infection. The diagnosis, which is an emergency, can be made by antigen detection, RT-PCR, virus isolation, electron microscopy and immunostaining of tissue, and IgM serology.

Further reading

Bannister B (2010) Viral haemorrhagic fevers imported into non-endemic countries: risk assessment and management. Br Med Bull 95: 193–225.

Chen LH, Wilson ME (2010) Dengue and chikungunya infections in travelers. Curr Opin Infect Dis 23: 438–44.

Dionisio D, Esperti F, Vivarelli A, et al. (2003) Epidemiological, clinical and laboratory aspects of sandfly fever. Curr Opin Infect Dis 16: 383–8.

Feldman H (2014) Ebola – a growing threat. N Engl J Med 371: 1375–8.

Gould EA, Solomon T (2008) Pathogenic flaviviruses. Lancet 371: 500–9.

Gould EA, Coutard B, Malet H, et al. (2010) Understanding the alphaviruses: recent research on important emerging pathogens and progress towards their control. Antiviral Res 87: 112–24.

Guzman MG, Harris E (2015) Dengue. Lancet 385: 453–65.

Halstead SB (2007) Dengue. Lancet 370: 1644–52.

Lindquist L, Vapalahti O (2008) Tick-borne encephalitis. Lancet 371: 1861–71.

Liu Q, He B, Huang S-Y, et al. (2014) Severe fever with thrombocytopenia syndrome, an emerging tick-borne zoonosis. Lancet Infect Dis 14: 763–72.

MacNeil A, Ksiazek TG, Rollin PE (2011) Hantavirus pulmonary syndrome, United States, 1993-2009. Emerg Infect Dis 17: 1195–201.

Markoff L (2013) Yellow fever outbreak in Sudan. N Engl J Med 368: 689–91.

Martines RB, Ng DL, Greer PW, et al. (2015) Tissue and cellular tropism, pathology and pathogenesis of Ebola and Marburg viruses. J Pathol 235: 153–74.

McMullan LK, Folk SM, Kelly AJ, et al. (2012) A new phlebovirus associated with severe febrile illness in Missouri. N Engl J Med 367: 834–41.

Monath TP (2007) Dengue and yellow fever – challenges for the development and use of vaccines. N Engl J Med 357: 2222–5.

Monath TP, Vasconcelos PEC (2014) Yellow fever. J Clin Virol 64: 160–73.

Overturf GD (2009) World Arboviruses. The *Bunyaviridae*. Pediatr Infect Dis J 28; 1014–15.

Peterson AT, Bauer JT, Mill JN (2004) Ecologic and geographic distribution of filovirus disease. Emerg Infect Dis 10: 40–7.

Simmons CP, Farrar JJ, van Vinh Chau N, et al. (2012) Dengue. N Engl J Med 366: 1423–32.

Soldan SS, Gonzalez-Scarano F (2005) Emerging infectious diseases: The *Bunyaviridae*. J Neurovirol 11: 412–23.

Towner JS, Amman BR, Sealy TK, et al. (2009) Isolation of genetically diverse Marburg viruses from Egyptian fruit bats. PLoS Pathog 5: e1000536.

Vinh DC, Embil JM (2009) Hantavirus pulmonary syndrome: a concise clinical review. South Med J 102: 620–5.

Yu XJ, Liang MF, Zhang SY, et al. (2011) Fever with thrombocytopenia associated with a novel bunyavirus in China. N Engl J Med 364: 1523–32.

Zacks M, Paessler S (2010) Encephalitis alphaviruses. Vet Microbiol 140: 281–6.

SECTION III
Bacteriology

CHAPTER 9

Bacteriology

Structure of bacteria

Bacteria are prokaryotes, i.e. they do not have a nuclear membrane. They have a cell wall, a cell membrane, a chromosome, and ribosomes. In addition, some bacteria have flagella (singular flagellum), which enables these bacteria to move, and some have fimbriae (singular fimbria), also called pili (singular pilus), which enable the bacteria to adhere to epithelial surfaces (see virulence factors, later in this chapter), and to pass genetic material to other bacteria (see conjugation, later in this chapter). The basic structure of a bacterial cell is shown in Fig. 9.1.

Bacteria have a single chromosome which is arranged in a circle, and which is supercoiled, which reduces its volume. DNA replication occurs in both directions. In addition, they may have pieces of extrachromosomal DNA, called plasmids. These may contain genes for virulence factors or antibiotic resistance. The ribosomes, where proteins are synthesized, consist of 30S and 50S subunits. These are targets of antibiotics that inhibit protein synthesis.

There are three main shapes of bacteria: round – coccus (cocci); rod-shaped – bacillus/rod (bacilli/rods); spiral – spirochete(s).

Cell wall (Fig. 9.2)

This is a rigid structure, composed of cross-linked carbohydrates, that gives the organism shape and prevents osmotic injury. Enzymes necessary for its synthesis are present in the cell membrane. These are called penicillin-binding proteins and are targets of β-lactam antibiotics. In Gram-positive bacteria, the cell wall consists of about 40 layers of peptidoglycan, while in Gram-negative bacteria, it consists of 3–4 layers of peptidoglycan, outside which there is a complex lipid membrane called the outer membrane. A component of this membrane is endotoxin. Channels called porins allow macromolecules to pass through the outer membrane.

Practical Medical Microbiology for Clinicians, First Edition. Frank E. Berkowitz and Robert C. Jerris.

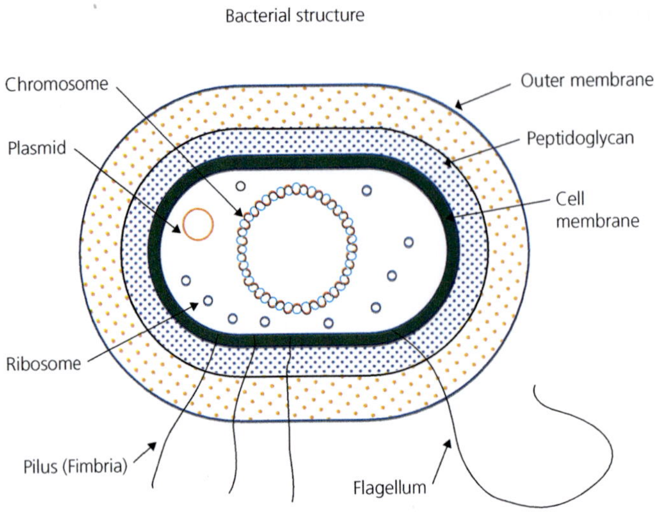

Fig. 9.1 Structure of a bacterial cell.

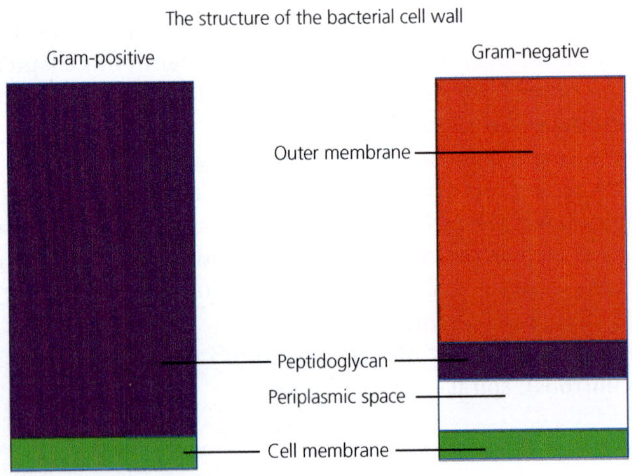

Fig. 9.2 Basic structure of Gram-positive and Gram-negative cell walls.

Genetic changes

Bacteria can change their genetic composition over a short period of time by the following mechanisms.

- Mutation: because bacteria have short generation times, mutations during replication can accumulate rapidly. Some of these may give the organisms a survival advantage.
- Acquisition of genetic material, by various mechanisms.
  - Conjugation: this is by the transfer of plasmids from a donor bacterial cell to a recipient cell. The plasmid DNA can replicate as does that of the chromosome. A plasmid that can be integrated into the chromosome is called an episome.

- Transformation: this is the acquisition of DNA from the environment. This phenomenon was first demonstrated when live non-encapsulated pneumococci were mixed in a medium with dead encapsulated pneumococci. The live pneumococci acquired the ability to make capsules, indicating their ability to take up genetic material from the medium. This observation contributed to the discovery of the structure of DNA.
- Transduction: this is the process by which a bacteriophage (a virus infecting a bacterium) introduces DNA into the bacterial cell. This DNA can be incorporated into the bacterial DNA (in which case it is called lysogenic), or it may remain in the cytosol, in which case it will replicate and cause lysis of the bacterial cell (lytic). An example of genetic material conferred by a bacteriophage is the gene for diphtheria toxin production in *Corynebacterium diphtheria* (see Chapter 12).

Pieces of genetic material that can move from a plasmid to a chromosome or to another plasmid are called transposons. Genes that can be carried in these genetic elements contain information that can convey antibiotic resistance and virulence. Groups of virulence genes constitute a pathogenicity island.

Bacterial virulence factors

These are bacterial properties that enable the organism to cause disease. The genes coding for these factors may be present on the chromosome, or on plasmids, and may have been derived from a bacteriophage of another bacterium. They can be categorized as follows.

- Facilitating bacterial adherence to host skin or mucosa, e.g. *Staphylococcus aureus*: fibronectin-binding proteins attaching to host fibronectin, and lipoteichoic acid; *E. coli*: fimbrial antigens facilitating attachment to uroepithelium.
- Promoting spread of the organism within host tissue, e.g. *Streptococcus pyogenes* hyaluronidase.
- Lysing tissue, e.g. *Clostridium perfringens* α-toxin.
- Damaging host leukocytes, e.g. *Staphylococcus aureus* α-toxin, Pantin-Valentine leukocidin (PVL); several *Streptococcus pyogenes* enzymes.
- Interfering with host immune responses, e.g. *Yersinia pestis* V antigen.
- Preventing phagocytosis, e.g. polysaccharide capsules of *Streptococcus pneumoniae, Haemophilus influenzae* type b, *Neisseria meningitidis, E. coli, Klebsiella pneumoniae, Bacillus anthracis,* and *Yersinia pestis*; *Streptococcus pyogenes* hyaluronic acid and M-protein.
- Exotoxins, with specific effects. These generally have a structure of a heavy component, which attaches to the target, and a light component, which exhibits the toxic activity. Examples include tetanus toxin and botulinum toxin. Some bacterial exotoxins are ADP-ribosyl transferases, whose effect on the target is shown in the equation.

$$\begin{array}{c}\text{Target}+\text{NAD}\rightarrow\text{Target}-\text{ADPR}+\text{N}\\ \uparrow\\ \text{ADP ribosyl transferase}(\text{exotoxin})\end{array}$$

where NAD is nicotinamide adenine dinucleotide and ADP is adenine diphosphate.

These include diphtheria toxin and *Pseudomonas aeruginosa* exotoxin A, which ADP-ribosylate elongation factor 2 on the ribosome, inhibiting protein synthesis, and cholera toxin. Enterotoxins are exotoxins which have an effect on the intestine, e.g. cholera toxin, *E. coli* heat-labile enterotoxin, and staphylococcal enterotoxins.

- Superantigens: these are exotoxins that can cause non-specific activation of large numbers of T-lymphocytes, resulting in a large outpouring of cytokines. Examples include streptococcal pyrogenic toxins and staphylococcal toxic shock syndrome toxin.
- Endotoxin (lipid A): this is a component of the lipopolysaccharide of the outer membrane of Gram-negative bacteria. It causes activation of the host systemic inflammatory response, through its interaction with Toll-like receptor 4, playing a major role in the pathogenesis of the sepsis syndrome caused by Gram-negative bacteria, e.g. meningococcemia.
- Antigenic variation: the ability of some bacteria to change their surface coats rendering them immune to antibodies that have developed against them, as occurs in *Borrelia recurrentis.*
- Changes in the expression of some virulence factors when moving from colonizing a mucosal surface to invading tissue, e.g. *Neisseria meningitidis,* or when moving from one type of host to another, as occurs when *Yersinia pestis* moves from a flea to a mammal.

Detection and identification of bacteria

This is addressed in detail in Chapter 2.

Resistance to antibacterial agents

See section on antimicrobial resistance in Chapter 1.

Mechanisms of resistance

There are five basic mechanisms by which bacteria can exhibit resistance to antimicrobial agents. More than one of these mechanisms can work simultaneously and synergistically.

- Failure of the drug to enter the organism in order to reach its target (barrier to entry): this barrier is posed by the cell wall and the cell membrane of the bacterium. This is particularly important in Gram-negative bacteria. Macromolecules enter Gram-negative bacteria through channels in their membranes called porins. Mutations in the porins can result in resistance. Examples of this include resistance of Gram-negative rods to aminoglycosides and β-lactams.
- Efflux mechanism: this is an energy-dependent mechanism whereby the drug is pumped out of the bacterial cell, before it reaches its target. This is important in resistance to tetracyclines and fluoroquinolones.
- Enzymatic alteration of the drug, rendering it inactive. This is probably the best known mechanism. Examples include β-lactamases (of which there are several hundred) and aminoglycoside-modifying enzymes.

- Alteration of the target, so that it no longer binds to the drug. Examples of this include alteration of penicillin-binding proteins in penicillin-resistant *Streptococcus pneumoniae* and methicillin-resistant staphylococci, and changes in the ribosome preventing binding of streptomycin.
- Creation of an alternative biochemical pathway for a particular function, such that the drug's target is no longer necessary for the organism. Examples of this include a different pathway for folate synthesis causing sulfonamide resistance, and use of D-ala-D serine (instead of D-ala-D-ala) for cross-linkages in peptidoglycan, causing vancomycin resistance.

Antibacterial agents

Antibacterial agents have revolutionized the management of infectious diseases. The first such agent was Prontosil, a synthetic sulfur compound, and the precursor of a group of agents called the sulfonamides. The strict meaning of the term antibiotic is an antimicrobial agent derived from a microbe. The majority are derived from fungi or species of Streptomyces. The first antibiotic was penicillin, which was discovered by Alexander Fleming, who noted inhibition of staphylococcal growth on agar plates which had become contaminated by a penicillium mold. This was purified and brought to industrial production by Florey and Chain. The history of discovery of newer agents and the chemical modification of these agents is closely linked to the development of resistance to them. In this chapter, the terms antibiotic and antibacterial agent are used synonymously.

From a clinical perspective, the susceptibility of a bacterial organism to a particular antimicrobial agent is determined by the ability of the agent to inhibit growth of the organism at a concentration achievable in the tissue, without causing toxicity. The measure of the susceptibility of an organism to a particular agent is the minimal inhibitory concentration (MIC), which is a property of the organism in relation to a particular drug. It is the lowest concentration of the drug that inhibits growth of the organism. The lower the MIC, the more susceptible the organism is to that drug. The MIC is usually measured in μg/mL or mg/L (the values are the same) (see Chapter 2).

The following antibacterial agents are considered here: penicillins, cephalosporins, carbapenems, aztreonam, aminoglycosides, tetracyclines, tigecycline, macrolides, glycopeptides, lipoglycopeptides, daptomycin, chloramphenicol, rifamycins, fluoroquinolones, linezolid, streptogramins, clindamycin, sulfonamides, other folate antagonists, metronidazole, polymyxins, fosfomycin, and urinary antiseptics. Their sites of action are shown in Fig. 9.3.

β-Lactams

These agents inhibit the growth of or kill bacteria by binding to their penicillin-binding proteins (PBPs), which are transpeptidases, carboxypeptidases, and endopetidases located on the bacterial cell membrane, and which are important in formation of cross-linkages in the bacterial cell wall. The active part of the drug is the β-lactam bond. The basic structures of these drugs have been modified chemically by the addition of various side-chains to produce drugs with different antimicrobial spectra and pharmacokinetics from the parent drug. Hydrolysis of the β-lactam bond is accomplished by bacterial

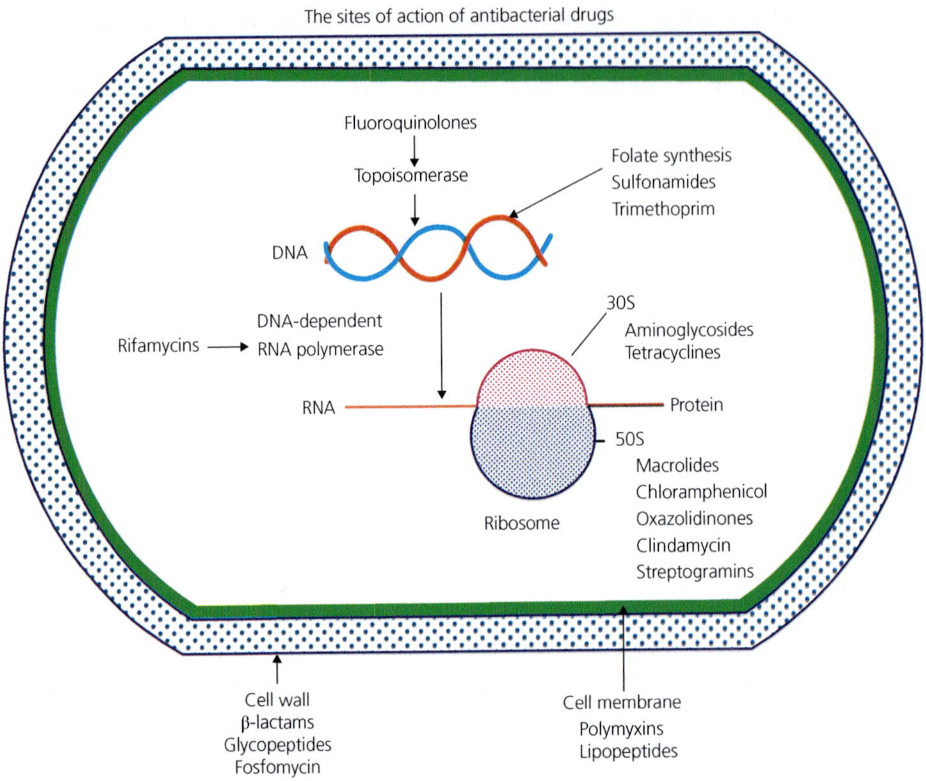

Fig. 9.3 Sites of action of different classes of antibacterial agents. DNA, deoxyribose nucleic acid; RNA, ribose nucleic acid; 30S, 30S ribosomal subunit; 50S, 50S ribosomal subunit.

β-lactamases, which render the drug inactive. Other mechanisms of bacterial resistance to these drugs are mutations of the penicillin-binding proteins, the drugs' targets, and a barrier to drug entry. (see section on drug-resistance) β-Lactams consist of four groups of drugs, namely penicillins, cephalosporins, monobactams, and carbapenems.

Penicillins

Penicillin (natural penicillin)

This drug, whose basic structure is shown in Fig. 9.4, in its various forms is active against most streptococci (including *Streptococcus pneumoniae*, all *Streptococcus pyogenes* and *Streptococcus agalactiae*), enterococci, *Neisseria meningitidis*, *Treponema pallidum*, Actinomyces, *Streptobacillus moniliformis*, most clostridia (including *Clostridium perfringens* and *Clostridium tetani*), *Corynebacterium diphtheriae*, many oral Gram-negative rods, and some oral anaerobes. Although it was previously active against staphylococci, these are almost all resistant nowadays.

It can be given intravenously (penicillin G, which has a short half-life and requires administration every 4–6 hours), intramuscularly (penicillin G, procaine penicillin G – which has a longer half-life and can be given every 12 hours, and benzathine penicillin, which has a very long half-life and results in low blood concentration for

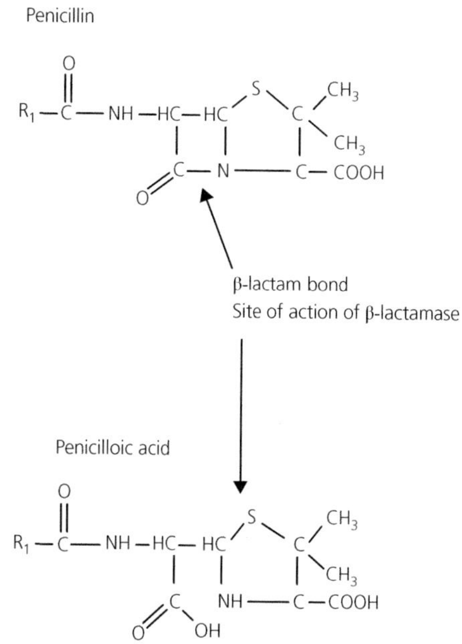

Fig. 9.4 The basic structure of penicillins.

about 3 weeks), and orally (penicillin V). It is a very safe drug, the main adverse effect being allergy. In very large dosages, it can cause seizures and hemolytic anemia.

Isoxazolyl penicillins (oxacillin, cloxacillin, flucloxacillin) and nafcillin

These are active against staphylococci that produce β-lactamase but, by definition, not against "methicillin-resistant staphylococci" whose mechanism of resistance is not mediated by β-lactamase. They are also active against *Streptococcus pyogenes*, but not against enterococci. They are the most narrow-spectrum antibacterial agents available. Although resistance to these drugs among staphylococci is called "methicillin resistance," methicillin represents the class and has not been in clinical use for many years due to its nephrotoxicity.

Oxacillin and nafcillin must be administered parenterally, while the others can be given by mouth.

β-Lactamase inhibitors

Because much of the bacterial resistance to penicillins is mediated by β-lactamases, inhibitors of these enzymes have been developed. They do not have much intrinsic antimicrobial activity, but they inhibit many β-lactamases. When combined with a penicillin, they inhibit the β-lactamase, allowing the penicillin to attack the organism. Four penicillin/inhibitor combinations are available in the USA: ampicillin/sulbactam, amoxicillin/clavulanic acid, piperacillin/tazobactam, and ticarcillin/clavulanic acid, as well as two cephalosporin/inhibitor combinations (see later).

Aminopenicillins

These drugs are modifications of penicillin accomplished by the addition of an amino side-chain.

Ampicillin

This has a similar antibacterial spectrum and adverse effect profile to those of penicillin, but it has additional activities against *Listeria monocytogenes* and some Gram-negative rods, in particular *Haemophilus influenzae, Escherichia coli,* and *Proteus mirabilis.* Although *Salmonella* spp. and *Shigella* spp. were susceptible to this agent, resistance is now widespread among these bacteria. Resistance among *H. influenzae* and *E. coli* strains, due to β-lactamase, is also common. It can be administered parenterally and orally.

Amoxicillin

This is very similar to ampicillin and has the same antimicrobial spectrum. However, it is better absorbed from the intestine than is ampicillin. It is the most active oral agent against penicillin-resistant *Streptococcus pneumoniae.* When amoxicillin or ampicillin is combined with a β-lactamase inhibitor, their antimicrobial spectrum is extended to methicillin-susceptible staphylococci, *H. influenzae, E. coli, Klebsiella* spp., *Moraxella catarrhalis* and many anaerobes, including *Bacteroides* spp.

Carboxypenicillins: carbenicillin and ticarcillin

In these penicillins, a carboxy side-chain has been added to penicillin. Their main value is their activity against *Pseudomonas aeruginosa, Enterobacter* spp., indole-positive species of Proteus (*P. vulgaris, P. rettgeri,* and *Morganella morganii*), and *Providencia* spp. They are not active against *Klebsiella* spp. In the USA, carbenicillin is not available, and ticarcillin is available only in combination with clavulanic acid. This broadens its spectrum to many Gram-negative rods, including *Klebsiella* spp., and anaerobes.

Ureidopenicillins

These penicillins are modifications of aminopenicillins, by addition of a chain to the amino group by a ureido bond. These are mezlocillin, azlocillin, and piperacillin (piperacillin has a piperazine side-chain). Piperacillin is active against most bacteria against which penicillin is active, in addition to several Gram-negative rods, in particular *Pseudomonas aeruginosa* and *Klebsiella* spp. In the USA, it is available only in combination with tazobactam, a β-lactamase inhibitor. This broadens the spectrum to include *Enterobacter* spp., many anaerobic bacteria, including *Bacteroides* spp., and methicillin-susceptible staphylococci.

Cephalosporins

The first cephalosporin was discovered in a fungus from the Mediterranean Sea. The basic structure is shown in Fig. 9.5. Many modifications have been made to this basic structure, and variations are often considered as belonging to first, second, third, fourth or fifth generations. The modifications occur at two sites on the molecule: those at the C7 site increase stability to β-lactamase, thus expanding the antimicrobial spectrum, while those at the C3 site affect its half-life. The early cephalosporins, cephaloridine and cephalothin, are no longer in use. In general, the cephalosporins are active against streptococci (but not enterococci), methicillin-susceptible staphylococci, *E. coli, Klebsiella* spp., and *Proteus mirabilis.* Their activity against Gram-negative bacteria expands with advancing generations. They are not active against *Listeria monocytogenes,* and, with the exceptions of the cefamycins, they have limited activity against anaerobes.

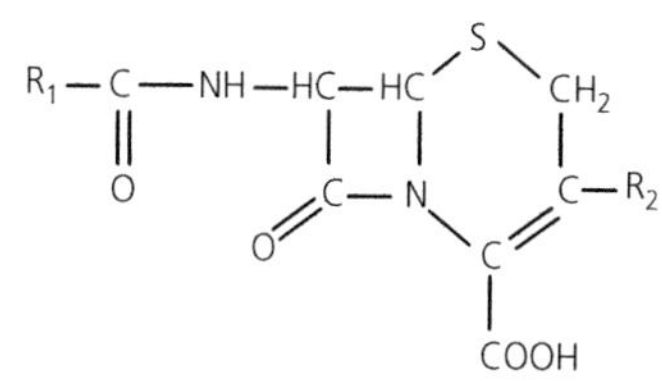

Fig. 9.5 The basic structure of cephalosporins.

First-generation cephalosporins

These are active against streptococci (but not against enterococci nor penicillin-resistant *Streptococcus pneumoniae*), *E. coli*, *Klebsiella pneumoniae*, and *Proteus mirabilis*. They include cefazolin, which is administered parenterally, and cephalexin which is used orally.

Second-generation cephalosporins

In addition to activity against those bacteria for which the first-generation cephalosporins are active, they are active against *H. influenzae*, and fairly active against penicillin-resistant *Streptococcus pneumoniae*.

Cefuroxime is available for parenteral and oral use.

The cephamycins, cefoxitin and ceftizoxime, are considered together with second-generation cephalosporins. They have a methoxy group at the C7 site, which gives them activity against anaerobes.

Third-generation cephalosporins

These are "extended-spectrum" cephalosporins, and include cefotaxime, ceftriaxone, and ceftazidime. They are very active against many Gram-negative rods. Only ceftazidime is active against *Pseudomonas aeruginosa*. Cefotaxime and ceftriaxone are very active against streptococci, including penicillin-resistant *Streptococcus pneumoniae*, but ceftazidime is not. Ceftriaxone has a very long half-life, so that it can be administered only once per day. Orally available drugs in this category include cefixime, which has poor activity against Gram-positive organisms, and cefdinir.

Fourth-generation cephalosporins

The agent within this category is cefepime, which is resistant to the hydrolytic activity of the inducible broad-spectrum β-lactamases elaborated by **S***erratia marcescens*, **P***seudomonas aeruginosa*, **I***ndole-positive Proteus* spp., **C***itrobacter freundii, and* **E***nterobacter* spp. (so-called "SPICE" organisms).

Fifth-generation cephalosporins

Cefobiprole and ceftaroline are active against methicillin-resistant *Staphylococcus aureus*, and ceftozolane/tazobactam is active against Gram-negative rods.

Carbapenems

These have a β-lactam structure but instead of a sulfur atom, they have an oxygen atom. They are active against a broad range of bacteria, including methicillin-susceptible staphylococci, streptococci, enterococci, a broad range of Gram-negative rods, including *Pseudomonas aeruginosa* (except for ertapenem), and anaerobic bacteria. They include the following.

Imipenem

This was the first clinically available carbapenem. It is a small molecule, which enables it to penetrate the cell membrane of Gram-negative bacteria, and it is resistant to β-lactamases by virtue of having its hydroxyethyl side-chain in the trans rather than the cis configuration. Because it is rapidly hydrolyzed by the renal tubular enzyme dihydropeptidase I, which results in nephrotoxic metabolites, it is combined with cilastatin, which inhibits this enzyme.

Meropenem and doripenem

These are very similar in spectrum and usage. They are active against *Burkholderia cepacia* (which imipenem is not), but not against *Rhodococcus equi* (which imipenem is).

Ertapenem

This has the advantage of a long half-life, so that it can be administered once per day. However, it is much less active against *Pseudomonas aeruginosa* and *Acinetobacter baumanni* than the other carbapenems.

Monobactam

Aztreonam has a single ring. It is active against only aerobic Gram-negative bacteria. It does not enter cerebrospinal fluid well, so it is not appropriate for use in patients with meningitis.

Aminoglycosides

These are aminocyclitol-aminoglycoside sugars. They have multiple side-chains, consisting of hydroxyl or amino groups. They act by irreversibly binding to the 30S subunit of the ribosome, causing inhibition of protein synthesis. This binding cannot occur under anaerobic or acidic conditions. Therefore these drugs are inactive within cells or pus, or against anaerobic bacteria. They do not cross membranes well, and are widely distributed in extracellular fluid. They are active against aerobic Gram-negative rods, staphylococci, and some mycobacteria, and they are synergistic with cell wall active agents against enterococci, enabling killing of these organisms. Their killing of bacteria is concentration dependent, and they have a prolonged postantibiotic effect (i.e. the antibacterial effect continues after the concentration of drug drops below the MIC.) Their main adverse effects are 8th cranial nerve toxicity, which is irreversible, and renal toxicity, which is usually reversible. Because they have a narrow therapeutic index, blood concentrations should be monitored in most cases.

Resistance can be due to the following: enzymatic alteration of the molecule (the side-chains can be changed by bacterial phosphotransferases, adenyltransferases or acetyltransferases, specific for each site on the molecule); mutation of porins, preventing entry of the drug into the bacterial cell; and mutation of the ribosome (which occurs mainly with streptomycin).

Streptomycin

This was the first aminoglycoside in clinical usage. It was the first antibiotic used for the treatment of tuberculosis. Its main use now is for the treatment of patients with tuberculosis, and for synergy with ampicillin or vancomycin for the treatment of patients with severe enterococcal infections, such as infective endocarditis, where the organism does not exhibit synergy with gentamicin.

Kanamycin

This is not widely used, except for resistant tuberculosis.

Gentamicin

This is the most commonly used of the aminoglycosides. It is active against staphylococci and a wide range of Gram-negative rods, including *Pseudomonas aeruginosa*. It is not active against streptococci, except for synergy with a cell wall active agent (see Streptomycin earlier). It is also used for the treatment of patients with plague (*Yersinia pestis* infection), tularemia (*Francisella tularensis* infection), and brucellosis (caused by *Brucella* spp.). Tobramycin is similar to gentamicin, but is slightly more active against *Pseudomonas aeruginosa*. Netilmicin is also similar to gentamicin, and possibly less toxic.

Amikacin

This has similar activity to that of gentamicin, but is active against some bacteria which are resistant to gentamicin, because it is not susceptible to as many drug-modifying enzymes as are the other aminoglycosides. It is therefore used mainly in hospitals where resistant Gram-negative rods are prevalent. It is also used for its antimycobacterial activity and its activity against many strains of *Nocardia* spp.

Neomycin

This is too toxic for systemic usage, but it is widely used in topical preparations.

Tetracyclines

These antibiotics have broad-spectrum activity against spirochetes, rickettsiae, ehrlichiae, chlamydiae, and mycoplasmas, as well as against Gram-positive and Gram-negative bacteria. Their mechanism of action is by inhibition of protein synthesis at the 30S ribosomal subunit. Resistance among pyogenic bacteria is widespread and is due to mutations of the ribosomal target, and an efflux mechanism. Their main adverse effects are photosensitivity, gastrointestinal disturbances, bacterial or fungal overgrowth on mucosae, and staining of permanent teeth if used in children younger than 8 years of age. Therefore, they are contraindicated in pregnant women and children younger than 8 years, unless they have rickettsial or ehrlichial infection. They can also cause pseudotumor cerebri.

The main tetracyclines in use are doxycycline, tetracycline, and minocycline. They are used mainly for patients suspected of having rickettsial infection, ehrlichial infection, chlamydial infection (genital *Chlamydia trachomatis* infections, *Chlamydophila psittaci*, and *Chlamydophila pneumoniae* infections), mycoplasma and ureaplasma infections, leptospirosis, Lyme borreliosis, and specifc non-tuberculous mycobacterial infections. They are also frequently used for the control of *Propionbacterium acnes*, which plays a role in the pathogenesis of acne vulgaris. Minocycline is also used for *Nocardia* spp. infections.

Glycylcycline

Tigecycline is a derivative of minocycline, but it is not susceptible to the mechanisms of bacterial resistance to which the other tetracyclines are susceptible, namely efflux and ribosomal protection. This renders it active against a wide variety of Gram-positive and Gram-negative bacteria, including anaerobes, and those that are resistant to other drugs. Its value lies in the treatment of patients with multidrug-resistant infections, particularly caused by Gram-negative rods.

Macrolides

This is a group of antibiotics consisting of a 14-member lactone ring with two sugar moieties attached. They inhibit protein synthesis by attaching to the 50S ribosomal subunit, preventing elongation of the peptide chain. Resistance is due to alteration of the target by methylation, to efflux, or to drug inactivation (mainly among Gram-negative bacteria). Methylation of the target also results in resistance to clindamycin and streptogramin B (see later in this chapter). Target methylation can be constitutive or inducible (see D-test in section on staphylococci, Chapter 10).

The macrolides are active mostly against Gram-positive bacteria, such as streptococci, staphylococci, and *Corynebacterium diphtheriae*, but also against some Gram-negative bacteria, such as *Neisseria* spp., *Bordetella pertussis*, *Legionella* spp., *Bartonella henselae*, *Campylobacter* spp., and against bacteria that lack cell walls. They are used mainly for the treatment of patients with infections caused by *Mycoplasma pneumoniae*, *Chlamydia trachomatis*, *Chlamydophila pneumoniae*, *Bordetella pertussis*, *Legionella* spp., *Campylobacter* spp., and some non-tuberculous mycobacteria. They are used as alternatives to penicillins in patients allergic to penicillin with infections caused by *Streptococcus pyogenes*, *Streptococcus pneumoniae*, and syphilis. They can be used in patients with staphylococcal infection, although resistance is common, especially among methicillin-resistant staphylococci. The main macrolides in use are as follows.

Erythromycin

This was the first discovered macrolide and is named for its source, *Streptomyces erythreus*. It is available as several different salts. Its main adverse effects are abdominal pain and diarrhea. If used in neonates younger than 2 weeks, it predisposes to the development of hypertrophic pyloric stenosis. Erythromycin estolate can cause cholestatic jaundice. Intravenous administration of erythromycin (seldom indicated) can cause severe arrhythmias. It can be associated with several drug–drug interactions.

Clarithromycin

This is used mainly for treating patients with pneumonia, *Helicobacter pylori* infection, and some non-tuberculous mycobacterial infections.

Azithromycin

This is an azalide (it has a 15-member ring). It has largely replaced erythromycin because it is much better tolerated. Because it enters cells well, it can be used for several infections in which the bacteria remain within macrophages, such as some non-tuberculous mycobacterial infections, and salmonella infections. It can be used for treating patients with shigellosis and campylobacteriosis. It has a long half-life, so it can be administered once per day and, in most situations, for only a few days. Clarithromycin and azithromycin are active against *H. influenzae*, which erythromycin is not.

Glycopeptides

Vancomycin

This is derived from the organism *Streptomyces orientalis*. It is active against a wide variety of Gram-positive bacteria. Being a large molecule, it cannot penetrate the outer membrane of Gram-negative bacteria. Its mechanism of action is by combining with the D-ala-D-ala of the cross-linking portion of peptidoglycan, preventing formation of the bacterial cell wall. One of the mechanisms that vancomycin-resistant bacteria utilize for resistance is creating these side-chains with D-ala-D-serine or D-ala-D-lactose.

Vancomycin is active against staphylococci (methicillin susceptible and resistant), streptococci (penicillin and cephalosporin susceptible and resistant), enterococci, *Corynebacterium* spp., *Bacillus* spp., and most *Clostridium* spp. Notable Gram-positive bacteria that are inherently resistant to vancomycin are *Leuconostoc* spp., *Pediococcus* spp., and *Erysipelothrix rhusiopathiae*. Its main uses are for the treatment of patients with severe infections caused by methicillin-resistant staphylococci or ampicillin-resistant enterococci, meningitis caused by β-lactam-resistant *Streptococcus pneumoniae*, antibiotic-associated colitis (see later), and patients with severe infections and with significant allergies to β-lactams, who would otherwise be treated with one of these drugs.

Vancomycin is not absorbed from the gut and therefore must be administered intravenously. The one exception to this is in the treatment of patients with *Clostridium difficile* colitis, in which case the drug is administered orally. Because it can be toxic (ototoxic and nephrotoxic), especially when used in combination with an aminoglycoside, it has a narrow therapeutic index. Therefore, in many circumstances, blood concentrations should be monitored.

Vancomycin-resistant staphylococci are an emerging problem, and vancomycin-resistant enterococci (VRE) (mostly *Enterococcus faecium*) have become a serious healthcare-associated problem.

Teicoplanin

This drug has a similar antimicrobial spectrum to that of vancomycin. Because it has a long half-life, it can be administered once per day (whereas vancomycin should be administered 2–4 times per day). Some VRE are susceptible to teicoplanin.

Lipoglycopeptides

These drugs, namely dalbovancin, oritavancin, and telavancin, are glycopeptides with a lipophilic side-chain. This side-chain enables them to attach to the cell membrane, enhancing their ability to kill the target bacteria. They are active against only Gram-positive bacteria, including anaerobes. They are active against methicillin-susceptible and -resistant *Staphylococcus aureus*, vancomycin-intermediate *Staphylococcus aureus*, and some VRE.

Lipopeptides

Daptomycin is a cyclic lipopeptide. It is active against only Gram-positive bacteria, including those resistant to other antimicrobial agents, being unable to penetrate the outer membrane of Gram-negative bacteria. Its mechanism of action is by causing potassium efflux from the bacterial cell. Its main use is for the treatment of patients with severe infections caused by methicillin-resistant staphylococci and VRE. However, it is not effective in pulmonary infections. It must be administered intravenously, and its main adverse effect is myopathy.

Chloramphenicol

This drug's mechanism of action is inhibition of protein synthesis by attaching to the 50S subunit of the ribosome It has broad-spectrum activity against Gram-positive and Gram-negative bacteria, including anaerobes, and some rickettsiae. Its advantages are that it is very well absorbed from the intestinal tract, it penetrates cerebrospinal fluid and brain tissue well, and it is very cheap. Its disadvantages are that it can cause aplastic anemia, which is rare, and, in newborns, a syndrome called "gray baby" syndrome. This is due to toxicity resulting in shock caused by mitochondrial dysfunction. Because of the potential for causing aplastic anemia, it has gone out of favor in the USA, and the oral form is no longer available in this country. Its main value has been for the treatment of *Haemophilus influenzae* infections, including meningitis, brain abscess, typhoid fever (resistance is now common), and *Streptococcus pneumoniae* infections.

Rifamycins

These drugs inhibit DNA-dependent RNA polymerase. They are active mainly against Gram-positive bacteria and mycobacteria, and their main value is for the treatment of patients with tuberculosis, leprosy, and some other mycobacterial infections. A very important adverse effect is their ability to induce cytochrome oxidase enzymes, which metabolize many other drugs, resulting in a decrease in the effects of those drugs.

Rifampin

This is the most frequently used rifamycin. In addition to being a mainstay of antituberculous therapy, it is used (always together with another active drug) for treatment of patients with staphylococcal infection on foreign bodies such as prosthetic heart valves. In addition to causing drug–drug interactions, it can cause hepatotoxicity, thrombocytopenia, and an influenza-like illness if used intermittently.

Rifabutin

This causes less drug–drug interaction than rifampin, and is used for patients with HIV infection and tuberculosis, who require protease inhibitors and antituberculous therapy. One of its important adverse effects is uveitis.

Rifapentine

This drug has a very long half-life and is used, in combination with isoniazid, for short-course treatment of latent tuberculosis infection.

Fluoroquinolones

This is a group of antimicrobials derived from fluorination of position 6 of nalidixic acid. Their mechanism of action is by inhibiting bacterial topoisomerases II (DNA gyrase) and IV, the enzymes that regulate the supercoiled state of bacterial DNA. This causes cessation of DNA synthesis.

Although, overall, they have broad antimicrobial spectra, their spectra are not the same. The older drugs, e.g. ciprofloxacin, have little antistaphylococcal and antipneumococcal activity and better antipseudomonal activity compared with the newer ones (e.g. levofloxacin); the newest ones (e.g. moxifloxacin) have good antianaerobic activity. They can enter cells, including macrophages, contributing to their value as antimycobacterial, antilegionella, and antisalmonella agents. Most have excellent bioavailability when administered via the oral route. Because early work showed that beagle puppies given a fluoroquinolone developed joint disease, there was, for many years, a reluctance for these drugs to be used in children. However, the canine problem does not seem to affect humans, and there is substantial experience of these drugs being used safely in children.

The main adverse effects are gastrointestinal disturbances and headache. A rare but dramatic effect is tendon rupture. Resistance is due to mutations in the topoisomerases or to efflux pumps.

Fluoroquinolones have been widely used for the treatment of individuals with *Neisseria gonorrhoeae* infection. However, resistance is becoming widespread.

Ciprofloxacin

This was the first fluoroquinolone in clinical usage. It is used widely for the treatment of patients with urinary tract infections and other Gram-negative rod infections.

Levofloxacin

This is the L-enantiomer of ofloxacin, and is used mainly for the treatment of patients with respiratory tract infections, because it is active against *Streptococcus pneumoniae, Mycoplasma pneumoniae, Chlamydophila pneumoniae,* and *Legionella* spp. It is also used in the treatment of multidrug-resistant *Mycobacterium tuberculosis* infection.

Moxifloxacin

This has activity against anaerobes, which the other fluoroquinolones do not.

Oxazolidinones

Linezolid inhibits protein synthesis by attaching to the surface of the 50S ribosomal subunit, preventing the subunit's attachment to the 30S subunit. It is active against Gram-positive bacteria and mycobacteria. Its main value is in the treatment of patients with infections caused by staphylococci, streptococci, and enterococci resistant to more conventional therapy such as β-lactams or vancomycin. It is also of value in the treatment of patients with multidrug-resistant tuberculosis. Its advantage is that it has excellent bioavailability after oral administration. It has several important adverse effects, notably myelosuppression, and peripheral and optic neuritis.

Streptogramins

Quinuprisitine/dalfopristine is a 30:70 mixture of quinupristine, a derivative of streptogramin B, and dalfopristine, a derivative of streptogramin A. It is active against Gram-positive bacteria only. Its main value has been for the treatment of patients with infections caused by vancomycin-resistant *Enterococcus faecium*. It must be administered intravenously and it has significant drug–drug interactions, due to its metabolism by cytochrome P_{450}3A enzymes.

Lincosamide

Clindamycin is a derivative of lincomycin (no longer in use), which was isolated from *Streptomyces lincolnensis*. It inhibits protein synthesis by combining with the 50S ribosomal subunit. It is active against Gram-positive bacteria, including streptococci, staphylococci (not enterococci), *Actinomyces* spp., some clostridia, *Bacillus* spp., and Gram-positive and Gram-negative anaerobes. Resistance, which has become common among anaerobes, is due to changes in the ribosome target. It is well absorbed from the intestinal tract. Its major adverse effect is diarrhea, and it is an important cause of *Clostridium difficile* colitis. It is useful in the treatment of patients with methicillin-resistant *Staphylococcus aureus* infections, or with penicillin allergy.

Folic acid antagonists

Sulfonamides

These derivatives of an azo dye containing a sulfonamide group were the first antibacterial agents in use (1932). Their mechanism of action is by inhibition of the formation of folate, which is necessary for the synthesis of purines and pyrimidines. The chemistry is shown in Fig. 9.6. They are well absorbed from the intestine and are widely distributed in tissue, including the cerebrospinal fluid. Although many categories of bacteria were once susceptible to these drugs, resistance is now widespread.

In most circumstances, sulfonamides are used in a fixed combination with trimethoprim, another folate synthesis inhibitor, as trimethoprim/sulfamethoxazole (cotrimoxazole). This drug combination is active against staphylococci, including those resistant to methicillin, *Streptococcus pneumoniae*, *Nocardia* spp., and many Gram-negative rods, including *E. coli*, *Klebsiella* spp., *Enterobacter* spp., *Salmonella* spp., *Shigella* spp., *Serratia marcescens*, *Stenotrophomonas maltophilia* (for which it is the first drug of choice), *Burkholderia cepacia*, some fungi, notably *Pneumocystis jiroveci*, and some protozoa, namely *Cyclospora cayetanensis* and *Cystoisospora belli* (sulfadiazine, in combination with pyrimethamine, is used for treating patients with *Toxoplasma gondii* infection). Cotrimoxazole is not active against *Pseudomonas aeruginosa* and has limited activity against *Streptococcus pyogenes*. Sulfonamides and cotrimoxazole are well absorbed from the intestine. Although these are fairly safe drugs, they can be associated with several types of hypersensitivity reactions, and cotrimoxazole can be associated with liver and kidney disease.

Dapsone, a sulfone, also inhibits the synthesis of dihydrofolate. It is used for the treatment of patients with Hansen disease (leprosy), as well as some non-infectious bullous skin diseases.

Metronidazole

This is a synthetic agent, which exerts its effect by entering cells and being metabolized by reduction into several bactericidal products. It is very active against many anaerobic

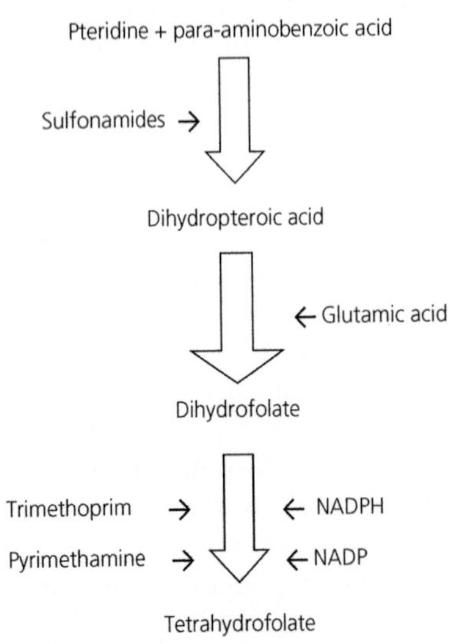

Fig. 9.6 The mechanisms of action of folate synthesis antagonists.

bacteria, as well as protozoa, including *Giardia intestinalis*, *Entamoeba histolytica*, and *Trichomonas vaginalis*. It is well absorbed from the intestine and well distributed in all tissues, including the brain. It is used for treating patients with severe anaerobic infections, including intraabdominal and intracerebral infections, infections caused by the above-mentioned parasites, and bacterial vaginosis.

Polymyxins

These are derived from *Bacillus polymyxa*. Two polymyxins are in use: polymyxin B and polymyxin E (colistin). Their mechanism of action is by attaching to the bacterial cell membrane and causing leakage of ions. They also have antiendotoxin activity. They must be administered intravenously or by inhalation (in patients with cystic fibrosis). They are active against Gram-negative rods (most members of the Enterobacteriaceae, excluding *Serratia* spp., *Proteus* spp., and *Providencia* spp.) and *Pseudomonas aeruginosa*, but not against *Burkholderia cepacia* or *Burkholderia pseudomallei*. Their main adverse effects are nephrotoxicity (the reason for very limited usage) and neurotoxicity. Their main use is for the treatment of patients with severe infections caused by multi-resistant Gram-negative rods.

Fosfomycin

This is an epoxide antibiotic derived from phosphonic acid and obtained from *Streptomyces* spp. It interferes with synthesis of bacterial cell walls at a stage before that affected by β-lactams and by glycopeptides. It has broad-spectrum activity against both Gram-positive bacteria, including staphylococci and streptococci, and Gram-negative bacteria, including *E. coli*. Being a small molecule that has a low degree of protein binding, it is widely distributed in tissue fluids, including the cerebrospinal fluid. In the USA, it is approved for use only for the treatment of patients with urinary tract infections, while in other parts of the world it is used for other types of infections as well.

Urinary antiseptics

Nitrofurantoin

This is a nitrofuran that is enzymatically reduced to an active form, bacterial enzymes being many-fold more active in this than are host enzymes. The metabolites exert their antibacterial effect by causing DNA damage.

Methenamine

This drug is converted, in an acidic environment, to water and formaldehyde, which is antibacterial. It is active mainly against *E. coli* and enterococcus.

Antimycobacterial drugs

These are discussed in Chapter 17 on mycobacteria (Chapter 17).

Further reading

Alvarez-Elcoro S, Enzler MJ (1999) The macrolides: erythromycin, clarithromycin, and azithromycin. Mayo Clin Proc 74: 613–34.

Carpenter CF, Chambers HF (2004) Daptomycin: another novel agent for treating infections due to drug-resistant Gram-positive pathogens. Clin Infect Dis 38: 994–1000.

Dotis J, Iosifidis E, Ioannidou M, et al. (2010) Use of linezolid in pediatrics: a critical review. Int J Infect Dis 14: e638–e648.

Edson RS, Terrell CL (1999) The aminoglycosides. Mayo Clin Proc 74: 519–28.

Garrison MW, Kawamura NM, Wen MM (2012) Ceftaroline fosamil: a new cephalosporin active against resistant Gram-positive organisms including MRSA. Expert Rev Anti Infect Ther 10: 1087–103.

Guskey MT, Tsuji BT (2010) A comparative review of lipoglycopeptides: oritavancin, dalbavancin, and telavancin. Pharmacotherapy 30: 80–94.

Hellinger WC, Brewer NS (1999) Carbapenems and monobactams: imipenem, meropenem, and aztreonam. Mayo Clin Proc 74: 420–34.

Kassamali Z, Rotschafer J, Jones RN, et al. (2013) Polymyxins: wisdom does not always come with age. Clin Infect Dis 57: 877–83.

Kasten MJ (1999) Clindamycin, metronidazole, and chloramphenicol. Mayo Clin Proc 74: 825–33.

Marshall WF, Blair JE (1999) The cephalosporins. Mayo Clin Proc 74: 187–95.

Nathwani D (2005) Tigecycline: clinical evidence and formulary positioning. Int J Antimicrob Agents 25: 185–92.

Popovic M, Steinort D, Pillai S, et al. (2010) Fosfomycin: an old, new friend? Eur J Clin Microbiol Infect Dis 29: 127–42.

Roberts MC (2003) Tetracycline therapy: update. Clin Infect Dis 36: 462–7.

Smilack JD (1999) The tetracyclines. Mayo Clin Proc 74: 727–9.

Smilack JD (1999) Trimethorprim-sulfamethoxazole. Mayo Clin Proc 74: 730–4.

Van Bambeke F, van Laethem Y, Courvalin P, et al. (2004) Glycopeptide antibiotics. From conventional molecules to new derivatives. Drugs 64: 913–36.

Walker RC (1999) The fluoroquinolones. Mayo Clin Proc 74: 1030–7.

Wilhelm MP, Estes L (1999) Vancomycin. Mayo Clin Proc 74: 928–35.

Wright AJ (1999) The penicillins. Mayo Clin Proc 74: 290–307.

Zhanel G, Calic D, Schweizer F, et al. (2010) New lipoglycopeptides: a comparative review of dalbovancin, oritavancin and telavancin. Drugs 70: 859–86.

CHAPTER 10

Gram-positive cocci

The main kinds of Gram-positive cocci (Fig. 10.1) causing disease in human beings are staphylococci and streptococci. Their major features are compared in Table 10.1, and Fig. 10.2 summarizes how these are identified in the laboratory.

Catalase test

Catalase is an enzyme that converts hydrogen peroxide to oxygen and water. The catalase test is performed by making a smear from a colony of the test organism on a microscope slide, and adding a drop of 3% hydrogen peroxide to the smear. Catalase-positive organisms cause bubbling (oxygen), while catalase-negative organisms do not (Fig. 10.3).

Coagulase test

This tests for the presence of a prothrombin-like protein that is used to differentiate between *Staphylococcus aureus* (coagulase positive) and other staphylococci (coagulase negative). It depends on the ability of the organism to cause clotting of rabbit plasma. A slide coagulase test (Fig. 10.4) detects coagulase bound to the organism, while a tube coagulase test (more specific) detects free coagulase.

Staphylococci

There are two main groups of staphylococci affecting humans: *Staphylococcus aureus* (coagulase positive) and coagulase-negative staphylococci. They are differentiated by the coagulase test (see earlier).

Staphylococcus aureus is a very common human pathogen. It colonizes many individuals on the anterior nares and the skin, particularly in the perineum and axillae. It has many virulence factors, and it elaborates several enzymes and exotoxins that contribute to its pathogenicity. It has the tendency to cause abscesses. It causes infections related to the skin and its structures, such as impetigo (a superficial infection), furuncles (boils), and cellulitis (a diffuse infection of the dermis), and it is the most common cause of wound infections, including surgical wound infections. It can

Practical Medical Microbiology for Clinicians, First Edition. Frank E. Berkowitz and Robert C. Jerris.

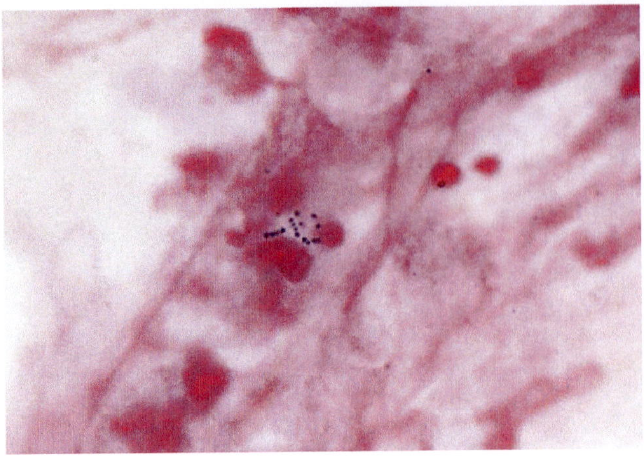

Fig. 10.1 Gram-positive cocci in chains, in this case *Streptococcus pyogenes*.

Table 10.1 Gram-positive cocci, their main characteristics, diseases they cause, and antimicrobial susceptibilities.

Organism	Characteristic	Diseases	Susceptibilities
Staphylococcus aureus	Coagulase positive	Skin, abscess, wounds, pneumonia, bacteremia, TSS, IE, HCAI	oxa,[1] ceph,[1] clinda, macrol,[1] vanc, TMP/S, genta, linezolid, dapto, tetra, rif
Coagulase –ve staphylococci	Coagulase negative	HCAI, foreign material	vanc
Streptococcus pyogenes (Group A)	β-hemolytic Bacitracin sensitive	Pharyngotonsillitis, impetigo, cellulitis, abscess, sepsis, TSS, rheumatic fever, glomerulonephritis	pen, amp, clinda, ceph, macrol, vanc, linezolid
Streptococcus agalactiae (Group B)	β-hemolytic	Neonatal infection Infection in diabetics	pen, amp, ceph, vanc
Streptococcus dysgalactiae (Group C)	β-hemolytic	Pharyngotonsillitis	pen, amp, ceph, vanc
Streptococcus pneumoniae	α-hemolytic Optochin sensitive	Pneumonia, otitis media, sinusitis, conjunctivitis, bacteremia, meningitis, HUS	pen,[2] amp,[2] ceftriax,[3] cefotax,[3] macrol,[1] tetra,[1] TMP/S,[1] clinda,[1] vanc, linezolid
Streptococcus anginosus group	Any type	Visceral abscesses	pen, amp, ceph, vanc
Enterococcus faecalis	γ-hemolytic	UTI, HCAI, IE	amp, vanc amino synergy linezolid
Enterococcus faecium	γ-hemolytic	UTI, HCAI, IE	amp, vanc amino synergy linezolid, quin/dalf
Rothia mucilaginosa	Adherent to agar	Sepsis in neutropenic hosts	May be penicillin resistant
Pediococcus spp.	Catalase negative	HCAI	Vancomycin resistant
Leuconostoc spp.	α/non-hemolytic	Central line infections	Vancomycin resistant
Aerococcus spp.	Catalase negative	UTI, IE	Penicillin and vancomycin susceptible

amino, aminoglycoside; amp, ampicillin; ceph, cephalosporins; clinda, clindamycin; dapto, daptomycin; genta, gentamicin; IE, infective endocarditis; HCAI, healthcare-associated infection; HUS, hemolytic-uremic syndrome; macrol, macrolides; oxa, oxacillin; pen, penicillin; quin/dalf, quinupristine/dalfopristine; rif, rifampin; tetra, tetracycline; TSS, toxic shock syndrome; TMP/S, trimethoprim/sulfamethoxazole; vanc, vancomycin.

[1] Resistance is common.

[2] Relative resistance is fairly common; it can often be overcome by high dosages, except in meningitis.

[3] Resistance occurs, and is important in meningitis.

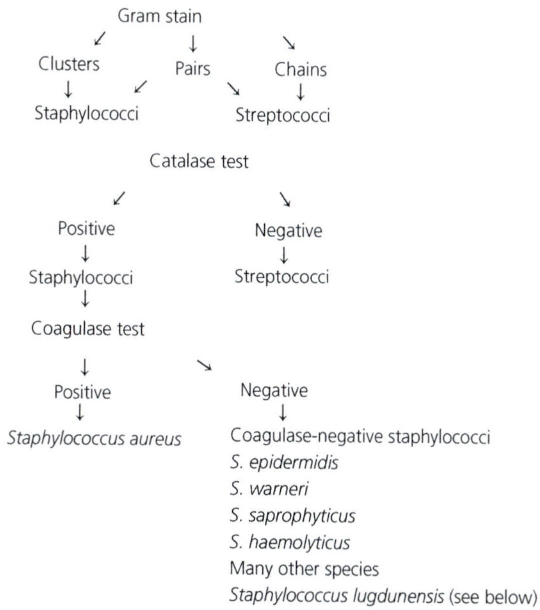

Fig. 10.2 Diagram showing how Gram-positive cocci are differentiated.

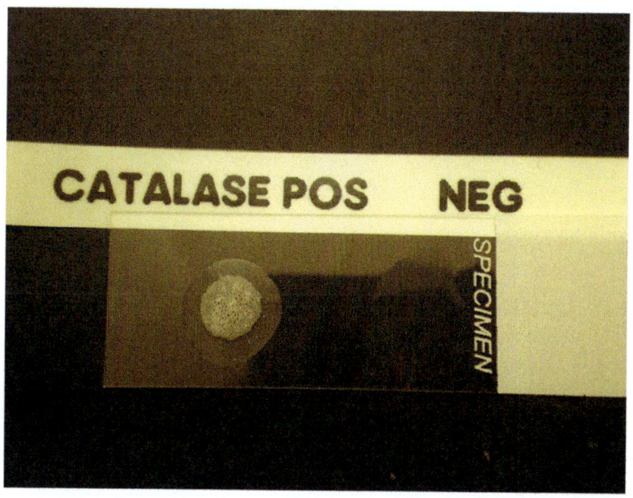

Fig. 10.3 A positive catalase test (bubbles) on the left (staphylococcus), and a negative test (no bubbles) on the right (streptococcus).

spread to deeper structures, such as the fascia and muscle, and to the blood. Thence it can spread to any part of the body, causing sepsis syndrome, bone and joint infections, visceral abscesses, and infective endocarditis. It can cause pneumonia by two routes: hematogenous, complicating bacteremia, and aerogenous (via the airway), often complicating a viral respiratory tract infection such as influenza or measles.

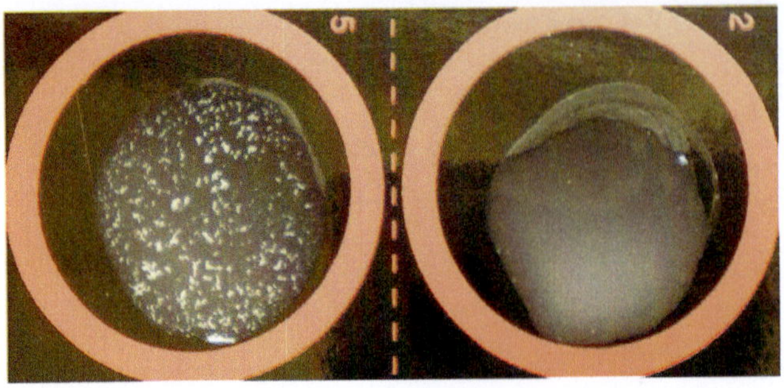

Fig. 10.4 A slide coagulase test to differentiate coagulase-positive *Staphylococcus aureus* (*left-hand panel*) from coagulase-negative staphylococci (*right-hand panel*).

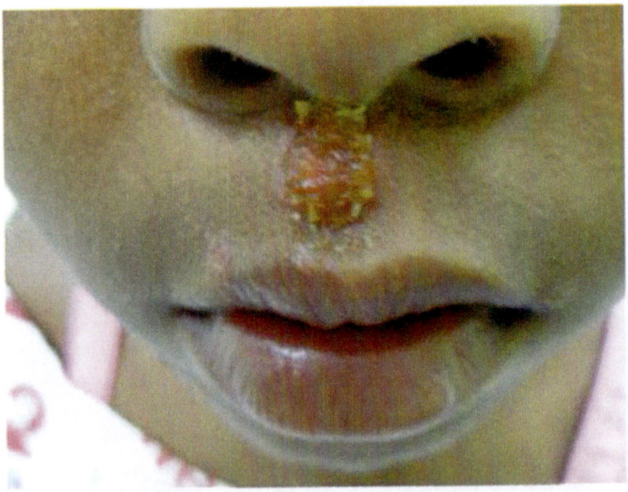

Fig. 10.5 A child with impetigo caused by *Staphylococcus aureus*.

A relatively recently recognized syndrome is rapidly progressive bacteremia, mainly in teenage boys, often associated with deep venous thrombosis.

This organism also causes several specific syndromes due to the elaboration of specific toxins. These include acute food poisoning (due to enterotoxins), toxic shock syndrome (due to toxic shock syndrome toxin), and scalded skin syndrome (due to exfoliative toxins A and B) (Figs 10.5–10.9).

Staphylococcus aureus can be readily cultured in the laboratory on blood agar (see Box 10.1).

Therapy of infected patients depends on surgical drainage of abscesses and antimicrobial therapy. The emergence of antimicrobial resistance in *Staphylococcus aureus* over time exemplifies that of antimicrobial resistance in general. When penicillin was first used, during the 1940s, it was active against the organism. Resistance, due to the production by the organism of a β-lactamase, was recognized in the early 1950s. This was overcome by the isoxazolyl penicillins (methicillin and

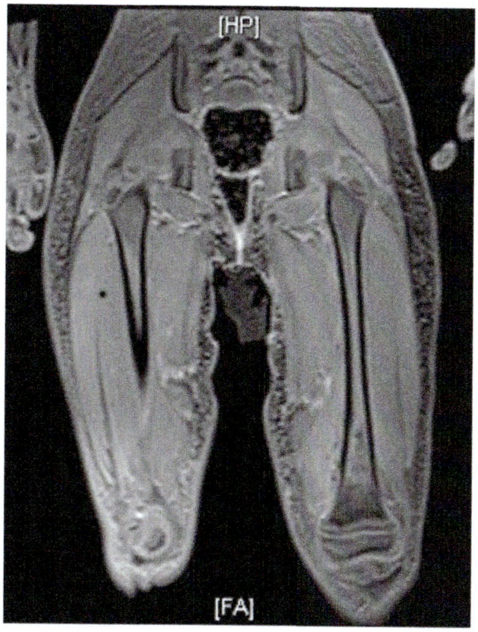

Fig. 10.6 Child with left distal femoral osteomyelitis caused by *Staphylococcus aureus*.

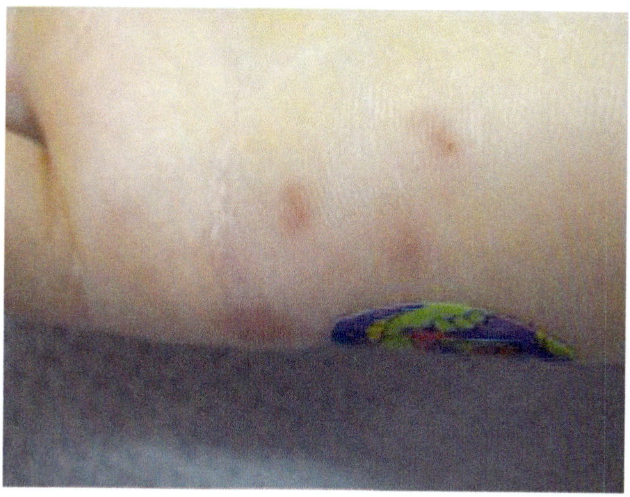

Fig. 10.7 Child with infective endocarditis caused by *Staphylococcus aureus*. Note the embolic lesions on the sole.

cloxacillin) and nafcillin. However, resistance to these drugs ("methicillin-resistant *Staphylococcus aureus*" – MRSA) was recognized in the 1960s. Although this was initially a problem mainly in hospitals, this is no longer the case. In many cities in the USA, currently up to 70–80% of community-acquired *Staph. aureus* cases are methicillin resistant.

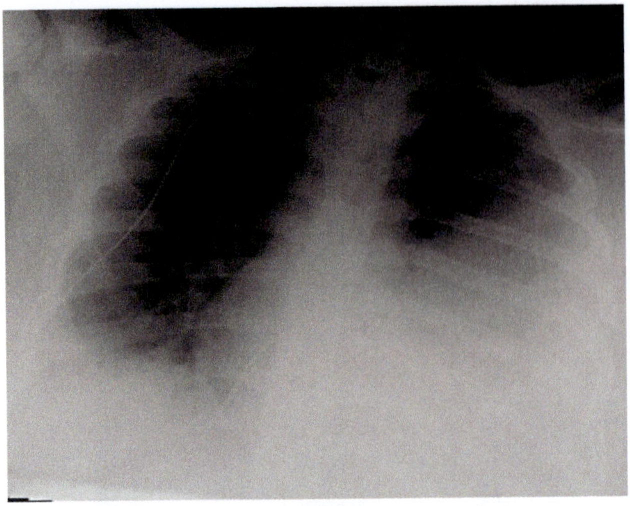

Fig. 10.8 Radiograph of a child with pneumonia and pleural effusion caused by *Staphylococcus aureus*, complicating influenza.

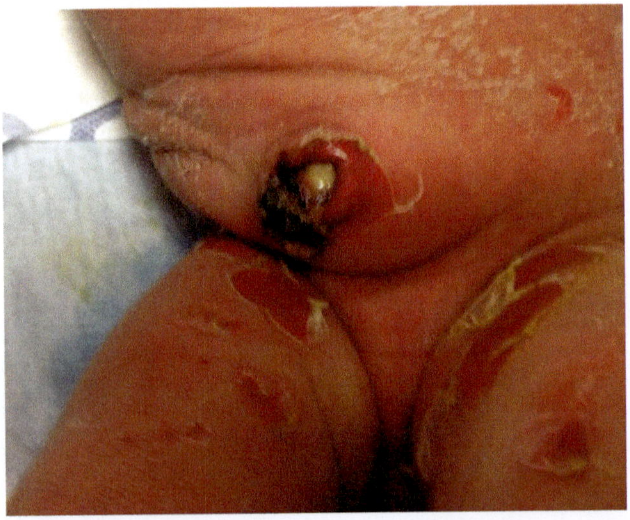

Fig. 10.9 Newborn infant with staphylococcal scalded skin syndrome.

Box 10.1 Case study

A 10-month-old boy presented with fever and swelling of his left thigh. On examination, he was febrile, and he had marked erythema and swelling of the left thigh. He was treated with vancomycin for the possibility of methicillin-resistant *Staphylococcus aureus* cellulitis. An MRI of the thigh showed swelling of the soft tissues of the thigh, with edema of muscle and fascia. Although his general condition improved, the thigh developed a pustule (Fig. 10.10). This was drained and the pus was sent to the laboratory for Gram stain, which showed Gram-positive cocci in clusters, and culture, which showed golden colonies on blood agar (Figs 10.11 & 10.12).

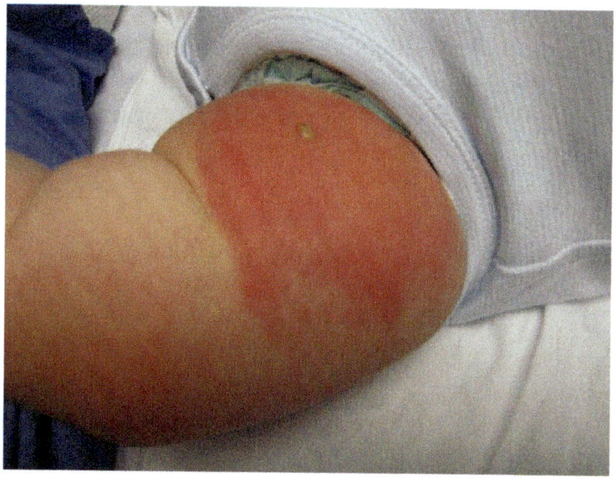

Fig. 10.10 Infant with thigh infection.

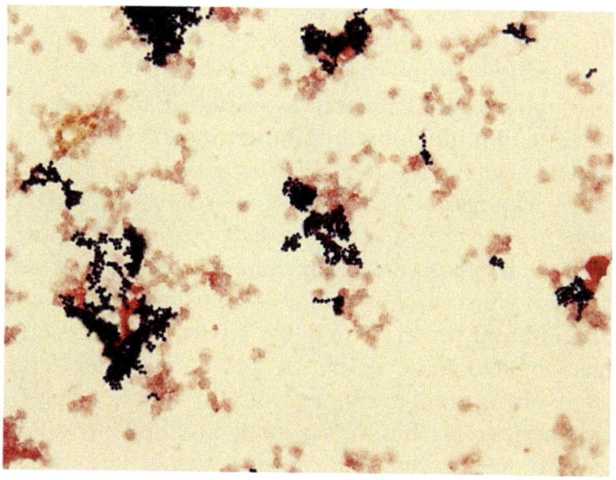

Fig. 10.11 Gram stain showing Gram-positive cocci in clusters.

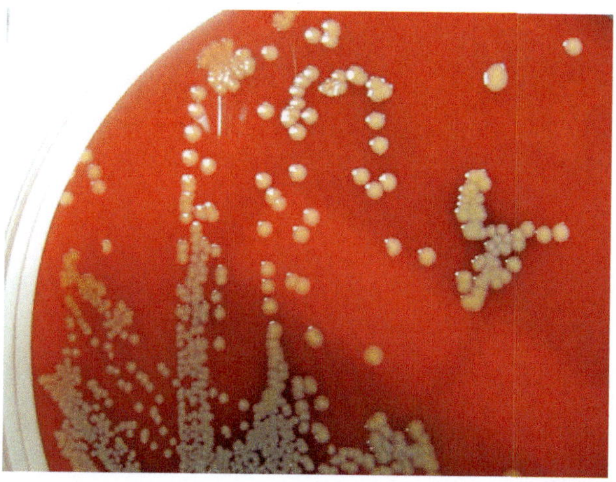

Fig. 10.12 Culture showing golden (aureate) colonies produced by *Staphylococcus aureus*.

The following drugs can be used empirically to treat patients with staphylococcal infections:

- if the prevalence of MRSA is low, nafcillin, oxacillin, cloxacillin, first-generation cephalosporins, or macrolides (for mild infections) can be used
- if the prevalence of MRSA is high, for severe infections vancomycin, linezolid or daptomycin should be used
- for less severe infections, clindamycin, trimethoprim/sulfamethoxazole, or tetracycline can be used.

Fusidic acid, which is not available in the USA, is also active against MRSA. Rifampin, which is active against staphylococci, should never be used alone, due to the rapid emergence of resistance to it.

Erythromycin resistance in staphylococci can be due to two mechanisms: an efflux mechanism and a target enzyme alteration. The latter may be inducible. If this occurs, it renders the organism resistant to clindamycin as well as erythromycin. This might not be apparent when clindamycin susceptibility is tested. Therefore, an isolate that is resistant to erythromycin but susceptible to clindamycin should be further tested for inducible resistance. This is done with the "D-test": after an agar plate is seeded with the test organism, erythromycin and clindamycin disks are placed about 2 cm apart on the plate. If there is induction of clindamycin resistance by the erythromycin, there will be a narrower zone of inhibition around the clindamycin disk on the side closer to the erythromycin disk, giving the zone a D-shape (Fig. 10.13).

Coagulase-negative staphylococci contain many species, including *Staph. epidermidis*, *Staph. warneri*, *Staph. haemolyticus*, and *Staph. saprophyticus*. These are normal flora of the skin, and they cause infections mainly in circumstances in which a medical device, especially plastic, which constitutes a foreign body, is inserted through the skin or is implanted in an organ, and remains there for a period of time ranging from days to years. The organisms colonize the plastic and embed themselves in a biofilm, rendering them relatively resistant to antimicrobial therapy. Examples include vascular catheters, ventriculoperitoneal shunts, hip implants, and prosthetic heart valves. These organisms are therefore common causes of healthcare-associated infections.

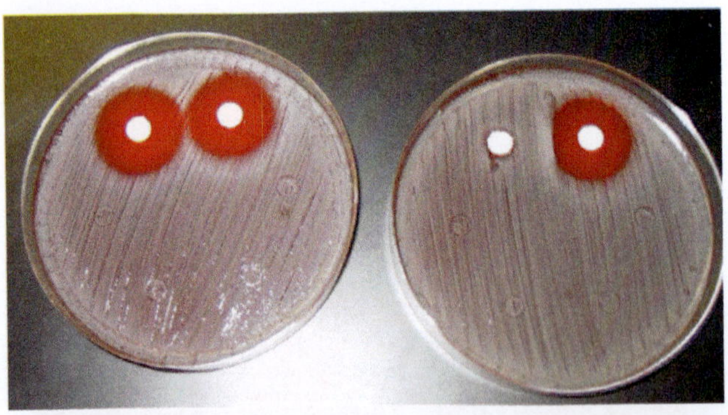

Fig. 10.13 The "D" test. In each plate the erythromycin disk is on the left, and the clindamycin disk is on the right. The plate on the left shows both erythromycin and clindamycin susceptibility (large zones of inhibition); the plate on the right shows erythromycin resistance (no zone of inhibition), and inducible clindamycin resistance (a flatter zone of inhibition on the side of the clindamycin disk closer to the erythromycin disk).

When the infection affects an endovascular foreign body, bacteremia usually occurs. The organisms tend to be resistant to many antimicrobial agents, including methicillin. Therapy often requires that the foreign body be removed and drugs such as vancomycin be administered.

Because coagulase-negative staphylococci are present on normal skin, they are frequent contaminants of blood cultures. There is no absolute way to determine whether or not a culture isolate is a contaminant or not. One must take into account the clinical situation and the risk factors for an isolate of this organism. For normal children who have blood cultures taken because of a febrile illness and whose blood cultures are reported to be growing Gram-positive cocci in clusters (suggesting a staphylococcus), we recommend managing the child for presumed *Staphylococcus aureus* only if there is clinical evidence of a focal infection or the child is ill-appearing. Otherwise we recommend waiting for determination of whether this is coagulase negative (much more common) or *Staph. aureus.*

Staphylococcus lugdunensis

This staphylococcus, named for Lyon, France, where it was first recognized, has some features of *Staphylococcus aureus,* and some of coagulase-negative staphylococci. Although it does not excrete coagulase, and therefore tests as a coagulase-negative staphylococcus, some strains have bound coagulase, which can be detected by slide agglutination. It can be differentiated from *Staph. aureus* by a few simple biochemical tests. It can cause a variety of infections, notably native valve endocarditis. Its antimicrobial susceptibilities are more like those of *Staph. aureus,* and the recommended breakpoints for antimicrobial resistance are the same as those for *Staph. aureus.*

Streptococci

Many different streptococci are important human pathogens. There are several different classification systems of streptococci which are overlapping. These are based on (a) the type of hemolysis they produce on blood agar, and (b) their cell wall carbohydrate antigen (Lancefield classification). Not all streptococci fit into these classification systems.

β-Hemolytic streptococci

These produce complete (clear) hemolysis (Fig. 10.14). The most important of these are (according to the Lancefield grouping), Group A (*Streptococcus pyogenes*), Group B (*Strep. agalactiae*), Group C (*Strep. dysgalactiae*), and Group G.

Group A streptococcus is identified by its sensitivity to bacitracin, which is shown by a zone of inhibition around a bacitracin disk (Fig. 10.14). It can also be identified, as can the other groups, by agglutination with group-specific antiserum (Fig. 10.15).

Streptococcus pyogenes (Group A streptococcus)

This is a common cause of infection, especially in children. It has several virulence factors, the most important of which is M-protein, which prevents opsonization of the organism by phagocytes. M-protein determines the serotype of the organism. There are many different M-proteins, and thus many different serotypes, between which there is no cross-immunity. The organism also produces several enzymes and extracellular proteins which play a role in pathogenesis.

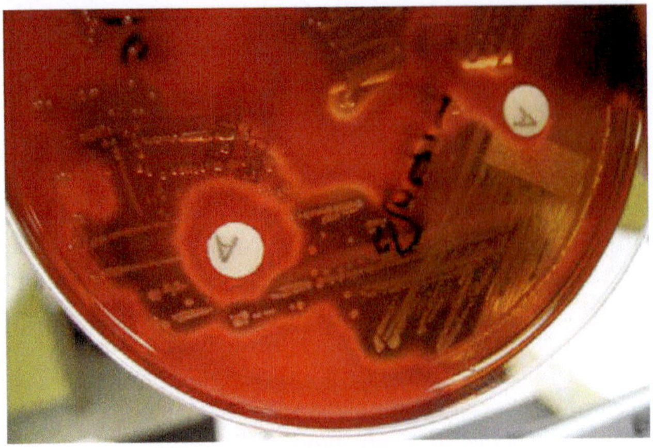

Fig. 10.14 β-Hemolysis (complete); note the zone of inhibition around the A disk, which is bacitracin, which identifies this isolate as *Streptococcus pyogenes*.

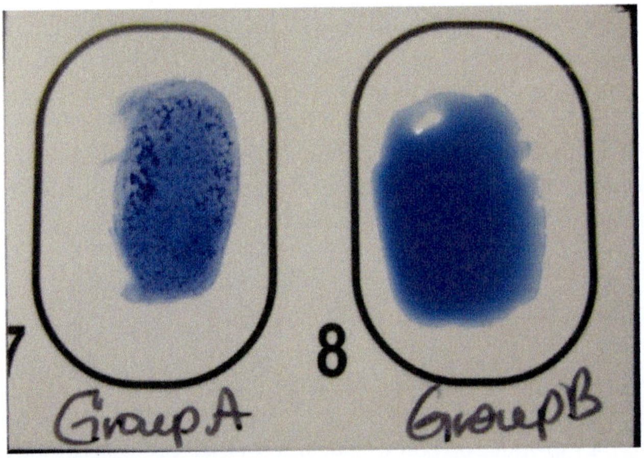

Fig. 10.15 Agglutination with Group A antiserum.

The clinical diseases caused by this organism are considered as follows.

- Skin and soft tissue infections: erysipelas (superficial), impetigo, cellulitis, wound infection, necrotizing fasciitis and myositis.
- Pharyngotonsillitis, from which infection can spread to cause peritonsillar abscess, cervical lymphadenitis, and otitis media. When associated with a diffuse, fine, erythematous rash, the illness is called scarlet fever. It is due to the elaboration of a pyrogenic exotoxin.
- Systemic infection (bacteremia), which can complicate any focal infection, but usually follows wound and soft tissue infections, rather than pharyngeal infections. This may, in turn, cause focal metastatic infection, including septic arthritis, osteomyelitis, meningitis, visceral abscesses, and infective endocarditis (see Box 10.2).
- Toxic shock syndrome.
- Immune (postinfective) complications: (a) acute rheumatic fever, which is now uncommon in industrialized countries, but very serious, and which may lead to

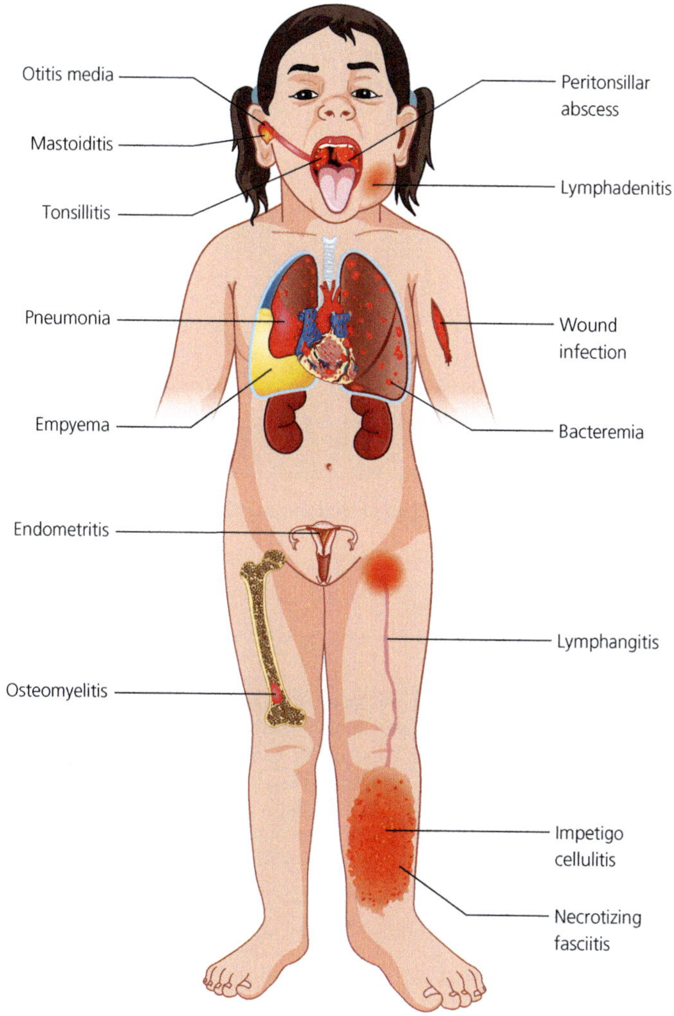

Fig. 10.16 Sites of infection caused by *Streptococcus pyogenes*.

chronic rheumatic heart disease. It is a major cause of acquired heart disease in developing countries;(b) post-streptococcal acute glomerulonephritis; (c) reactive arthritis. The sites of infection caused by *Strep. pyogenes* are summarized in Fig. 10.16.

This organism is readily cultured from body fluids and pus, and readily identified (see earlier). Because pharyngotonsillitis is a frequent reason for children to present for medical care, and has the potential to lead to acute rheumatic fever, rapid diagnostic tests, based on detection of Group A antigen, and which can be performed in the clinic, are widely used (Fig. 10.17).

Streptococcus pyogenes is susceptible to penicillin. In contrast to many bacteria of medical interest, which have developed resistance to penicillin since its introduction to clinical medicine (1940s), this organism has not manifested such resistance yet! It is also susceptible to cephalosporins, clindamycin, macrolides, vancomycin, and linezolid, but is not very susceptible to trimethoprim/sulfamethoxazole.

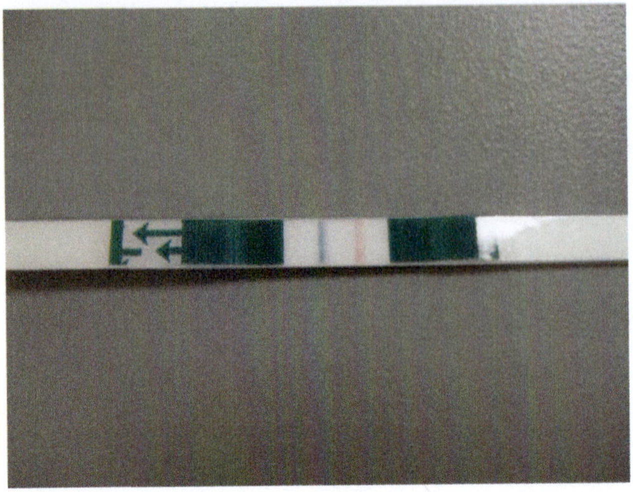

Fig. 10.17 A rapid *Streptococcus pyogenes* antigen detection test. The blue line indicates a positive test and the red line is the negative control.

Box 10.2 Case study

A 4-year-old boy presented with fever and rash for a few days. On examination, he was ill-appearing with a temperature of 39°C. His eyes showed conjunctival injection and his lips were red, as were his palms. He would not open his mouth due to pain. He had a vague lacy macular rash on his limbs, and swelling of his submandibular area with overlying erythema of the skin (Fig. 10.18).

Blood cultures grew out Gram-positive cocci in chains. These showed β-hemolysis and were susceptible to bacitracin which, together with agglutination with Group A antiserum, confirmed the identification as *Streptococcus pyogenes*. He was treated successfully with penicillin.

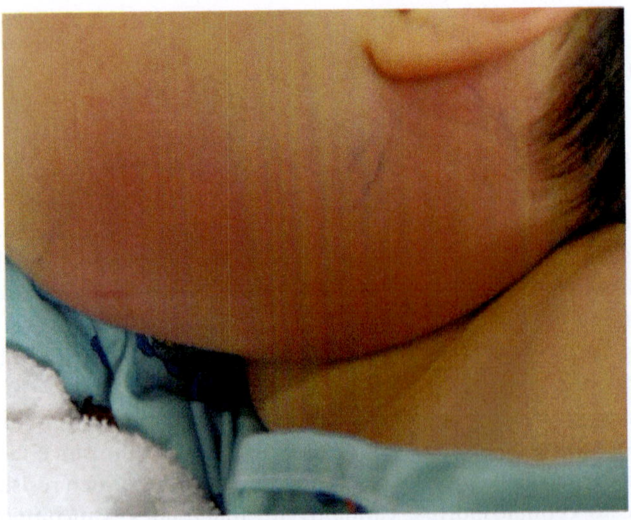

Fig. 10.18 Patient showing submandibular cellulitis.

Streptococcus agalactiae (Group B streptococcus)

This is normal flora of the vagina of many women, and of the intestinal tract. It is currently the most common cause of neonatal sepsis and meningitis in Europe and North America. Currently recommended practice is for testing of pregnant women close to term for the presence of the organism in the vagina and rectum, using enrichment media or molecular methods, and treating those who are colonized with penicillin or ampicillin during delivery. This has significantly reduced the incidence of early-onset infection in the newborn (i.e. earlier than 1 week of age), but not of infection occurring after that. Early-onset infection usually manifests as pneumonia, sepsis syndrome, or meningitis, while late-onset disease, occurring as late as 4 months, often manifests as bacteremia, meningitis, or skeletal infection.

The organism appears on Gram stain as do other streptococci (Gram-positive cocci in pairs or chains; Fig. 10.19) and can be readily identified in culture, as β-hemolytic colonies (narrower zone than that produced by Group A streptococcus; Fig. 10.20) that agglutinate with Group B antiserum. It should be suspected in the appropriate clinical situation. It is susceptible to penicillins and cephalosporins.

β-Hemolytic streptococci that have the Group C antigen (e.g. *Strep. dysgalactiae* subsp. *equisimilis*) and the Group G antigen (e.g. *Strep. canis*) are occasional causes of pharyngitis and other focal infections in humans.

α-Hemolytic streptococci

α-Hemolysis (green hemolysis) is not caused by a hemolysin elaborated by the organism, but by the release of hydrogen peroxide, which causes changes in the red cells in the agar.

γ-Hemolytic streptococci

These do not cause hemolysis. There is so much overlap between α-hemolytic streptococci and non- (γ)hemolytic streptococci that they are generally grouped together. They are divided into two groups, according to their susceptibility to a cuprous

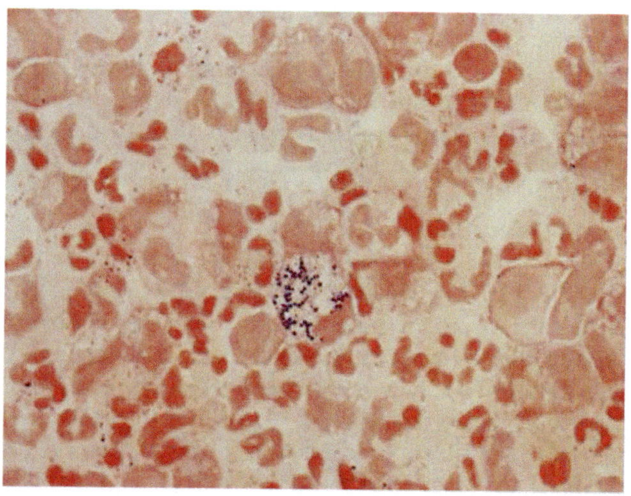

Fig. 10.19 A Gram-stained smear of cerebrospinal fluid of a 4-month-old infant, showing many leukocytes and Gram-positive cocci in pairs and chains, which were *Streptococcus agalactiae*.

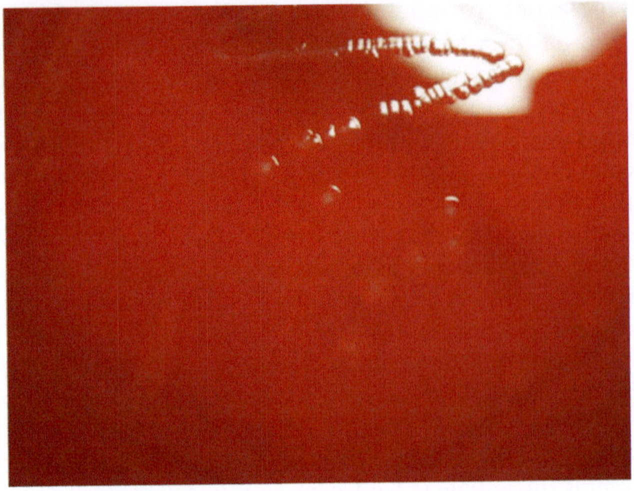

Fig. 10.20 Colonies of *Streptococcus agalactiae* on blood agar. Note the small zone of β-hemolysis.

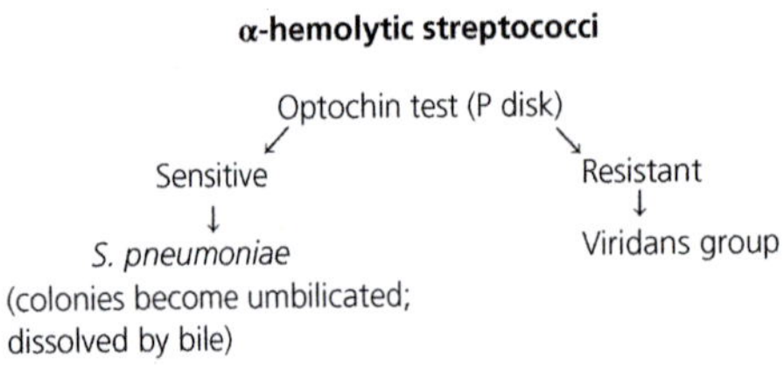

Fig. 10.21 Diagram showing how α-hemolytic streptococci are differentiated.

compound called optochin. The optochin test is performed by placing a paper disk (P disk) impregnated with optochin on an agar plate, which has been inoculated with the test organism. A zone of inhibition demonstrates susceptibility (Figs 10.21 & 10.22). If they are susceptible to optochin, they are *Strep. pneumoniae*. If they are resistant to optochin, they are thrown into the large basket of streptococci called viridans (viridans is the Latin word for green) streptococci, discussed later in this chapter. **This is not a species of streptococcus.** It contains many species, some of which also have Lancefield groups, e.g. *Strep. durans*.

Streptococcus pneumoniae

Streptococcus pneumoniae (also called pneumococcus) is a common inhabitant of the nasopharynx, especially in young children, and a very important pathogen. Its major virulence factor is the polysaccharide capsule, which prevents opsonization by phagocytes. There are more than 80 different types of such capsules, each determining the organism's respective serotype. Immunity is type specific. The specific type can be determined in the laboratory by specific antiserum, using the Quellung ("capsular swelling") test.

On Gram stain, the organism appears as Gram-positive diplococci (its former name was *Diplococcus pneumoniae*) (Fig. 10.23) and it grows on regular media, producing α-hemolytic colonies on blood agar. These are sensitive to optochin (see earlier) and are dissolved by bile. Older colonies develop a concavity at the top, giving the colony an

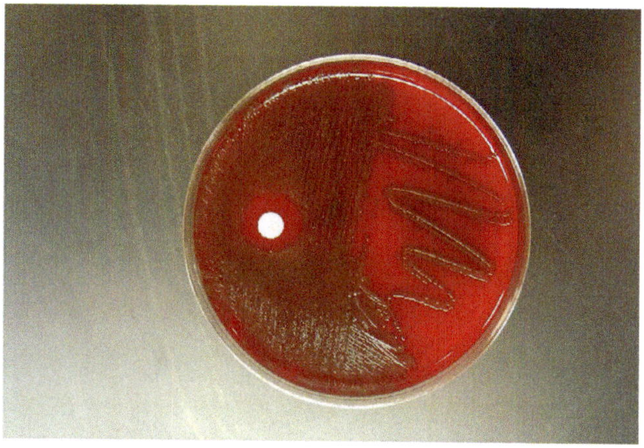

Fig. 10.22 α-Hemolytic streptococci, with a zone of inhibition around the optochin disk (optochin susceptibility), indicating that this is *Streptococcus pneumoniae.*

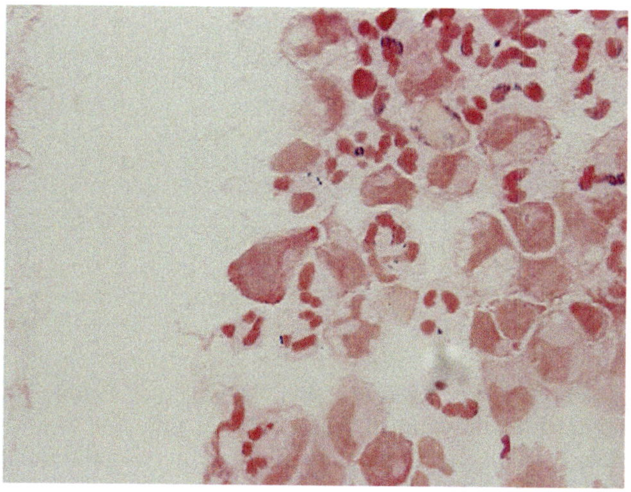

Fig. 10.23 Gram stain of cerebrospinal fluid, showing Gram-positive cocci in pairs (*Streptococcus pneumoniae*).

umbilicated appearance (Fig. 10.24). These features distinguish them from viridans streptococci.

Streptococcus pneumoniae causes a wide variety of infections, which are considered in the following categories.

- Acute otitis media, acute sinusitis, and conjunctivitis.
- Pneumonia: *Strep. pneumoniae* is an important cause of pneumonia in all age groups. This can be diffuse or lobar, and it can be complicated by empyema and lung abscess.
- Bacteremia – this may occur in young children in the absence of an obvious focus of infection ("occult bacteremia"). However, it can lead to metastatic infection, the most important of which is meningitis. Other such focal infections include septic arthritis, osteomyelitis, pericarditis, and primary peritonitis.
- Hemolytic-uremic syndrome.

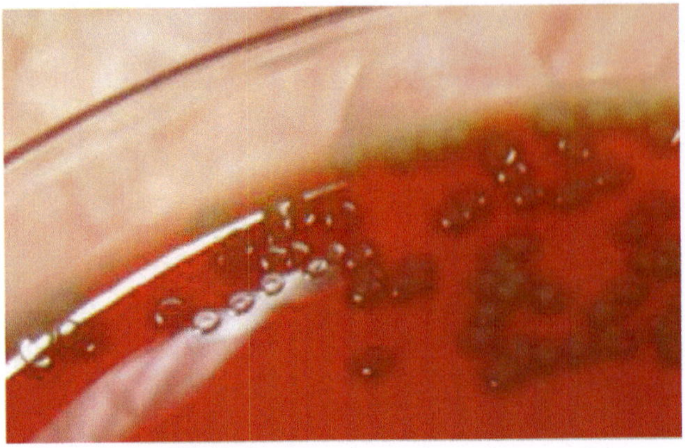

Fig. 10.24 α-Hemolytic colonies, showing central umbilication (*Streptococcus pneumoniae*).

Box 10.3 Case study

A 4-month-old infant who had been treated with amoxicillin and then ceftriaxone for an ear infection presented with fever and irritability. On examination, he had neck stiffness and a bulging anterior fontanelle. A diagnosis of bacterial meningitis was suspected. Cerebrospinal fluid revealed 406 leukocytes per mm^3 with 80% neutrophils. The protein and glucose concentrations were 88 and 41 mg/dL respectively. The Gram stain is shown in Fig. 10.23.

The culture grew out α-hemolytic streptococci, which were umbilicated (Fig. 10.24) and sensitive to optochin, identifying them as *Streptococcus pneumoniae*.

This organism was, historically, susceptible to penicillins. However, over the past few decades, penicillin resistance, as well as resistance to other antibiotics (such as chloramphenicol, macrolides, clindamycin, and tetracycline), has become progressively more prevalent. Third-generation cephalosporins (ceftriaxone and cefotaxime) are active against most strains of pneumococcus, but resistant strains occur. This is important particularly in the case of meningitis, when there is no room for inadequate treatment. In this situation, one should assume that the causative organism is resistant to penicillin and third-generation cephalosporins. Vancomycin should be used in addition to ceftriaxone or cefotaxime, until the susceptibilities are known. Other drugs active against penicillin-resistant pneumococci are carbapenems, linezolid, and levofloxacin.

Viridans streptococci

The viridans streptococci are normal flora of the mouth and intestine. There are six main groups, *mutans, salivarius, anginosus, sanguinis, mitis,* and *bovis,* each of which contains several species. Some play a role in the pathogenesis of dental caries, one of the most common afflictions of humans. However, they can cause more serious visceral disease, such as sinusitis and brain abscess, and they constitute one of the most common causes of infective endocarditis. They can be differentiated from one another by biochemical tests, MALDI-TOF MS, or molecular methods, but in most clinical situations this is not necessary. The *Strep. anginosus* and *Strep. bovis* groups are described in more detail below.

Box 10.4 Case study

A 13-year-old girl with Di George syndrome and complex congenital heart disease, including a prosthetic aortic valve, presented with a history of fever for 1 week. Several blood cultures grew out a viridans group streptococcus. A diagnosis of infective endocarditis was made, and she was treated successfully with intravenous penicillin.

Streptococcus anginosus group

This group of streptococci, formerly called, *Strep. milleri*, consists of three species, namely *Strep. anginosus*, *Strep. constellatus*, and *Strep. intermedius*. Although they are considered to belong to the viridans streptococci, they may be α-, β-, or γ-hemolytic. Sometimes they have Group A, C, G, or F Lancefield antigens. They are normal flora of the mouth, intestine, and female genital tract, but they can cause a wide variety of pyogenic infections in viscera, such as liver abscess, brain abscess, pneumonia and lung abscess, and infective endocarditis. They are often present together with anaerobic bacteria.

They can be readily cultured, and grow optimally under microaerophilic or anaerobic conditions. Their colonies have a characteristic caramel or butterscotch odor. They can be differentiated from other streptococci by their ability to produce acetoin (Voges–Proskauer test), their ability to hydrolyze arginine, and their inability to ferment sorbitol.

Streptococcus bovis

This group, which expresses the Lancefield Group D antigen (as do enterococci), consists of three biotypes: I (*Strep. gallolyticus* subsp. *gallolyticus*), II/1 (*Strep. infantarius*), and II/2 (*Strep. gallolyticus* subsp. *pasteurianus*). They cause bacteremia, visceral abscess, and infective endocarditis. Bacteremia and infective endocarditis caused by biotype I are associated with colonic cancer.

Enterococci

Enterococci were formerly grouped within the genus Streptococcus, but are now classified as a separate genus. There are two main species affecting humans: *E. faecalis* and *E. faecium*. They appear morphologically as streptococci, and share the streptococcal Group D antigen. When grown on blood agar, the colonies are usually non-hemolytic (Fig. 10.25). They are present in the intestinal tract and cause disease mainly related to the intestine and urinary tract. They are also important causes of healthcare-associated infections. They are distinguished from streptococci by the following: positive reaction with pyrrolidonyl arylamidase (PYR) (Fig. 10.26) (*Strep. pyogenes* is also PYR positive), growth on bile-esculin, and growth in 6.5% saline.

Although they are usually susceptible to cell wall active agents, namely penicillin, ampicillin, and vancomycin, (not cephalosporins), all these agents can only inhibit growth of the enterococci, not kill them. When killing is required, as in cases of infective endocarditis, the addition of an aminoglycoside, such as streptomycin or gentamicin, is necessary for synergistic activity and killing. (Aminoglycosides alone are not active against these organisms.) However, the following antibiotic

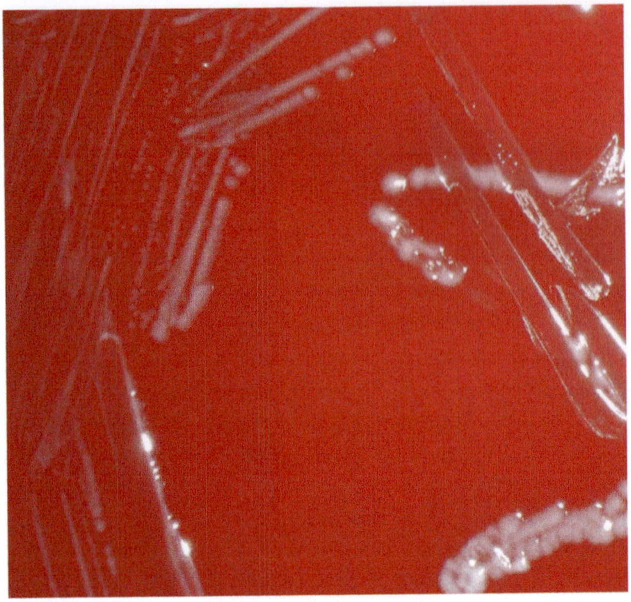

Fig. 10.25 Enterococci growing on blood agar. Note the absence of hemolysis.

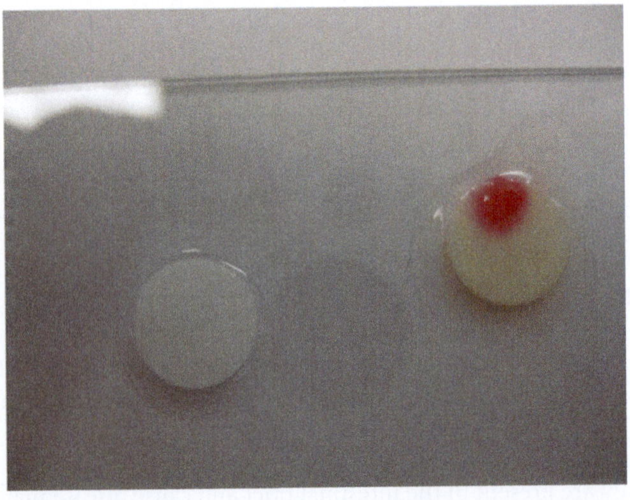

Fig. 10.26 The pyrrolidonyl arylamidase (PYR) test for identifying enterococci. The disk on the left is negative, and that on the right is positive.

resistance patterns have emerged: (a) resistance to penicillin and ampicillin, with retention of vancomycin susceptibility; (b) failure of synergy with aminoglycosides; (c) resistance to vancomycin, which occurs primarily in *E. faecium*. Vancomycin resistance is usually also associated with penicillin-ampicillin resistance.

Alternative therapy for patients with infections caused by such organisms is very limited, mainly to linezolid, quinupristine/dalfopristine (only in cases of *E. faecium* infection), daptomycin, and tigecycline.

Box 10.5 Case studies

A 13-year-old girl who had multiple medical problems (liver, small bowel, and pancreas transplant, and a neurogenic bladder), and who was receiving trimethoprim/sulfamethoxazole and nitrofurantoin, presented with a urinary tract infection. There were many leukocytes in the urine, and the culture grew out small colonies that hydrolyzed bile-esculin, which were identified as *Enterococcus faecalis*.

A 10-year-old girl presented with a history of headache for a few days, followed by dragging of her left foot. On examination, she was alert but had a mild left hemiparesis. A brain CT scan (Fig. 10.27) revealed an epidural abscess, which was drained surgically. Gram stain of the pus obtained at surgery is shown in Fig. 10.28. Although organisms that had the appearance of streptococci were present on the Gram stain, they did not grow on blood agar.

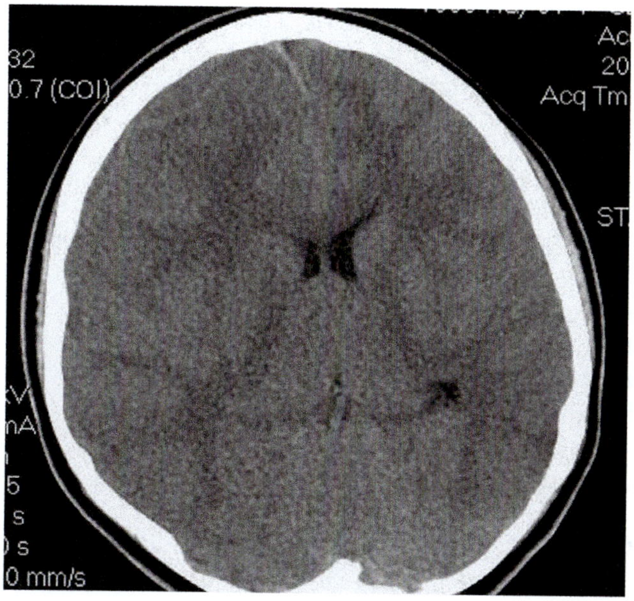

Fig. 10.27 CT scan showing an epidural abscess in the right frontal area.

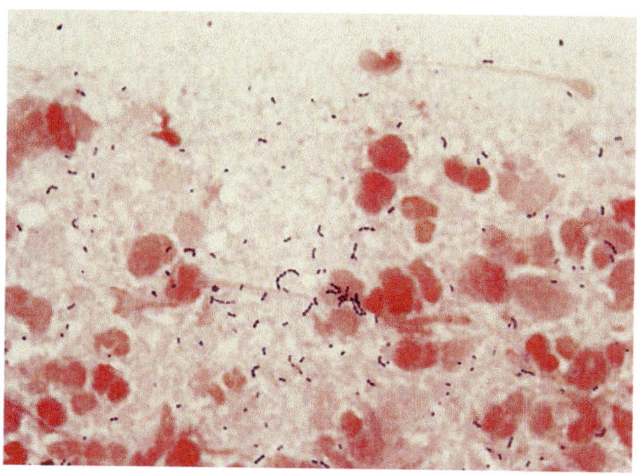

Fig. 10.28 Pus and many Gram-positive cocci in chains.

When organisms are seen on Gram stain, but fail to grow in culture, there are two possible reasons for this.

- The patient had received antimicrobial therapy, which had killed or inhibited growth of the organism.
- The culture environment was not conducive to growth, due to either lack of necessary nutrients or the presence of inhibitors (e.g. oxygen in the case of strict anaerobes).

In this case, when the agar was supplemented with cysteine, the organisms grew. These belong to a group of bacteria formerly called nutritionally deficient or nutritionally variant streptococci. They require cysteine or pyridoxal supplementation of agar for their growth. They are part of the normal oral flora, and belong to the following genera and species: *Abiotrophia defectiva* and *Granulicatella adiacens, G. elegans,* and *G. para-adiacens,* and *Gemella* spp. They cause endocarditis and other infections, and are relatively resistant to penicillins. It is difficult to test their antimicrobial susceptibilities.

Other Gram-positive cocci

Rothia mucilaginosa (formerly called *Stomatococcus mucilaginosus*)

This is part of the normal oral flora. It causes severe infections in neutropenic and other immunocompromised individuals. The following characteristics should alert one to the possibility of this organism, when Gram-positive cocci are seen:

- non-hemolytic colonies
- adherence of colonies to the agar surface
- catalase test – weakly positive or negative
- coagulase test – negative
- oxidase test – negative
- failure to grow in 6.5% saline.

Identification of this organism is important because it may be resistant to penicillins. The drug to which it is most susceptible is rifampin. However, vancomycin, gentamicin, ampicillin, and other β-lactams have variable activity against it.

Pediococcus spp.

These organisms have the following characteristics: Gram-positive cocci, catalase-negative, oxidase-negative, α- or non-hemolytic, type with Group D antiserum, susceptible to penicillin and ampicillin, but, of note, **resistant to vancomycin.**

They are present in various foods and on mucosal surfaces. The main species causing opportunistic infections in humans are *P. acidilactici* and *P. pentosaceus.*

Leuconostoc spp.

These are Gram-positive cocci that resemble streptococci. The main species are *L. mesenteroides* (of which there are several subspecies), *L. paramesenteroides, L. citreum,* and *L. lactis.* They may be present in food such as dairy products. They are opportunistic pathogens, causing disease mainly in individuals with abnormal intestinal tracts, and central venous lines. They appear as Gram-positive cocci in pairs and short chains, and are α-hemolytic or non-hemolytic on blood agar. The main significant laboratory characteristic that should alert one to the possible identification of these organisms is their resistance to vancomycin. This is very important because most bacteremic patients at risk for infection caused by these organisms are likely to be treated empirically with vancomycin. They are susceptible to cefotaxime, but relatively resistant to penicillin.

Aerococcus spp.

These are Gram-positive cocci that form tetrads and clusters, resembling staphylococci, but they are catalase negative. They are rare causes of urinary tract infections and infective endocarditis, especially in the elderly. They are susceptible to vancomycin and penicillin, and their susceptibilities to other antimicrobial agents are variable.

Further reading

Barton LL, Rider ED, Coen RW (2001) Bacteremic infection with Pediococcus: vancomycin-resistant opportunist. Pediatrics 107: 775–6.

Bisno AL, Brito MO, Collins CM (2003) Molecular basis of group A streptococcal virulence. Lancet Infect Dis 3: 191–200.

Cole JN, Barnett TC, Nizet V, et al. (2011) Molecular insight into invasive group A streptococcal disease. Nat Rev Microbiol 9: 724–36.

Committee on Infectious Diseases and Committee on Fetus and Newborn (2011) Recommendations for the prevention of perinatal Group B streptococcal (GBS) disease. Pediatrics 128: 611–16.

DeLeo FR, Otto M, Kreiswirth BN, et al. (2010) Community-associated methicillin-resistant Staphylococcus aureus. Lancet 375: 1557–68.

Doern CD, Burnham CAD (2010) It's not easy being green: the viridans group streptococci, with a focus on pediatric clinical manifestations. J Clin Microbiol 48: 3829–35.

Florescu D, Hill L, Sudan D, et al. (2008) *Leuconostoc* bacteremia in pediatric patients with short bowel syndrome: case series and review. Pediatr Infect Dis J 27: 1013–19.

Frank KL, del Pozo JL, Patel R (2008) From clinical microbiology to infection pathogenesis: how daring to be different works for *Staphylococcus lugdunensis*. Clin Microbiol Rev 21: 111–133.

Iwen PC, Mindru C, Kalil AC, et al. (2012) *Pediococcus acidilactici* endocarditis successfully treated with daptomycin. J Clin Microbiol 50: 1106–8.

Krishna S, Miller LS (2012) Host-pathogen interactions between the skin and *Staphylococcus aureus*. Curr Opin Microbiol 15: 28–35.

Liu C, Bayer A, Cosgrove SE, et al. (2011) Clinical practice guidelines by the Infectious Diseases Society of America for the treatment of methicillin-resistant *Staphylococcus aureus* infection in adults and children. Clin Infect Dis 52: 1–38.

Lowy FD (2011) How *Staphylococcus aureus* adapts to its host. N Engl J Med 364: 1987–90.

Senn L, Entenza JM, Greub G, et al. (2006) Bloodstream and endovascular infections due to Abiotrophia defectiva and Granulicatella species. BMC Infect Dis 6: 1–6.

Shelton-Dodge K, Vetter EA, Kohner PC, et al. (2011) Clinical significance and antimicrobial susceptibilities of *Aerococcus sanguinicola* and *Aerococcus urinae*. Diagn Microbiol Infect Dis 70: 448–51.

Steer AC, Danchin MH, Carapetis JR (2007) Group A streptococcal infections in children. J Paediatr Child Health 43: 203–13.

Watkins RR, David MZ, Salata RA (2012) Current concepts on the virulence mechanisms of methicillin-resistant Staphylococcus aureus. J Med Microbiol 61(pt 9): 1179–93.

CHAPTER 11
Gram-negative cocci

Neisseria

Neisseria are Gram-negative cocci, belonging to the family Neisseriaceae. This genus contains two very important human pathogens, *N. meningitidis* and *N. gonorrhoeae*, in addition to several other species that are normal inhabitants of the oropharynx (e.g. *N. lactamica* and *N. sicca*). The latter are important in that they must be distinguished from the pathogens and that they occasionally cause disease.

Neisseria meningitidis (also called meningococcus) causes rapidly progressive sepsis, which carries a very high fatality rate, and meningitis, of which it is one of the most important causes. The sepsis syndrome (meningococcemia) is characterized by fever and rapid progression to shock over a few hours. These patients may not have meningitis. A variety of rashes occurs, including petechial, macular, and hemorrhagic (Figs 11.1 & 11.2). Severe vascular occlusions may occur as well as adrenal hemorrhage, a condition called the Waterhouse–Friderichsen syndrome (Figs 11.3 & 11.4). The manifestations of meningitis are the same as for other types of bacterial meningitis. However, the presence of petechiae would suggest that this is the cause (Fig. 11.5).

There are six clinically significant serogroups (based on the polysaccharide capsule, an important virulence factor) – A, B, C, W135, X, and Y, and multiple serotypes, based on the outer membrane protein. The polysaccharides and outer membrane proteins are important epidemiologically and immunologically. The infection occurs sporadically, and in epidemics. There is a particularly high rate of disease in a band of sub-Saharan Africa from Senegal to Ethiopia. In the USA, the rate of infection has declined from about 1.5 to about 0.2 cases per 100,000 in the past 10 years.

On Gram stain, the organism has a characteristic appearance. It is bean shaped, often two together (diplococci), with their long axes parallel to one another (as opposed to streptococci, whose long axes are in a straight line). In cerebrospinal fluid, most of the leukocytes lack organisms, and a few contain large numbers of organisms (Fig. 11.6).

N. meningitidis can be cultured on blood agar. Blood culture bottles containing sodium polyanethol sulfonate (SPS) can inhibit its growth. Skin lesion aspirates should

Practical Medical Microbiology for Clinicians, First Edition. Frank E. Berkowitz and Robert C. Jerris.

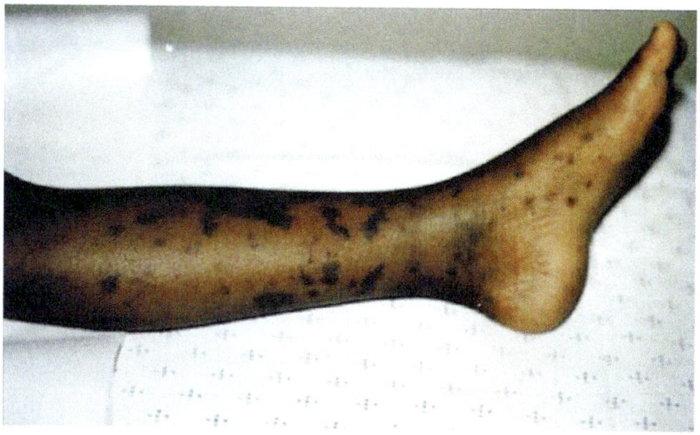

Fig. 11.1 Child with meningococcemia, showing a hemorrhagic rash. Copyright © 2007 Frank E. Berkowitz. Reprinted with permission of Cambridge University Press.

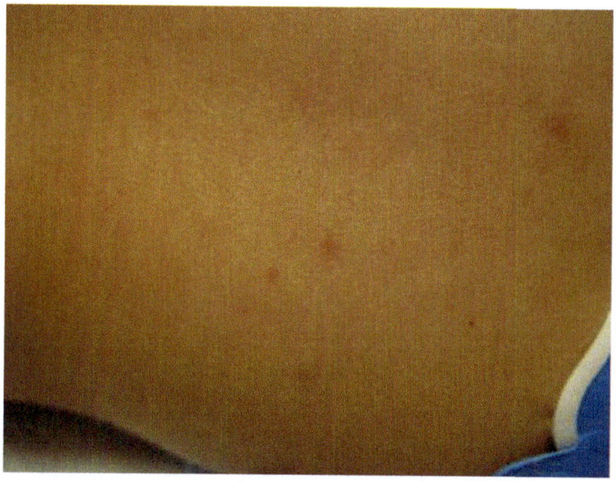

Fig. 11.2 Child with meningococcal meningitis showing a macular rash.

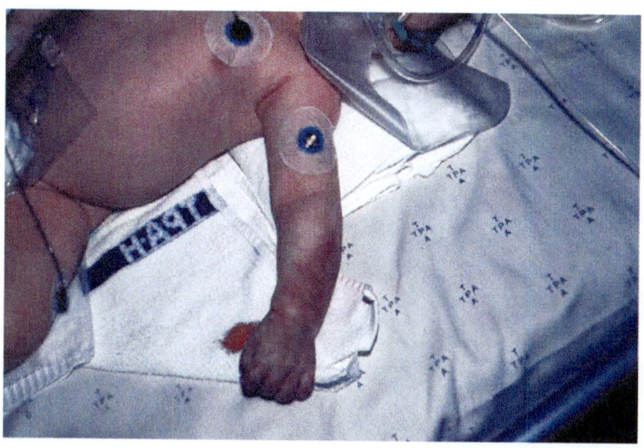

Fig. 11.3 Child with meningococcemia and vascular occlusion affecting the left upper limb. (These affected the right upper limb and both lower limbs as well.)

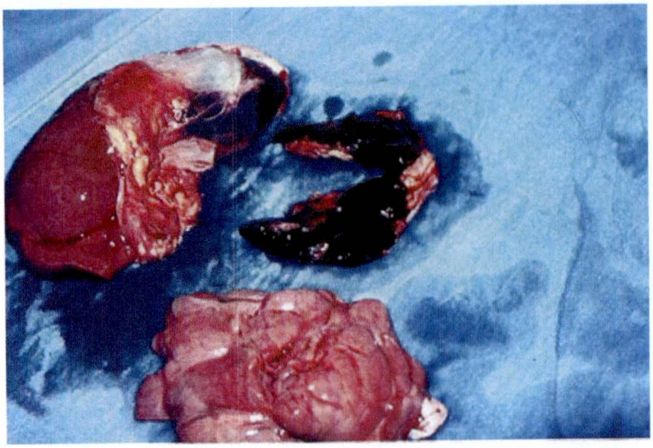

Fig. 11.4 Autopsy specimen of a child dying of meningococcemia, showing adrenal hemorrhage (Waterhouse–Friderichsen syndrome). Copyright © 2007 Frank E. Berkowitz. Reprinted with permission of Cambridge University Press.

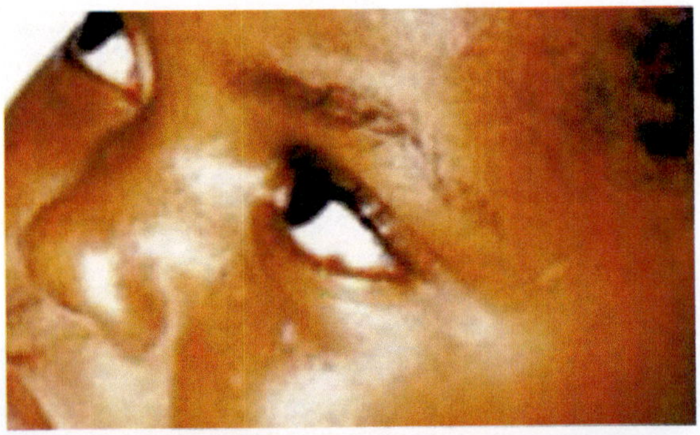

Fig. 11.5 Child with meningococcal meningitis, showing a subconjuctival petechia. Copyright © 2007 Frank E. Berkowitz. Reprinted with permission of Cambridge University Press.

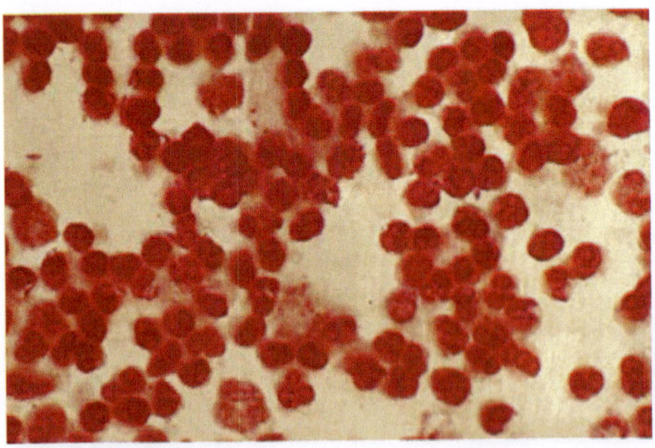

Fig. 11.6 Gram stain of cerebrospinal fluid showing leukocytes containing *Neisseria meningitidis*. Copyright © 2007 Frank E. Berkowitz. Reprinted with permission of Cambridge University Press.

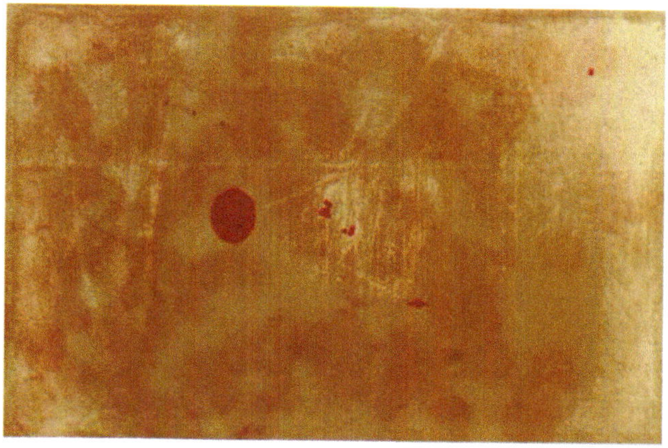

Fig. 11.7 Gram stain of a skin aspirate, showing Gram-negative cocci (*Neisseria meningitidis*).

be Gram stained (Fig. 11.7). If screening for pharyngeal carriers is performed (not routinely indicated), a selective medium must be used (see culture for gonococcus below).

Decades ago, this organism was susceptible to sulfonamides but this is no longer the case. Although it is universally susceptible to penicillin using clinical cut-offs, its minimum inhibitory concentrations (MICs) to penicillin have risen. It is susceptible to third-generation cephalosporins. Ceftriaxone is widely used for treatment because it is active against other causes of bacterial meningitis, and it eliminates the carrier state.

Several meningococcal vaccines, both polysaccharide, and polysaccharide-protein conjugate vaccines, against serogroups A, C, W-135, and Y, are in use. A vaccine against serogroup B, made from surface proteins, not from the polysaccharide capsule, is licensed in Europe. Although this serogroup causes about one-third of cases of infection in the USA, the vaccine is not in use here, other than in outbreak situations.

Neisseria gonorrhoeae (also called gonococcus, or GC) is transmitted sexually, causing an infection called gonorrhea, which manifests mainly as urethritis in males and cervicitis in females. However, it can cause infection in other sites such as the rectum and pharynx, and occasionally it can spread systemically, affecting especially joints, tendon sheaths, and the skin. In the newborn, it can cause a severe conjunctivitis, which can lead to keratitis and destruction of the eye (Figs 11.8 & 11.9). Because the organism is spread so readily, it poses a very important public health problem.

Because gonococcus is usually present on mucosae, where other organisms are also present, culture requires suppression of these other organisms. Therefore culture swabs should be inoculated on chocolate agar that includes vancomycin, colistin, and nystatin (Thayer–Martin agar). In cases of suspected gonorrhea, a culture should be performed. However, this is not adequately sensitive. Therefore, one of the nucleic acid amplification tests (NAATs) should also be performed. Because obtaining swabs from the cervix or urethra is time-consuming, screening of high-risk populations can be done more readily using urine tests, which detect DNA. (They are often combined with tests for *Chlamydia trachomatis*, which is frequently a co-pathogen.)

Because resistance to penicillin and fluoroquinolones is so widespread, and cephalosporin resistance is emerging, current recommended therapy for patients infected with this organism is ceftriaxone plus either azithromycin or doxycycline. Cefixime plus either azithromycin or doxycycline is considered second-line therapy.

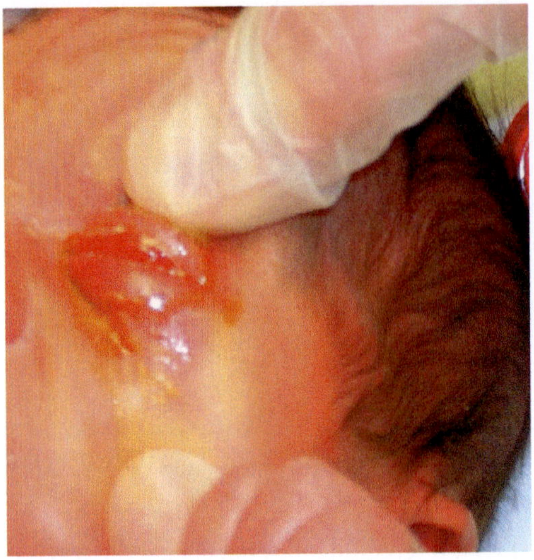

Fig. 11.8 Eye of a newborn infant showing severe conjunctivitis caused by *Neisseria gonorrhoeae*. Copyright © 2007 Frank E. Berkowitz. Reprinted with permission of Cambridge University Press.

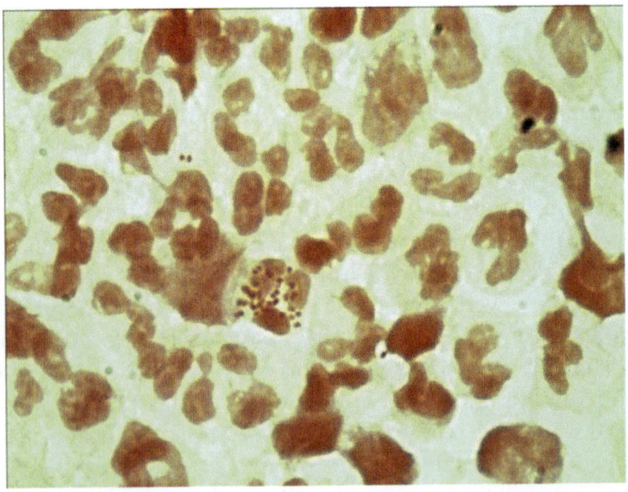

Fig. 11.9 Gram-stained smear from the eye of the patient in Fig. 11.8. It shows characteristic Gram-negative cocci, both inside and outside leukocytes. Copyright © 2007 Frank E. Berkowitz. Reprinted with permission of Cambridge University Press.

Moraxella catarrhalis (formerly called *N. catarrhalis*, and then *Branhamella catarrhalis*) is a member of the Neisseriaceae, with the same appearance on Gram stain as *N. meningitidis* and *N. gonorrhoeae*. It is normal flora of the pharynx, and plays a role in causing acute otitis media and sinusitis. It frequently elaborates β-lactamase, rendering it resistant to amoxicillin, but not to amoxicillin/clavulanate.

The main laboratory methods for differentiating between the different Gram-negative cocci are shown in Table 11.1.

Table 11.1 Key laboratory tests to differentiate the most common Gram-negative cocci.

Organism	Growth on media		Fermentation of carbohydrate				Butyrate hydrolysis	Other
	Thayer–Martin*	Basic agar†	Glucose	Maltose	Lactose	Sucrose		
Neisseria gonorrhoeae	+	–	+	–	–	–	–	
N. meningitidis	+	+	+	+	–	–	–	
N. lactamica	+	+	+	+	+	–	–	
Moraxella catarrhalis	+/–	+	–	–	–	–	+	Hockey-puck like movement when picking colony

*, or other similar media for selective growth of *N. gonorrhoeae* (e.g. Martin Lewis). Note that *Kingella* species may also grow on these media, hence Gram stain is important to differentiate among the species.
†, Nutrient agar, Mueller Hinton, trypicase soy (no blood) at 35°C.

Further reading

Brigham KS, Sandora TJ (2009) Neisseria meningitidis: epidemiology, treatment and prevention in adolescents. Curr Opin Pediatr 21: 437–43.

Centers for Disease Control and Prevention (2010) Sexually transmitted disease treatment guidelines. MMWR 59 (RR-12): 49–56.

Centers for Disease Control and Prevention (2010) Update to sexually transmitted disease treatment guidelines, 2010: oral cephalosporins no longer a recommended treatment of gonococcal infections. MMWR 61: 590–4.

Centers for Disease Control and Prevention (2013) Prevention and control of meningococcal disease. Recommendations of the Advisory Committee on Immunization Practices (ACIP). MMWR 62(RR02): 1–22.

Kirkcaldy RD (2012) New Treatment Guidelines for Gonorrhea: Antibiotic Change. Centers for Disease Control and Prevention. Expert Commentary. www.medscape.com/viewarticle/768883

Murphy TF, Parameswaran GI (2009) *Moraxella catarrhalis*, a human respiratory tract pathogen. Clin Infect Dis 49: 124–31.

Vidakovics MLP, Riesbeck K (2009) Virulence mechanisms of *Moraxella catarrhalis* in the pathogenesis of infection. Curr Opin Infect Dis 22: 279–85.

CHAPTER 12

Gram-positive rods

This is a heterogenous group of bacteria, both sporogenous and non-sporogenous.

Sporogenous Gram-positive rods

Genus: Bacillus

The Gram-positive aerobic sporogenous rods constitute the genus Bacillus. (Sporogenous anaerobic Gram-positive rods constitute the genus Clostridium – see section on anaerobic bacteria, Chapter 14). This contains many species, the most important of which, from a medical and veterinary viewpoint, is *B. anthracis*, the cause of anthrax.

Anthrax in humans occurs in four forms: cutaneous, the most common form, inhalation anthrax (characterized by severe, rapidly progressive mediastinitis), ingestion anthrax causing gastrointestinal disease, and systemic disease, including meningitis. The name is derived from the black color of the cutaneous eschar, "anthrax" being the Greek word for coal (Fig. 12.1). The organism is present in soil and can contaminate the hides of animals, which is the usual source of infection in humans. It has also been used as a form of bioterrorism. The virulence factors are a polygluconic acid capsule, which is antiphagocytic, and three synergistic toxins, lethal toxin, protective factor, and edema factor.

Bacillus anthracis can be grown on regular media. It is critical to communicate directly with the laboratory when cultures are obtained from suspected cases of anthrax. The role of the routine clinical laboratory is to rule it out or to refer suspicious organisms to a reference laboratory.

Direct Gram stain from specimens demonstrates large Gram-positive rods 1–1.5 × 3–5 μm in size, often in short chains of 2–4 cells and frequently with a halo (capsule) around the cells.

Plates demonstrating growth should be immediately placed in a biological safety cabinet, where all further manipulation should be performed. Colonies from blood agar plates are non-hemolytic, have a "ground-glass" morphologic appearance and may have a "Medusa-head" characteristic (Fig. 12.3). A Gram-stained preparation from

Practical Medical Microbiology for Clinicians, First Edition. Frank E. Berkowitz and Robert C. Jerris.

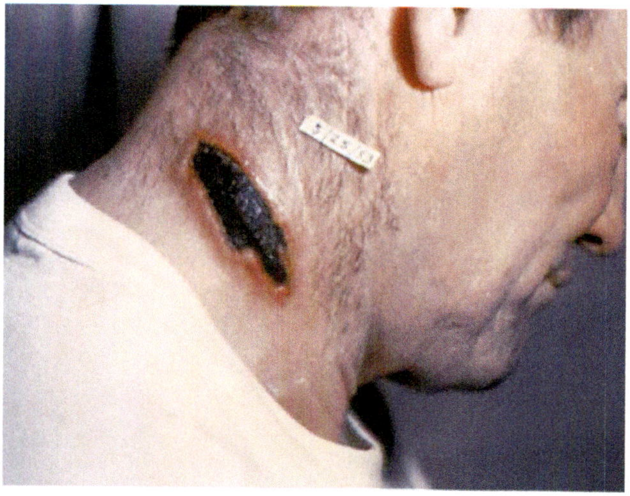

Fig. 12.1 Cutaneous anthrax. Courtesy of PHIL, CDC.

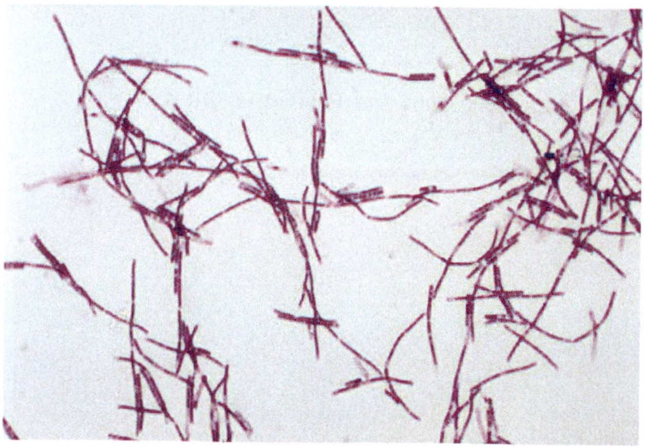

Fig. 12.2 *Bacillus anthracis* on Gram stain. Spores can be seen. Courtesy of PHIL, CDC.

colonies generally shows long chains of Gram-positive rods, and spores, if present, are located centrally or subterminally. Rarely are capsules seen from primary isolation plates (Fig. 12.2).

The key morphologic and phenotypic tests on which to base the decision to refer isolates for further work-up to rule out *B. anthracis* are:

- Colony: non-β-hemolytic
- Catalase: positive
- Motility: negative.

Although it is usually susceptible to penicillin, it may produce β-lactamase. Suspected cases of anthrax should be treated with a combination of a fluoroquinolone and linezolid. If meningitis is suspected, then meropenem should also be used.

The other members of the genus Bacillus are widespread in the environment, and on skin. Therefore, they are common contaminants of cultures of specimens such as blood.

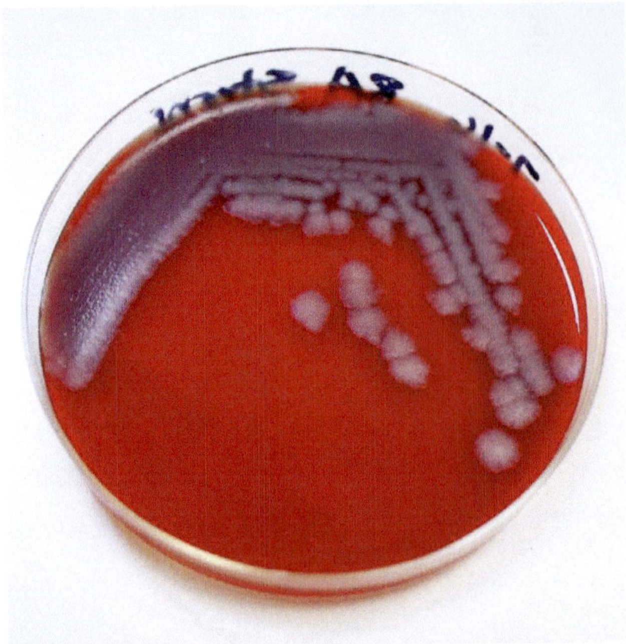

Fig. 12.3 *Bacillus anthracis* growing on blood agar. Courtesy of PHIL, CDC.

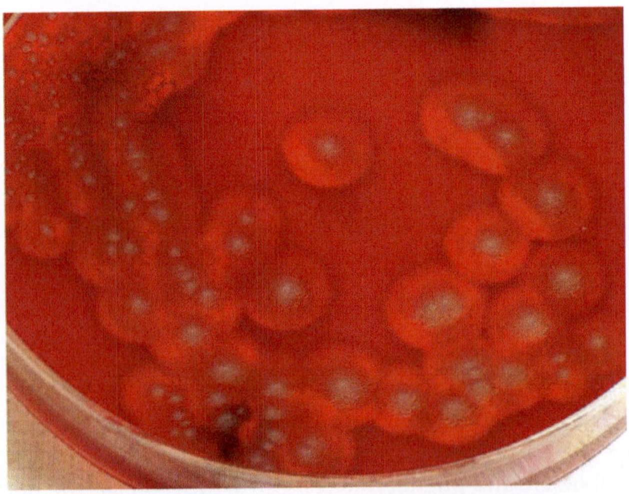

Fig. 12.4 *Bacillus cereus colonies* growing on blood agar showing hemolysis. Courtesy of PHIL, CDC.

However, they can cause real infections, particularly on plastic foreign bodies such as vascular catheters. Most clinical laboratories do not identify them to a species level, as long as *B. anthracis* has been excluded. Most are resistant to penicillin but susceptible to clindamycin, vancomycin, and fluoroquinolones. Of this group, the most important is *B. cereus* (Fig. 12.4). It can cause a wide variety of infections, including those of the skin and soft tissue, and systemic infections. Infection of foreign bodies is particularly important. When present in wound infections, it must be differentiated from *Clostridium perfringens*. Demonstration of a capsule, using an India ink preparation, would indicate *C. perfringens*. *B. cereus* can cause two forms of acute food poisoning due to the elaboration of enterotoxins. *B. cereus* is resistant to β-lactams and β-lactam/β-lactam inhibitor combinations. However, it is susceptible to vancomycin, clindamycin, macrolides, chloramphenicol, tetracycline, aminoglycosides, fluoroquinolones, linezolid, rifampin, and daptomycin.

Spores of the organism *B. stearothermophilus* are used to test that autoclaves are functioning properly, as sterilization should kill spores.

Non-sporogenous Gram-positive rods

Genus: Listeria

Listeria monocytogenes is an aerobic Gram-positive rod that is an important pathogen in neonates, the elderly, and individuals with impaired cell-mediated immunity, in whom it causes bacteremia and meningitis (Fig. 12.5). It can also cause acute diarrhea. The organism is present in the environment, in unpasteurized dairy products, meat, and vegetables, and can grow in the cold. It can live within macrophages, requiring cell-mediated immunity for its elimination. It is ingested and colonizes the intestine and vagina, and can cause bacteremia, which, in normal individuals, may not be of great significance. However, a pregnant woman can transmit the organism transplacentally to the fetus, causing an intrauterine infection called granulomatosis infantiseptica. Most cases of neonatal infection are acquired intrapartum.

It can be grown on blood agar, on which it forms small gray colonies, surrounded by a small zone of β-hemolysis. In a hanging drop preparation it demonstrates tumbling motility.

It is susceptible to penicillin and ampicillin, although it is tolerant to these drugs (they are bacteristatic rather than bactericidal). It is also susceptible to vancomycin and trimethoprim/sulfamethoxazole. Treatment consists of ampicillin with or without gentamicin. Of particular importance is the fact that it is resistant to cephalosporins, especially third generation.

Genus: Corynebacterium

The genus Corynebacterium, whose name is derived from the Greek for a small club, contains many species, which are aerobic or facultatively anaerobic. The most important is *C. diphtheriae*, the cause of diphtheria. Diphtheria is a very important infection, mainly in children. In countries where widespread vaccination against this organism has been carried out (for decades), it has become very rare. The organism is spread from the throat (the main site of carriage and infection) by droplets. It is not an invasive organism, but causes disease by elaboration of an exotoxin, which is an ADP-ribosylase (see Chapter 9 Bacteria: general features). This inhibits elongation factor 2 in the ribosome, preventing protein synthesis. Clinically, the disease manifests as a severe pharyngotonsillitis with the formation of a pseudomembrane. This can extend into the larynx, causing airway obstruction and death. There may be a bloody nasal discharge. The toxin also causes a toxic cardiomyopathy, which can manifest with myocardial failure and complete heart block, the other main cause of death. A diffuse peripheral neuropathy and renal failure can also occur. The organism can also infect the vagina and skin.

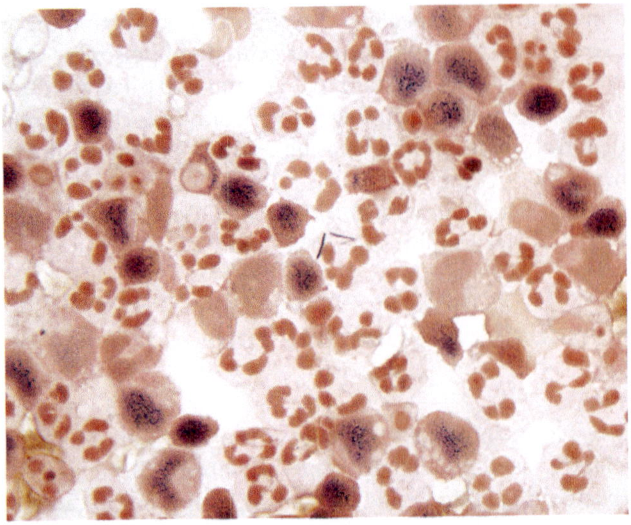

Fig. 12.5A Gram stain of cerebrospinal fluid of a patient with Listeria monocytogenes meningitis, showing numerous leukocytes, and Gram-positive rods. Courtesy of Marylyn Addo M.D. Ph.D., Division of Infectious Diseases, Massachusetts General Hospital and Harvard Medical School, and Judith Holden, MT (ASCP) MPH, Micropathology Laboratory, Massachusetts General Hospital, Boston, MA. This image was first published on Partners' Infectious Disease Images web site, whose content is copyrighted by Partners Healthcare System, Inc., and is used with permission. All rights reserved.

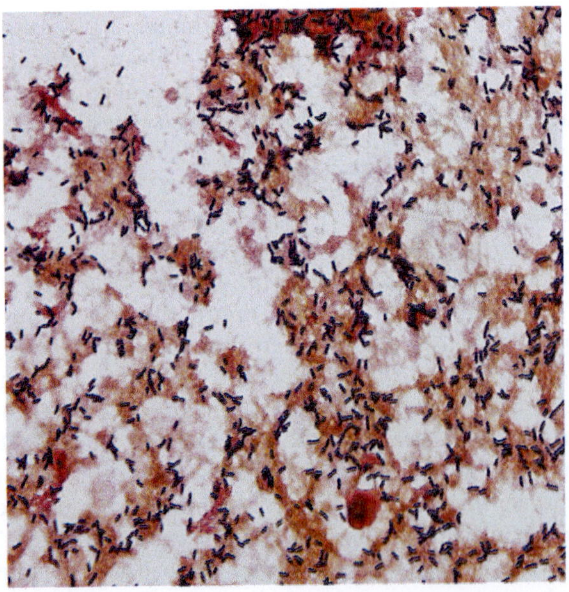

Fig. 12.5B Gram stain of *Listeria monocytogenes* growing in blood culture. Courtesy of Eileen Burd, PhD, D(ABMM),Director, Clinical Microbiology, Emory University Hospital.

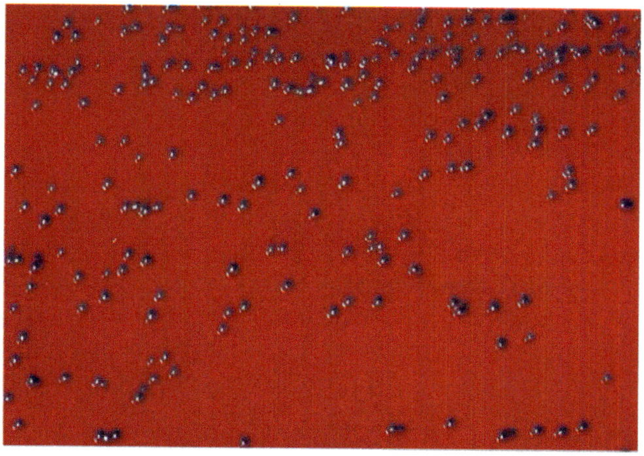

Fig. 12.6 *Corynebacterium diphtheriae* growing on agar containing tellurite. Courtesy of PHIL, CDC.

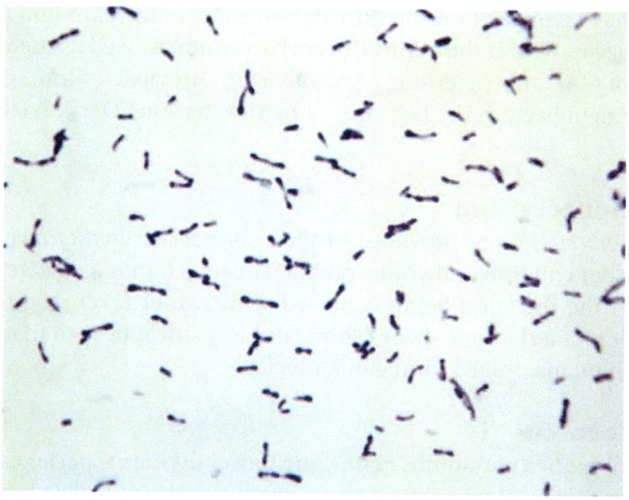

Fig. 12.7 *Corynebacterium diphtheriae* showing bipolar enhancement of staining. Courtesy of PHIL, CDC.

Because the throat contains many commensal organisms, culture of *C. diphtheriae* requires selective media. These include Loeffler's medium and tellurite-containing agar, on which it produces black colonies (Fig. 12.6). When stained with methylene blue, the organism has metachromatic ends (Fig. 12.7). If cultured, the isolate should be tested for toxin production, using an immunodiffusion test (Elek plate). Because this infection is so rare in the USA, clinical laboratories do not routinely culture throat swabs for this organism. Therefore, when suspected, specimens should be sent to a reference or public health laboratory.

Treatment consists of supportive care, often requiring an artificial airway, antimicrobial therapy, and antitoxin. *C. diphtheriae* is susceptible to penicillin and macrolide antibiotics. Because it has significant public implications, it is reportable to public health authorities.

Diphtheria vaccine is a toxoid. It is produced by treatment of the toxin with formaldehyde that renders it non-toxic but with retention of its immunogenicity.

Other corynebacteria

Non-lipophilic Corynebacterium species include *C. ulcerans* (which can produce diphtheria toxin), *C. xerosis, C. minutissimum* (which causes erythrasma, a superficial skin infection), *C. striatum,* and *C. pseudodiphtheriticum*. Lipophilic Corynebacterium species include *C. jeikeium* and *C. urealyticum* (which is associated with urinary tract infection in the elderly). Because they are present in the environment and on skin, they are often "contaminants" of cultures such as blood cultures. However, they can cause opportunistic infections, especially hospital-acquired infections, and infections on foreign bodies such as vascular catheters, ventriculoperitoneal shunts, and prosthetic heart valves. Clinical laboratories seldom identify these organisms to species level, nor do they perform antimicrobial susceptibility tests. Many are resistant to β-lactam antibiotics. Antimicrobial therapy usual requires the administration of vancomycin.

Coryneform bacteria

This is a group of Gram-positive rods that are morphologically similar to *C. diphtheriae*, and are therefore often referred to as "diphtheroids." They are widespread in the environment, and may be present on the skin and mucosae of humans and animals. They include other species within the genus Corynebacterium, as well as members of other genera, including Arcanobacterium, Arthrobacter, Brevibacterium, Cellulomonas, Dermabacter, Exiguobacterium, Leifsonia, Microbacterium, Oerskovia, Rothia, and Turicella.

Genus: Arcanobacterium

Arcanobacterium haemolyticum, previously called *Corynebacterium haemolyticum*, causes pharyngitis in older children and young adults. The infection is associated with a rash, which results in the infection being confused with scarlet fever. It grows better on human or rabbit blood than on sheep blood, and it is susceptible to many antibiotics, including penicillin, macrolides, and clindamycin.

Genus: Rhodococcus

There are several species within this genus, the most important species causing disease in humans being *R. equi* (formerly called *Corynebacterium equi*). This is present in the environment and is acquired by inhalation. It causes a chronic pneumonia, primarily in individuals with impaired cell-mediated immunity, such as AIDS. It resembles *Mycobacterium tuberculosis* in the following ways: it causes chronic pneumonia, sometimes with cavity formation; it multiplies within macrophages; and it grows in Lowenstein–Jensen medium, after which it stains acid fast with Kinyoun stain (a modified acid-fast stain). It is resistant to β-lactam antibiotics but is susceptible to macrolides, rifampin, clindamycin, vancomycin, aminoglycosides, and fluoroquinolones.

Genus: Actinomyces

These are branching filamentous Gram-positive rods consisting of several species that are part of the normal human flora, primarily in the mouth, intestine, and female genital tract (Fig. 12.8). They are classified as anaerobes but are often microaerophilic. The species include *A. israelii, A. odontolyticus, A. meyeri,* and *A. naeslundii.* They can cause chronic infections that originate from the mucosal surfaces where they are present, sometimes due to minor trauma. They can cross tissue planes, suggesting the diagnosis of cancer. They can cause infections complicating dental infection, toothpick injuries,

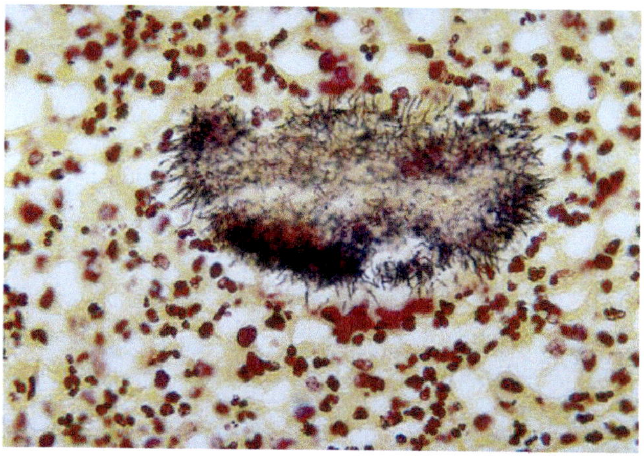

Fig. 12.8 Inflammation and branching rods of Actinomyces. Courtesy of PHIL, CDC.

and intrauterine contraceptive devices, and they can also spread hematogenously. Foci of inflammation can become mineralized, so that granules, called "sulfur" granules, form in the pus, giving it a gritty texture. They grow on regular media, requiring prolonged incubation times, and they are susceptible to many antibiotics, including penicillin, macrolides, and clindamycin. Therapy must, however, be prolonged.

Genus: Nocardia

These are also branching Gram-positive rods. The genus contains many species, some of which have been divided into complexes according to their antimicrobial drug susceptibilities. These include the following: *N. abscessus*, *N. brevicatena/paucivorans* complex, *N. nova* complex, *N. transvalensis* complex, *N. farcinica* complex, *N. asteroides* complex, *N. cyriacigeorgica*, *N. brasiliensis*, *N. pseudobrasiliensis*, and *N. otitidiscaviarum*. They are present in the environment (soil, vegetation, and water), and are not part of the normal flora. With the exception of *N. brasiliensis*, they are opportunists, affecting individuals with impaired cell-mediated immunity. They enter the host via the lungs so the most common infection is pneumonia. They have the predilection to spread via the bloodstream to the brain and to subcutaneous tissue. *N. brasiliensis* can be inoculated from vegetation into the skin, causing subcutaneous infection and lymphangitis, similar to that occurring in sporotrichosis (see Chapter 20). Nocardia species can have the same appearance on Gram stain as do Actinomyces species (branching, beaded, filamentous Gram-positive rods). However, they may also stain red with a modified acid-fast stain (Kinyoun stain) (Fig. 12.9). Their antimicrobial susceptibilities are very variable, and testing of isolates is therefore very important. However, they are usually susceptible to trimethoprim/sulfamethoxazole, amikacin, and linezolid. For patients with severe infection, initial empiric therapy should consist of a combination of trimethoprim/sulfamethoxazole, amikacin and imipenem or ceftriaxone.

Erysipelothrix rhusiopathiae

This organism is widespread in the environment, and colonizes many animals and fish. Therefore, individuals working with animal and fish carcases are at risk of becoming infected. It causes a painful cellulitis and, rarely, infective endocarditis.

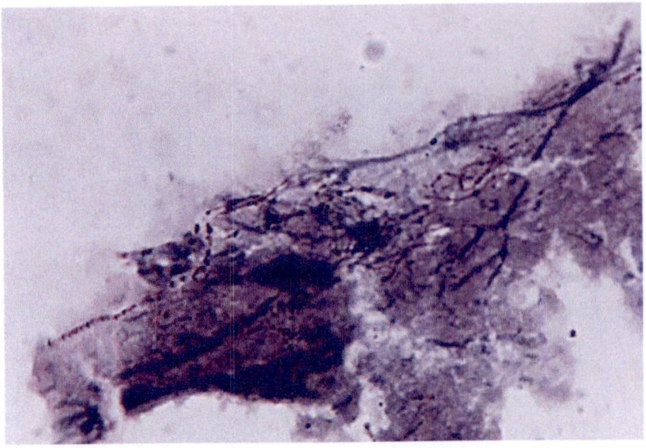

Fig. 12.9 *Nocardia* spp. in lung tissue, stained with the Kinyoun stain. Courtesy of PHIL, CDC.

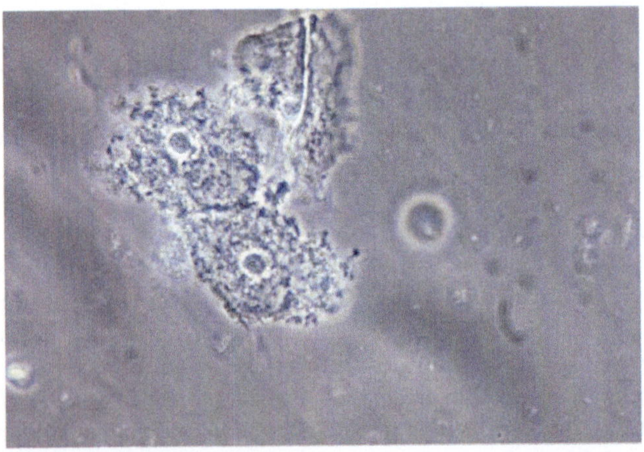

Fig. 12.10 "Clue cells." Courtesy of PHIL, CDC.

It is susceptible to penicillin, carbapenems, and fluoroquinolones but, notably, is resistant to vancomycin. This is of great importance because many other Gram-positive rods, such as the coryneform bacteria and *Bacillus* spp., are resistant to β-lactams and susceptible to vancomycin.

Gardnerella vaginalis

This is a Gram-positive rod that is facultatively anaerobic. It plays an important role in the pathogenesis of bacterial vaginosis. In this disease, numerous bacteria covering vaginal squames give these cells a granular appearance. Such cells, called "clue cells," are used for the diagnosis of bacterial vaginosis (Fig. 12.10).

Tropheryma whipplei

This causes Whipple disease, a chronic infection resulting in arthritis, chronic diarrhea, and disease of other organs. Although the disease had been recognized as an infection for many years, the organism was characterized by nucleic acid methods before it could be cultivated *in vitro*. The diagnosis is made by PCR of body fluids or tissue. Intestinal biopsy shows macrophages containing material that stains with the periodic acid-Schiff stain (also used for staining fungi and glycogen).

Other Gram-positive rods

These occasionally cause human disease, especially in compromised hosts, and include members of the genera *Gordonia*, *Tsukamurella*, and *Streptomyces*.

Further reading

www.asm.org/index.php/guidelines/sentinel-guidelines

Adalja AA, Toner E, Inglesby TV (2015) Clinical management of potential bioterrorism-related conditions. N Engl J Med 372: 954–62.

Bernard K (2005) *Corynebacterium* species and coryneforms: an update on taxonomy and diseases attributed to these taxa. Clin Microbiol Newsl 27: 9–18.

Bottone EJ (2010) *Bacillus cereus*, a volatile human pathogen. Clin Microbiol Rev 23: 382–98.

Braden CR (2003) Listeriosis. Pediatr Infect Dis J 22: 745–6.

Brown-Elliot BA, Brown J, Conville PS, et al. (2006) Clinical and laboratory features of the *Nocardia* spp. based on current molecular taxonomy. Clin Microbiol Rev 19: 259–82.

Kadirova R, Kartoglu HU, Strebel PM (2000) Clinical characteristics and management of 676 hospitalized diphtheria cases, Kyrgyz Republic, 1995. J Infect Dis 181(Suppl 1): S110–S115.

Muzny CA, Schwebke JR (2013) *Gardnerella vaginalis*: still a prime suspect in the pathogenesis of bacterial vaginosis. Curr Infect Dis Rep 15: 130–5.

Puechal X (2013) Whipple's disease. Ann Rheum Dis 72: 797–803.

Stenfors Arnesen LP, Fagerlund A, Granum PE (2008) From soil to gut: Bacillus cereus and its food poisoning toxins. FEMS Microbiol Rev 32: 579–606.

Sullivan DC, Chapman SW (2010) Bacteria that masquerade as fungi. Proc Am Thorac Soc 7: 216–21.

Wilson JW (2012) Nocardiosis: updates and clinical overview. Mayo Clin Proc 87: 403–7.

CHAPTER 13

Gram-negative rods

General features

This large group of bacteria includes many families, genera, and species. Most clinical isolates belong to the family Enterobacteriaceae. It may be convenient (although taxonomically not necessarily accurate) to consider them in the following main groups:

- Enterobacteriaceae
- non-Enterobacteriaceae acquired from the environment
- non-Enterobacteriaceae acquired from humans or animals.

Most Gram-negative rods have the same appearance on Gram stain, which is therefore not useful in distinguishing between them. Their identification depends on their growth characteristics on different media, their biochemical activities, and other phenotypic characteristics (see Chapter 2).

Media

The usual media used are blood agar, MacConkey agar, chocolate agar (heated blood agar) and, sometimes selective media (to enhance the growth of a particular organism, while inhibiting the growth of others).

Colonies may be fairly characteristic on blood agar (e.g. large, gray colonies, which may be mucoid, if the organism has a large capsule). MacConkey agar inhibits the growth of Gram-positive cocci. It also contains lactose and a pH indicator which turns pink in acid. Thus, on this medium, an organism that ferments lactose produces acid, making the colonies pink (Fig. 13.1). The recognition of these "lactose-fermenting" organisms largely limits the identification of the organism to the genera Escherichia, Klebsiella, or Enterobacter.

Biochemical tests

Oxidase test

This tests the ability of the organism to utilize the cytochrome oxidase system. The test entails smearing a colony of the bacteria on a piece of filter paper on which has been placed a drop of the reagent tetramethyl-p-phenylenediamine dihydrochloride, which

Practical Medical Microbiology for Clinicians, First Edition. Frank E. Berkowitz and Robert C. Jerris.

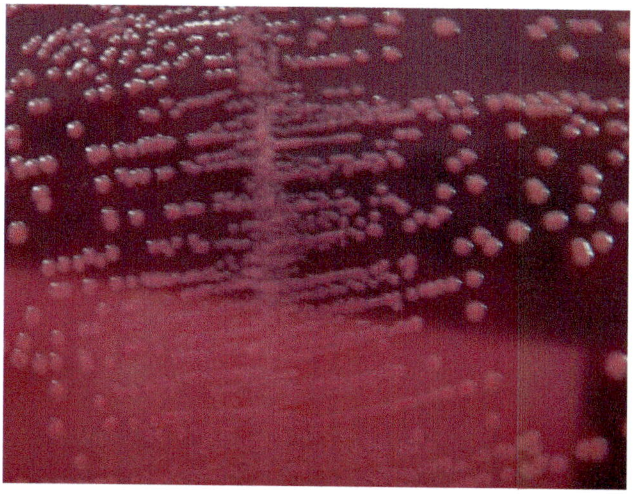

Fig. 13.1 Pink colonies of *E. coli* on MacConkey agar due to lactose fermentation.

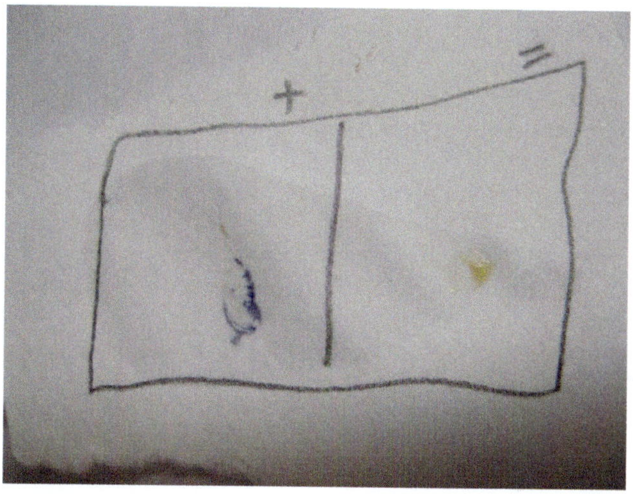

Fig. 13.2 Oxidase test. The panel on the left shows a positive test (*Pseudomonas aeruginosa*), and that on the right is a negative test (*E. coli*).

is colorless. A lavender/purple reaction indicates a positive test (Fig. 13.2). The main value of this test is to differentiate all members of the Enterobacteriaceae (negative) from *Pseudomonas aeruginosa* (positive).

Other biochemical tests

These involve fermentation of various sugars (all members of the Enterobacteriaceae ferment glucose) including lactose (see earlier in this chapter) and maltose; metabolism of organic acids, such as citrate, lysine, and ornithine; ability to produce gas; and ability to reduce hydrogen sulfide. These tests were formerly performed in test tubes ("long sugars") but are now performed in small multichamber plastic containers (see Chapter 2).

Different bacteria have different combinations of biochemical reactions. The reactions of a particular isolate can be compared with those of large numbers of isolates with a known identification in a large database, which enables one to identify them.

Some species of bacteria, such as *Escherichia coli* (*E. coli*), are normal flora in feces, but certain strains are pathogenic by virtue of possessing certain virulence factors. These pathogenic strains cannot be identified using the usual methods but require methods that detect the virulence factor (e.g. Shiga toxin in enterohemorrhagic *E. coli*), or the gene responsible for producing the virulence factor.

Motility in semi-solid agar

A tube containing semi-solid agar is stabbed with a platinum wire inoculated with bacteria from a colony. Lateral spread from the original stab indicates motility.

Enterobacteriaceae

Members of the family Enterobacteriaceae are characterized by:
- aerobic and facultatively anaerobic growth
- fermentation of glucose (and sometimes other sugars)
- reduction of nitrate to nitrite.

Many of these bacteria live normally (or are pathogenic) in the gastrointestinal tract and in the environment (which is often contaminated by animal feces), but they may cause disease in other parts of the body. Their cell surfaces have many antigens, some of which are used to identify species or serotypes of the organisms. The somatic antigen is termed the "O" antigen, and for motile organisms (which have flagella), the flagellar antigen is termed the "H" antigen. For those organisms that are encapsulated, the capsular antigen is termed the "K" antigen. So, for example, the predominant serotype of enterohemorrhagic *E. coli* is O 157: H 7.

The members of the Enterobacteriaceae of medical importance, their usual sources, and the infections they cause are shown in Table 13.1, and their usual antimicrobial susceptibilities are shown in Table 13.2.

Escherichia coli

This organism has provided much of the material on which our knowledge of biochemical pathways is based. It has been used widely for the study of genetics. It forms part of the normal flora of human and animal large bowel. However, when it gets out of the bowel it can cause disease, such as peritonitis after perforation of the bowel. It is the most common cause of urinary tract infection, and an important cause of neonatal sepsis and meningitis, and of healthcare-associated infection, such as ventilator-associated pneumonia. Certain strains possess specific virulence factors that enable them to cause specific disease syndromes, e.g. adhesins which enable them to attach to uroepithelium, and toxins that enable them to cause diarrhea. The following are the different types of pathovars causing diarrhea: enterotoxigenic *E. coli* (ETEC), enteroinvasive *E. coli* (EIEC) – which are really Shigellae, enteropathic *E.coli* (EPEC), enteroaggregative *E. coli* (EAEC), enterohemorrhagic *E. coli* (EHEC), and diffusely adherent *E. coli* (DAEC). Although the virulence factors cannot be readily detected in clinical laboratories, there is an increasing ability for their genes

Table 13.1 Enterobacteriaceae most commonly encountered, according to tribe (sometimes used in their taxonomy), their usual sources, and the infections they cause.

Organism	Source	Infections
Tribe I: Escherichieae		
Genus: Escherichia		
E. coli	A, H, End	Diarrhea, HUS, UTI, neonatal infection, peritonitis, HCAI
Genus: Shigella	H	Diarrhea, dysentery, encephalopathy, HUS
S. dysenteriae		
S. flexneri		
S. sonnei		
S. boydii		
Tribe II: Edwardsielleae		
Genus: Edwardsiella	Env (water)	Wound, other foci
E. tarda		
Tribe III: Salmonelleae		
Salmonella enteritidis	A	Diarrhea, bacteremia
>2400 non-typhi	H	Enteric (typhoid) fever
S. typhi		
Tribe IV: Citrobactereae		
Citrobacter freundii	End	Neonatal infection (HCA) – meningitis, brain abscess, UTI
C. koseri		
Tribe V: Klebsielleae		
Klebsiella pneumoniae	End, H	Pneumonia, UTI, HCAI
K. oxytoca	End, H	Pneumonia, UTI, HCAI
K. ozaenae	H	Atrophic rhinitis
K. granulomatis	H	Granuloma inguinale
K. rhinoscleromatis	H	Nasal granuloma
Enterobacter cloacae	End, H	UTI, HCAI
E. aerogenes		
Pantoea agglomerans	End	HCAI
Cronobacter sakazakii	Baby formula	Neonatal HCAI
Hafnia alvei	H	OI, diarrhea
Serratia marcescens	Env	HCAI, CGD
S. liquifaciens		
Raoultella ornitholytica	W, fish	Wound, sepsis
Tribe VI: Proteeae		
Proteus mirabilis	H	UTI, HCAI
P. vulgaris		
Providencia stuartii	H	UTI, HCAI
P. rettgeri		
Morganella morganii	H	UTI, HCAI
Tribe VII: Yersinia		
Yersinia pestis	A	Plague
Y. enterocolitica	A	Diarrhea
Y. pseudotuberculosis	A	Diarrhea
Several other genera and species		

A, animal; CGD, chronic granulomatous disease; End, endogenous; Env, Environment; H, human; HCAI, healthcare-associated infection; HUS, hemolytic-uremic syndrome; OI, opportunistic infection; UTI, urinary tract infection.

Table 13.2 Enterobacteriaceae and their usual antimicrobial susceptibilities. With increasing antimicrobial resistance, there may be resistance to some of these drugs.

Organism	Susceptibilities
E. coli	Ampicillin, cefazolin, TMP/S, 3rd gen. ceph, ticar/clav, pip/tazo, aminoglycosides, fluoroquinolones, polymyxin
Shigella spp.	Ampicillin,† TMP/S,† 3rd gen. ceph, cefixime, azithromycin
Klebsiella spp.	Cefazolin, 3rd gen. ceph, ticar/clav, pip/tazo, aminoglycosides, fluoroquinolones, polymyxin
Enterobacter spp.	3rd gen. ceph,* TMP/S, ticar/clav,* pip/tazo,* aminoglycosides, fluoroquinolones, polymyxin
Citrobacter spp.	3rd gen. ceph,* ticar/clav,* pip/tazo,* aminoglycosides, fluoroquinolones, polymyxin
Serratia spp.	3rd gen. ceph,* TMP/S, ticar/clav,* pip/tazo,* aminoglycosides, fluoroquinolones, polymyxin
Proteus mirabilis	Ampicillin, cefazolin, 3rd gen. ceph, ticar/clav, pip/tazo, aminoglycosides, fluoroquinolones, polymyxin
Proteus vulgaris	3rd gen. ceph,* ticar/clav,* pip/tazo,* aminoglycosides, fluoroquinolones, polymyxin
Providencia spp.	3rd gen. ceph,* ticar/clav,* pip/tazo,* aminoglycosides, fluoroquinolones, polymyxin
Morganella morganii	3rd gen. ceph,* ticar/clav,* pip/tazo,* aminoglycosides, fluoroquinolones, polymyxin
Salmonella spp.	Ampicillin,† TMP/S, 3rd gen. ceph., chloramphenicol, fluoroquinolones, azithromycin
Yersinia enterocolitica	Gentamicin, streptomycin, TMP/S, gentamicin, 3rd gen. ceph

3rd gen. ceph, third-generation cephalosporins; TMP/S, trimethoprim/sulfamethoxazole (cotrimoxazole); ticar/clav, ticarcillin/clavulanate; pip/tazo, piperacillin/tazobactam.

*These bacteria frequently harbor the genes for inducible resistance to all β-lactams except cefepime and carbapenems.

†High rates of resistance.

to be detected. Of *E.coli* strains that cause extraintestinal disease, two pathovars have been recognized, namely uropathogenic *E. coli* (UPEC) and neonatal meningitis *E. coli* (NMEC). About 80% of *E. coli* strains causing neonatal meningitis are of the K1 capsular type.

Many of the EHEC do not ferment sorbitol, which most other strains of *E. coli* do. This provides a screening tool for such strains.

Escherichia coli has been susceptible to a wide variety of antimicrobial agents, including ampicillin. However, over time the prevalence of resistance strains has increased, especially to β-lactam agents. There are now strains resistant to extended-spectrum cephalosporins, due to their ability to make extended-spectrum β-lactamases (ESBLs).

Genus: Shigella

This genus is very closely related to *E. coli*. There are 47 serotypes (based on their O-antigen), which are distributed among four species, namely *S. dysenteriae, S. flexneri, S. sonnei*, and *S. boydii*. They are non-motile (as opposed to *E. coli* and *Salmonella* spp.). They cause intestinal infection resulting in diarrhea, which is initially watery and may become bloody (dysentery). Infection is associated with high fever and sometimes with seizures. The infective dose is very low (10^2–10^3 organisms). Although the organisms can be readily detected in the laboratory in stool cultures, they are fastidious, and their numbers in stool decrease later in the course of the illness.

Antibiotic therapy is beneficial for infected patients. Ampicillin and trimethoprim/sulfamethoxazole were formerly effective in treating patients, but resistance to these drugs is now common. Therefore, azithromycin, ciprofloxacin, or ceftriaxone should be used, unless susceptibility to other agents is known.

Klebsiella pneumoniae

This causes severe pneumonia in debilitated individuals and urinary tract infections, and is a very important cause of healthcare-associated infections. It produces a large capsule, is lactose fermenting and non-motile, which make it relatively easy to recognize in the laboratory. It is always resistant to ampicillin. Although it was formerly susceptible to first-generation cephalosporins, now many strains are resistant. Some strains are resistant to third-generation cephalosporins due to the production of extended-spectrum β-lactamases(see *E.coli* earlier); in addition, strains that are resistant to carbapenems due to the production of carbapenemases (*Klebsiella pneumoniae* carbapenemase – KPC) have been recognized. These strains are detected by a modified Hodge test (Figs 13.3 & 13.4).

Genus: Salmonella

The genus Salmonella has two species: *S. enterica* and *S. bongari*. *S. enterica* accounts for almost all medical disease, and will be discussed further here. It contains more than 2500 different serovars, all of which are transmitted from animals, with the exception of *S. enterica* serovars *typhi* and *paratyphi* A, which are confined to humans. For convenience, the serovar name is used widely as the species name, e.g. *S. enterica* serovar *typhimurium* is called *S. typhimurium*. The non-typhi salmonellae have interesting names such as the towns where they were first reported (e.g. *S. mbandaka*) or their animals of origin (e.g. *S. pullorum*).

Salmonellae are transmitted by the fecal–oral route. They can invade cells and live inside macrophages, and thus require cell-mediated immunity for their elimination.

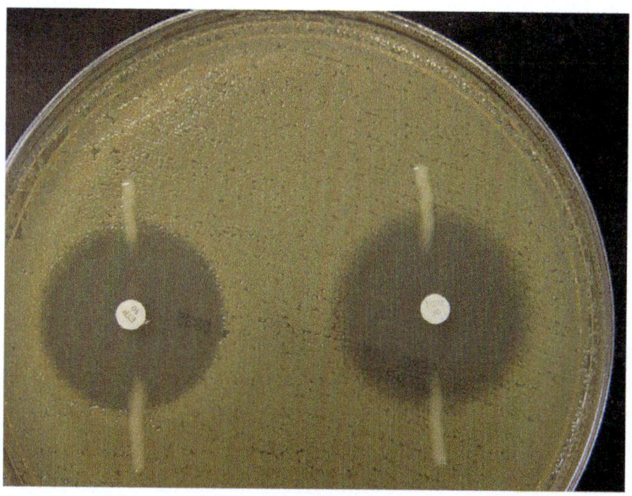

Fig. 13.3 A negative Hodge test (left disk is ertapenem, right disk is meropenem). Note that there is no growth close to the disk. Compare this with Fig. 13.4.

Fig. 13.4 A positive Hodge test (left disk is ertapenem, right disk is meropenem). Note growth up to the disk. Compare this with Fig. 13.3.

The non-typhi salmonellae cause acute gastroenteritis and dysentery, and occasionally systemic infection, especially in young infants, individuals with defects in cell-mediated immunity, and those with sickle cell disease. *S. typhi* and *S paratyphi* A cause typhoid (enteric) fever, which is a systemic infection affecting the terminal ileum and the reticuloendothelial system.

Salmonellae can readily be identified in the laboratory by their biochemical features, and *S. typhi* can be readily differentiated from the other salmonellae. Serotyping and identification of specific strains, which are important epidemiologically, are usually performed by reference laboratories.

Serratia marcescens

This organism is seldom a pathogen in normal hosts, but it is an important cause of healthcare-associated infections. The production of red pigment, prodigiosum, is a characteristic that has facilitated detection of the organism. The organism possesses the gene for inducible amp C β-lactamase, which is a broad-spectrum β-lactamase. Therefore β-lactam drugs, except carbapenems, should not be used for treating patients with severe infections caused by this organism. The other important species of this genus is *S. liquifaciens.*

Proteus (P. mirabilis, P. vulgaris), Providencia (P. stuartii, P. rettgeri), and *Morganella morganii*

This is a group of Enterobacteriaeae that cause mainly urinary tract infections, especially in individuals with indwelling urinary catheters. They also cause other healthcare-associated infections. When *Proteus* causes neonatal meningitis, it has the tendency to cause brain abscesses. Proteus species produce urease, which converts urea in urine to ammonia, rendering the urine alkaline. Finding a high pH in infected urine is a diagnostic clue to the presence of these organisms.

Proteus mirabilis and *P. vulgaris* can be readily detected on agar because they cause swarming. This has the appearance of the lines left by waves on a beach (Fig. 13.5).

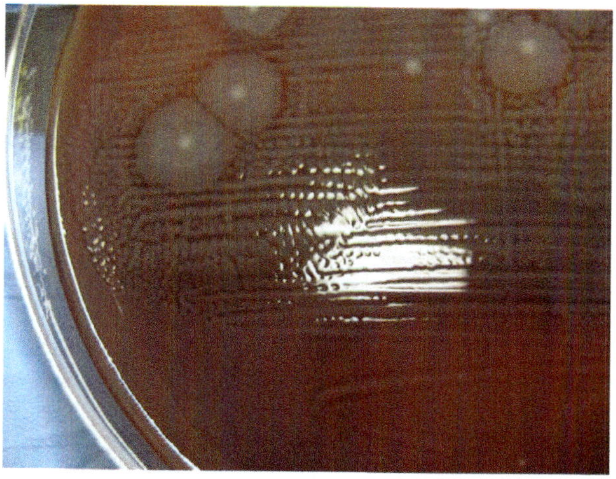

Fig. 13.5 *Proteus mirabilis* swarming on agar.

P. mirabilis, the most frequent isolate within this group, is usually susceptible to ampicillin, while *P. vulgaris* is always resistant to this drug. *P. vulgaris, P. rettgeri,* and *M. morganii,* a group often referred to as "indole-positive Proteus," possess the gene for inducible amp C β-lactamase. Therefore β-lactams are not suitable for the treatment of patients infected with them. The indole positivity is due to the action of the enzyme tryptophanase, by which the organism hydrolyzes and deaminates tryptophan, releasing indole, pyruvic acid, and ammonia. It is detected by a colorimetric test.

Genus Yersinia

This contains three species, namely *Y. pestis, Y. pseudotuberculosis,* and *Y. enterocolitica.*

Y. pestis, the cause of plague, is a genetic descendant of *Y. pseudotuberculosis,* an intestinal pathogen. It has been associated with three pandemics (the Justinian plague, 5th–7th centuries; the Black Death, 13th–15th centuries; and the modern plague, 1870s to the present). Although the disease has had a world-wide distribution, most recent cases have occurred in Africa and Asia. The distribution of cases in the USA in the past 42 years is shown in Fig. 13.6.

This infection is transmitted by fleas. The organism causes a high bowel obstruction in the flea, so that when it feeds, it regurgitates organisms into the host. These travel to the local lymph nodes, where they cause inflammation (bubonic plague), and then systemically, causing bacteremia, which can be very rapidly progressive (septicemic plague) (Figs 13.7 & 13.8). When they spread to the lung, they cause pneumonia (pneumonic plague) which is the only form that is contagious from human to human. The infection carries a very high fatality rate.

The organism is a small Gram-negative rod with bipolar staining, which grows readily on regular medium. Its virulence factors allow multiplication in the flea intestine, and they promote dissemination and prevent phagocytosis in the mammalian host. The laboratory staff should be informed about specimens to be submitted from a suspected case of plague, because it poses a significant hazard. The antimicrobial agent with which there is most experience is streptomycin. However, gentamicin seems as effective. Doxycycline and chloramphenicol are alternatives.

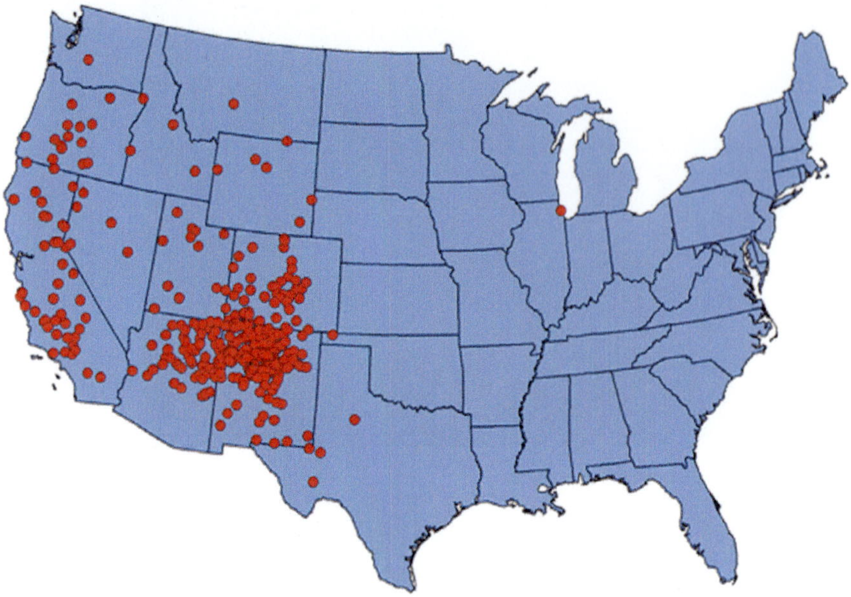

Fig. 13.6 Reported cases of plague in the USA, 1970–2012. Courtesy of CDC.

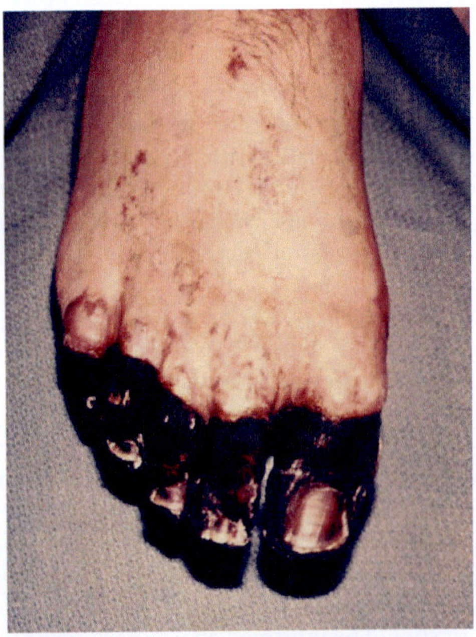

Fig. 13.7 Patient with septicemic plague, showing peripheral gangrene. Courtesy of PHIL, CDC.

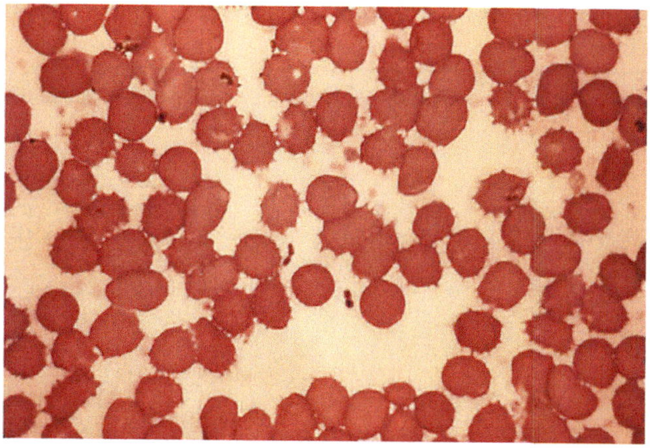

Fig. 13.8 Blood smear of patient with septicemic plague, showing Gram-negative rods with bipolar staining, characteristic of *Yersinia pestis*. Courtesy of PHIL, CDC.

Yersinia enterocolitica is an important cause of acute infectious diarrhea. This is usually watery diarrhea, but it can become bloody. The organism can also cause mesenteric lymphadenitis, which can be confused with appendicitis. Young infants and individuals with iron overload are particularly susceptible to bacteremia. It can grow in the cold, so that blood collected for transfusion and stored under refrigeration, if contaminated with the organism at the time of collection, can cause infection in the recipient. It can be acquired from several different animals, the most important of which is the pig. There are several different serotypes and biotypes. It can be detected in stool culture using various selective media. Whether patients with diarrhea benefit from antimicrobial therapy is unclear. Patients with bacteremia should be treated with gentamicin or trimethoprim/sulfamethoxazole.

Non-Enterobacteriaceae Gram-negative rods from the environment

These bacteria, the clinical diseases they cause, and their usual antimicrobial susceptibilities are summarized in Table 13.3.

Genus: Pseudomonas

This contains many species, of which only a small proportion are human pathogens. The most important by far is *P. aeruginosa*. Others which have been reported to cause disease in humans include *P. fluorescens, P. alcaligenes, P. luteola, P. mendocina, P. mosselli, P. oryzihabitans, P. otitidis, P. putida*, and *P. stutzeri*.

Pseudomonas aeruginosa

This organism lives in the environment, especially aquatic environments, and can cause severe infections. It has many virulence factors including hydrolytic enzymes, and toxins, and it produces several pigments (Figs 13.9 & 13.10), which can suggest its presence. It is important especially as a nosocomial pathogen, as a cause of pulmonary infection in individuals with cystic fibrosis, and of bacteremia in those with neutropenia. When a Gram-negative rod is cultured and found to be oxidase positive, this is the most likely suspect. It possesses many mechanisms for antimicrobial resistance,

Table 13.3 Environmental Gram-negative rods, the clinical diseases they cause, and their usual antimicrobial susceptibilities.

Organism	Clinical	Antimicrobial therapy
Pseudomonas aeruginosa	Pneumonia, sepsis, HCAI, UTI	ceftaz, pip/tazo, ticar/clav, carbap, fluoroq, aminoglyc
Burkholderia cepacia complex	Pneumonia	carbap, TMP/S, minocycline
Burkholderia pseudomallei	Melioidosis (pneumonia, sepsis, focal infections)	ceftaz, carbap, TMP/S, doxy
Stenotrophomonas maltophilia	HCAI, pneumonia	TMP/S, ticar/clav
Acinetobacter baumannii	HCAI, pneumonia	3rd gen. ceph, aminoglyc, fluoroq
Aeromonas	Wound	3rd gen. ceph, carbap, gent, fluoroq
Achromobacter xylosoxidans	OI	ceftaz, carbap, TMP/S
Legionella spp.	Pneumonia	fluoroq, azithromycin
Plesiomonas shigelloides	Diarrhea, sepsis	ceph, aminoglyc, fluoroq, TMP/S, chloro, doxy
Chromobacterium violaceum	Wound, sepsis	fluoroq, TMP/S, chloro, carbap
Elizabethkingia meningosepticum	Meningitis, sepsis	vanc + rif, fluoroq, pip/tazo
Vibrio cholerae	Cholera	azithro, doxy, fluoroq, TMP/S
Vibrio parahaemolyticus	Diarrhea	fluoroq, doxy
Vibrio vulnificus	Septicemia, wound	ceftaz + doxy

3rd gen. ceph, third-generation cephalosporins; aminoglyc, aminoglycosides; carbap, carbapenem; ceftaz, ceftazidime; doxy, doxycycline; fluoroq, fluoroquinolone; HCAI, healthcare-associated infection; OI, opportunistic infection; pip/tazo, piperacillin/tazobactam; rif, rifampin; TMP/S, trimethoprim/sulfamethoxazole (cotrimoxazole); ticar/clav, ticarcillin/clavulanate; UTI, urinary tract infection; vanc, vancomycin.

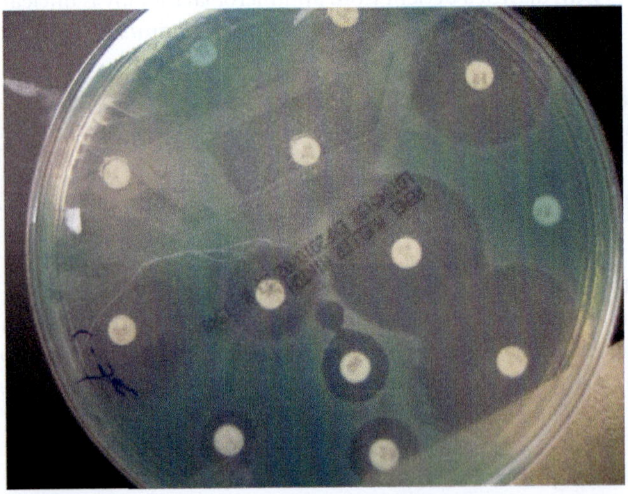

Fig. 13.9 Agar plate showing the blue pigment (pyocyanin) produced by *P. aeruginosa*.

rendering it resistant to several antimicrobial agents. Drugs that are appropriate for treating patients with infections caused by this organism include ceftazidime, cefepime, ticarcillin, piperacillin, aminoglycosides (gentamicin, tobramycin, and amikacin), carbapenems (except ertapenem), and fluoroquinolones.

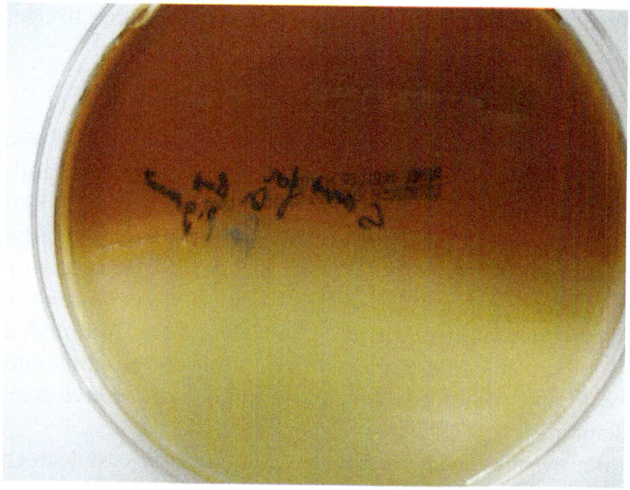

Fig. 13.10 Agar plate showing red pigment (pyorubrin) produced by *P. aeruginosa*.

Genus: Burkholderia

Burkholderia cepacia complex

This is a group of organisms that are important pathogens in individuals with cystic fibrosis, and their acquisition in such patients often portends a rapidly downhill clinical course. They are also organisms to which individuals with chronic granulomatous disease are susceptible. The group consists of the following species: *B. cepacia, B. cenocepacia, B. multivorans, B. stabilis, B. vietnamiensis, B. dolosa, B. ambifaria, B. anthina,* and *B. pyrrocinia.*

They are resistant to many antimicrobial agents, including some to which *P. aeruginosa* is susceptible. They are usually susceptible to carbapenems, minocycline, and trimethoprim/ sulfamethoxazole, and may also be susceptible to piperacillin/ tazobactam and chloramphenicol. Combinations, such as meropenem plus minocycline, might improve antibacterial activity.

Burkholderia pseudomallei

This organism is present in soil and water in South East Asia, northern Australia, parts of South America, and Puerto Rico. It causes melioidosis, an infection characterized by bacteremia, pneumonia, or focal infections, often after exposure to these environmental sources. It is particularly well recognized in Thailand. It is associated with a very high fatality rate. Recognition of the organism in the laboratory is difficult, if it is not suspected. Antimicrobial therapy should include ceftazidime or a carbapenem for 10–14 days, followed by a course of oral therapy for up to 20 weeks with a combination of trimethoprim/sulfamethoxazole and doxycycline. Chloramphenicol may also be used. The rate of relapse is high.

Stenotrophomonas maltophilia (previously called *Pseudomonas maltophilia*, then *Xanthomonas maltophilia*)

This is primarily a nosocomial pathogen, causing pneumonia, bacteremia, and other infections. It is usually resistant to all β-lactams and aminoglycosides, and always resistant to carbapenems. It is usually susceptible to trimethoprim/sulfamethoxazole, which is the mainstay of therapy of patients infected with it, and to tigecycline, and it might be susceptible to ticarcillin/clavulanate.

Acinetobacter spp.

This is a Gram-negative coccobacillus, which belongs to the family Moraxellaceae. It is oxidase negative (which distinguishes it from *Pseudomonas aeruginosa* and *Neisseria* spp.), and it is a strict aerobe, which distinguishes it from members of the Enterobacteriaceae. Although its taxonomy is complex, for practical purposes (particularly antimicrobial susceptibilities) there are considered to be two species: *A. calcoaceticus-baumannii* complex and *A. lwoffii*. These organisms are widespread in the environment, and they cause a wide variety of infections, particularly nosocomial infections. Although they grow well in the laboratory and are able to use many different carbon sources for growth, they do not react in many biochemical tests. They may stain Gram positive (see pitfalls of the Gram stain in Chapter 2).

Acinetobacter calcoaceticus-baumannii has the potential for extreme degrees of antimicrobial resistance by several different mechanisms, including the production of amp C broad-spectrum β-lactamase, aminoglycoside-modifying enzymes, porin mutations, and efflux mechanisms.

Aeromonas spp.

These are widespread in nature, including fresh and estuarine water. They are oxidase positive and ferment glucose. There are two groups of aeromonas: those that are motile and mesophilic (they grow at mammalian body temperatures), and infect humans, and those that are non-motile, psychrophilic (they grow at low temperatures), and infect fish. Their presence in food probably reflects contamination of the food with infected water. *Aeromonas* spp. cause infections of wounds acquired in water, and related to leech bites (including iatrogenic), bacteremia, and diarrhea.

Most are resistant to ampicillin, first-generation cephalosporins, ticarcillin and carbenicillin, and susceptible to second- and third-generation cephalosporins, carbapenems, aminoglycosides, trimethoprim/sulfamethoxazole, fluoroquinolones, and tetracyclines.

The most commonly isolated species are *A. hydrophila*, *A. veronii* biovar *sobria*, and *A. caviae*.

Genus: Legionella

This is a group of bacteria that were first recognized after an outbreak of pneumonia in legionnaires at a convention in Philadelphia in 1976. The outbreak was traced to water in air-conditioning units. Currently several species are recognized, the most important of which is *L. pneumophila*. Other species affecting humans include *L. micdadei* (after McDade, who first recognized Legionella), *L. bozemanii*, *L. dumoffii*, *L. gormanii*, and *L. longbeachae*. The organism lives in bodies of water and in water systems such as

Box 13.1 Case study

A 17-month-old boy sustained multiple injuries, including skull fractures, during an earthquake. After neurosurgical management at a field hospital, he developed meningitis due to *Acinetobacter baumannii*. He was treated with meropenem, and recovered from this infection.

cooling towers. It can live within or outside amebae in the biofilm in water pipes, and in hot water systems, including in hospitals. It can therefore spread by aerosol. In the host, it lives inside macrophages. Therefore, cell-mediated immunity is necessary for its elimination. Although it causes pneumonia mainly in immunocompromised individuals and smokers, there is a growing recognition of its role in causing community-acquired pneumonia. It can cause an influenza-like illness that lasts for a few days and is self-limiting, called Pontiac fever.

Although it is a Gram-negative rod, it does not stain well with the Gram stain. It does not grow on regular media, but can be cultured on special media (see later). Laboratory diagnosis is based on the following:

- urinary antigen detection for *L. pneumophila* serotype 1, which accounts for most cases of Legionnaires' disease
- culture on buffered charcoal yeast extract agar. This takes up to 10 days (Fig. 13.11)
- direct fluorescent antibody staining on sputum (Fig. 13.12)
- serology (a four-fold rise in acute to convalescent titers, to at least 1:128).

Antibiotic therapy should consist of azithromycin or a fluoroquinolone.

Achromobacter xylosoxidans

Achromobacter xylosoxidans (formerly *Alkaligenes denitrificans* subspp. *xylosoxidans*) is a Gram-negative rod that is oxidase positive, catalase positive, and motile. It causes infections mainly in hospitalized and in immunocompromised patients. It is usually susceptible to antipseudomonal β-lactams (ceftazidime, piperacillin/tazobactam, and ticarcillin/clavulanate), trimethoprim/sulfamethoxazole, and meropenem, and resistant to fluorquinolones and often to aminoglycosides.

Ochrobactrum anthropi

This is a motile, non-lactose-fermenting, oxidase-positive and urease-positive Gram-negative rod, formerly classified within the genus Achromobacter, that is a rare cause of infections on foreign bodies. It is resistant to β-lactam antibiotics, but susceptible to aminoglycosides, fluoroquinolones, and trimethoprim/sulfamethoxazole.

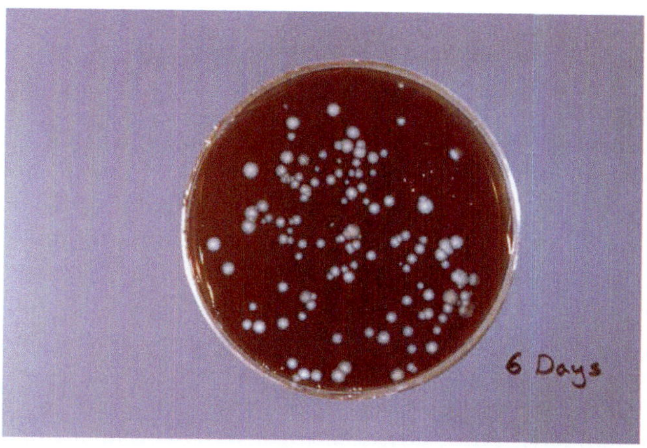

Fig. 13.11 A 6-day-old culture of *Legionella pneumophila* on charcoal-yeast extract agar. Courtesy of PHIL, CDC.

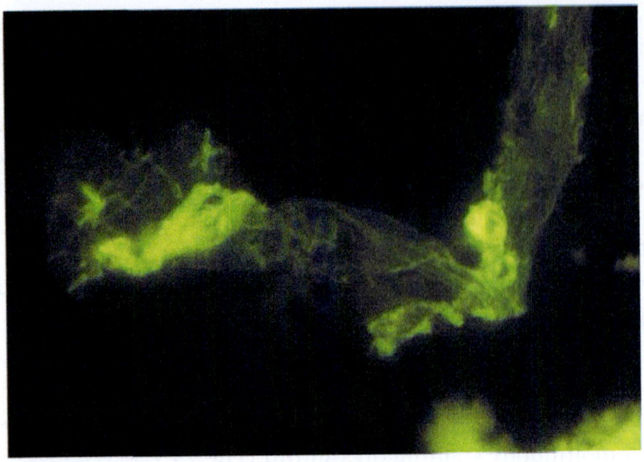

Fig. 13.12 Indirect fluorescent antibody stain showing *Legionella pneumophila* in sputum. Courtesy of PHIL, CDC.

> **Box 13.2** Case study
>
> A 9-year-old boy with common variable immunodeficiency had repeated hospital admissions for massive lymphadenopathy and drainage from cervical lymph nodes. The cultures repeatedly grew out *A. xylosoxidans*. Although he improved with appropriate antimicrobial therapy, the infections relapsed, often being resistant to the therapy that he had previously received.

> **Box 13.3** Case study
>
> A 3-year-old child with chronic granulomatous disease, who enjoyed playing in mud, presented in septic shock. Blood cultures grew out a Gram-negative rod identified as *Chromobacterium violaceum*. The cultures are shown in Figs 13.13 and 13.14.

Chromobacterium violaceum

This is a Gram-negative rod present in the environment, that is a rare cause of infections associated with submersion in water or soil. It causes wound and severe systemic infections. Individuals with chronic granulomatous disease are particularly susceptible to infections caused by this organism. It is recognized in the laboratory by its black color on blood agar. Appropriate antibiotics for treating infected patients are ciprofloxacin, trimethoprim/sulfamethoxazole, chloramphenicol, and imipenem.

Elizabethkingia meningoseptica

This organism (formerly called *Cryseobacterium meningosepticum*, and before that *Flavobacterium meningosepticum*) is an environmental agent that can cause bacteremia and meningitis, especially in newborns. It grows on regular media, producing a yellow pigment (hence "flavo"), and is oxidase positive. Although it is a Gram-negative

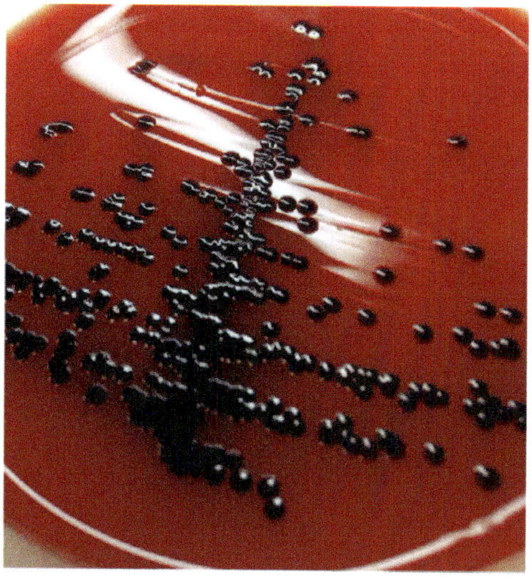

Fig. 13.13 *Chromobacterium violaceum* on blood agar. Courtesy of Dr Jumi Yi.

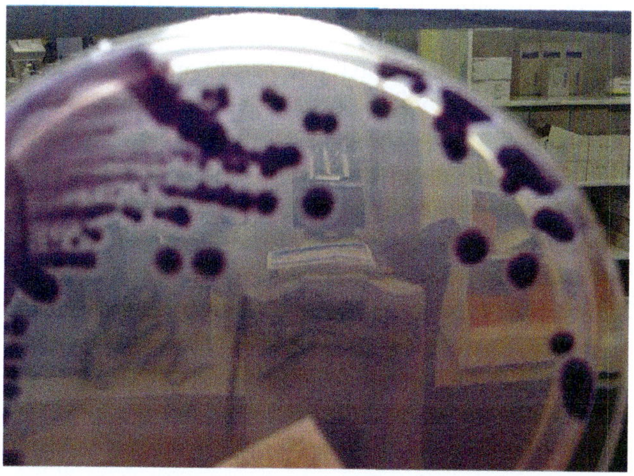

Fig. 13.14 *Chromobacterium violaceum* on MacConkey agar. Courtesy of Dr Ann Chahroudi.

rod, its antibiotic susceptibilities are, to some extent, more like those of Gram-positive organisms. Notably it is usually susceptible to vancomycin, rifampin, and linezolid. Optimum antimicrobial therapy is unclear but the following drugs should be considered: vancomycin plus rifampin, vancomycin plus ciprofloxacin, piperacillin/tazobactam, and linezolid.

Plesiomonas shigelloides

This is a Gram-negative rod present in aqueous environments. It is motile and oxidase positive. It has been implicated in travelers' diarrhea and foodborne outbreaks of diarrhea, both watery and bloody. It is usually susceptible to cephalosporins, aminoglycosides, chloramphenicol, trimethoprim/sulfamethoxazole, doxycycline, ciprofloxacin, and carbapenems. It is resistant to ampicillin, but susceptible to amoxicillin/clavulanate.

Vibrios

These are Gram-negative, comma-shaped organisms that are facultatively anaerobic, oxidase positive, and motile, with polar flagella. The most important is *V. cholerae*, the cause of cholera, an infection characterized by severe diarrhea.

Vibrio cholerae possesses virulence factors that enable it to attach to intestinal mucosa, where it elaborates cholera toxin. This toxin causes an increase in intestinal cell adenyl cyclase and secretion of electrolytes and water, resulting in diarrhea, which can be profound. The infection is associated with a high morbidity and mortality if rehydration therapy is not given. There have been seven recorded pandemics, the most recent starting in 1961. The most recent large outbreak started in Haiti in 2010, in circumstances that facilitated the spread of this organism, namely crowding together with inadequate sanitation. The taxonomy of this organism is as follows.

- Serogroup: according to O-antigen (>200 serogroups).
- O1 and one of its derivative serogroups, O139, are toxigenic, and associated with epidemics.
- O1 has two biotypes, classical and El Tor (the biotype associated with the seventh pandemic).
- Each biotype has three serotypes: Inaba, Ogawa, Hikojima (unstable).

Strains of *V. cholerae* that do not type with O1 or O139 antiserum are called non-O1 non-O139 strains (formerly called non-agglutinating [NAG] strains). These can cause extraintestinal infection as well as diarrhea.

Although this organism grows on regular media, it is detected readily on thiosulfate-citrate-bile salt (TCBS) agar on which it forms yellow colonies (Fig. 13.15).

Treatment consists of hydration therapy and antimicrobial therapy. The antibiotics that are suitable include azithromycin, doxycycline, fluoroquinolones, and trimethoprim/sulfamethoxazole. (The strain causing the 2010 epidemic in Haiti is resistant to trimethoprim/sulfamethoxazole.)

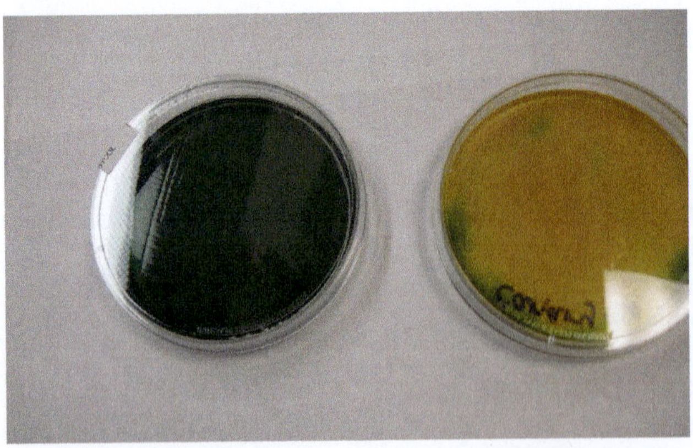

Fig. 13.15 Colonies of *Vibrio cholerae* on thiosulfate-citrate-bile salt (TCBS) agar. The left-hand plate has not been inoculated.

The non-cholera vibrios are halophilic (grow best in high salt concentrations). They are therefore found in seawater, and are associated with shellfish.

Vibrio parahaemolyticus

This causes diarrhea, primarily in areas close to the sea. It is a very important cause of diarrhea in Japan. In addition to hydration therapy, patients should be treated with a fluoroquinolone or doxycycline.

Vibrio vulnificus

This causes three syndromes: acute gastroenteritis, wound infection associated with seawater, and sepsis. Eating uncooked bivalves, e.g. oysters, which filter large volumes of water, retaining bacteria, is a risk factor for acquiring this organism. Sepsis occurs mainly in individuals who are immunocompromised or have diabetes mellitus, liver disease, or iron overload. This infection is rapidly progressive, and is characterized by severe necrotizing skin infection.

It is susceptible to a variety of antibiotics. Patients should be treated with ceftazidime plus doxycycline.

Non-Enterobacteriaceae Gram-negative rods from humans or animals

Gram-negative rods from humans or animals, their sources, the diseases they cause, and their antimicrobial susceptibilities are shown in Table 13.4.

Genus: Haemophilus

This contains several species, including *H. influenzae* and *H. parainfluenzae*. The most important is *H. influenzae*. Some strains have a polysaccharide capsule, which is the basis for serotyping (types a–f), and most strains are non-encapsulated ("non-typeable"). The non-typeable strains are part of the normal pharyngeal flora, and are an important cause of otitis media, sinusitis, and exacerbations of bronchitis in individuals with chronic obstructive airway disease. Of the typeable strains, the most important is type b. This was a common cause of bacteremia and the most common cause of acute bacterial meningitis in young children. It also caused septic arthritis, osteomyelitis, facial cellulitis, and epiglottitis. However, as a result of the widespread use of a vaccine against the polysaccharide capsule, an important virulence factor of the organism, these infections have largely disappeared in areas where the vaccine has been used. The first vaccine, made from the polysaccharide capsule, was not immunogenic in young children, the main group at risk. The current vaccines, which are made of the polysaccharide chemically conjugated to a protein, are effective in young infants, and have been very successful.

Members of the Haemophilus genus have specific growth requirements, which help in the identification of the organisms. *H. influenzae* requires factor V (nicotinamide adenine dinucleotide) and factor X, an iron protoporphyrin. These are present in heated blood agar (chocolate), which is an ideal medium for isolation of this organism. On Gram stain, they appear as small Gram-negative bacilli or coccobacilli (Fig. 13.16).

Table 13.4 Gram-negative rods from humans or animals, their sources, diseases they cause, and their antimicrobial susceptibilities.

Organism	Source	Disease	Antimicrobials
Haemophilus influenzae non-typeable	H, End	Otitis media, sinusitis, pneumonia	amp,* amox/clav, 3rd gen. ceph., fluoroq, azithro
Type b		Bacteremia, meningitis	amp,* 3rd gen. ceph, chloro
Haemophilus ducreyi	H	Genital ulcer	doxy, ceftriaxone
Bordetella pertussis	H	Pertussis	erythromycin, azithro, TMP/S
Brucella	A	Brucellosis	gentamicin, doxy, TMP/S, rifampin
Francisella tularensis	A	Tularemia	gentamicin
Pasteurella multocida	A	Bite infection	amox/clav, 3rd gen. ceph, TMP/S
Bartonella bacilliformis	Fly	Oroya fever	pen
B. henselae	Cat	Cat scratch disease	azithro, TMP/S, fluoroq, doxy
B. quintana	Louse	Trench fever	pen, 3rd gen. ceph, azithro, doxy, gent
Capnocytophaga canimorsus	Dog bite	Wound infection	amox/clav
HACEK	H-End	IE, bites, skeletal	amox/clav, 3rd gen. ceph
Streptobacillus moniliformis	Rat	Bacteremia – fever, rash, arthritis, IE	pen, doxy
Coxiella burnetii	A	Pneumonia, hepatitis, IE	doxy
Campylobacter	A	Diarrhea, sepsis	fluoroq, azithro

3rd gen. ceph, third-generation cephalosporins; A, animal; amox/clav, amoxicillin/clavulanate; amp, ampicillin; azithro, azithromycin; chloro, chloramphenicol; End, endogenous; H, human; fluoroq, fluoroquinolone; gent, gentamicin; IE, infective endocarditis; pen, penicillin; TMP/S, trimethoprim/sulfamethoxazole.

Haemophilus ducreyi

This is a sexually transmitted infection, causing painful genital ulcers associated with inguinal lymphadenopathy (Fig. 13.17). The diagnosis is mostly based on clinical features and exclusion of other conditions causing a similar clinical picture; culture requires a special medium that is not generally available, and PCR is not generally available. Treatment consists of a macrolide, ceftriaxone, or ciprofloxacin.

Haemophilus aphrophilus and *H. paraphrophilus*

These have been renamed, as a single species, *Aggregatibacter aphrophilus* (see "HACEK" organisms, later in this chapter).

Genus Bordetella

This consists of *B. pertussis, B. parapertussis, B. holmesii*, and *B. bronchiseptica*. *B. pertussis*, and occasionally *B. parapertussis*, cause pertussis or whooping cough, a very severe bronchial infection, affecting mostly young infants. The organism is spread from human to human by droplets, generated by coughing, sneezing or speaking. It causes severe damage to the respiratory epithelium. Globally, and within the USA, pertussis is an important cause of morbidity and mortality in infants, due to complicating pneumonia and apnea.

A nasopharyngeal swab is the specimen of choice for demonstrating the organism. It requires special media for its growth and to select it out from other pharyngeal flora.

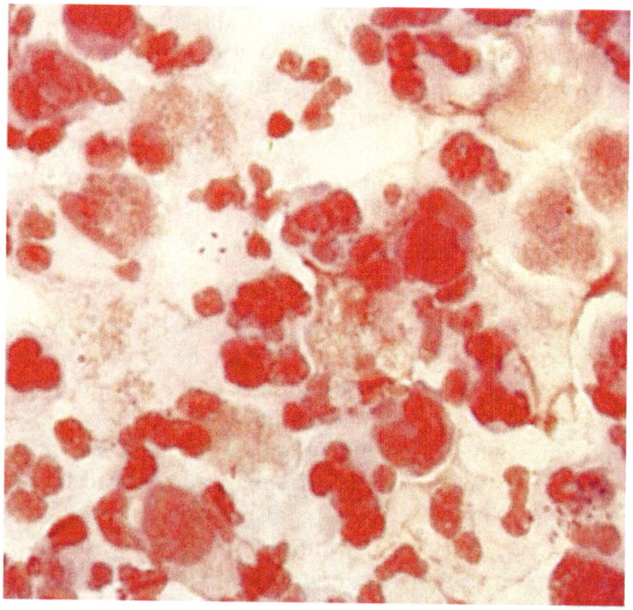

Fig. 13.16 *Haemophilus influenzae* in cerebrospinal fluid (note small Gram-negative coccobacilli).

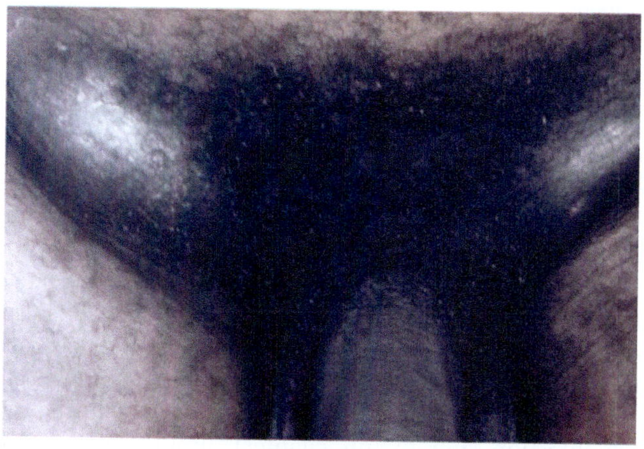

Fig. 13.17 Inguinal lymphadenopathy due to *Haemophilus ducreyi*. Courtesy of PHIL, CDC.

Such media include Bordet–Gengou agar. Because growth takes several days, more rapid diagnostic tests are necessary. Other tests include a direct fluorescent antibody test and PCR. The latter is the test of choice for diagnosing this infection.

Although treatment is primarily supportive, antimicrobial therapy reduces infectivity. Therefore, the child should be treated with a macrolide (erythromycin or azithromycin), as should family members, who might be infected with the organism. This infection can be largely prevented by vaccination. The current vaccines in use (which are given in combination with diphtheria and tetanus vaccines) are a killed whole cell vaccine (no longer used in the USA) and acellular vaccines, made from a combination of several antigens, including pertussis toxin, pertactin, and fimbrial agglutinin. The latter, which is currently used in the USA, does not seem to provide as long a duration of immunity as did the whole cell vaccine. Older children and adults, in whom immunity has waned, are an important source of infection in infants.

Genus: Brucella

This genus, which causes brucellosis, consists of several species, the most important being *B. melitensis* (named for the island of Malta), *B. abortis*, *B. suis*, and *B. canis*. These organisms are pathogens of various animals, and affect humans through their contact with infected animals or their excretions or secretions. They are present in several animal tissues which have high concentrations of erythritol, a preferred carbohydrate substrate for their growth. Although widespread in many countries, Brucella is now rare in the USA. The organism lives in macrophages, and cell-mediated immunity is necessary for its elimination. It causes an acute, subacute or chronic infection, which is initially bacteremic but becomes focal, such as infective endocarditis and osteomyelitis.

Although it grows on regular media, growth is slow, so that agar plates require incubation for 2–3 weeks. It poses a hazard to laboratory staff if aerosolized. Serologic tests are useful for diagnosis, and molecular tests, which are sensitive and specific, have been developed.

Drugs used for treatment are streptomycin, doxycycline, trimethoprim/sulfamethoxazole, and rifampin.

Genus: Francisella

Francisella tularensis (which includes subspecies *F. t. holartica* and *F. t. novicida*) is the cause of tularemia, a localized or systemic infection transmitted from an infected animal by direct inoculation through the skin (e.g. through abrasions incurred while skinning an infected animal), by ingestion, by inhalation, or by the bite of an infected arthropod, such as a tick or fly. It occurs across the northern hemisphere, and its distribution in the USA is shown in Fig. 13.18. The organism has a polysaccharide capsule, and can live within macrophages. There are six main clinical forms of the infection, namely ulceroglandular (the most common), oculoglandular (see also *Bartonella henselae* infection later in this chapter), oropharyngeal, gastrointestinal, pneumonic, and typhoidal. Systemic spread can complicate all of these forms. The organism requires cysteine and cystine for its growth, so it does not grow on blood agar (unless supplemented), but it does grow on chocolate agar. It poses a hazard to laboratory staff if aerosolized. It can be cultured from blood, lymph node aspirates, or other tissue. Gram stain of specimens has low sensitivity, and a direct fluorescent antibody test or PCR can be useful, if available. Serologic tests can also be used, but might be negative early in the infection. Antimicrobial therapy consists of gentamicin, but ciprofloxacin and doxycyline are alternatives.

This organism is considered a potential bioterrorism threat. The laboratory personnel must be informed when a specimen suspecting of containing this organism is submitted.

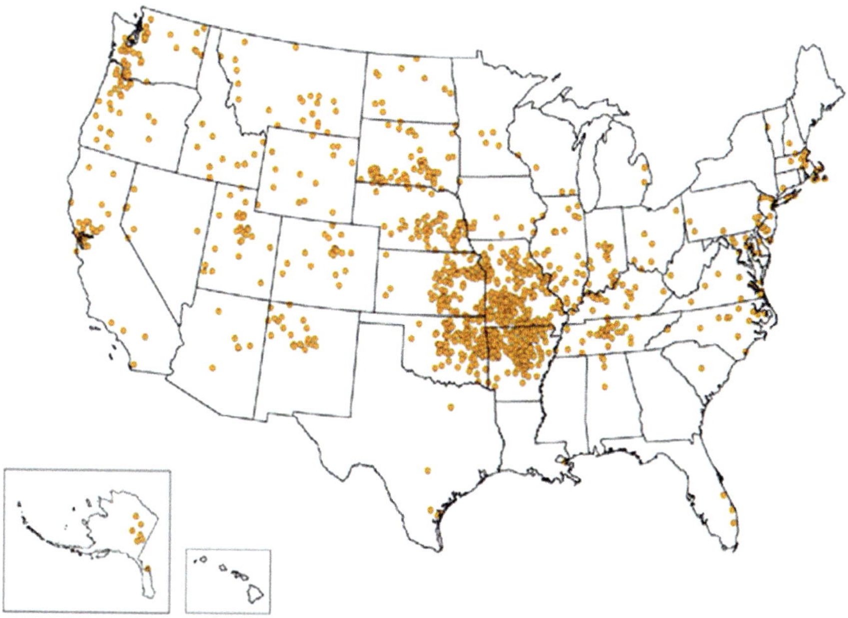

1 dot placed randomly within country of residence for each reported case

Fig. 13.18 Distribution of reported cases of tularemia within the USA from 2001 to 2010. Courtesy of CDC.

Genus: Pasteurella

Pasteurella multocida is normal flora of some animal mouths, especially members of the cat family, and is therefore an important cause of infections complicating bites. It grows on routine media, and is susceptible to penicillin, ampicillin, second- and third-generation cephalosporins, tetracyclines, and fluoroquinolones.

Genus: Bartonella

The first recognized species within this genus was *B. bacilliformis*, which causes Oroya fever. This is a fly-transmitted infection, occurring within specific altitudes in the Andes, in Peru and Ecuador. The organism infects erythrocytes. The early stage of the infection is characterized by hemolytic anemia, which is followed by the development of wart-like skin lesions called "verruca peruana." It can occur in outbreaks, for example in road construction work crews.

The diagnosis is based on visualization of the organism within erythrocytes on a Giemsa-stained blood smear.

A more common infection is cat scratch disease, caused by *B. henselae*. This organism infects cats, in particular kittens, without causing illness in them. It is inoculated into the human skin by a cat scratch, where it may cause a local infection. This may spread to the local lymph node, causing a lymphadenitis (Fig. 13.19).

Occasionally it spreads systemically, causing infection in the viscera, bone, and eye. It can also cause an encephalopathy.

This organism also causes a hemangioma-like lesion, called bacillary angiomatosis, and peliosis hepatis in immunocompromised patients such as those with AIDS. This, and the verruca peruana, suggest that members of this genus produce angiogenic factors.

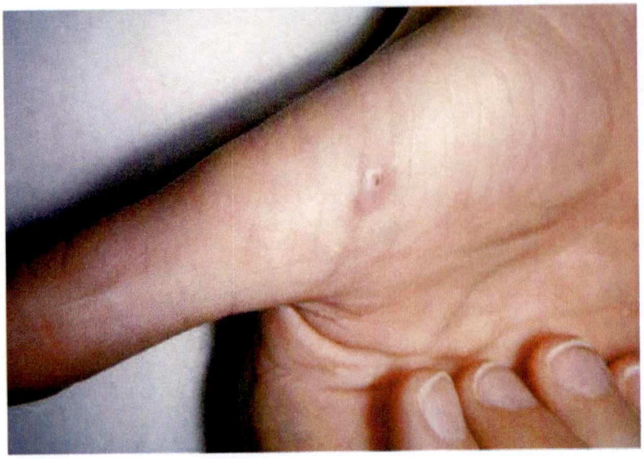

Fig. 13.19 Inoculation site in a case of cat-scratch disease (Courtesy of PHIL/CDC).

Although *B. henselae* can be recovered in culture, this is rarely accomplished. It can be detected by PCR performed on tissue or pus but diagnosis of the disease is usually made clinically, and is sometimes confirmed serologically. Although it is susceptible to several different antibiotics, including macrolides, gentamicin, trimethoprim/sulfamethoxazole, fluoroquinolones, and tetracycline, no adequate controlled trial has demonstrated benefit of a specific agent in patients with cat scratch disease. Macrolide treatment of patients with bacillary angiomatosis is beneficial.

Bartonella quintana causes trench fever, a louse-borne infection, occurring mainly in homeless individuals. The illness is characterized by fever, headache, myalgias, shin pain, conjunctival injection, splenomegaly and rash, that relapses in 5-day intervals (quintan fever). It can cause prolonged bacteremia, infective endocarditis, and bacillary angiomatosis. Diagnosis can be made by blood culture, serology, and PCR. It is susceptible to many antibiotics, including penicillin, third-generation cephalosporins, macrolides, tetracycline, and gentamicin.

Genus: Capnocytophaga

This is a group of organisms that are part of the normal oral flora of animals and humans. Their name is derived from their growth requirement for carbon dioxide. *Capnocytophaga canimorsus* is an important cause of infection complicating dog bites, which is especially severe in immunocompromised and asplenic individuals. Other species include *C. ochracea*, *C. sputigena*, *C. gingivalis*, and *C. cynodegmi*. Although many strains are susceptible to β-lactam antibiotics, the acquisition of β-lactamases has rendered them resistant to many of these drugs. They are susceptible to β-lactam/inhibitor combinations, carbapenems, clindamycin, and tetracycline.

Genus: Coxiella

Coxiella burnetii, a small Gram-negative coccobacillus, causes Q fever, an infection that derives its name from "query" – the search for an answer. The organism was formerly considered to be related to rickettsiae, but is more closely related to *Legionella pneumophila*. It exists in two variant forms: a small cell variant, which is inert and can remain viable in the environment for a long time, and a large cell variant, which infects monocytes and macrophages. It can live and multiply within these cells. It is present in the excretions and secretions of cattle, sheep, and goats, and contaminates the dust. It is transmitted among animals by ticks, and can be transmitted to humans in this manner. However, humans usually acquire the infection by inhalation of contaminated dust. The infection causes fever, headache, myalgia, pneumonia, and subclinical hepatitis, but all organs can be affected. It can cause infective endocarditis, which is chronic and blood culture negative. Many cases of infection are asymptomatic.

The diagnosis is made serologically, best by an indirect immunofluorescent test. For acute cases, antibodies to phase 2 should be tested while in chronic cases (endocarditis), antibodies to phase 1 should be tested.

The treatment for acute Q fever is doxycycline. For cases of endocarditis, treatment should consist of a combination of doxycyline and hydroxychloroquine. The hydroxychloroquine raises the pH in the phagolysosome, which increases the activity of the doxycycline.

"HACEK" organisms

This is a group of bacteria, namely *Haemophilus aphrophilus* (now called *Aggregatibacter aphrophilus*), *Actinobacillus actinomycetemcomitans* (now called *Aggregatibacter actinomycetemcomitans*), *Cardiobacterium hominis*, *Eikenella corrodens*, and *Kingella kingae*, that are grouped in this fashion because they cause "culture-negative" infective endocarditis. They are all part of normal oropharyngeal flora but they belong to different groups from a genetic relatedness viewpoint. Kingella belongs to the family Neisseriaceae (together with *Neisseria meningitidis* and *N. gonorrhoeae*). Although they may be susceptible to penicillin, their acquisition of β-lactamase has rendered them often resistant to this drug. They are susceptible to β-lactam/β-lactamase inhibitor combinations and to third-generation cephalosporins. *Kingella kingae* is an important cause of bone and joint infections in young children.

Genus: Campylobacter

There are many species, the most important of which are *C. jejuni*, *C. coli*, *C. upsaliensis*, and *C. concisus*, which are very important causes of acute intestinal infection in humans, and *C. fetus*, which causes bacteremia and other systemic infections in neonates and immunocompromised hosts. They are harbored in the intestines of poultry and other animals. Intestinal infection affects the small and large bowel causing fever, watery diarrhea, and sometimes dysentery. An important postinfectious complication is Guillain–Barré syndrome. Detection of these organisms in the laboratory requires special media, and incubation at 42°C. Antibiotic therapy of patients with intestinal infection accelerates recovery. The drugs most frequently used for treatment of such patients are azithromycin and ciprofloxacin, but resistance to ciprofloxacin is becoming more common. *C fetus* is susceptible to ampicillin, aminoglycosides, and carbapenems.

Genus: Helicobacter

Helicobacter pylori is an important cause of peptic and gastric ulcers, and plays a role in the pathogenesis of gastric carcinoma and gastric mucosa-associated lymphoid tissue (MALT) lymphoma. It is highly prevalent in poor communities, without necessarily causing illness. It produces urease, which enables it to survive in the acidic gastric mucosa. This property is used diagnostically. Gastric or duodenal biopsies are placed in media containing urea and a pH indicator. The conversion of urea to ammonia, and the consequent elevation of the pH and color change of the indicator, indicate the presence of the organism. Another diagnostic test utilizes this in the patient. The patient drinks a liquid containing urea with radio-labeled carbon. The radio-labeled ammonia is exhaled and quantitated. The presence of the organism can be detected by a fecal antigen test. Serologic tests can be used to show that the patient has been infected by the organism, but they are not useful in diagnosing disease.

The organism is susceptible to several antimicrobial agents. Standard therapy of affected patients consists of a combination of a proton pump inhibitor and two antimicrobial agents, such as amoxicillin and clarithromycin. However, the emergence of resistance has led to the use of other therapeutic regimens.

Genus: Streptobacillus

Streptobacillus moniliformis is a slender Gram-negative rod that is part of the normal flora of rats. Genetically, it is related to the genus Fusobacterium. It is transmitted by a bite or scratch, and can also be transmitted by contamination of food. Transmission from animals other than rats has occurred, including from mice, gerbils, and squirrels. It causes one of the forms of rat-bite fever, which is characterized by fever, rash, and arthritis. It can cause infective endocarditis and other severe visceral infections. It can also cause foodborne infection and outbreaks (Haverhill fever). It grows on artificial media supplemented with serum or ascitic fluid. In liquid media, it forms characteristic "puff-balls." It is inhibited by polyanethol sulfonate, which is present in many blood culture systems. Therefore, it is difficult to detect in regular culture, and when suspected, one should consult with the laboratory staff. It can be grown from cutaneous pustules that are part of the clinical illness (see Box 13.4). It is susceptible to penicillin and tetracycline.

Box 13.4 Case study

An 8-year-old girl presented with a 4-day history of fever, rash, and joint pain (Figs 13.20 & 13.21). She had two pet rats, which she kissed. Fluid from a pustule was cultured on C and S blood agar. After 48 hours, tiny dewdrop-like colonies appeared. The Gram stain from these colonies demonstrated Gram-negative rods in chains (Fig. 13.22). She responded well to doxycyline therapy, but she relapsed, and she was therefore treated successfully with penicillin.

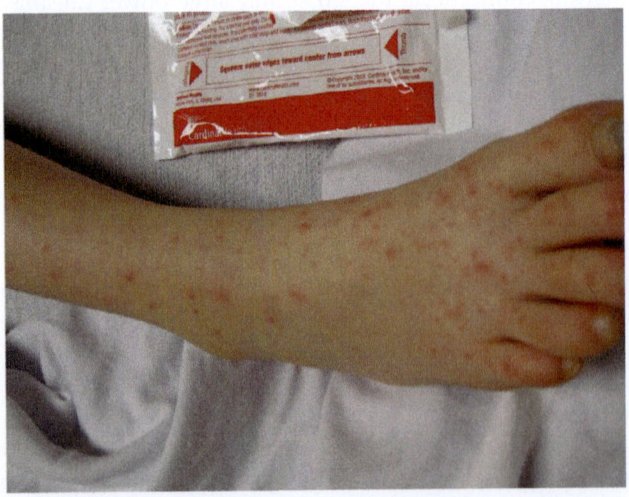

Fig. 13.20 The rash of the patient with rat-bite fever in Fig. 13.19.

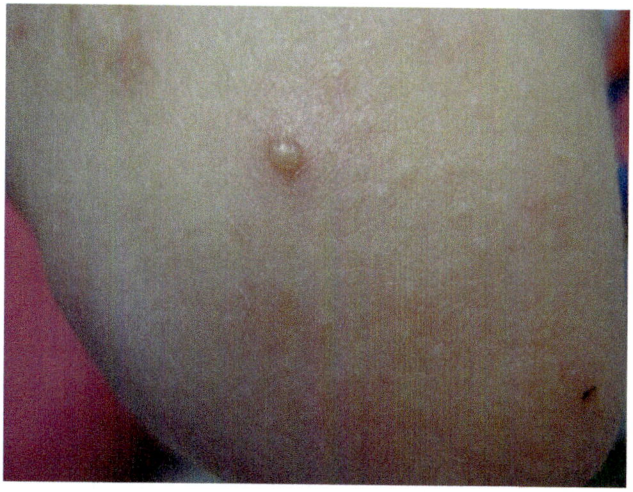

Fig. 13.21 A pustule on the patient with rat-bite fever in Fig. 13.19.

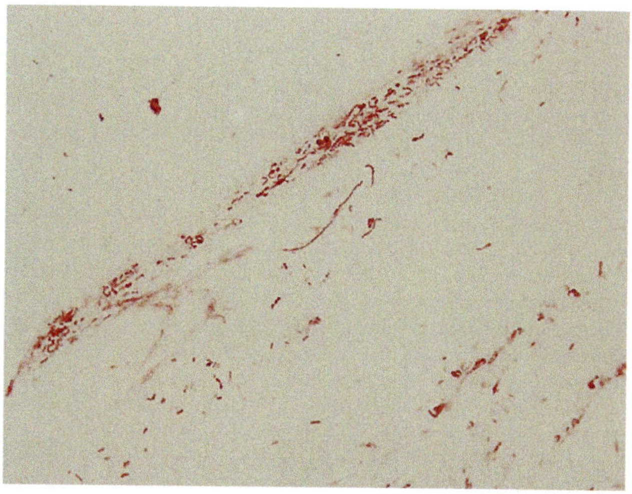

Fig. 13.22 Gram stain of colonies from culture of material from the pustule in Fig. 13.21, showing chains of Gram-negative rods, which were identified as *Streptobacillus moniliformis*.

Further reading

Adalja AA, Toner S, Inglesby TV (2015) Clinical management of potential bioterrorism-related conditions. N Engl J Med 372: 954–62.

Aisenberg G, Rolston K, Safdar A (2004) Bacteremia caused by *Achromobacter* and *Alcaligenes* species in 46 patients with cancer. Cancer 101: 2134–40.

Al Dahouk S, Nockler K (2011) Implications of laboratory diagnosis on brucellosis therapy. Expert Rev Anti Infect Ther 9: 833–45.

Briceno DF, Quinn JP, Villegas MV (2010) Treatment options for multidrug-resistant nonfermenters. Expert Rev Anti Infect Ther 8: 303–15.

Carratala J, Garcia-Vidal C (2010) An update on Legionella. Curr Opin Infect Dis 23: 152–7.

Cherry JD (2012) Epidemic pertussis in 2012 – the resurgence of a vaccine-preventable disease. N Engl J Med 367: 785–7.

Croxen MA, Finlay BB (2010) Molecular mechanisms of *Escherichia coli* pathogenicity. Nat Rev Microbiol 8: 26–38.

Di Pentima MC, Mason EO, Kaplan SL (1998) In vitro antibiotic synergy against *Flavobacterium meningosepticum*: implications for therapeutic options. Clin Infect Dis 26: 1169–76.

Feasey NA, Dougan G, Kingsley RA, et al. (2012) Invasive non-typhoidal salmonella disease: an emerging and neglected tropical disease in Africa. Lancet 379: 2489–99.

Foley JE, Nieto NC (2010) Tularemia. Vet Microbiol 140: 332–8.

Gaastra W, Boot R, Ho HTK, et al. (2009) Rat bite fever. Vet Microbiol 133: 211–28.

Gilligan PH (2014) Infections in patients with cystic fibrosis. Diagnostic microbiology update. Clin Lab Med 34: 197–217.

Jolivet-Gougeon, A, Sixou J-L, Tamanai-Shacoori Z, et al. (2007) Antimicrobial treatment of *Capnocytophaga* infections. Int J Antimicrob Agents 29: 367–73.

Kosek M, Penataro P, Yori P, et al. (2010) Shigellosis update: advancing antibiotic resistance, investment empowered vaccine development, and green bananas. Curr Opin Infect Dis 23: 475–80.

Lan R, Reeves PR, Octavia S (2009) Population structure, origins and evolution of major *Salmonella enterica* clones. Infect Genet Evol 9: 996–1005.

Looney WJ, Narita M, Mühlemann K (2009) *Stenotrophomonas maltophilia*: an emerging opportunist human pathogen. Lancet Infect Dis 9: 312–23.

McColl KEL (2010) Helicobacter pylori infection. N Engl J Med 362: 1597–604.

Ming Man S (2011) The clinical importance of emerging Campylobacter species. Nature Rev Gastroenterol Hepatol 8: 669–85.

Munoz-Price LS, Weinstein RA (2008) Acinetobacter infection. N Engl J Med 358: 1271–81.

Paddock CD, Sanden GN, Cherry JD, et al. (2008) Pathology and pathogenesis of fatal *Bordetella pertussis* infection in infants. Clin Infect Dis 47: 328–38.

Parker JL, Shaw JG (2006) *Aeromonas* spp. clinical microbiology and disease. J Infect 62: 109–18.

Parker N, Barralet JH, Bell AM (2006) Q fever. Lancet 367: 679–88.

Peacock SJ (2006) Meliodosis. Curr Opin Infect Dis 19: 421–8.

Pitout JDD (2008) Multiresistant Enterobacteriaceae: new threat of an old problem. Expert Rev Anti Infect Ther 6: 657–69.

Prentice MB, Rahalison L (2007) Plague. Lancet 369: 1196–207.

Rolain JM, Brouqui P, Koehler JE, et al. (2004) Recommendations for treatment of human infections caused by *Bartonella* species. Antimicrob Agents Chemother 48: 1921–33.

Romero Gomez MP, Sobrino Daza JA, Saez Nieto JA, et al. (2004) Prosthetic mitral valve endocarditis due to *Ochromobactrum anthropi*: case report. J Clin Microbiol 42: 3371–3.

Sack D, Sack RB, Nair GB, et al. (2004) Cholera. Lancet 363: 223–33.

Sader HS, Jones RN (2005) Antimicrobial susceptibility of uncommonly isolated non-enteric Gram-negative bacilli. Int J Antimicrob Agents 25: 95–109.

Sanchez-Vargas FM, Abu-El-Haija MA, Gomez-Duarte OG (2011) *Salmonella* infections: an update on epidemiology, management, and prevention. Travel Med Infect Dis 9: 263–77.

Sirinavin S, Techasaensiri C, Benjaponpitak S, et al. (2005) Invasive *Chromobacterium violaceum* infection in children: case report and review. Pediatr Infect Dis J 24: 559–61.

Wain J, Hendriksen RS, Mikoleit ML, et al. (2015) Typhoid fever. Lancet 385: 1136–45.

Weitkamp J-H, Tang Y, Haas D, et al. (2000) Recurrent *Achromobacter xylosoxidans* bacteremia associated with persistent lymph node infection in a patient with hyper-immunoglobulin M syndrome. Clin Infect Dis 31: 1183–7.

CHAPTER 14

Anaerobic bacteria

General properties of anaerobes

Many bacteria grow anaerobically, including members of the Enterobacteriaceae and some streptococci. However, the term "anaerobic bacteria" refers to those bacteria that do not grow aerobically (i.e. in the presence of oxygen). There are various degrees of aerotolerance among anaerobes, i.e. some are more sensitive to the presence of oxygen than others. Of importance for their growth is a low redox potential. As with other bacteria, they exhibit a wide variety of mechanisms of virulence, including exotoxins, enzymes, and polysaccharide capsules.

Many of these bacteria constitute the bulk of the normal mucosal flora, and are present in quantities of up to 10^{12}/mL of fluid. Therefore, their presence in an infection usually suggests that the infection arose from a mucosal surface. For the same reason, infections involving anaerobes are often polymicrobial. They are important pathogens in chronic sinusitis and its intracranial complications such as epidural empyema and brain abscess, peritonsillar abscess, jugular septic thrombophlebitis (Lemierre syndrome), aspiration pneumonia, lung abscess and pleural empyema, intraperitoneal infections, and infections of the female genital tract.

The presence of anaerobes should be suspected in the following circumstances:

- the infection site, or material, e.g. pus, from the infection, is malodorous
- organisms are seen on Gram stain, but do not grow in regular culture.

Specific identification of anaerobes is often difficult and usually not of clinical importance.

The reasons for knowing that anaerobes are present in a particular infection are that:

- the infection likely arose from a mucosal surface
- the infection might be polymicrobial (Fig. 14.1)
- antimicrobial therapy active against anaerobes should be used.

Successful isolation of anaerobic bacteria from clinical specimens depends on transportation of the specimen in an anaerobic medium, and use of anaerobic conditions

Practical Medical Microbiology for Clinicians, First Edition. Frank E. Berkowitz and Robert C. Jerris.

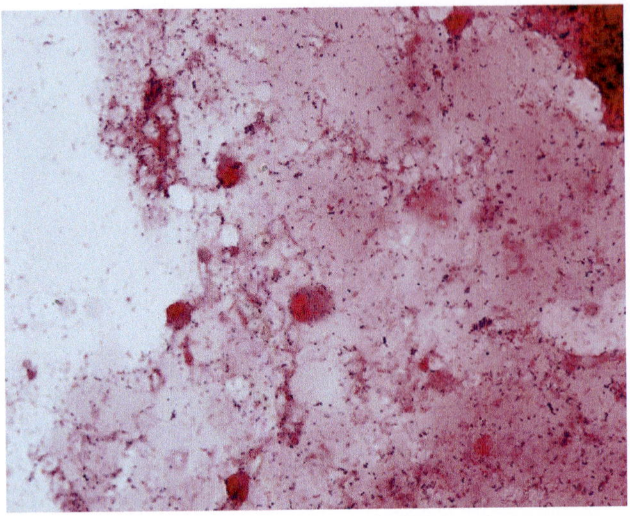

Fig. 14.1 Gram stain of peritoneal fluid from a child with peritonitis complicating a perforated appendix. Note the large Gram-negative rods, small Gram-negative coccobacilli, and Gram-positive cocci in pairs.

for culture. There are several systems available for generating such conditions in the laboratory. Once there is growth, identification depends on biochemical tests and, in some laboratories, on gas-liquid chromatography of fatty acids and MALDI-TOF MS.

The anaerobes will be considered here in the following groups: sporulating Gram-positive rods; non-sporulating Gram-positive rods; Gram-positive cocci; Gram-negative rods; and Gram-negative cocci.

Sporulating gram-positive rods

These constitute the genus Clostridium, which contains many species. They are present in the intestine of humans and animals, and some are widespread in environment. Some cause specific clinical syndromes, and others cause a wide variety of pyogenic infections.

Clostridial syndromes

Tetanus

This is caused by *C. tetani,* which elaborates a very potent exotoxin, tetanus toxin (also called tetanospasmin). This is a well-characterized toxin consisting of a heavy and a light polypeptide chain linked by a disulfide bridge. The organism enters the body via a wound (or the umbilical stump in a neonate) where it elaborates the toxin. The toxin ascends along axons to reach the central nervous system, where it inhibits inhibitory neurons. This results in muscle excitation, causing muscle spasms. Although it usually affects the whole body, it can be localized. These spasms prevent normal respiration, and cause facial spasms (risus sardonicus) and arching of the back (Figs 14.2 & 14.3).

The diagnosis is made clinically. Although the organism can be cultured in the laboratory, it is extremely sensitive to oxygen, so specimens must be collected anaerobically. It has characteristic terminal spores (Fig. 14.4).

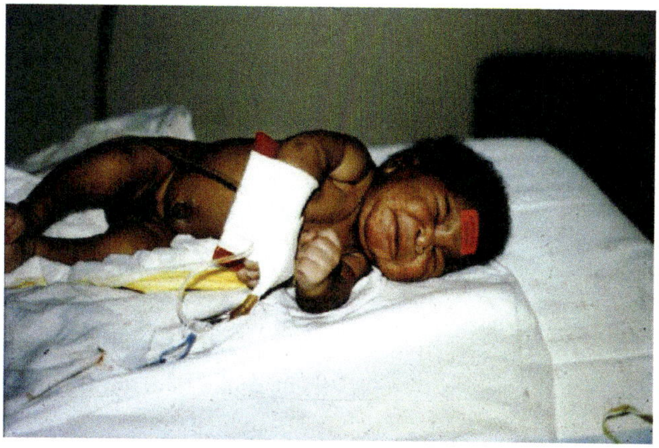

Fig. 14.2 A 5-day-old infant with tetanus, showing clenching of the fist and grimacing. Copyright © 2007 Frank E. Berkowitz. Reprinted with the permission of Cambridge University Press.

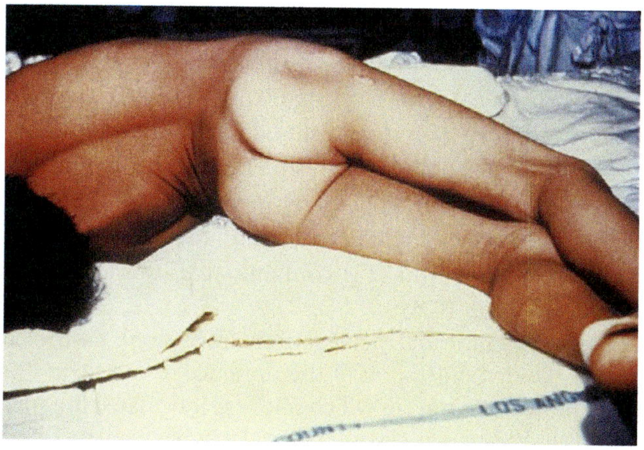

Fig. 14.3 Opisthotonus due to tetanus. Courtesy of PHIL, CDC.

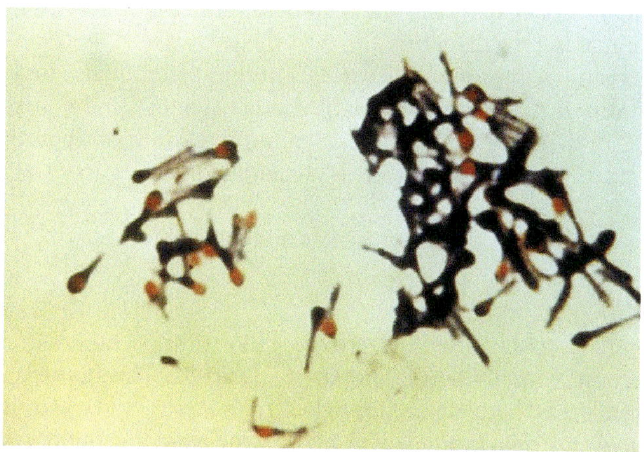

Fig. 14.4 Micrograph showing *Clostridium tetani* with terminal spores. Courtesy of PHIL, CDC.

The treatment is mainly supportive. This often requires ventilatory support for several weeks. In addition, penicillin and tetanus antitoxin (tetanus immune globulin) should be administered.

Tetanus can be prevented by active immunization with tetanus toxoid, which is made by treating the toxin with formaldehyde. Neonatal tetanus is a major health problem in some developing countries, and accounts for most cases of tetanus worldwide. It can be prevented by good midwifery practices, and immunization of the mother, whose antibodies (IgG) will cross into the fetus.

Botulism

This is caused by *C. botulinum*, which produces a toxin, botulinum toxin, which is structurally very similar to tetanus toxin. However, its mode of action, namely inhibition of release of acetylcholine from presynaptic nerve terminals, results in floppiness and paralysis, which is generalized. In classic botulism (see later in this chapter), the paralysis is typically descending and includes ptosis. There are seven different antigenic types of botulinum neurotoxin, A through G. Only A, B, E, and F affect humans. In addition to *C. botulinum*, three other species of clostridium can produce botulinum toxin: *C. butyricum* (type E toxin), *C. baratii* (type F toxin), and *C. argentinense* (type G toxin).

There are three different clinical types of botulism:

- classic botulism, in which the toxin is ingested. This occurs when food, contaminated with spores, is improperly prepared, and stored under anaerobic conditions during which the spores germinate and the organism elaborates the toxin
- infant botulism, in which an infant ingests food, such as honey, containing the organism, which elaborates the toxin inside the infant (Fig. 14.5)
- wound botulism, in which a wound is contaminated by spores, which germinate and elaborate the toxin (Fig. 14.6).

The diagnosis depends on culturing the organism from stool, gastric contents or the suspected food, or by the demonstration of the presence of the toxin in food, blood or stool. This is done by mouse bioassay, in which mice are inoculated intraperitoneally with test material. Some mice are inoculated with material that has been mixed with botulinum antitoxin, while others are inoculated with untreated material. If the mice injected with the treated specimen survive but those injected with untreated specimen develop botulism, the initial suspect material contained botulinum toxin. Electromyography may also be helpful in making the diagnosis.

Treatment is mainly supportive. However, antitoxin should also be used when the diagnosis is suspected. An equine antitoxin against types A and B toxin is available through the CDC in the USA and, for infants, a human antitoxin (botulinum immune globulin), also against types A and B, is available through the California Health Department.

Botulism is a disease that could be used as a biological weapon.

Gas gangrene (myonecrosis)

This is a dramatic infection caused by histotoxic clostridia. It occurs in areas of tissue damage that become contaminated with these clostridia. The buttock and thigh are commonly affected areas, because they are close to the anus, and clostridia may be present in feces. These clostridia produce various enzymes, the most important of which is α-toxin, a lecithinase, which lyzes cell membranes. This results in lysis of muscle tissue

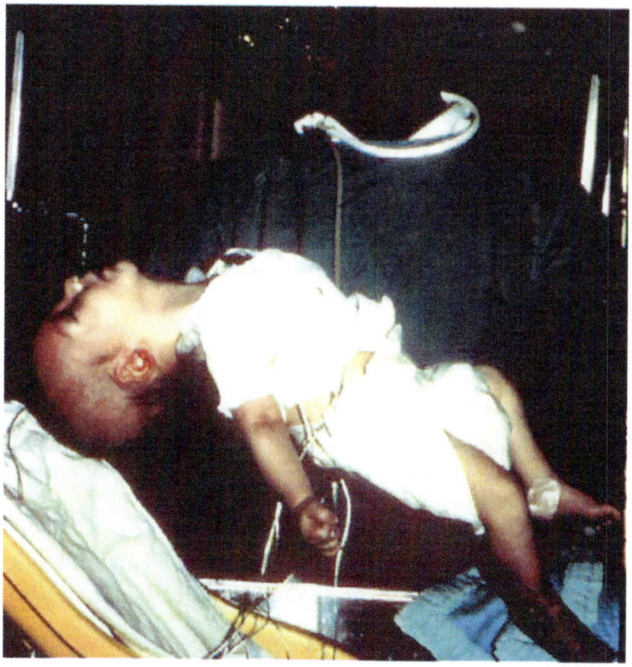

Fig. 14.5 An infant with marked hypotonia, due to botulism. Courtesy of PHIL, CDC.

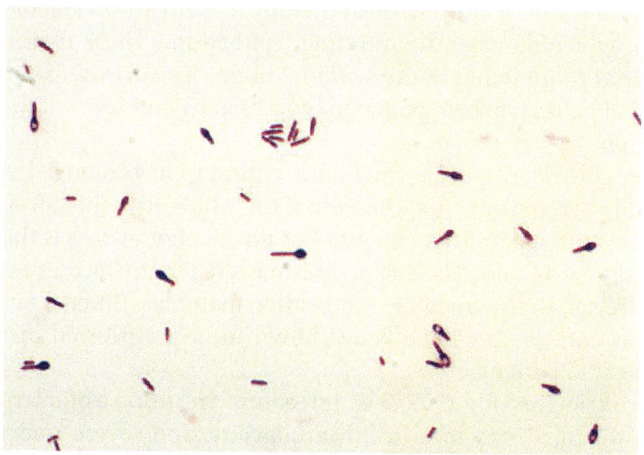

Fig. 14.6 Gram stain of *Clostridium botulinum*. Note the spores. Courtesy of PHIL, CDC.

and red cells. The infection is characterized by the rapid progression of swelling and dark discoloration of the affected area, with the appearance of purplish-brown bullae. Crepitations over the area may be detected. The bullae contain bacteria but there is a paucity of leukocytes, because they are lyzed (Fig. 14.7).

The most common clostridial species causing this disease is *C. perfringens*, but other causative histotoxic species include *C. septicum*, *C. histolyticum*, *C. sordellii*, and *C. novyi*. These organisms can be grown in the laboratory under anaerobic conditions. They can also be tested for the production of lecithinase.

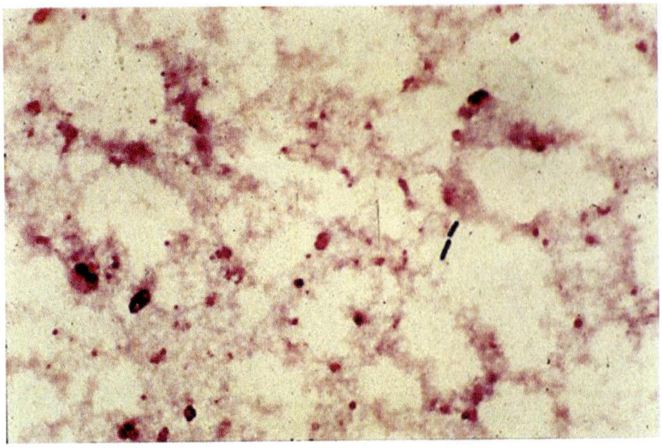

Fig. 14.7 Gram-positive rod in an aspirate from a bulla in a child with gas gangrene. This was identified as *Clostridium septicum*. Copyright © 2007 Frank E. Berkowitz. Reprinted with the permission of Cambridge University Press.

Treatment consists of aggressive surgical debridement and an antibiotic active against clostridium. Penicillin, metronidazole, clindamycin, and carbapenems are suitable. The role of hyperbaric oxygen therapy is controversial.

Pseudomembranous (antibiotic-associated) colitis

This is caused by *C. difficile*. The disease results from an overgrowth of this organism in the large bowel as a result of antimicrobial therapy. The organism elaborates two cytotoxins, A and B, which damage the intestinal epithelium. These toxins are unstable and may degrade at room temperature within 2 hours. The disease is characterized by crampy abdominal pain, diarrhea, which may be bloody, and fever. It may progress to colonic perforation.

As with other clostridia, spore formation is critical to its prolonged survival in the environment and is responsible for the ease with which it is spread. Key to control efforts is the spores' susceptibility to bleach, but not alcohol, which is the antibacterial agent found in common hand gels. The organism is shed in the feces, so surfaces (toilet seats), devices (rectal thermometers), and other materials (linens) may serve as a reservoir for infection. Spores have been shown to be transferred on the hands of healthcare workers to patients.

The classic risk factors for *C. difficile* infections are prior antimicrobial therapy, advanced age, prolonged stay in a healthcare facility, and severe underlying illness. Additional noted risk factors include inflammatory bowel disease, gastrointestinal surgery, gastric acid suppression, and immunosuppression.

The severity of this infection appears to be increasing with increased morbidity and mortality. There now appears to be an increased prevalence in "low-risk" populations that have no exposure to antimicrobials, no hospitalization, nor any underlying disease. In addition, new hypervirulent strains have emerged throughout the United States, Canada, and Europe. Such a strain, known as NAP1 (by pulse field gel electrophoresis), BI (by restriction endonuclease analysis), or 027 (by PCR ribotyping), produces up to 16-fold more toxin A, up to 23-fold more toxin B, than other strains, and a binary toxin of uncertain function. These strains demonstrate enhanced sporulation and resistance to fluoroquinolones.

Diagnostic tests include assays that culture the organism from the stool and those that test for toxin in the stool. Toxigenic culture, which is considered to be the gold standard, entails culturing the stool for the organism and then testing the cultured organism for toxin production. This is done by inoculating two tissue culture monolayers of an appropriate cell line with the isolated organism, and protecting one with a specific neutralizing antitoxin, in order to detect a specific effect of the toxin (Fig. 14.8). In addition, enzyme immunoassays that target the toxin plus a common antigen among all *C. difficile* strains (glutamate dehydrogenase[GDH]), used together, provide good sensitivity (90–95%) and specificity (85–90%) for diagnosis. Enzyme immunoassays to detect toxin only should not be used.

The most sensitive (90–95%) and specific (95–96%) assays are nucleic acid amplification tests (NAATs). Conventional PCR and isothermal assays (e.g. loop-amplified isothermal amplification [LAMP] assays) are available. While these assays are slightly more expensive, they offer the advantage of rapid turn-around, with some having the ability to detect the NAP1 strain. This is important for epidemiologic reasons, and because this strain is resistant to fidaxomycin, one of the drugs used in treatment.

Colonoscopy demonstrates inflammation of the colon with pseudomembranes.

Treatment consists of discontinuing the incriminated antimicrobial agent, if possible, and the administration of specific antimicrobial therapy, namely metronidazole, or vancomycin orally. (This is the only indication for oral vancomycin therapy.) Fecal transplant has recently been used for recalcitrant cases.

Box 14.1 Case study

A 2-year-old boy with multiple medical problems, for which he had received antibiotics, developed severe diarrhea and fever. On examination, he had abdominal distension and tenderness. Because of persistent diarrhea and fever, and features of sepsis, he underwent CT scanning, which demonstrated thickened large bowel loops (Fig. 14.9). *C. difficile* toxin was demonstrated in his stool, and a diagnosis of antibiotic-associated colitis was made. He was treated successfully with metronidazole.

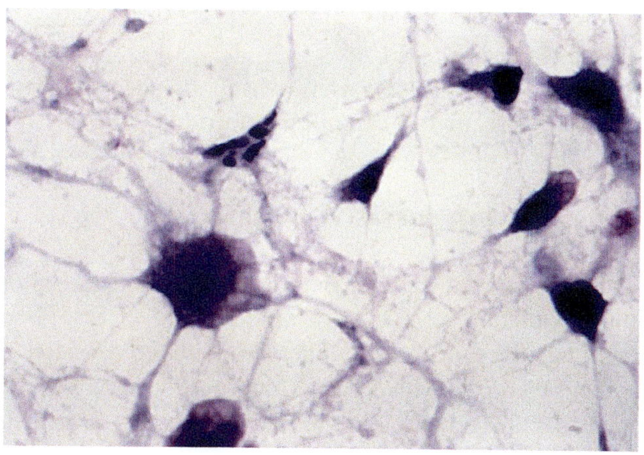

Fig. 14.8 The cytotoxic effect of *C. difficile* toxin in tissue culture. Courtesy of PHIL, CDC.

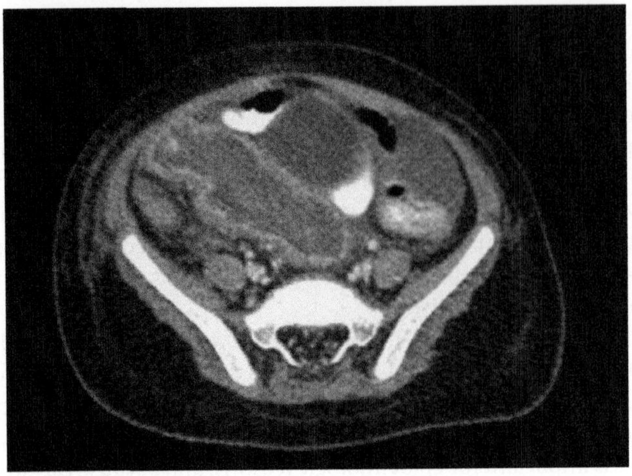

Fig. 14.9 CT scan showing thickened loops of large bowel.

Box 14.2 Case study

A 13-year-old boy was struck with the end of a bamboo stick on the bridge of his nose. He subsequently developed a severe cellulitis and abscess, and at surgery, multiple pieces of bamboo were removed (Fig. 14.10). Culture of the infected material grew out several Gram-negative rods, *Enterococcus casseliflavus*, and *Clostridium perfringens*, the Gram stain of the culture of which is shown in Fig. 14.11.

Food poisoning

Clostridium perfringens causes acute food poisoning. Because this organism is widespread in the environment, it can contaminate food readily. If spores contaminate a large mass of meat, which is allowed to cool slowly, the spores germinate and the vegetative organisms elaborate an enterotoxin. When the food is ingested, the toxin causes acute vomiting and diarrhea.

Soft tissue infections

The histotoxic clostridia can cause a variety of soft tissue infections, especially in immunocompromised hosts, and can complicate injuries.

Non-sporulating Gram-positive rods

This is a heterogenous group of bacteria that are uncommon causes of disease, and include the following genera.

- Actinomyces: although this is not a strict anaerobe, it is often considered together with other anaerobes. It is discussed with aerobic Gram-positive rods (see Chapter 12).
- Bifidobacterium: this is the dominant colonizer of the intestine in breastfeeding babies.
- Leptotrichia spp.
- Eubacterium (Eggerthella) spp.

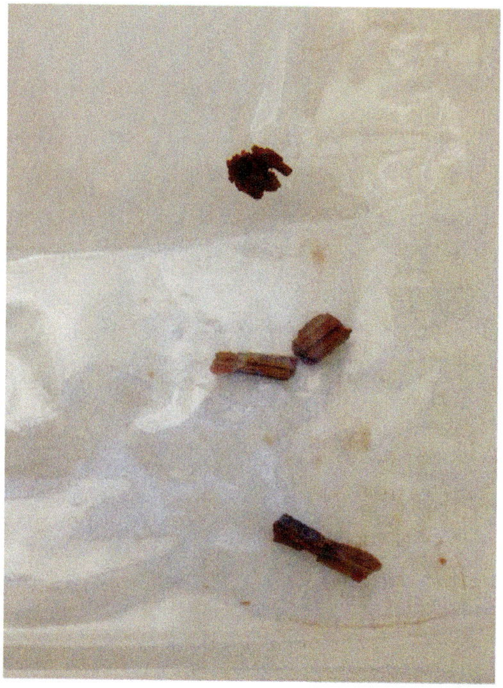

Fig. 14.10 The pieces of bamboo removed from the child's nasal bridge.

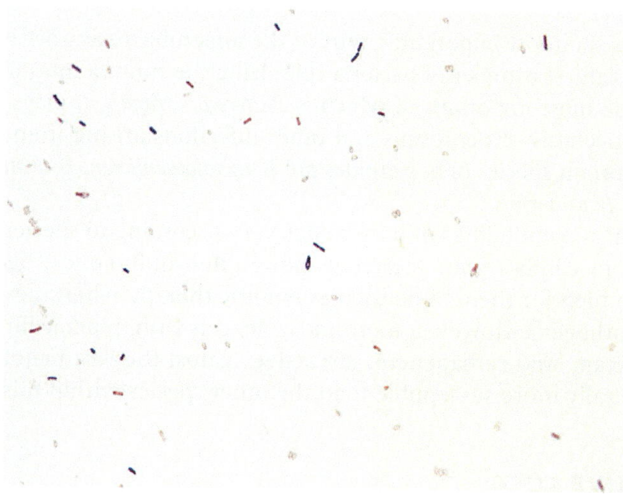

Fig. 14.11 Gram stain of colonies of *Clostridium perfringens* cultured from the wound. Note the Gram-variable rods, some with subterminal spores.

- Mobiluncus spp. (curved rod, associated with bacterial vaginosis)
- Lactobacillus spp.
- *Propionibacterium acnes*: this organism is present on skin, and plays a role in the pathogenesis of acne vulgaris. It is an important colonizer of plastic foreign bodies, such as ventriculoperitoneal shunts, vascular catheters, and prosthetic devices, such as lens implants (Fig. 14.12).

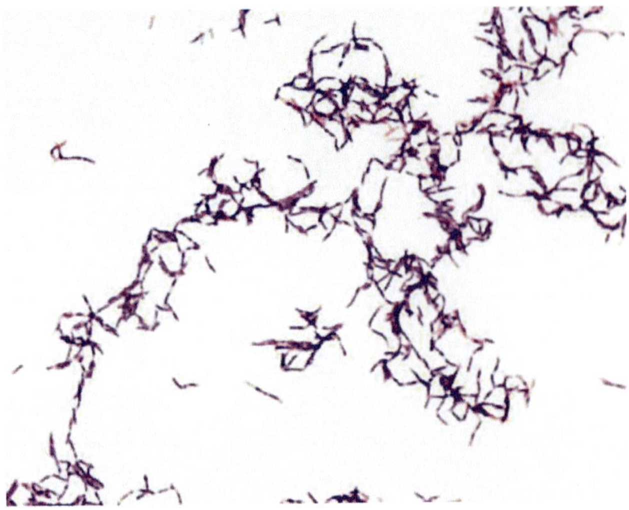

Fig. 14.12 *Propionibacterium granulosum* cultured from a brain abscess.

Gram-negative rods

These belong to several main genera, namely Bacteroides, Fusobacterium, Porphyromonas, Prevotella, Bilophilia, Sutterella, Selenomonas, and Campylobacter. Each of these genera contains many species.

Bacteroides is the most important genus of the anaerobic Gram-negative rods, and accounts for the largest number of bacteria colonizing the human intestine. It contains many species, the most important of which is *Bacteroides fragilis.* It plays an important role in the pathogenesis of peritonitis and other infections arising from the intestinal tract. Other important species of bacteroides are *B. thetaiotaomicron, B. ovatus, B. vulgatus, B. distasonis,* and *B. uniformis.*

Antimicrobial susceptibilities of bacteroides vary according to species. Overall, the resistance rates to clindamycin, cefotetan, newer fluorquinolones, tigecycline, and linezolid are too high for them to be used as empiric therapy when these bacteria are the suspected pathogens. However, metronidazole, cefoxitin, piperacillin/tazobactam, ticarcillin/sulbactam, and carbapenems are active against the vast majority of strains. *B. fragilis* is generally more susceptible than the other species within this group.

Gram-positive cocci

These include the genera *Peptococcus* and *Peptostreptococcus*. The latter has been split into several new genera, including *Peptostreptococcus, Anaerococcus, Finegoldia, Micromonas, Parvimonas,* and *Peptoniphilus*. They cause infections associated with mucosal surfaces, such as sinusitis, and brain abscess, and those related to the intestinal and female genital tracts. They are generally more susceptible to antibiotics than are the Bacteroides group, but the rates of resistance to penicillin, clindamycin, and metronidazole are variable. They are almost all susceptible to β-lactam/β-lactamase inhibitor combinations and carbapenems.

Box 14.3 Case study

A 15-year-old girl presented with a 4-day history of abdominal pain. On examination, she was hypotensive and had a distended abdomen, with lower abdominal tenderness. She was resuscitated with intravenous fluid and treated with piperacillin/tazobactam, for suspected sepsis arising from an abdominal source. A blood culture grew out Gram-negative rods in the anaerobic bottle only. This was identified as *Bacteroides fragilis*. An abdominal CT scan showed several large abscesses, which were drained (Fig. 14.13). The source was a perforated appendix, which was subsequently removed.

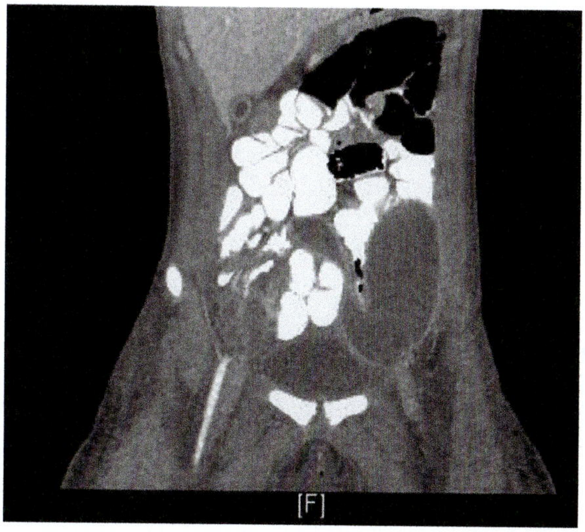

Fig. 14.13 CT scan showing a large intra-abdominal abscess.

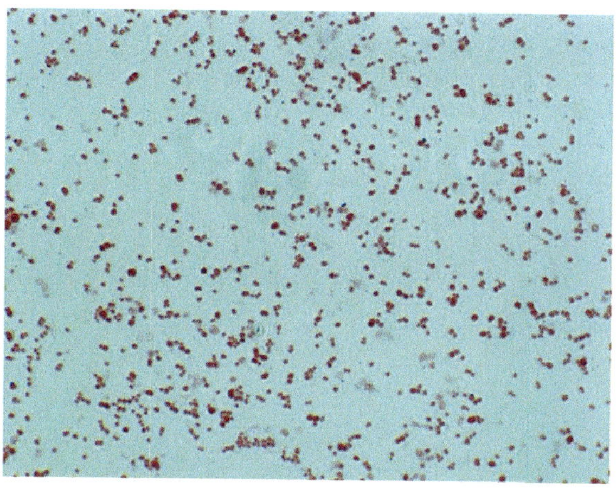

Fig. 14.14 Gram stain of *Veillonella parvula*, cultured from an infant's blood. Courtesy of Ann Chahroudi.

Gram-negative cocci

The only genus within this group is Veillonella. This is present in the pharynx. It is seldom isolated from clinical specimens (Fig. 14.14).

Further reading

Adalja AA, Toner E, Inglesby TV (2015) Clinical management of potential bioterrorism-related conditions. N Engl J Med 372: 954–62.

Ananthakrishnan AN (2011) Clostridium difficile infection: epidemiology, risk factors and management. Nat Rev Gastroenterol Hepatol 8: 17–26.

Brecher SM, Novak-Weekley SM, Nagy E (2013) Laboratory diagnosis of *Clostridium difficile* infections: there is light at the end of the colon. Clin Infect Dis 57: 1175–881.

Brook I (2007) The role of anaerobic bacteria in cutaneous and soft tissue abscesses and infected cysts. Anaerobe 13: 171–7.

Dembek ZF, Smith LA, Rusnak JM (2007) Botulism: cause, effects, diagnosis, clinical and laboratory identification, and treatment modalities. Disaster Med Public Health Prep 1: 122–34.

Du Pont HL, Garey K, Caeiro J-P, et al. (2008) New advances in *Clostridium difficile* infection: changing epidemiology, diagnosis, treatment and control. Curr Opin Infect Dis 21: 500–7.

Farrar JJ, Yen LM, Cook T, et al. (2000) Tetanus. J Neurol Neurosurg Psychiatry 69: 292–301.

Goonetilleke A, Harris JB (2004) Clostridial neurotoxins. J Neurol Neurosurg Psychiatry 75(Suppl III): iii35–iii39.

Hecht DW (2006) Anaerobes: antibiotic resistance, clinical significance, and the role of susceptibility testing. Anaerobe 12: 115–21.

Kachrimanidou M, Malisiovas N (2011) *Clostridiuim difficile* infection: a comprehensive review. Crit Rev Microbiol 37: 178–87.

Kufelnicka AM, Kirn T (2011) Effective utilization of evolving methods for laboratory diagnosis of Clostridium difficile infection. Clin Infect Dis 52: 1451–7.

Kuijper EJ, Coignard B, Brazier J, et al. (2007) Update of *Clostridium difficile*-associated disease due to PCR ribotype 027 in Europe. Euro Surveill 12(6): 163–6.

Loo VG, Poirier L, Miller M, et al. (2005) A predominantly clonal multi-institutional outbreak of *Clostridium difficile*-associated diarrhea with high morbidity and mortality. N Engl J Med 353: 2442–9.

McDonald LC, Owings M, Jernigan DB (2006) Clostridium difficile infections in patients discharged from US short-stay hospitals, 1996-2003. Emerg Infect Dis 12(3): 409–15.

Murphy EC, Frick I-M (2013) Gram-positive anaerobic cocci – commensals and opportunistic pathogens. FEMS Microbiol Rev 37: 520–53.

Nagy E (2010) Anaerobic infections. Update on treatment considerations. Drugs 70: 841–58.

Popoff MR (2014) Botulinum neurotoxins: more and more diverse and fascinating toxic proteins. J Infect Dis 209: 168–9.

Song Y, Liu C, Finegold SM (2007) Development of a flow chart for identification of Gram-positive anaerobic cocci in the clinical laboratory. J Clin Microbiol 45: 512–16.

Thwaites CL, Beeching NJ, Newton CR (2015) Maternal and neonatal tetanus. Lancet 385: 362–70.

Wexler HM (2007) *Bacteroides*: the good, the bad, and the nitty-gritty. Clin Microbiol Rev 20: 593–621.

CHAPTER 15

Mycoplasmas, Chlamydiae, Rickettsiae, and Ehrlichiae

Mycoplasmas and Ureaplasma

Mycoplasmas and Ureaplasma are mollicutes, that is, bacteria that lack cell walls. They are the smallest self-replicating organisms that can live outside a cell, and they are difficult to culture *in vitro*.

Mycoplasma pneumoniae

This organism is a common cause of respiratory tract infection in both children and adults. It is spread by respiratory droplets and because it has a long generation time (6 hours), it is associated with a long incubation period (weeks). It attaches to the respiratory epithelial cell, where it damages the cilia and causes inflammation. This results in pharyngitis, bronchitis, and pneumonia, which is usually bilateral and patchy (Fig. 15.1).

These infections can be complicated by disease of other organs, which may be immune mediated. They include disease of the skin (erythema multiforme and Stevens–Johnson syndrome), brain (encephalitis), and blood (autoimmune hemolytic anemia). Making the diagnosis of mycoplasma infection is difficult. Culture is rarely performed in routine laboratories, and it takes weeks for the organism to grow. Serology, using IgM, or seroconversion is the best way to confirm the diagnosis, but not practical regarding patient management; PCR is becoming more widely available. Nevertheless, when the infection is suspected, patients should be treated empirically. The organism is susceptible to tetracyclines and fluoroquinolones, the usual treatments given to adults, and macrolides, the usual treatment given to children younger than 8 years. It is resistant to sulfonamides and rifampin.

Mycoplasma hominis and *Ureaplasma urealyticum*

These are present in the genital tracts of many women. They attach to the epithelium and can cause inflammation. They can cause puerperal infections, and ureaplasma can play a role in preterm labor and chorioamnionitis. It can also cause infections of the newborn, including meningitis. These organisms can be cultured *in vitro* in cell-free medium (which

---

Practical Medical Microbiology for Clinicians, First Edition. Frank E. Berkowitz and Robert C. Jerris.

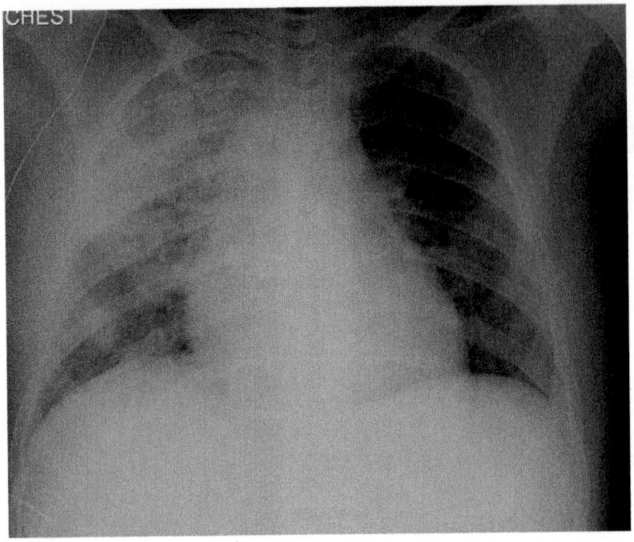

Fig. 15.1 Chest X-ray of a teenage girl with *Mycoplasma pneumoniae* pneumonia.

is also used for antimicrobial susceptibility testing), but this requires inoculation of the culture at the bedside, or transport to the laboratory in transport medium. They can be detected by PCR. *M. hominis* is susceptible to tetracycline, clindamycin, and fluoroquinolones, but resistant to macrolides, while *U. urealyticum* is susceptible to macrolides, tetracycline, and fluoroquinolones, but resistant to clindamycin.

Mycoplasma genitalium

This causes urethritis in males and cervicitis in females. The diagnosis is very difficult and, for practical purposes, depends on PCR, if available. Although it has been considered susceptible to macrolides, tetracyclines, and fluoroquinolones, a recent Danish study revealed a high rate of macrolide resistance.

Chlamydia and Chlamydophila

These are small bacteria that lack cell walls and cannot multiply extracellularly. Their life cycle is shown in Fig. 15.2.

The infectious particle is the elementary body. This is expelled in secretions from the source and infects an epithelial cell of the new host. Here it develops into a reticulate body, which grows, and subsequently divides up into many elementary bodies, which repeat the cycle.

Chlamydia trachomatis

This consists of many serovars, which are divided into three groups and which cause three different clinical diseases.

Oculogenital disease

This is caused by serovars B, D, E, F, G, Ga, H, I, Ia, J, and K. These organisms cause genital tract infection, characterized in males by urethritis and in females by vaginitis, cervicitis (Fig. 15.3), and salpingitis. When salpingitis occurs, secondary infection by

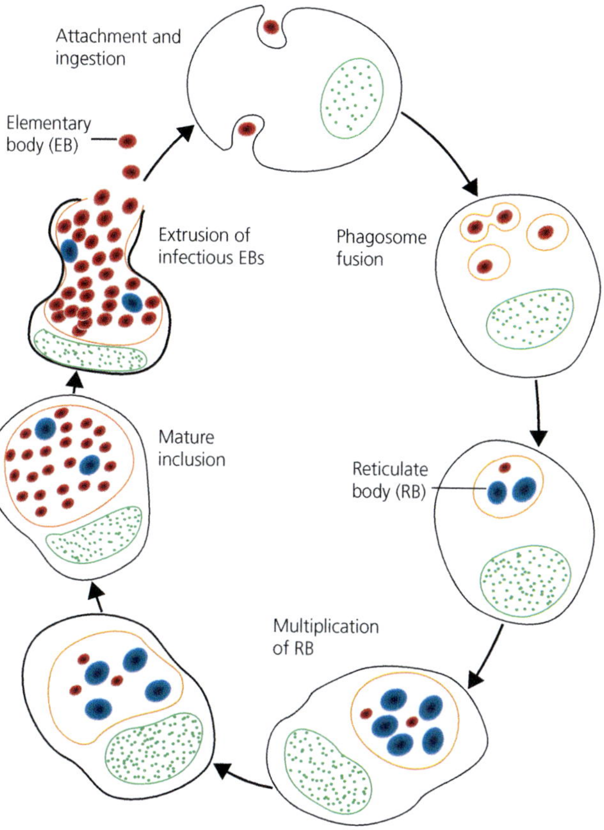

Fig. 15.2 Life cycle of chlamydiae.

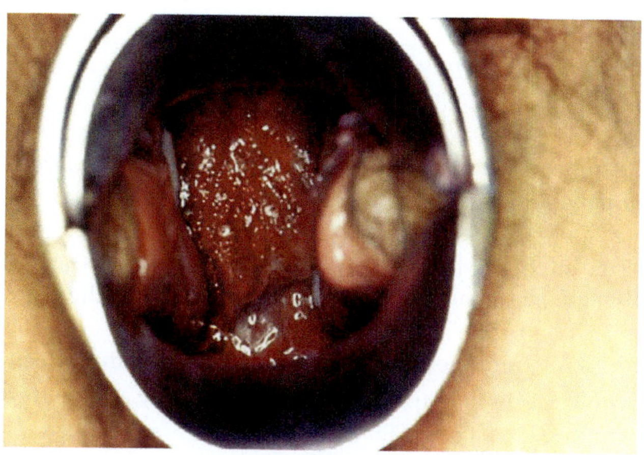

Fig. 15.3 Chlamydial cervicitis. Courtesy of PHIL, CDC.

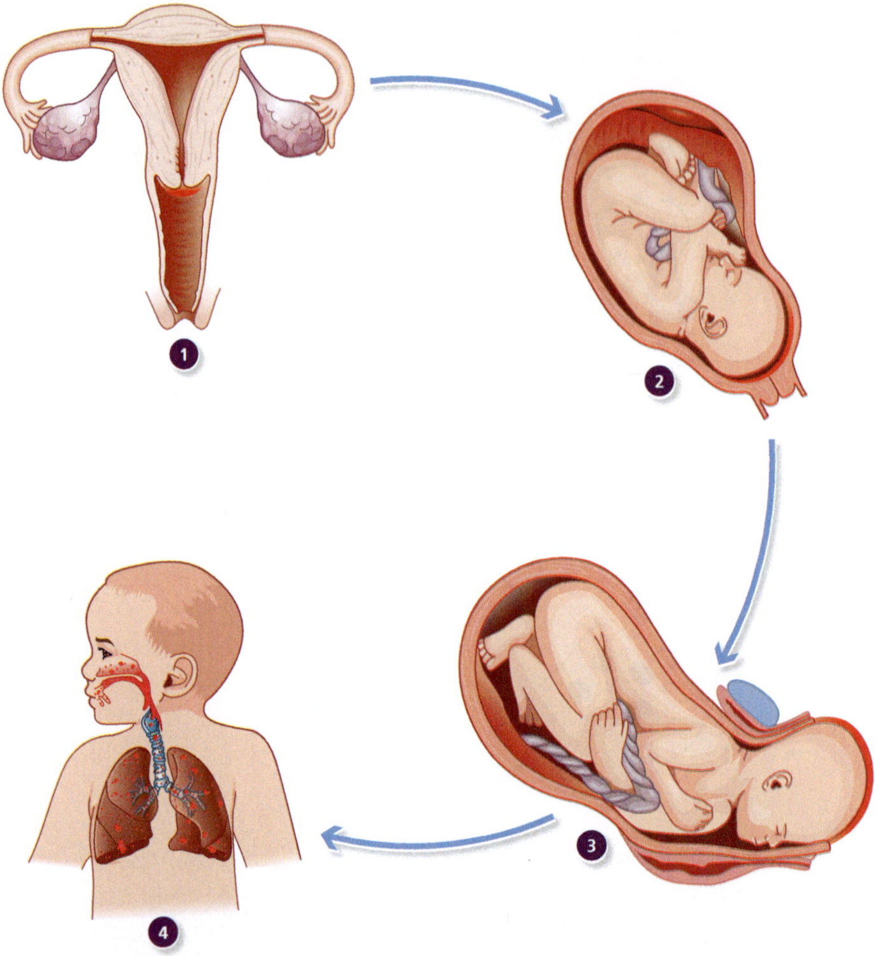

Fig. 15.4 Pathogenesis of intrapartum chlamydial infection. Red indicates area of chlamydial infection. 1. Cervical infection; 2. cervical infection, with fetus *in utero*; 3. infant passing through chlamydia-infected birth canal; 4. routes of spread of chlamydia in newborn infant.

other bacteria, such as enteric rods and anaerobes, may ensue and the infection may spread from the tubes into the peritoneal cavity. causing localized peritonitis, a syndrome called pelvic inflammatory disease. Proctitis can occur in individuals who have anal sex.

These organisms infect newborn infants as they pass through the birth canal, causing conjunctivitis, and, by spreading down from the pharynx to the lungs, pneumonia, as shown in Figs 15.4 and 15.5.

There are several different methods for diagnosing chlamydial infection.

- Culture: this requires growth in tissue culture, is expensive and not readily available, and is therefore not widely used (Fig. 15.6).

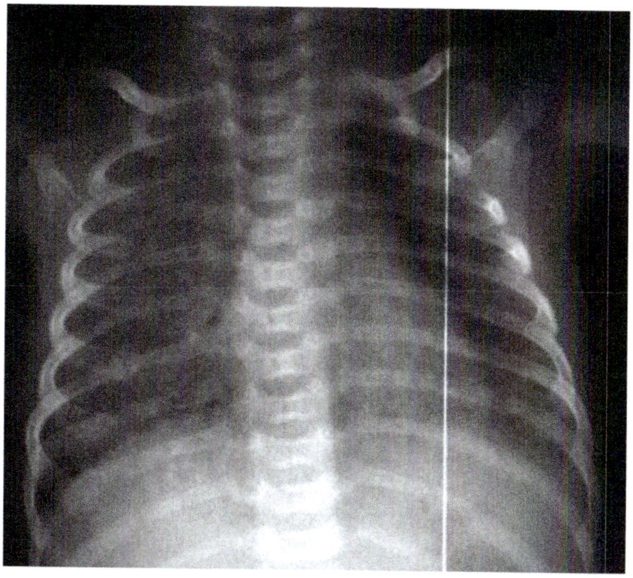

Fig. 15.5 Chest X-ray of a young infant with *Chlamydia trachomatis* pneumonia. Note the bilateral patchy infiltrates.

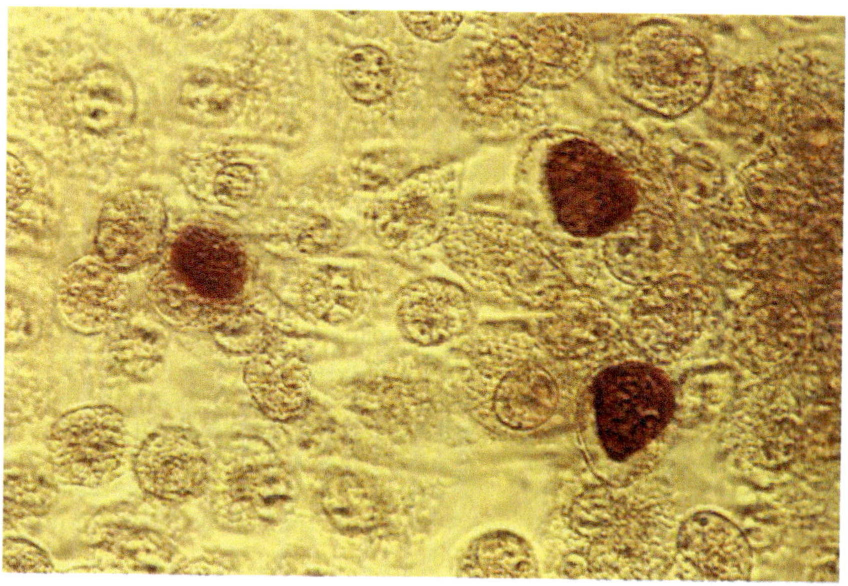

Fig. 15.6 *Chlamydia trachomatis* growing in tissue culture in McCoy cells, demonstrating inclusions stained with iodine. Courtesy of PHIL, CDC.

- Demonstration of chlamydial antigen or nucleic acid. Nucleic acid amplification tests (NAATs) are widely used. These can be performed on urine specimens, and are useful for screening, because the collection of genital specimens is not necessary.

Treatment of patients with chlamydial infection consists of a tetracycline or a macrolide.

Lymphogranuloma venereum (LGV)

This is caused by serovars L1, L2, L2a, and L3. It is characterized by a genital ulcer, followed by localized lymphadenitis. Anorectal sites can result in progressive tissue damage, and the lymph node can become purulent and form draining sinuses. Although this chlamydial infection can be diagnosed using the same tests as are used for the serovars causing oculogenital infection, testing for the specific serovars causing LGV is difficult. Treatment consists of a tetracycline or macrolide.

Trachoma

This is a severe eye infection caused by serovars A, B, Ba, and C. It is prevalent in countries with poor sanitation and, worldwide, it is the most common infectious cause of blindness. The organism is spread by hands and flies, and causes recurrent conjunctivitis. This results in scarring of the tarsal conjunctiva, with eventual entropion (in-turning of the eyelashes). This causes the eyelashes to rub on the cornea (trichiasis), causing scarring and blindness.

Microbiologic diagnosis is difficult in areas where the disease is prevalent.

Management consists of **S**urgical treatment of trichiasis, **A**ntimicrobial therapy with erythromycin or tetracycline, **F**acial cleanliness, and **E**nvironmental cleanliness (SAFE).

Chlamydophila pneumoniae

This organism causes bronchitis and pneumonia that clinically resemble illness caused by *Mycoplasma pneumoniae*. The diagnosis is difficult to make, and the therapy is the same as that for the mycoplasma infections.

Chlamydophila psittaci

This causes psittacosis (ornithosis), which is a zoonosis transmitted to humans from birds and some mammals. There are eight serovars, based on the major outer membrane protein. Six of these (A–F) occur in various birds, while WC and M56 occur in cattle and rodents respectively. It is transmitted to humans by the inhalation of the organism in dried feces or respiratory secretions. Although it causes mainly pneumonia, which can be very severe, it can affect any organ. Presentation is usually acute, with severe headache and respiratory symptoms. Large outbreaks have occurred in the turkey-producing industry.

The diagnosis depends largely on the history of exposure. Although the organism can be cultured, this requires a level 3 laboratory and is seldom performed. The diagnosis is usually confirmed by serologic tests. The complement fixation and ELISA tests do not distinguish between *C. psittaci* and *C. pneumoniae*. The best test currently is the microimmunofluorescence test. NAATs have been developed but are not generally available. The treatment is a tetracycline. Macrolides have good activity *in vitro* and are effective in an animal model. Fluoroquinolones (with the exception of norfloxacin) have good *in vitro* activity against the organism, but in an animal model only gatifloxacin was effective.

The different chlamydiae and the main clinical syndromes that they cause are shown in Table 15.1.

Table 15.1 Chlamydiae and Chlamydophilas, and their main clinical syndromes.

Species	Site	Syndrome
Chlamydia trachomatis		
Serovars B, Da, Ia, D–K	Genital	Cervicitis, urethritis, pelvic inflammatory disease
	Neonate	Conjunctivitis, pneumonia
Serovars L1, L2, L2a, L3	Genital	Lymphogranuloma venereum
Serovars A, B, Ba, C	Eye	Trachoma
Chlamydophila psittaci	Lung, systemic	Psittacosis
Chlamydophila pneumoniae	Respiratory tract	Bronchitis, Pneumonia

Rickettsiales

This order contains two families, the Rickettsiaceae and the Anaplasmataceae. The former contains the genera Rickettsia and Orientia, while the latter contains the genera Ehrlichia and Anaplasma.

Rickettsiaceae

These organisms are small obligate intracellular organisms that are classified with Gram-negative rods. They have a worldwide distribution. They infect endothelial cells and therefore cause generalized systemic infection. There are three main groups: the typhus group, the spotted fever group, and *Orientsia tsutsugamushi*.

Typhus group

There are two organisms in this group: *Rickettsia typhi* (also called *R. mooseri*), which causes endemic typhus, and *R. prowazekii*, which causes epidemic typhus. These infections are characterized by fever, severe headache, and generalized rash (Fig. 15.7).

Rickettsia typhi is transmitted by animal fleas. *R. prowazekii* is transmitted by the body louse *Pediculus humanus corporis* (see Chapter 26). This is prevalent in circumstances of poverty and crowding ("lousy" circumstances). The louse lays its eggs on the seams of clothing, and after the nits have hatched, the lice feed on the host's blood. The infection can progress to multiorgan failure and, if untreated, it carries a very high fatality rate. It has caused many millions of death and profoundly influenced the course of human history. It is the only rickettsiosis of which humans are the reservoir. It can reactivate within a host many years after the initial infection, a condition called Brill–Zinsser disease. If this occurs in conditions favorable for spread, an outbreak can occur.

Spotted fevers

These are arthropod-borne (mostly tick-borne) infections, and most areas of the world have their own particular rickettsial spotted fever. Table 15.2 lists some of the rickettsiae, their vectors, and their areas of prevalence.

The most important rickettsial disease in North America is Rocky Mountain spotted fever, which is caused by *R. rickettsii*. It is transmitted by the hard ticks *Dermacentor variabilis, Dermacentor andersoni, Rhipicephalus sanguineus,* and *Amblyomma cajennense.*

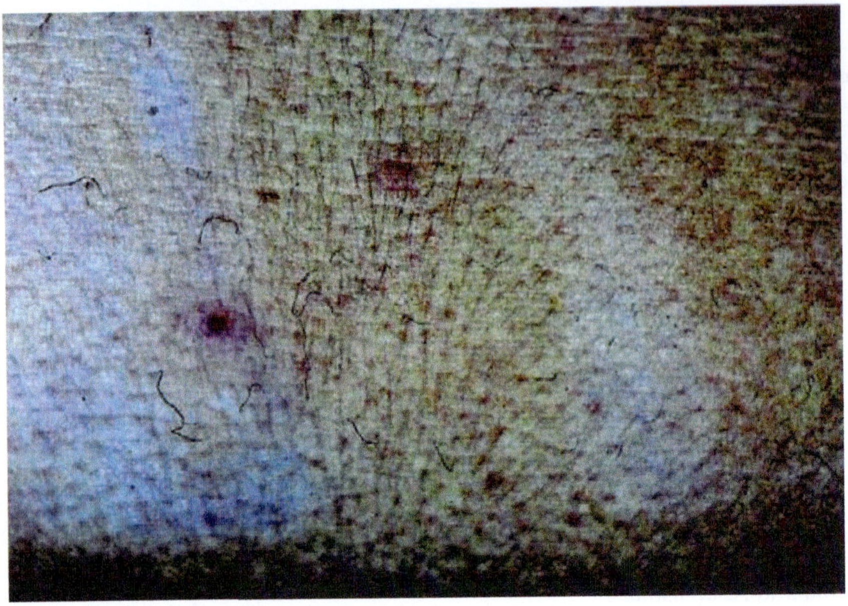

Fig. 15.7 Rash in an individual with epidemic typhus. Reproduced from Niang M, Brouqui P, Raoult D (1999) Epidemic typhus imported from Algeria. Emerg Infect Dis 5: 716–18, courtesy of CDC.

Table 15.2 Rickettsial infections, their vectors, and distribution.

Organism	Vector	Distribution
Typhus group		
R. typhi	Flea	Worldwide
R. prowazekii	Body louse	Worldwide
Spotted fever group		
R. rickettsii	Tick	Americas
R. conorii	Tick	Mediterranean, Africa
R. africae	Tick	Africa, West Indies
R. aeschlimannii	Tick	Africa
R. australis	Tick	Australia
R. sibirica	Tick	Russia, Asia
R. parkeri	Tick	N. America
R. mongolotimonae	Tick	Europe, Africa, Asia, Europe
R. akari	Mite	N. America
R. slovaka	Tick	Europe
R. japonica	Tick	Japan
R. honei	Tick	Tasmania
R. heilongjiangensis	Tick	China
R. felis	Flea	Worldwide
R. helvetica	Tick	Europe,Thailand
Orientia tsutsugamushi	Mite	E. Asia, Western Pacific

The infection is characterized by fever, severe headache, and a rash that develops 2–4 days after the onset of fever. Although the rash classically starts distally on the extremities, this is not always the case. It is initially macular, but becomes petechial (Fig. 15.8). If untreated, it can lead to multiorgan disease and death.

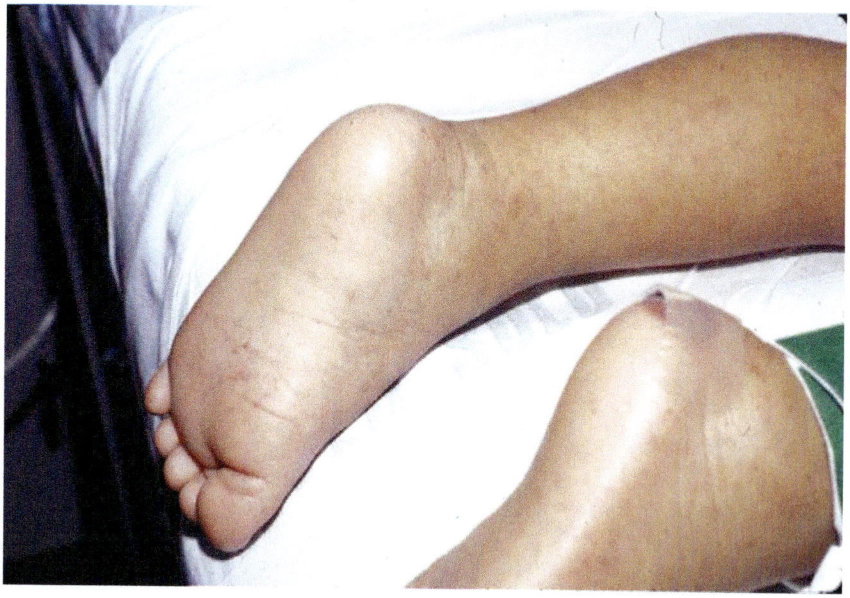

Fig. 15.8 Rash of a child with Rocky Mountain spotted fever. Copyright © 2007 Frank E. Berkowitz. Reprinted with the permission of Cambridge University Press.

The decision to treat should be based on the clinical picture, because there is no readily available, reliable test that can confirm or exclude the diagnosis. The tests that can be performed are serologic and are of value, in retrospect, only if acute and convalescent specimens are tested. The tests for different rickettsiae cross-react with one another. PCR tests on blood and skin biopsies, and immunohistochemical staining of skin biopsies can be performed but are not routinely available. Although rickettsiae can be cultured in tissue culture, this requires a level 3 laboratory and is not generally available.

The treatment is a tetracycline (usually doxycycline) irrespective of age.

Mediterranean spotted fever (Boutenneuse fever) and African tick bite fever are characterized by an eschar at the site of the tick bite (Fig. 15.9), and regional lymphadenopathy.

Anaplasmataceae

This family contains three organisms known to infect humans, namely *Ehrlichia chaffeensis*, the cause of human monocytic ehrlichiosis (HME), *Anaplasma phagocytophilum*, the cause of human granulocyctic anaplasmosis (HGA), and *Ehrlichia ewingii*. The ehrlichiae are named in honor of Paul Ehrlich, a German microbiologist and Nobel laureate, who discovered the precursor of sulfonamides. Recently, a new ehrlichial organism, related to *E. muris*, was reported in the northern US. Clinically, these infections cause non-specific manifestations, such as headache, fever, myalgias, and, occasionally, a rash (Fig. 15.10). They multiply within the respective leukocytes, forming inclusions called morulae (from Latin for mulberry). These stain with Romanowsky stains and can therefore sometimes be seen in a blood smear (Fig. 15.11). However, this is not sensitive enough to be relied upon as a diagnostic test for the infection. Other laboratory findings that suggest the possibility of ehrlichial infection are leukopenia, thrombocytopenia, and elevation of serum transaminases.

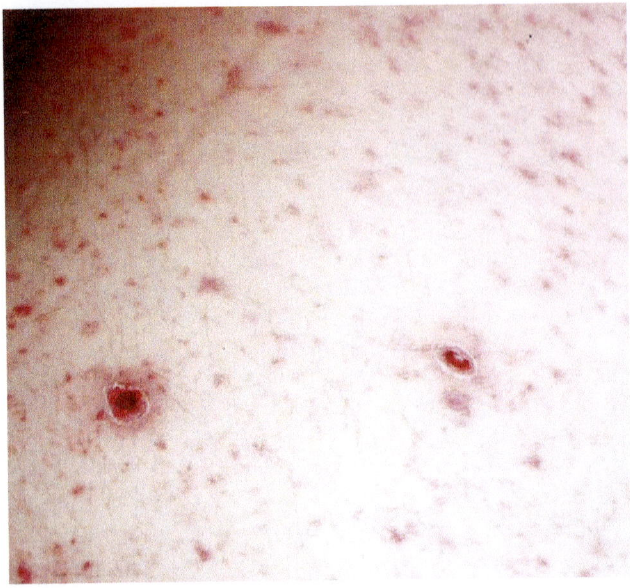

Fig. 15.9 Rash and eschars in a patient with African tick bite fever. Copyright © 2007 Frank E. Berkowitz. Reprinted with the permission of Cambridge University Press.

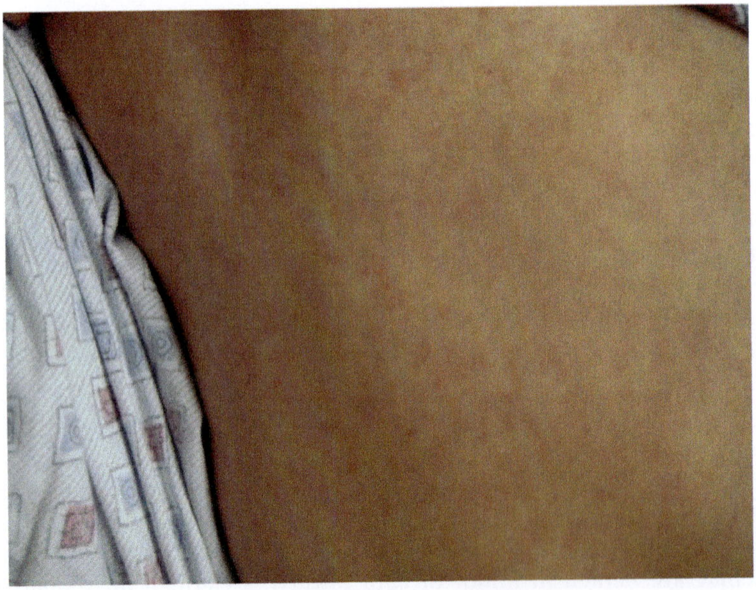

Fig. 15.10 Rash in a boy with ehrlichiosis. Courtesy of Dr Anita McElroy.

The diagnosis can be confirmed by serologic methods, using paired sera, or by PCR. This is not of value in decision making about therapy (see Rickettsiaceae).

Coxiella burnetii (see Chapter 13)

The different tick-borne diseases, their vectors, and their geographic areas of prevalence in the USA are shown in Table 15.3.

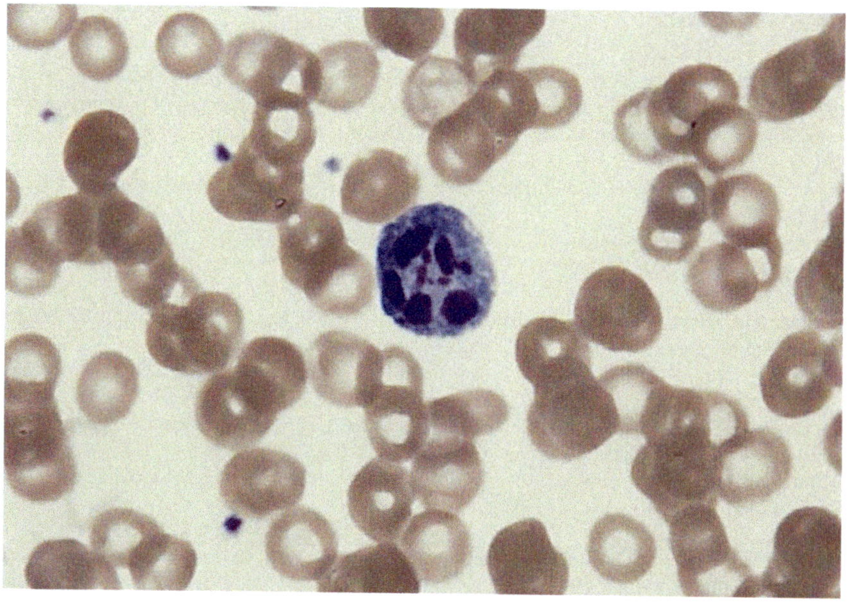

Fig. 15.11 Morulae in a mononuclear cell of the boy shown in Fig. 15.10. Courtesy of Dr Anita McElroy.

Table 15.3 Tick-borne diseases in America, their vectors, and distributions.

Organism	Vector	Distribution
Viruses		
Colorado tick fever	*Dermacentor andersoni*	Rocky Mountains
Powassan virus	*Ixodes* spp.	Northeast, Great Lakes
Heartland virus	*Amblyomma*	Central USA
Bacteria		
Borrelia burgdorferi	*Ixodes scapularis*	Northeast, Midwest
	Ixodes pacificus	West
Borrelia hermsii	*Ornithodorus* spp.	Western Mountains
Anaplasma phagocytophilum	*Ixodes scapularis*	Northeast, Midwest
	Ixodes pacificus	
Ehrlichia chaffeensis	*Amblyomma americanum*	South, Central
Ehrlichia ewingii	*Amblyomma americanum*	South, Central
Rickettsia rickettsii	*Dermacentor variabilis*	USA
	Dermacentor andersoni	Mexico
	Rhipicephalus sanguineus	Central, South America
	Amblyomma cajennense	
Francisella tularensis	*Dermacentor* spp., *Amblyomma americanum*, flies	Whole USA, mostly central and western
Protozoa		
Babesia spp.	*Ixodes scapularis* *Ixodes pacificus*	Northeast, Midwest, northern Pacific coast
Toxin		
Tick paralysis	Many types of ticks	Worldwide
Unknown organism		
Southern tick-associated rash illness (STARI)	*Amblyomma americanum*	Southern USA

Further reading

Bechah Y, Capo C, Mege J-L, et al. (2008) Epidemic typhus. Lancet Infect Dis 8: 417–26.

Centers for Disease Control and Prevention (2006) Diagnosis and management of tickborne rickettsial diseases: Rocky Mountain spotted fever, ehrlichiosis, and anaplasmosis – United States. MMWR 55(RR04): 1–27.

Centers for Disease Control and Prevention (2010) Sexually transmitted disease treatment guidelines. MMWR 59(RR12): 1–110.

Dantas-Torres F (2007) Rocky Mountain spotted fever. Lancet Infect Dis 7: 724–31.

Niang M, Brouqui P, Raoult D (1999) Epidemic typhus imported from Algeria. Emerg Infect Dis 5: 716–18.

Pritt BS, Sloan LM, Hoang Johnson DK, et al. (2011) Emergence of a new pathogenic Ehrlichia Species, Wisconsin and Minnesota, 2009. N Engl J Med 365: 422–9.

Rolain J-M, Jensenius M, Raoult D (2004) Rickettsial infections – a threat to travelers. Curr Opin Infect Dis 17: 433–7.

Salado-Rasmussen K, Jensen JS (2014) *Mycoplasma genitalium* testing pattern and macrolide resistance: a Danish nationwide retrospective survey. Clin Infect Dis 59: 24–30.

Sethi S, Singh G, Samantha P, et al. (2012) Mycoplasma genitalium: an emerging sexually transmitted pathogen. Indian J Med Res 136: 942–55.

Stewardson AJ, Grayson ML (2010) Psittacosis. Infect Dis Clin North Am 24: 7–25.

US Department of Health and Human Services, Centers for Disease Control and Prevention (2014) *Tickborne Diseases of the United States. A Reference Manual for Health Case Providers*, 2nd edn. Bethesda, Centers for Disease Control and Prevention.

Waites KB, Talkington DF (2004) *Mycoplasma pneumoniae* and its role as a human pathogen. Clin Microbiol Rev 17: 697–728.

Waites KB, Schelonka RL, Xiao L, et al. (2009) Congenital and opportunistic infections: Ureaplasma species and Mycoplasma hominis. Semin Fetal Neonatal Med 14: 190–9.

Watt G, Parola P (2003) Scrub typhus and tropical rickettsioses. Curr Opin Infect Dis 16: 429–36.

CHAPTER 16
Spirochetes

Spirochetes are spiral-shaped bacteria (Fig. 16.1). Some are part of the normal mucosal flora, but several are important pathogens. Four genera are pathogenic in humans, namely Treponema, Borrelia, Leptospira, and Spirillum. Their modes of transmission, the diseases they cause, and the tests used to diagnose them are summarized in Table 16.1.

Syphilis

Syphilis (formerly called lues) is caused by *Treponema pallidum* subspecies *pallidum*. It is a chronic systemic infection that is initially transmitted sexually in most cases. It can also be transmitted transplacentally to the fetus, and by blood transfusion. There are three clinical stages of the infection.

Primary syphilis

Characterized by a painless ulcer (chancre) at the site of inoculation (usually genitalia), and painless regional lymphadenopathy, appearing 2–3 weeks after infection (Fig. 16.2). If the infection follows transplacental transmission or blood transfusion, the primary stage does not occur.

Secondary syphilis

This is the systemic stage of the infection. Initially there is bacteremia with dissemination of the organism to any organ, including the vasa vasorum. There may be fever, rash, generalized lymphadenopathy, mucosal ulcers, perianal warts (condylomata lata), retinitis, meningitis, and hepatitis (Fig. 16.3).

Tertiary syphilis

This manifests many years after infection. It has several manifestations: gummata (focal, destructive granulomas), in viscera, bone, and nasal septum; tabes dorsalis (disease of the posterior columns of the spinal cord); general paralysis of the insane (a dementia); and aortitis, with aneurysm formation of the ascending aorta.

Practical Medical Microbiology for Clinicians, First Edition. Frank E. Berkowitz and Robert C. Jerris.

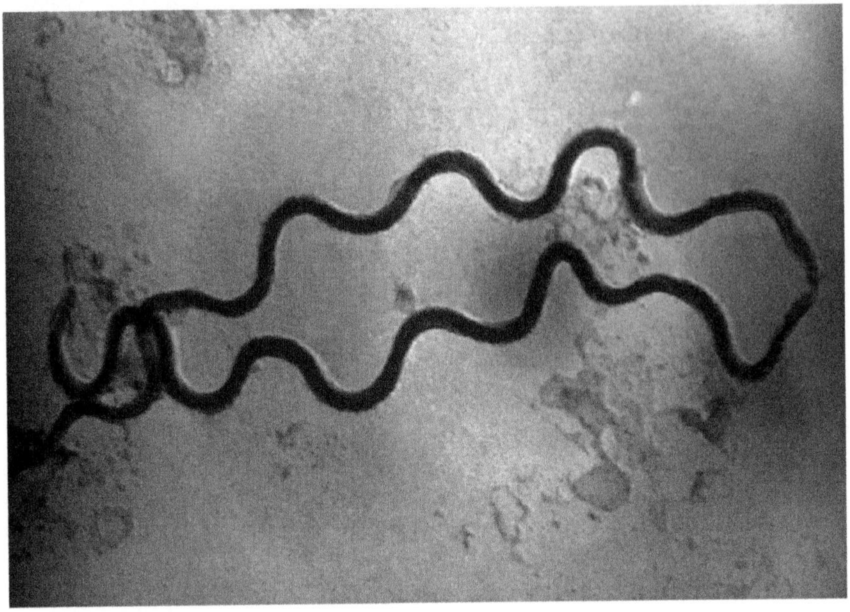

Fig. 16.1 *Treponema pallidum*, the cause of syphilis. Courtesy of PHIL, CDC.

Table 16.1 Spirochetes pathogenic in humans, modes of transmission, clinical syndromes, and diagnostic tests.

Spirochete	Transmission	Syndrome	Diagnostic tests
Treponema			
T. pallidum ssp. *pallidum*	Sexual, vertical	Syphilis	Dark-field microscopy, serology
T. pallidum ssp. *pertenue*	Contact	Yaws	Serology
T. pallidum ssp. *endemicum*	Contact	Bejel	Serology
T. pallidum ssp. *carateum*	Contact	Pinta	Serology
Leptospira			
L. interrogans Many serovars	Animal urine (often via water)	Leptospirosis	Culture, serology
Borrelia			
B. burgdorferi	Tick	Lyme disease	Serology
B. recurrentis	Louse	Relapsing fever	Blood smear, serology
B. duttoni	Tick	Relapsing fever	Blood smear, serology
B. hermsii, turicatae, parkeri	Tick	Relapsing fever	Blood smear, serology
Spirillum			
S. minus	Rat bite	Rat-bite fever (Asian)	Serology

Latent syphilis

The infection is present but there are no clinical manifestations. This is categorized as early latent infection (<1 year after infection) and late latent infection.

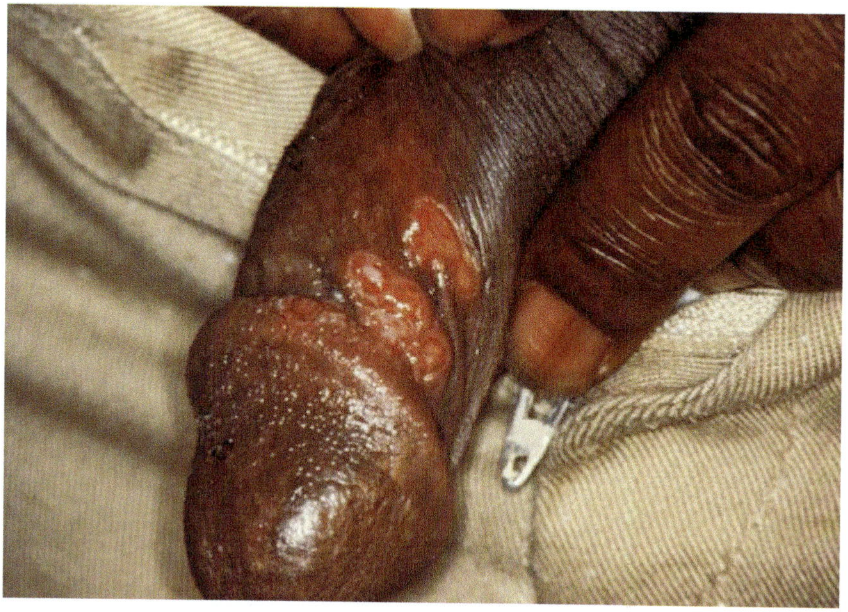

Fig. 16.2 Chancre due to primary syphilis. Courtesy of PHIL, CDC.

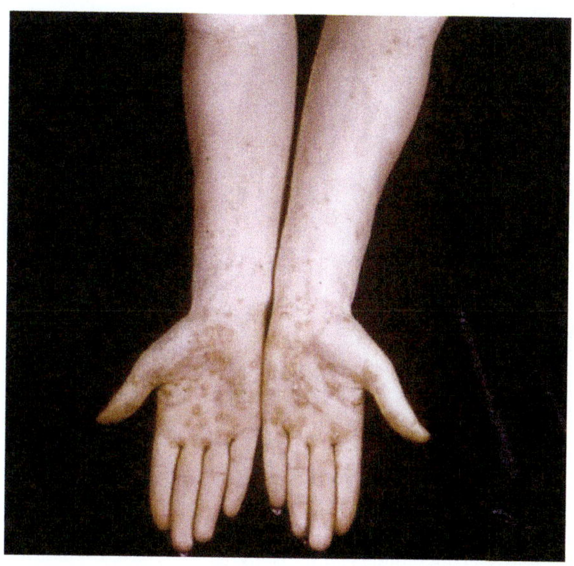

Fig. 16.3 Rash caused by secondary syphilis. Courtesy of PHIL, CDC.

Congenital syphilis

This is a form of secondary syphilis, following transplacental transmission. Although there may be manifestations at birth, these may become apparent only during the first few months of life. All organ systems can be affected.

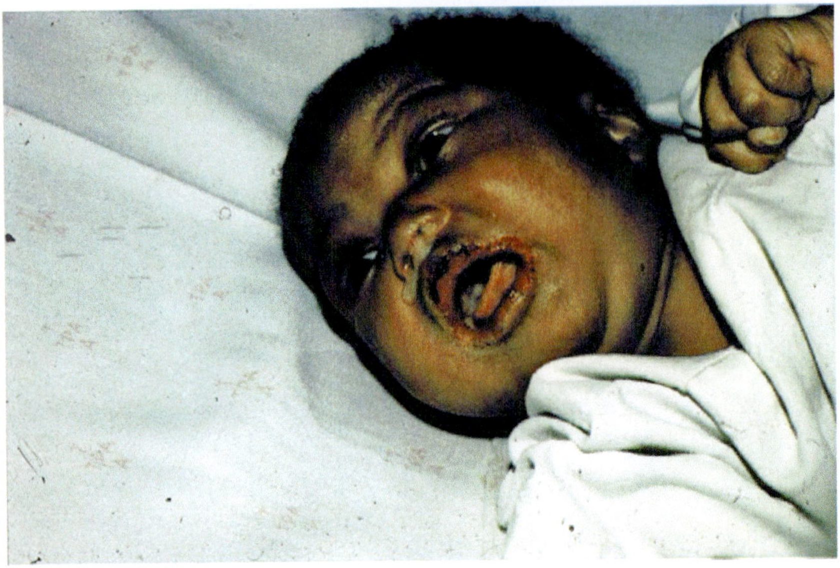

Fig. 16.4 Infant with snuffles due to congenital syphilis. Copyright © Frank E. Berkowitz. Reprinted with the permission of Cambridge University Press.

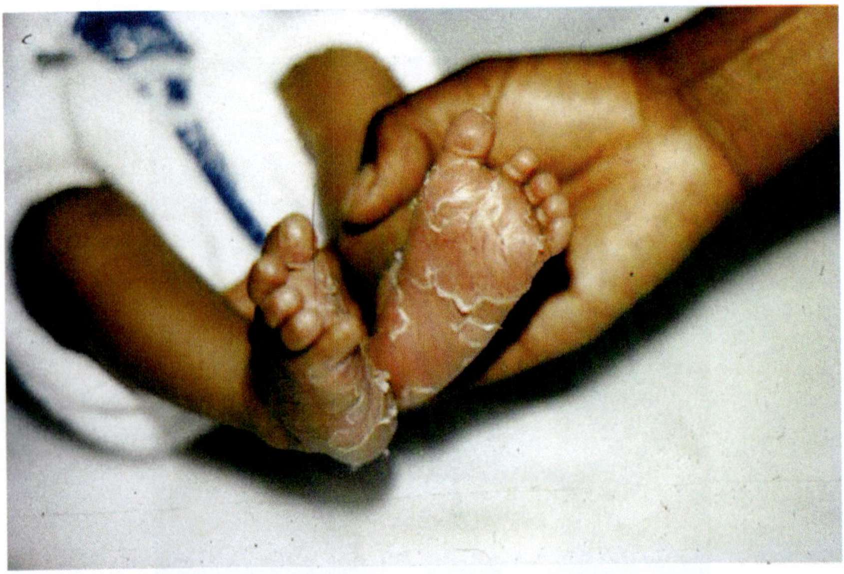

Fig. 16.5 Peeling of the soles in congenital syphilis. Copyright © Frank E. Berkowitz. Reprinted with the permission of Cambridge University Press.

- Nasal mucosa, resulting in bloody nasal discharge (snuffles) (Fig. 16.4)
- Rash with peeling of the palms and soles (Fig. 16.5)
- Kidneys: glomerulonephritis, leading to nephrotic syndrome (Fig. 16.6)
- Lymphoreticular system: anemia (hemolytic), generalized lymphadenopathy, hepatosplenomegaly, as well as hepatitis

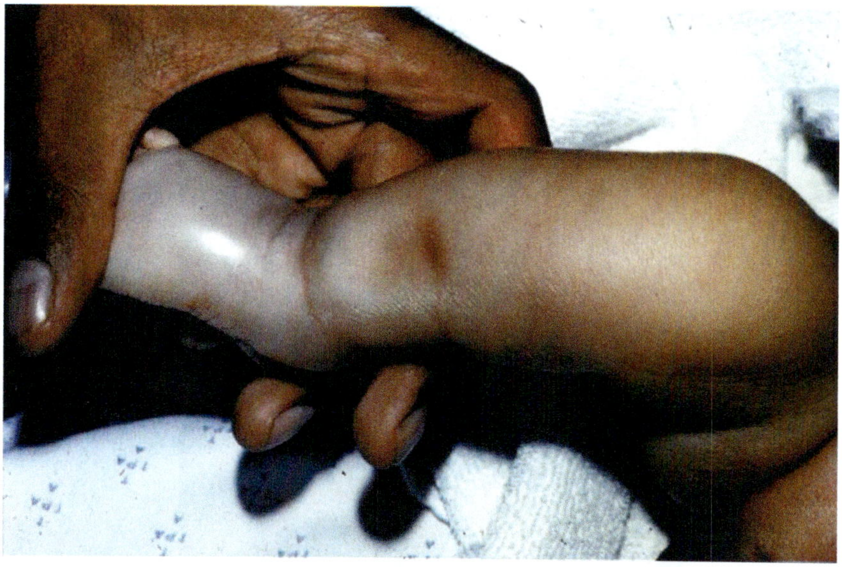

Fig. 16.6 Pitting edema due to the nephrotic syndrome caused by congenital syphilis. Copyright © Frank E. Berkowitz. Reprinted with the permission of Cambridge University Press.

- Bone: periostitis (Fig. 16.7)
- Brain: meningitis
- Lung: pneumonia alba, which can be fatal (Fig. 16.8)

If congenital syphilis is not treated, it may result in late congenital syphilis. This manifests as Hutchinson's triad (interstitial keratitis, sensorineural deafness, and peg-shaped upper central incisors) (Fig. 16.9), multicusped molars, and effusions in the knee joints (Clutton's joints) (Fig. 16.10). It may also progress to tertiary syphilis.

Syphilis in HIV-infected individuals

This is an important co-infection because the routes of acquisition of these infections are the same, namely sexual intercourse and through contaminated blood, such as by needle sharing. Specific issues in such patients are a higher rate of neurologic and ophthalmologic manifestations, a higher rate of therapeutic failure, and a higher rate of false-negative specific treponemal tests (see later in this chapter). HIV-infected individuals have a higher rate of false-positive non-treponemal tests than non-infected individuals.

Diagnosis of syphilis

Because *T. pallidum* cannot be cultured *in vitro*, alternative diagnostic methods are required. There are two main methods: direct visualization of the organism, by a fluorescent antibody test or dark-field microscopy; and serology. The direct visualization tests are not readily available. Dark-field microscopy is performed on fluid, such as serous fluid from skin lesions of primary or secondary syphilis, or mucosal lesions other than oral lesions. (The mouth may have commensal spirochetes that would give false-positive results.) This method requires a dark-field microscope and technical expertise.

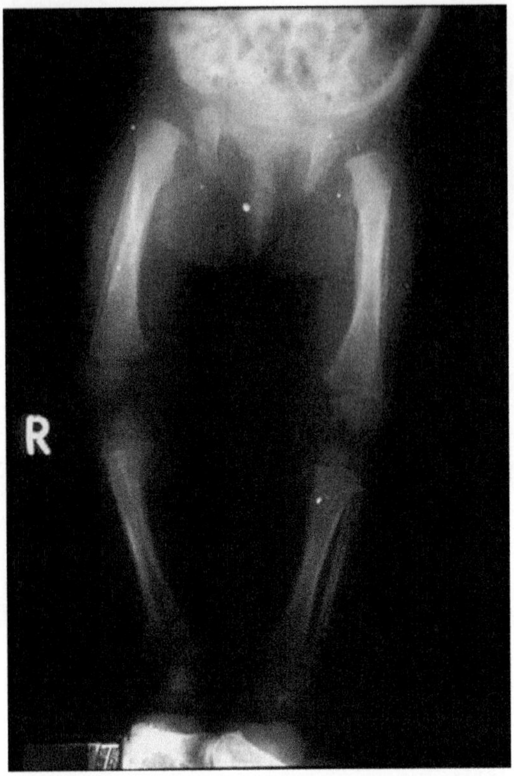

Fig. 16.7 Lower limb X-rays of a newborn infant showing periostitis due to congenital syphilis. Note the lucencies at the proximal ends of the tibias and the periosteal thickening of the femurs. Copyright © Frank E. Berkowitz. Reprinted with the permission of Cambridge University Press.

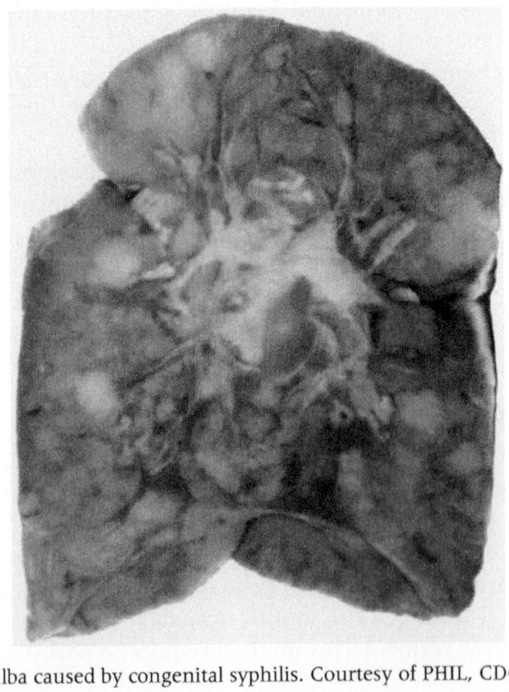

Fig. 16.8 Pneumonia alba caused by congenital syphilis. Courtesy of PHIL, CDC.

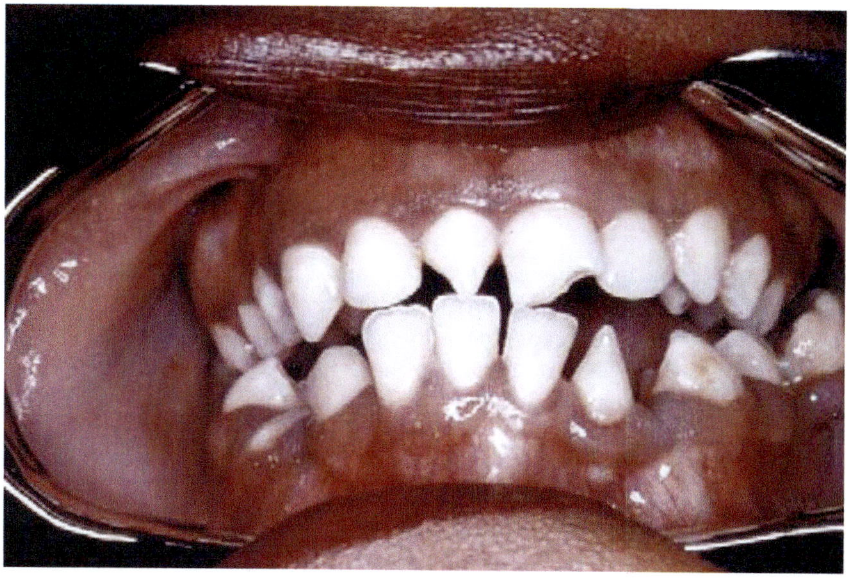

Fig. 16.9 Hutchinson teeth due to congenital syphilis. Courtesy of PHIL, CDC.

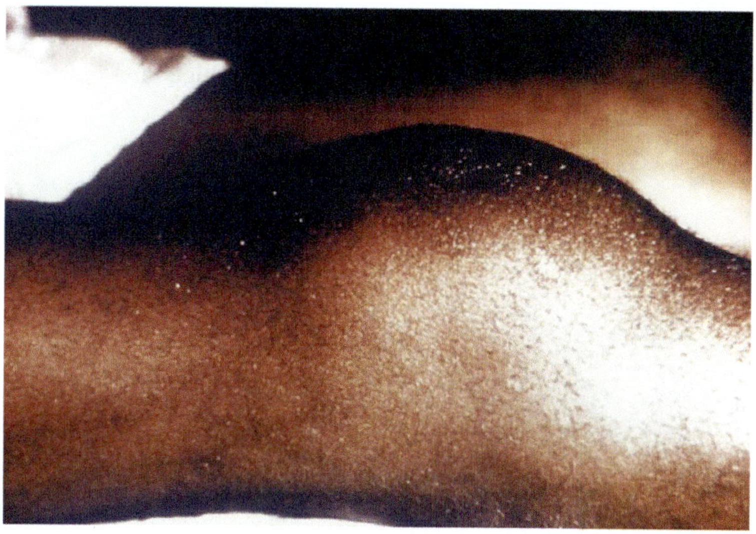

Fig. 16.10 Clutton's joints due to congenital syphilis. Courtesy of PHIL, CDC.

Serology

There are two basic types of tests: non-treponemal tests, in which the antigens used are not derived from the spirochete but contain cardiolipin, lecithins, and cholesterol, which flocculate when they react with IgG or IgM antibodies to *T. pallidum*; and specific treponemal tests.

Non-treponemal tests

These are the rapid plasma reagin tests (RPR), the Venereal Disease Research Laboratory test (VDRL), and toluidine red unheated serum test. Their advantages are that they are cheap, easily performed, and quantitative titers can be determined. These titers become positive 3–6 weeks after infection, and decrease with effective treatment. They can therefore be used for evaluating the effectiveness of treatment.

Treponemal tests

The older tests are the microhemagglutination assay for *T. pallidum* (MHA-TP), the *T. pallidum* particle agglutination (TPPA) test, the *T. pallidum* hemagglutination assay (TPHA), and the fluorescent treponemal antibody absorption test (FTA-ABS). Several enzyme immunoassays (EIA) and rapid tests have been developed. These antibodies persist despite effective treatment.

Factors that are important in interpretation of the tests, especially when used for screening, are the predictive values (which take into account the sensitivities, specificities, and prevalence of syphilis in the population), ease of performance, and cost. Syphilis testing algorithms, using treponemal and non-treponemal tests, are shown in Fig. 16.11.

The traditional algorithm detects active infection but may miss early primary and treated infections. In addition, there is a high rate of false positives that require a specific treponemal test for confirmation.

The reverse algorithm using EIA or chemiluminescence immunoassay (CIA) detects early primary and treated infection that may be missed with the traditional algorithm. These tests are not specific and also suffer from high rates of false positivity

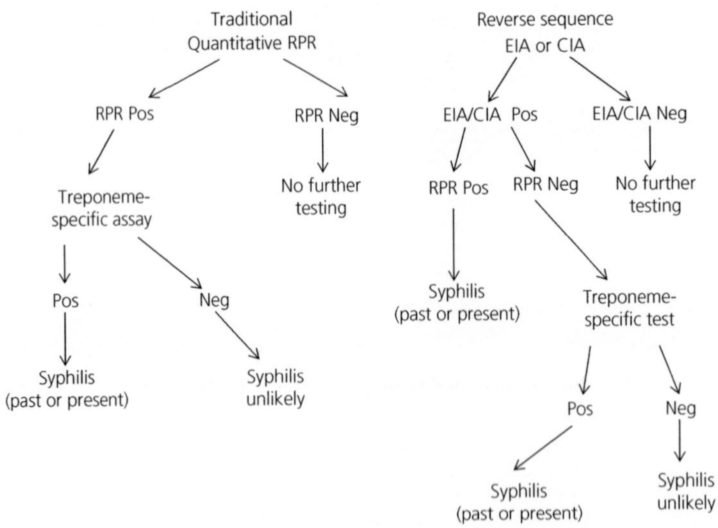

Fig. 16.11 Traditional and reverse sequence algorithms for syphilis testing. CIE, chemiluminescence immunoassay; EIA, enzyme immunoassay; RPR, rapid plasma reagin test. Adapted from Radolf JD, Bolan G, Park IU, et al. (2011) Discordant results from reverse sequence syphilis screening – five laboratories, United States, 2006–2010. MMWR 60:133–7.

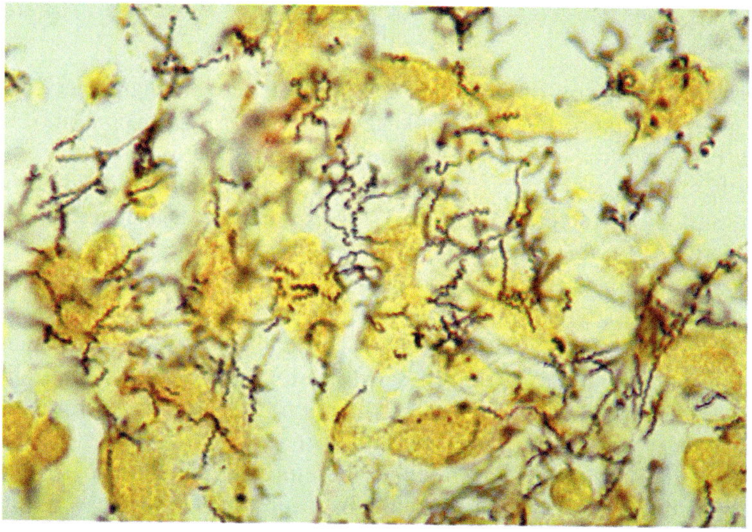

Fig. 16.12 Histology showing spirochetes (*Treponema pallidum*) stained with a silver stain. Courtesy of PHIL, CDC.

and therefore have high sensitivity and low specificity. If positive, a non-treponemal test is required to detect active infection. All positives by either RPR or EIA/CIA must be confirmed by specific treponemal antigen test. The advantage of the reverse algorithm is that testing is amenable to automation and is useful when testing large numbers of samples.

Although histology is not generally used for making the diagnosis, the organism can be seen in tissue stained with a silver impregnation technique (Fig. 16.12).

Treatment

The treatment of syphilis is penicillin, by injection. The number of doses depends on the stage of the infection. Treatment can be complicated by a Jarisch–Herxheimer reaction, in which there is an initial clinical deterioration characterized by increasing temperature and hypotension.

Endemic treponematoses

The organisms that cause these infections are indistinguishable from *Treponema pallidum* subspecies *pallidum*, and serologic tests give the same results as in syphilis. They are yaws (caused by *T. p.* subspecies *pertenue*), bejel (caused by *T. p.* subspecies *endemicum*) and pinta (caused by *T. p.* subspecies *carateum*). They occur in limited areas, mainly of the tropics and poor areas. Yaws occurs in South America, sub-Saharan Africa, and South and South East Asia, bejel occurs in West and Southern Africa and Saudi Arabia, and pinta occurs only in South and Central America. They are transmitted by close contact, and affect mainly children and teenagers. The initial manifestations are mostly cutaneous, but they can lead to secondary and tertiary disease, as in syphilis.

Leptospirosis

This is caused by organisms of the genus Leptospira. There are eight species, 23 serogroups (e.g. icterohaemorrhagiae, canicola, grippotyphosa, and pomona), and 230 serovars. It has a worldwide distribution, and is transmitted to humans from animal urine, mostly via bodies of water, such as lakes. It is a hazard for individuals working with meat, and also a recreational hazard, for example for triathletes who swim in lakes. Within the USA, the state with most cases is Hawaii.

The clinical manifestations are protean, because all organs can be affected. It causes a febrile illness which may be associated with many other clinical features. Although it may be biphasic, with an initial febrile illness and a second phase associated with multiple organ abnormalities, this is not necessarily the case. Additional features include myalgias, headache, meningitis, rash, jaundice, conjunctival suffusion, pulmonary hemorrhage, and renal failure.

The diagnosis cannot be made rapidly. Therefore the decision to treat must be made on clinical grounds, based on potential exposure and clinical features.

Diagnostic tests

Although the organism can be cultured, this requires special media and takes several weeks to grow (Fig. 16.13). It can be visualized by dark-field microscopy of body fluids such as serum, cerebrospinal fluid, and peritoneal dialysate, but the sensitivity and specificity of this are low. The diagnostic tests of choice are serologic, the gold standard being the microhemagglutination test (MAT). Seroconversion or a four-fold rise in titers is considered diagnostic, while a single titer of >1/400 is suggestive. PCR is not readily available.

Treatment, which consists of penicillin, ceftriaxone, or doxycyline, can be complicated by a Jarisch–Herxheimer reaction (see Syphilis).

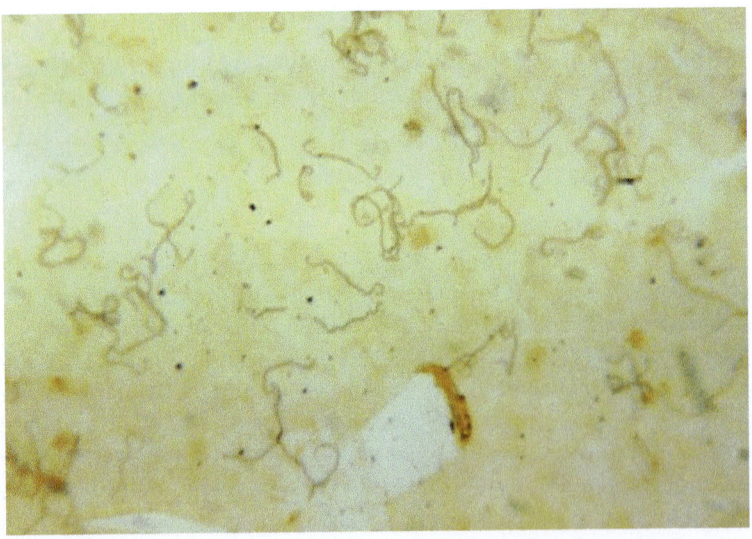

Fig. 16.13 Liver smear of a fatal case of leptospirosis stained with a silver impregnation stain and showing leptospira organisms. Courtesy of PHIL, CDC.

Borrelia

There are two groups of borrelias, those causing Lyme disease and those causing relapsing fever.

Lyme disease

This is caused by *Borrelia burgdorferi* (*Borrelia burgdorferi* sensu strictu, the only species causing disease in the USA) and related borrelias (*Borrelia burgdorferi* sensu lata, namely *B. afzelii* and *B. garinii*) which cause Lyme disease in Europe. The organism is transmitted by hard ticks. In the USA these are *Ixodes scapularis* and *Ixodes pacificus*, and in Europe, *Ixodes ricinus*.

The disease has early and late manifestations. The early manifestation, which develops 1–4 weeks after the tick bite, is erythema migrans, an erythematous localized rash occurring at the site of the tick bite. This can be characteristic enough for a clinical diagnosis to be made and for treatment to be instituted (Fig. 16.14).

Early disseminated disease occurs 3–5 weeks after inoculation. This may manifest as multiple skin lesions, cranial nerve palsies, especially facial, meningitis, and carditis.

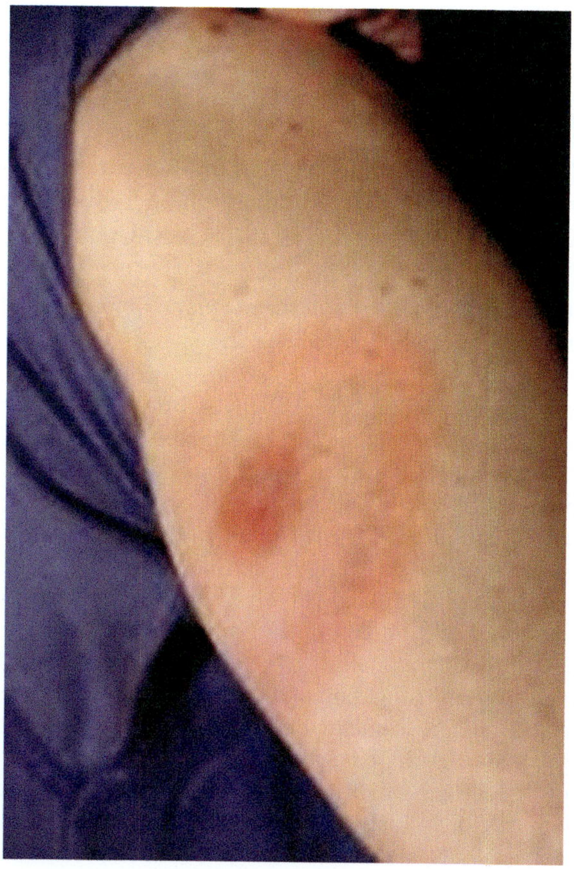

Fig. 16.14 Patient with erythema migrans. Courtesy of PHIL, CDC.

The latter can be in the form of pericarditis or myocarditis, especially heart block of different degrees. During this stage, non-specific constitutional symptoms may occur. After several weeks to months, patients may develop late disease characterized by arthritis, usually of large joints, and encephalitis.

The diagnosis is made clinically in the early stage and serologically in subsequent stages. Diagnostic testing is fraught with hazards in interpretation, and should not be undertaken lightly. There are many false positives, and patients who have vague symptoms might be led to ascribe all their symptoms to Lyme disease, which they do not have. Current recommendations from the CDC are that serologic testing should be done in two stages: an ELISA, which is sensitive, and, if positive, a Western blot, which is more specific. However, a recent study suggests that a single test utilizing the VlsE protein or its C6 peptide can be used. It has the additional advantage that it is useful for diagnosing cases acquired in the USA and in Europe. Testing should not be performed in individuals with a pretest probability of infection of less than 0.2. Although the organism can be cultured, e.g. from skin lesions, this is not generally available. PCR tests are not useful at this time.

Treatment consists of amoxicillin, tetracycline, or ceftriaxone, for variable periods, depending on the age of the patient (children younger than 8 years should not be treated with a tetracycline) and the stage of the infection.

Relapsing fever

This is a group of arthropod-borne infections characterized by periods of fever, lasting a few days, interspersed with asymptomatic periods also lasting a few days.

There are three main groups.

- *B. recurrentis*: this has a world-wide distribution, but currently occurs mainly in Sudan and Ethiopia. It is transmitted by the human body louse *Pediculus humanus humanus* (see Chapter 26), and carries the highest fatality rate of the relapsing fevers.
- *B. duttonii*: this is transmitted by the soft tick, *Ornithodoros moubata*, in Africa.
- A group of American borrelias named for the species of soft tick of the genus Ornithodorus that transmits them, e.g. *B. hermsii*, *B. turicatae*, and *B. parkeri*.

The relapsing fever caused by these organisms is due to the phenomenon called antigenic variation (see *Trypanosoma brucei* in Chapter 24). The surface coat of the organism, which contains variable membrane protein, changes periodically. As the host develops antibodies to it and the number of organisms in the blood declines, the coat changes and the antibodies are no longer active against it.

Clinically, in addition to the relapsing fever, there may be rashes, enlargement of liver or spleen, jaundice, and neurologic involvement.

The diagnosis is best made by demonstration of the organism in a stained thin blood smear (Fig. 16.15). Giemsa staining is the standard method, the sensitivity of which can be increased by centrifugation of the blood. Acridine orange and fluorescent stains can also be used. Phase contrast and dark-field microscopy of diluted blood (1:10) are other methods. Serology is not useful for immediate diagnosis. There are additional problems associated with the use of serology: the major antigens are the variable membrane proteins, which vary over time; antibodies persist after previous infections; and there are cross-reactions with other spirochetes.

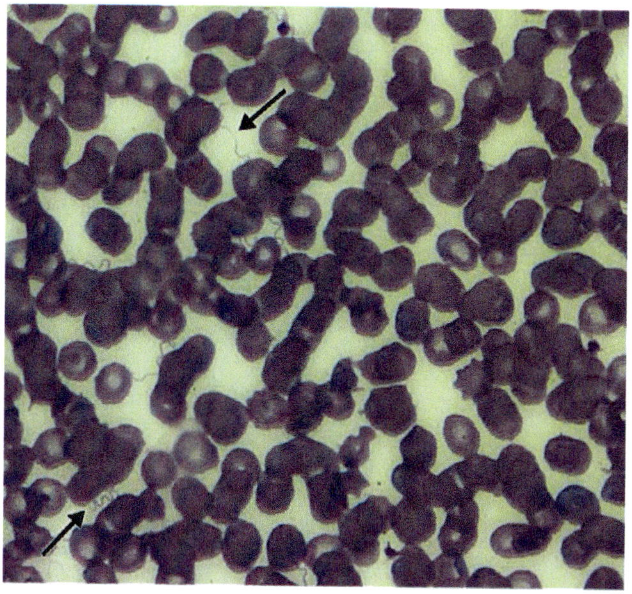

Fig. 16.15 Giemsa-stained blood smear showing *Borrelia hermsii*. Courtesy of PHIL, CDC.

Several different antimicrobial agents can be used for treatment, including doxycycline, amoxicillin, ceftriaxone, and penicillin. Treatment can be complicated by the Jarisch–Herxheimer reaction (see treatment of syphilis earlier in this chapter).

Spirillum minus

This organism causes a form of rat-bite fever called Sodoku. It occurs mainly in Asia, but has been reported in Brazil. It has not been cultured *in vitro*, and consequently little is known about it. It is motile by virtue of bipolar flagellae, and it can be visualized by dark-field microscopy. The clinical illness is similar to that caused by *Streptobacillus moniliformis* (see Gram-negative rods, Chapter 13), but it has a longer incubation period (2–3 weeks) and is often associated with a chancre at the bite site, and regional lymphadenopathy.

Further reading

Branda JA, Strle F, Strle K, et al. (2012) Performance of United States serological assays in the diagnosis of Lyme borreliosis acquired in Europe. Clin Infect Dis 57: 333–40.

Farnsworth N, Rosen T (2006) Endemic treponematosis: review and update. Clin Dermatol 24: 181–90.

Larsson C, Andersson M, Bergstrom S (2009) Current issues in relapsing fever. Curr Opin Infect Dis 22: 443–9.

Lau C, Smythe L, Weinstein P (2010) Leptospirosis: an emerging disease in travellers. Travel Med Infect Dis 8: 33–9.

Mitja O, Asiedu K, Mabey D (2013) Yaws. Lancet 381: 763–73.
Murray TS, Shapiro ED (2010) Lyme disease. Clin Lab Med 30: 311–28.
Pierce EF, Katz KA (2011) Darkfield microscopy for point-of-care syphilis diagnosis. Med Lab Obs 43(1): 30–1.
Ricaldi JN, Vintz JM (2006) Leptospirosis in the tropics and in travelers. Curr Infect Dis Rep 8: 51–8.
Sena AC, White BL, Sparling PF (2010) Novel Treponema pallidum serologic tests: a paradigm shift in syphilis screening for the 21st century. Clin Infect Dis 51: 700–8.
Wöhrl S, Geusau A (2007) Clinical update: syphilis in adults. Lancet 369: 1912–14.

CHAPTER 17
Mycobacteria

Taxonomy

The mycobacteria are a diverse, complex group of organisms. They are related to the Gram-positive rods such as Nocardia, Streptomyces, Rhodococcus, Tsukamurella, and Gordonia. They differ in that their cell walls contain N-glycolyl muramic acid (instead of N-acetyl muramic acid) and they have an extremely high lipid content. As a result, the dyes used in the Gram stain routinely fail to detect them, although the trained microscopist may see a "cell outline" if the organisms are present in abundance. Special staining procedures are necessary to visualize them. Because of the cell wall composition, they resist decolorization with 3% hydrochloric acid after being exposed to basic fuchsin dye for prolonged periods of time (or alternatively, heating of this dye on the slide). Therefore they are classified as "acid fast."

Organisms within the group can be separated based on growth rate and unique colony morphologies that further distinguish the species, as described below. Current updates on nomenclature/taxonomy can be found at www.bacterio.cict.fr/ or www.dsmz.de/home.html.

The most important species of the genus is *M. tuberculosis*, the cause of human tuberculosis. However, there are more than 100 species of mycobacteria, most of which are environmental bacteria that cause human disease mainly as opportunists. They are not spread like tuberculosis from human to human. These are often referred to as non-tuberculous mycobacteria (NTM), or mycobacteria other than tuberculosis (MOTT). They are discussed later in this chapter.

Mycobacterium tuberculosis consists of a complex of several organisms, some of which cause tuberculosis mainly in animals, but can infect humans. These are *M. tuberculosis*, *M. bovis* (bovines), *M. caprae* (goats), *M. africanum* (humans), *M. microti* (voles), *M. canetti*, and *M. pinnipedii* (seals).

Practical Medical Microbiology for Clinicians, First Edition. Frank E. Berkowitz and Robert C. Jerris.

Tuberculosis

This is a very important infection worldwide, with about 9 million new cases per year, of which about 10% occur in children. The incidence and prevalence are extremely variable across the globe; for example, the overall incidence in the USA is about 3/100,000 per year, while that in some parts of Africa and Asia is greater than 300/100,000 (Fig. 17.1).

Because the organism multiplies slowly (doubling time of about 24 hours, and up to 36 hours, compared with that of *E.coli* of about 30 minutes), the disease is slowly progressive and individuals can be contagious for a long time.

Pathogenesis

An infected individual (the source is almost always an adult) coughs up the organism which can remain suspended in the air for many hours on small droplet nuclei (<5 microns diameter); another individual (the victim) inhales the organism which enters the alveolus and is ingested by a pulmonary macrophage, in which it can survive. This is carried to the draining lymph nodes in the hilum of the lung, where the organism can spread and cause focal inflammation. The initial pulmonary focus is called a primary (Ghon) focus, and the combination of the pulmonary and lymph node foci is called a primary (Ghon) complex. This sequence of events, which is subclinical, is called primary

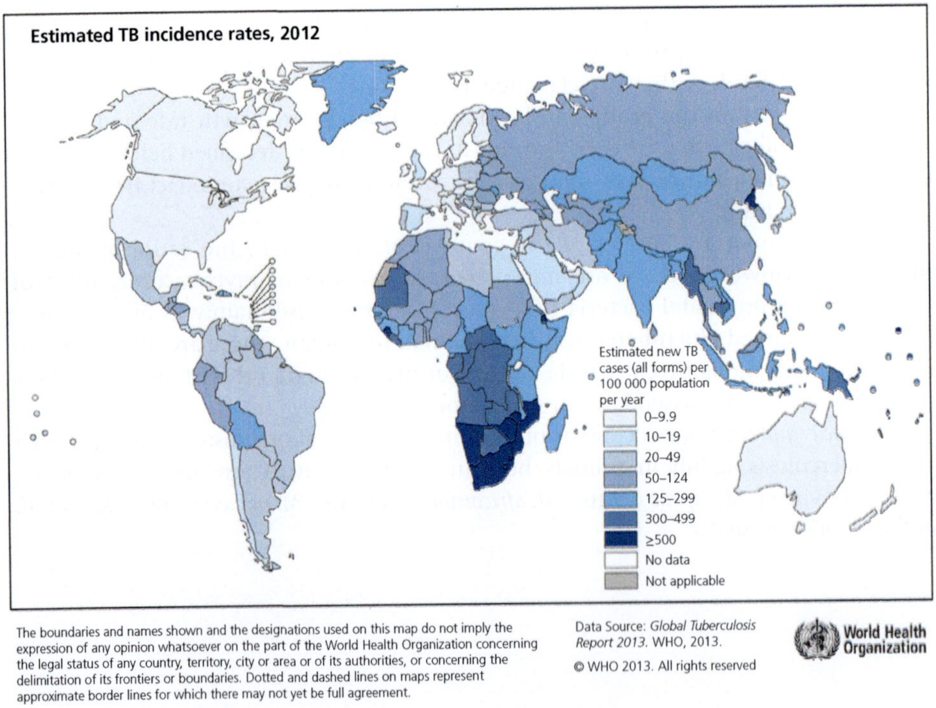

Fig. 17.1 World map showing the global distribution of tuberculosis, according to incidence (cases per 100,000 population) as of 2012. Source: Global Tuberculosis Report 2013, with permission from the World Health Organization.

tuberculosis, and is the usual course of events in children. The organism can spread into the blood and be carried to any part of the body, such as bone, kidney, and brain. In children, the enlarged hilar lymph nodes play a very important role in the clinical manifestations (Fig. 17.2). They can compress the adjacent bronchi, leading to distal collapse or pneumonia, and they can erode into the bronchus, causing endobronchial disease, which can spread along the whole of the bronchial tree to different parts of the lungs. Infection can spread within the primary focus of infection, causing progressive primary tuberculosis. When large numbers of bacteria enter the blood, as can occur if a blood vessel is eroded, multiple small foci of inflammation develop wherever these bacteria settle, a condition called miliary tuberculosis (named for millet seeds, which these foci resemble).

In most cases of infection, the organism is controlled by the host's immune system, so that no disease occurs. The fact that the individual had ever been infected can then be determined only by detecting an immune response by a tuberculin skin test or a γ-interferon release test (see later). However, the organism remains dormant and can become reactivated months, years, or decades later. Immunosuppression, in particular HIV infection, is an important reason for reactivation but the cause cannot always be determined. Reactivation and reinfection (i.e. infection in an individual who has previously been infected with the organism) is called postprimary tuberculosis. It differs

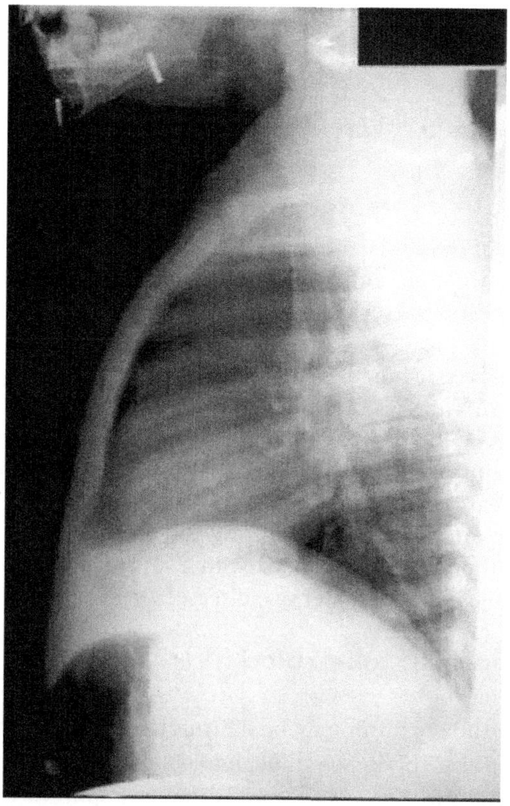

Fig. 17.2 Chest X-ray showing hilar lymphadenopathy due to tuberculosis in a young child.

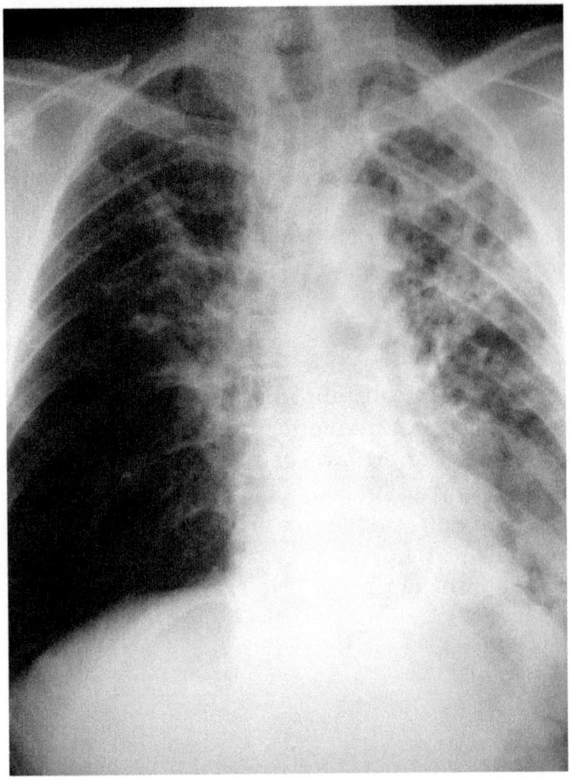

Fig. 17.3 Chest X-ray showing left upper lobe pulmonary infiltrate with a cavity. Courtesy of Dr Aliya Yamin.

from primary disease in that it usually occurs in the upper lobes of the lung, where it causes significant destruction and cavitation, which are associated with positive sputum smears and cultures (see later) (Fig. 17.3); the hilar lymph nodes are not enlarged.

Tuberculosis causes a wide range of clinical manifestations, mostly affecting the lung. Pleural disease can also occur (Figs 17.4 & 17.5). There may be non-specific manifestations of infection, such as fever, night sweats, and of chronic illness, such as weight loss (or, in children, failure to gain weight appropriately). Cough and hemoptysis are the main symptoms of lung disease in adults. The infection should be suspected in patients with symptoms of focal or multifocal disease lasting for several weeks, especially in high prevalence areas or if there has been contact with a suspected case. Pulmonary findings may not be present unless there is extensive disease.

Laboratory diagnosis of tuberculosis

This poses two challenges.

- Material containing the organism may be difficult to obtain. The main specimen that can be studied is sputum. This is very difficult to obtain in young children. In children, alternative specimens can be obtained from induced sputum and gastric material containing swallowed sputum. Bronchoalveolar wash specimens should be used if

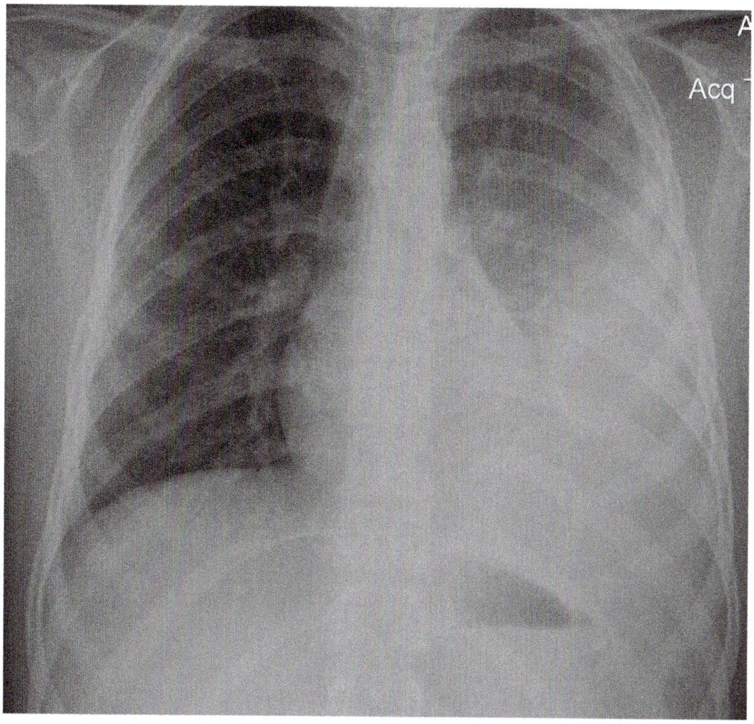

Fig. 17.4 Chest X-ray showing a tuberculous pleural effusion in a teenage boy.

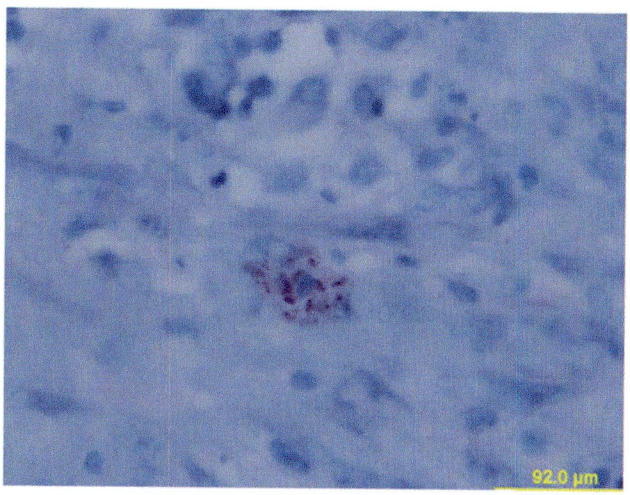

Fig. 17.5 Acid-fast bacilli within a macrophage in a pleural biopsy of the patient shown in Fig. 17.4.

obtained. However, whenever tissue is readily obtainable, biopsy or fine needle aspiration should be performed to obtain material for culture and histology.

- The organism is slow-growing, so that culture results may take several weeks to become positive. This poses problems not only in the diagnosis of the infection, but also in the determination of antimicrobial susceptibilities, which can take an additional few weeks.

Mycobacterium tuberculosis can be detected by microscopy, culture, and genome detection (nucleic acid amplification with optimal identification from acid-fast smear-positive specimens). In addition, the host's response to infection can also be detected, in the laboratory and in the patient.

Any type of specimen can be subjected to microscopy for mycobacteria. However, specimens for culture require decontamination and digestion if they are from non-sterile sites, such as sputum, the most common specimen.

Stains for microscopy

The classic stain is the Ziehl–Neelsen stain. This entails staining the specimen on a slide with heated carbol-fuchsin, a red stain, and washing it with a decolorizing agent, acidified alcohol. The slide is then counterstained with a blue or green stain. Mycobacteria retain the stain after the wash, hence the name "acid-fast" (Fig. 17.6). Because examination of slides stained with Ziehl–Neelsen stain requires oil-immersion lens examination (×1000), it is very labor intensive. This has therefore been replaced by use of a fluorescent stain, auramine, which stains DNA, and also requires an acid-alcohol wash. The auramine-stained slides can be examined under lower power (×450), but this requires a fluorescent microscope (Fig. 17.7).

Culture for isolation of mycobacteria (*M. tuberculosis* and non-tuberculous mycobacteria) (see later)

The key to success in laboratory detection is to have a high index of suspicion and communication with the laboratory personnel. Most of the organisms, with the exception of the "rapid grower" group (see later), have a prolonged time of incubation for detection (as many as 6–8 weeks, mean time for all groups 3 weeks). In addition,

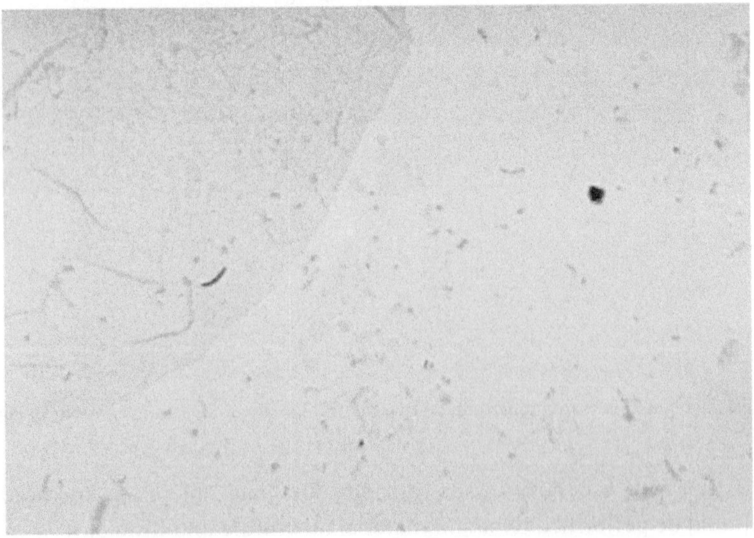

Fig. 17.6 Ziehl–Neelsen showing an acid-fast bacterium. Courtesy of PHIL, CDC.

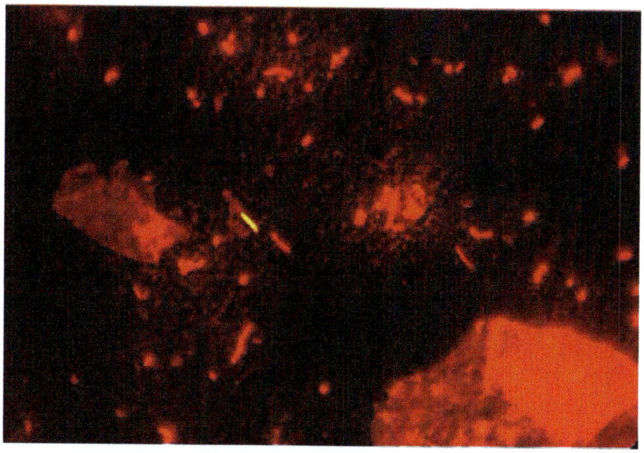

Fig. 17.7 An acid-fast bacillus stained with aureamine, appearing yellow. Courtesy of PHIL, CDC.

specific organisms within the group may require specialized media/growth factors or modification of temperature, e.g. *M. haemophilum* requires hemin (hemoglobin); *M. marinum* and *M. ulcerans* prefer lower temperatures (30–33°C). Mycobacteria prefer an environment of increased CO_2 of 5–10%.

Detection and identification of mycobacteria require fastidious laboratory techniques. Taking advantage of the unique cell wall of the mycobacteria specimens that may contain normal microbiota (respiratory sources), procedures for pretreatment prior to culture are used to eliminate non-mycobacterial isolates. These pretreatments (referred to as digestion-decontamination) may include N-acetyl-L-cysteine-sodium hydroxide, zephiran-trisodium phosphate, oxalic acid, or cetylpyridinium chloride-sodium chloride. Specimens are then plated on specialized media. With the exception of the fastidious organisms noted above, most strains grow on Lowenstein–Jensen media (coagulated whole eggs, salts, glycerol, potato flour, and malachite green [as inhibitor to routine bacterial growth]) and Middlebrooks 7H11 plates (defined salts, vitamins, co-factors, albumin, catalase, glycerol, casein hydrolysate, and malachite green). These organisms also grow in a variety of liquid media, which are now recommended as a routine because of faster growth and detection in automated instruments. They include BBL MGIT (mycobacteria growth indicator) tube for use in the BACTEC MGIT automated system, BactAlert Mycobacteria bottle, and TREK ESP culture system. Once detected in culture (Fig. 17.8), organisms may be identified by conventional phenotypic tests, molecular techniques, and MALDI-TOF MS. Note that key morphologic growth characteristics are extremely helpful in presumptive identification, such as the "cording phenomenon" with *M. tuberculosis* complex.

Major applications of molecular techniques have made detection and identification much faster. These include direct detection from clinical specimens (PCR); culture confirmation from growth media; DNA sequencing and DNA fingerprinting for epidemiology. GeneXpert (Cepheid) is a molecular test that can be used on sputum to detect both *M. tuberculosis* and rifampin resistance. This is of particular importance, because rifampin resistance is a surrogate marker for multidrug resistance. Processing takes about 90 minutes.

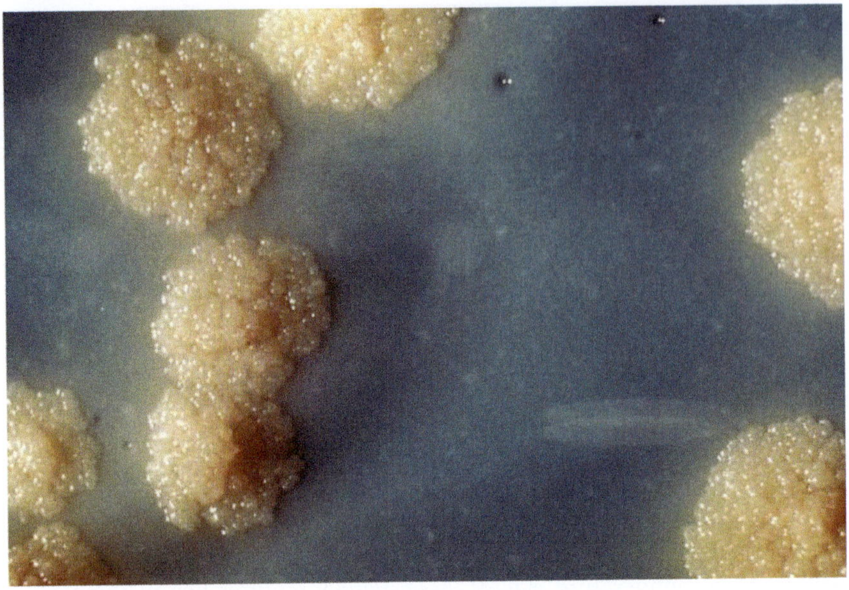

Fig. 17.8 Colonies of *Mycobacterium tuberculosis* growing on a solid medium. Courtesy of PHIL/CDC.

Tests of the host response to infection include the following.

- Tuberculin skin test (TST): this is the traditional test, which has been used for many years. It depends on the delayed-type hypersensitivity of the host to mycobacterial antigens (purified protein derivative – PPD) injected intradermally on the volar aspect of the forearm. This produces an area of induration around the area of the original injection, caused by edema and the infiltration of lymphocytes and macrophages. After 48–72 hours the transverse diameter of the area of induration is measured in millimeters with a flexible tape, and positivity is decided according to the pretest probability/risk of infection (Fig. 17.9). The disadvantages of the test are the following: placing and reading it requires skill, the edges of induration can be difficult to determine, it requires a second visit for the reading, and it is not specific for *M. tuberculosis* (patients who have previously been infected with non-tuberculous mycobacteria, or who have received BCG – tuberculosis vaccine made from attenuated *M. bovis* – can react).
- Interferon γ release assays (IGRAs): these tests, performed *in vitro*, depend on the fact that lymphocytes sensitized to mycobacterial antigens release interferon γ when reexposed to such antigens. The presence of the interferon is measured. There are two main types of IGRAs: (a) in-tube: blood is collected into a tube containing the antigens. After an incubation period, interferon γ is measured. (b) ELISPOT: a fixed number of peripheral blood mononuclear cells (lymphocytes) are stimulated in microtiter wells, whose floors have been coated with anti-interferon γ antibodies; the lymphocytes settle on the bottoms of the wells, and any interferon γ produced is bound by the underlying antibodies. After the lymphocytes have been washed off, another anti-interferon γ antibody, labeled with an enzyme, is added, followed by addition of a substrate of the enzyme that becomes colored when acted upon by the

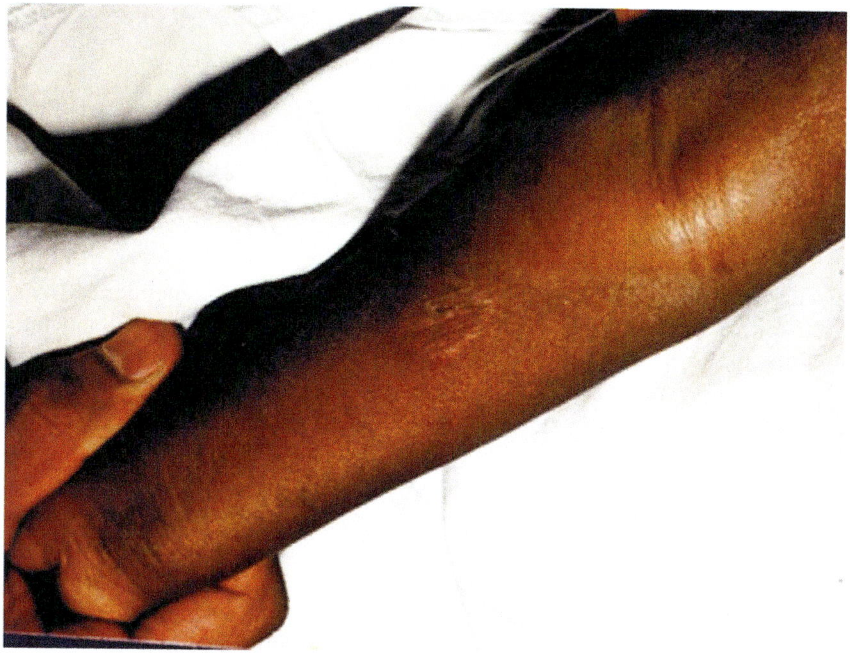

Fig. 17.9 A reactive tuberculin skin test (PPD).

enzyme. Wherever an interferon-producing lymphocyte had been present, a spot is produced. The number is quantified. (See the ELISA test, Chapter 2.)

The advantages of such tests are that the antigens used are specific for *M. tuberculosis*, and a repeat visit is not necessary. Unfortunately, these tests have not been shown to be more sensitive than the TST, especially in children younger than 2 years.

Treatment of tuberculosis

Mycobacterium tuberculosis exists in the host in several different states: rapidly dividing organisms, which are present in cavities in large numbers, slowly dividing organisms present in caseous material and within macrophages, and dormant organisms. Because mutations to resistance occur, at varying rates, resistance can emerge during therapy. This is more likely to occur if only a single drug is used, compared with multiple drugs, because resistance is not likely to develop to more than one drug. (The probability of this occurring is the product of the probability of this occurring to each drug.) The main principles of therapy should entail the following: (a) multiple drugs, to which the organism is believed to be susceptible (susceptibility is not known at the time that therapy is instituted, unless the susceptibility of the isolate from the contact is known); (b) prolonged (6 months), if optimal therapy can be used, longer (up to 18 months), if not; (c) directly observed (i.e. a healthcare worker administers the drug to the patient); (d) never add a single drug to a failing regimen.

Antituberculous drugs are considered in the categories of first line (isoniazid, rifampin, pyrazinamide, ethambutol, and streptomycin) and second line, used when the first-line drugs cannot be used, because of either patient intolerance or organism resistance (ethionamide, amikacin and kanamycin, fluoroquinolones, capreomycin and viomycin, cycloserine and terizidone, linezolid, and paraaminosalicyclic acid [PAS]). Their main adverse effects are summarized in Table 17.1.

Isoniazid (INH)

Isoniazid is a prodrug that is converted to an active drug by a bacterial enzyme, encoded by the katG gene. It interferes with mycolic acid synthesis. Resistance to the drug can be caused by a mutation of the katG gene.

Rifamycins

These drugs (rifampin, rifabutin, and rifapentine) inhibit DNA-dependent RNA polymerase. They are active mainly against Gram-positive bacteria and mycobacteria, and their main value is for the treatment of patients with tuberculosis, leprosy, and some other mycobacterial infections. A very important adverse effect is their ability to induce cytochrome oxidase enzymes, which metabolize many other drugs, resulting in a decrease in the effects of those drugs.

Rifampin

This is the most frequently used rifamycin. In addition to being a mainstay of antituberculous therapy, it is used for treatment of patients with staphylococcal infection on foreign bodies such as prosthetic heart valves. In addition to causing drug–drug interactions, it can cause hepatotoxicity, thrombocytopenia, and an influenza-like illness if used intermittently.

Rifabutin

This causes less drug–drug interaction than rifampin, and is used for patients with HIV infection and tuberculosis, who require protease inhibitors and antituberculous therapy. One of its important adverse effects is uveitis.

Rifapentine

This drug has a very long half-life and is used, in combination with isoniazid, for short-course treatment of latent tuberculosis infection.

Table 17.1 Antituberculous drugs and their major adverse effects.

Drug	Major adverse effects
Isoniazid	Hepatotoxicity, peripheral neuropathy (vitamin B6 responsive)
Rifampin, rifapentine	Orange urine and tears, hepatotoxicity, drug interactions, thrombocytopenia, "flu-like" illness, rash
Rifabutin	Myelosuppression, uveitis, corneal opacity
Pyrazinamide	Hepatotoxicity, hyperuricemia
Ethambutol	Optic neuropathy, retinopathy
Streptomycin	8th nerve damage
Amikacin, kanamycin	8th nerve damage, nephrotoxicity
Capreomycin, viomycin	Ototoxicity, nephrotoxicity, neuromuscular blockade
Ethionamide	Gastrointestinal intolerance, endocrine disturbances
Cycloserine, terizidone	Neurotoxicity
PAS	Gastrointestinal intolerance, hypothyroidism
Fluoroquinolones	Gastrointestinal disturbances, headache, tendon rupture
Linezolid	Myelosuppression, peripheral and optic neuropathy

Pyrazinamide

This is also a prodrug that is converted to pyrazinoic acid, which interferes with cell membrane synthesis. It functions in an acid environment, including within cells. *M.bovis* and Bacille Calmette–Guerin (BCG), the vaccine derived from it, are naturally resistant to this drug.

Ethambutol

This interferes with cell wall synthesis. Its main value lies in protecting the other drugs in the therapeutic regimen from the development of resistance.

Aminoglycosides

Streptomycin, amikacin, and kanamycin interfere with protein synthesis. They must be injected, and are used in the treatment of resistant cases.

Capreomycin and viomycin

These inhibit protein synthesis by binding to the 30S and 50S subunits of the ribosome. They must be administered by injection.

Ethionamide

This acts in the same manner as INH, and there is cross-resistance, depending on the gene mutation of katG causing INH resistance.

Cycloserine and terizidone

These inhibit cell wall synthesis.

Two new drugs have recently been approved for the treatment of multidrug-resistant tuberculosis: bedaquiline, which inhibits ATP synthetase, and delaminid, which inhibits mycolic acid synthesis.

Prevention

Because tuberculosis is highly contagious, prevention of spread depends largely on public health measures. This involves reporting of cases so that contacts can be traced, investigated for infection, and treated if indicated. BCG vaccine is used in most countries where tuberculosis is prevalent. Its effectiveness is much lower than that of vaccines used to prevent other infections.

Non-tuberculous mycobacteria

These have been classified by Runyon as follows.

A Slow growers (that take longer than 1 week to produce colonies)

1 Those producing pigment after exposure to light (photochromogens), e.g. *M. kansasii, M. marinum, M. simiae, M. genavense, M. asiaticum*
2 Those producing pigment in the dark (scotochromogens), e.g. *M. scrofulaceum, M. gordonae, M. szulgai, M. xenopi, M. celatum, M. flavescens*

3 Those not producing pigment (non-chromogens), e.g. *M. avium-intracellulare* complex, *M. malmoense, M. paratuberculosis, M. terrae, M. gastri, M. trivial, M. hemophilum, M. ulcerans*

B Rapid growers (that produce colonies in less than 1 week): *M. fortuitum, M. chelonei, M. abscessus,* and *M. smegmatis* (Fig. 17.10)

Non-tuberculous mycobacteria cause disease in normal hosts and individuals with immune system impairment such as transplantation, AIDS, and some primary immunodeficiencies, such as chronic granulomatous disease. Pulmonary infection accounts for the majority of cases. However, they also cause cervical lymphadenitis (mainly in otherwise normal children), skeletal infections, usually complicating penetrating trauma, and skin and soft tissue infections, also complicating trauma, which might have been minor, including surgery. A particularly severe type of skin infection, which progresses to bone infection, is Buruli ulcer, caused by *M. ulcerans*. This is prevalent in tropical areas of the world.

The main differential diagnoses of these infections are tuberculosis, fungal infections, and cancer.

The diagnosis depends on obtaining material from the infected tissue, optimally by biopsy, but fine needle aspiration is also useful (other than for lung tissue). Microscopy should be performed after staining with an acid-fast stain, and culture on liquid and solid media should be performed at 28–30°C and at 35–37°C. Identification of the species may be done by high-performance liquid chromatography of mycolic acids, and by various molecular methods, especially gene sequence analysis and MALDI-TOF MS.

The antimicrobial susceptibilities of these organisms are very variable; they are not susceptible to the drugs used for patients with tuberculosis, with the exception of the rifamycins and ethambutol. Treatment depends on the location of disease and the type

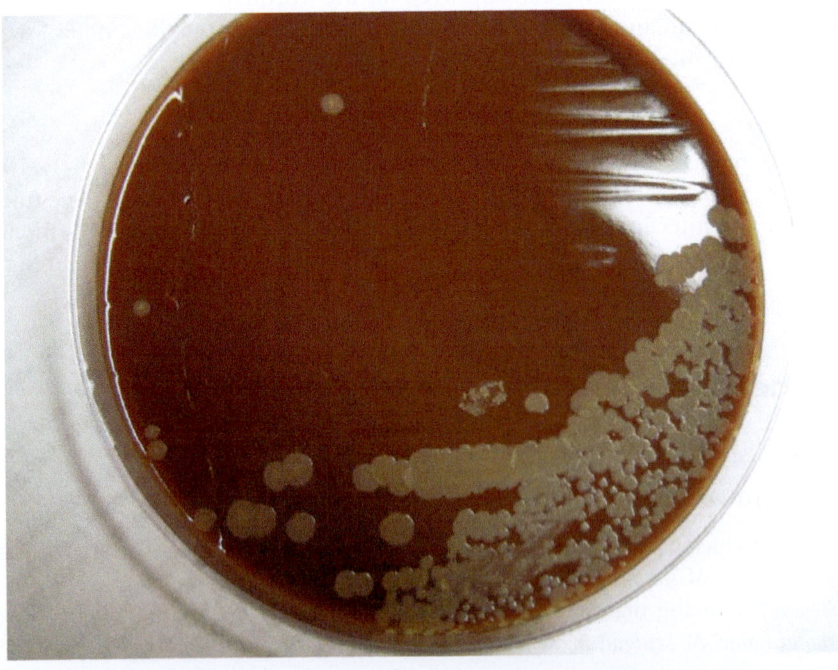

Fig. 17.10 Colonies of a rapidly growing mycobacterium on chocolate agar.

of host. For lymphadenopathy in a normal host, it is usually surgical, sometimes augmented by antimicrobial therapy. On the other hand, the treatment for a patient with AIDS and disseminated *M. intracellulare* complex infection is a combination of a macrolide (clarithromycin or azithromycin) and ethambutol.

Leprosy (Hansen's disease)

This is a very important mycobacterial infection with a world-wide distribution, caused by *M. leprae*, which was the first bacterium demonstrated to be a human pathogen (1873). It has been associated with significant social stigma and, in the past, prior to the advent of effective antimicrobial therapy (see later), infected individuals were banished from society or confined to leprosaria.

The organism is transmitted by close contact, mainly from infected nasal secretions. In the southern USA, it is also transmitted by armadillos. The incubation period is 2–12 years. There is likely a genetic influence on susceptibility to this infection, most individuals being resistant. The organism grows best at 35°C, and the organs most affected are the skin and peripheral nerves. The nerve disease, caused directly by the infection, as well as, in some cases, the immune reaction to it, results in deformities (Fig. 17.11). There are multiple skin manifestations, including hypopigmented or erythematous patches that are anesthetic (Fig. 17.12). The immune response to the infection varies widely, and has a profound effect on the clinical manifestations. On

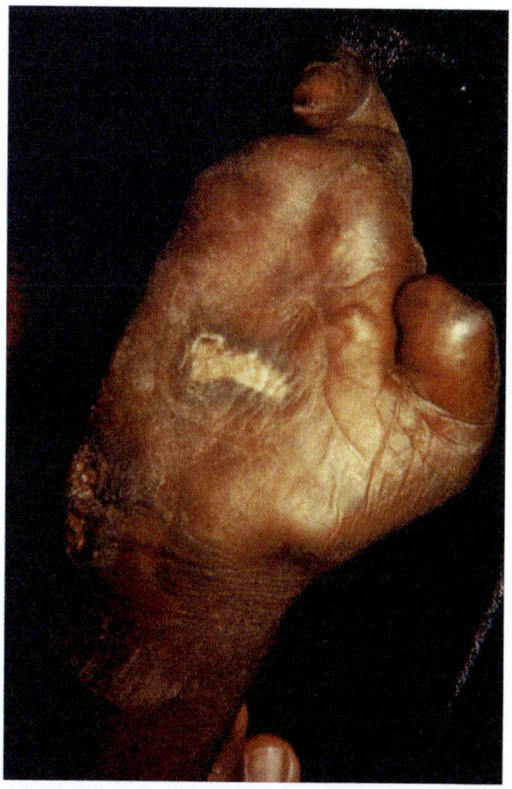

Fig. 17.11 Hand deformity resulting from leprosy. Courtesy of PHIL, CDC.

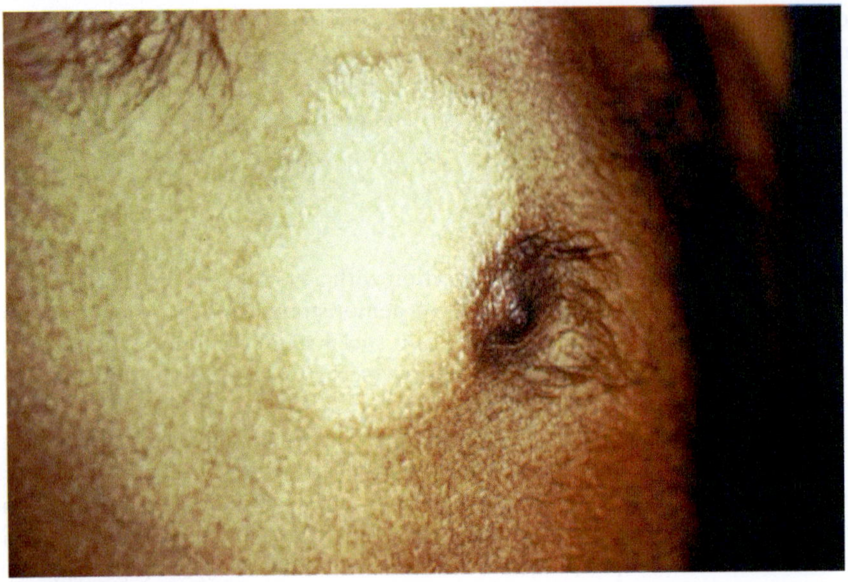

Fig. 17.12 Hypopigmented skin lesion of leprosy. Courtesy of PHIL, CDC.

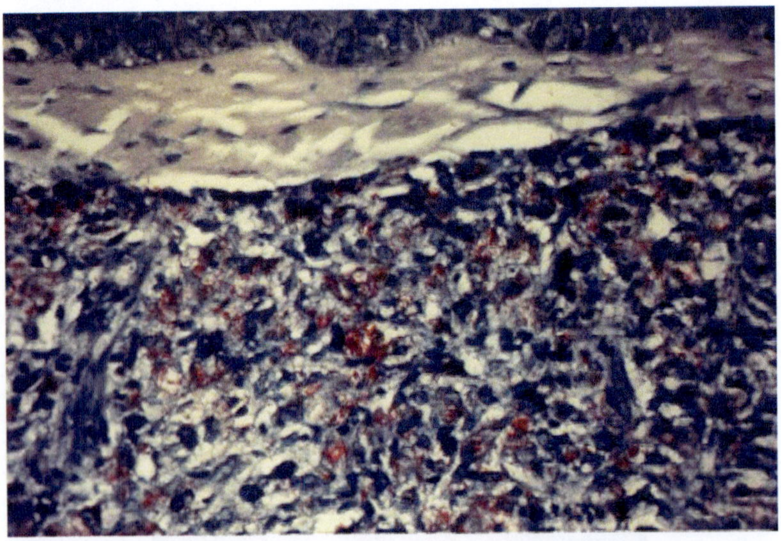

Fig. 17.13 Tissue specimen stained with an acid-fast stain showing *Mycobacterium leprae*. Courtesy of PHIL, CDC.

the one extreme (tuberculoid) are the cases in which there is a strong cell-mediated response, resulting in marked inflammation, and very few organisms (paucobacillary), and on the other extreme (lepromatous) are the cases in which there is minimal cell-mediated response, resulting in mild inflammation, and many organisms.

Because the organism cannot be cultured *in vitro*, the diagnosis is based on clinical manifestations, biopsy, and visualization of the organism after staining with an acid-fast stain (Ziehl–Neelsen on smears, Fite–Faraco on tissue). Specimens are obtained from biopsies of skin, nerve, or other affected tissue (Fig. 17.13). Also a slit-skin smear obtained from affected areas of skin, as well as from the ear lobes, is stained.

Because the infection has such a long incubation period, attempts have been made to diagnose the infection in the latent stage, analogous to the case with tuberculosis, using, for example, antibodies to phenolic glycolipid-1, a cell wall component. These have not proved useful. PCR on tissue shows increased sensitivity compared with slit-skin smears.

Treatment has been very effective. The mainstay of therapy is dapsone, a sulfone drug, rifampin (see Antituberculous drugs), and clofazamine. Ofloxacin, clarithromycin, and minocycline are also used in therapy.

Further reading

American Thoracic Society/Centers for Disease Control and Prevention/Infectious Diseases Society of America (2003) Treatment of tuberculosis. Am J Respir Crit Care Med 167: 603–62.

Banerjee S, Biswas N, Das NK, et al. (2011) Diagnosing leprosy: revisiting the role of the slit-skin smear with critical analysis of the applicability of polymerase chain reaction in diagnosis. Int J Dermatol 50: 1522–7.

Boehme CC, Nabeta P, Hillemann D, et al. (2010) Rapid molecular detection of tuberculosis and rifampin resistance. N Engl J Med 363: 1005–15.

Dockrell HM (2011) Keep the faith – leprosy still needs new diagnostic tools and laboratory research. Lepr Rev 82: 340–3.

Griffith DE, Aksamit T, Brown-Elliott BA, et al. (2007) An official ATS/IDSA statement: diagnosis, treatment, and prevention of nontuberculous mycobacterial diseases. Am J Respir Crit Care Med 175: 367–416.

Herrera V, Perry S, Parsonnet, et al. (2011) Clinical application and limitations of interferon-γ release assays for the diagnosis of latent tuberculosis infection. Clin Infect Dis 52: 1031–7.

Kar HK, Gupta R (2015) Treatment of leprosy. Clin Dermatol 33: 55–65.

Lawn SD, Zumla AI (2011) Tuberculosis. Lancet 378: 57–72.

Pai M, Denkinger CM, Kik SV, et al. (2014) Gamma interferon release assays for detection of *Mycobacterium tuberculosis* infection. Clin Microbiol Rev 27: 3–20.

Piersimoni C, Scarparo C (2009) Extrapulmonary infections associated with nontuberculous mycobacteria in immunocompetent persons. Emerg Infect Dis 15: 1352–8.

Piris A, Lobo AZC, Moschella SL (2010) Global dermatopathology: Hansen's disease – current concepts and challenges. J Cutan Pathol 37(suppl 1): 125–36.

Rodrigues LC, Lockwood DNJ (2011) Leprosy now: epidemiology, progress, challenges, and research gaps. Lancet Infect Dis 11: 464–70.

Schaaf HS, Marais BJ (2011) Management of multidrug-resistant tuberculosis in children: a survival guide for paediatricians. Paediatr Respir Rev 12: 31–8.

Talhari C, Talhari S, Penna GO (2015) Clinical aspects of leprosy. Clin Dermatol 33: 26–37.

Whitelow AC, Sturm WA (2009) Microbiological testing for Mycobacterium tuberculosis. In: Schaaf HS, Zumla AI (eds) *Tuberculosis. A Comprehensive Clinical Reference*. St Louis: Saunders Elsevier, pp. 169–78.

SECTION IV
Mycology

CHAPTER 18
Fungi

General properties of fungi

Fungi constitute one of the largest taxa of living things. They reproduce asexually and sexually. Very few of them cause disease in humans. However, several that are not pathogens under normal circumstances can cause disease in individuals who are severely immunocompromised.

A simple classification divides fungi into yeasts (e.g. candida), molds (e.g. aspergillus), and those that are yeasts at human body temperature and molds at environmental temperatures, i.e. dimorphic (e.g. histoplasma).

Structure of fungi

Fungi are eukaryotes (i.e. they have a nuclear membrane), that are aerobes. The cell membrane contains ergosterol, which is important as a target of some antifungal agents. External to the cell membrane is a cell wall, consisting of glucan, mannan, and chitin. The steps in the synthesis of ergosterol, an important component of the cell membrane, and the antifungal agents interfering with them are shown in Fig. 18.1. The main groups of antifungal agents, their targets in different components of the cell, and mechanisms of resistance to them are shown in Table 18.1.

Laboratory perspectives for mycelial fungi

While a comprehensive review of procedures is not practical for this text, certain fundamentals are important.

Direct smear

All specimens submitted for fungal isolation and identification should be accompanied by a direct smear if quantity is sufficient. The most common routine stains are the periodic acid-Schiff (PAS) and Calcofluor white stains, both of which target the chitin in the cell wall. The PAS appears as deep red when visualized under the light

Practical Medical Microbiology for Clinicians, First Edition. Frank E. Berkowitz and Robert C. Jerris.

Squalene

↓ Squalene epoxidase ← Allylamines

Lanosterol

↓ C-14-α-demethylase ← Triazoles

Ergosterol

Fig. 18.1 Steps in the synthesis of ergosterol.

Table 18.1 Mechanisms of action of antifungal agents, and mechanisms of resistance to them.

Agent	Target	Mechanism of action	Mechanism of resistance
Griseofulvin	Mitotic spindle	Inhibits formation	Unknown
Polyenes	Cell membrane sterols	Forms pores Leakage of contents Oxidative damage	Decreased ergosterol in membrane; accumulation of other sterols; decreased oxidative damage
Azoles	C-14-α demethylase	Inhibits ergosterol synthesis (needed for cell membrane integrity)	Efflux pump upregulation; decreased affinity for target; increased production of altered target; alternate pathway
Echinocandins	1,3 β-D-glucan synthase	Inhibits cell wall synthesis	Mutations in target enzyme complex
Allylamines	Squalene epoxidase	Inhibits ergosterol synthesis (needed for cell membrane integrity)	Unknown
5-Fluorocytosine (5-FC)	Converted to 5-FU; 5-FU →5-FUMP 5-FUMP→5-FUTP; 5-FUTP replaces UTP in RNA → inhibits protein synthesis 5-FUMP inhibits thymidylate synthase → inhibits cell replication	Inhibits nucleic acid synthesis	Decreased drug uptake; decreased conversion of 5-FC to 5-FU, or 5-FU to 5-FUMP (active form of the drug)

5-FU, 5-fluorouracil; 5-FUMP, 5-fluorouracil monophosphate; 5-UTP, 5-uracil triphosphate; RNA, ribose nucleic acid; UTP, uracil triphosphate.

microscope. The Calcofluor white stain requires a fluorescent microscope to visualize the hyphal elements which, depending on the filter combination, will stain white or apple green (Fig. 18.2).

Additionally, methylene blue or India ink may be used to visualize capsules (especially for *Cryptococcus neoformans/gattii*). Also commonly used for direct visualization from clinical specimens is a KOH preparation, in which a drop of 10% KOH is admixed with the specimen. The KOH clears cellular material, allowing enhanced visualization of hyphae. A silver impregnation stain (e.g. Gomori–Grocott stain) stains fungal walls black. These are used mainly for cytological and histological preparations (Fig. 18.3).

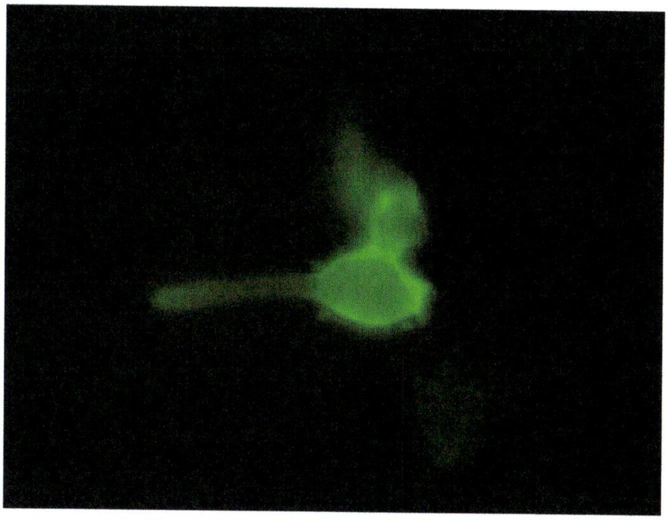

Fig. 18.2 *Aspergillus* spp. stained with Calcofluor white.

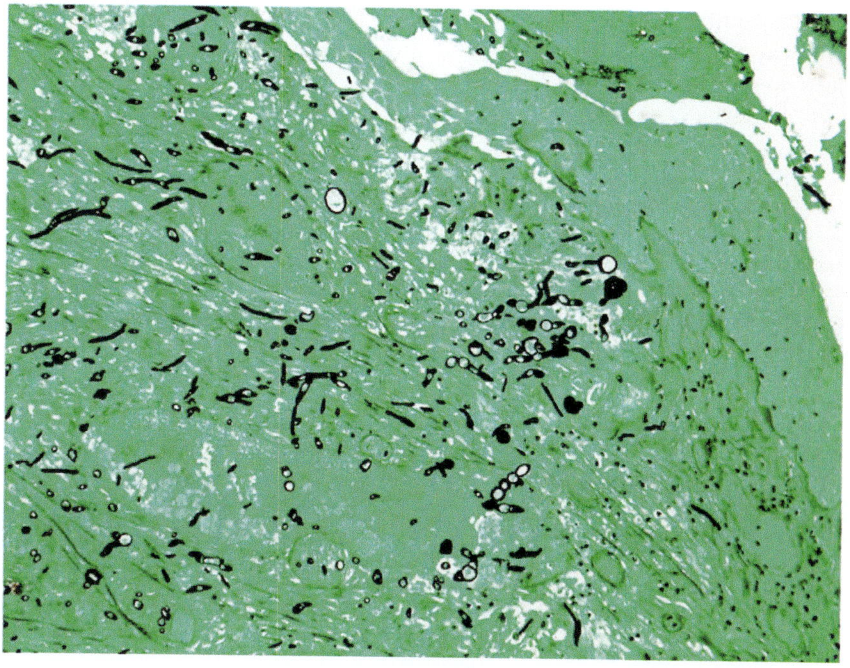

Fig. 18.3 Silver impregnation stain of sinus tissue, showing mold (*Bipolaris* spp.).

Key morphological features that can be observed on direct smear, and that can be helpful in identifying fungi, are as follows.

- Hyaline, septate hyphae are impossible to identify/differentiate among the many molds (e.g. *Aspergillus, Fusarium*).
- Dematiaceous septate-brown pigmented hyphae (e.g. *Bipolaris, Curvularia*).

- Sclerotic bodies (brown, round to pleomorphic thick, septated, walled cells) (e.g. *Fonsecaea, Phialophora*).
- Granules (different sizes, colors, compositions), e.g. 200–300 μm soft white, without cement-like matrix is typical of *Acremonium* while a 200–300 μm soft black granule without a cement-like matrix is typical of *Exophiala.*

Note that interpretation requires a skilled microscopist.

India ink preparations are also useful to demonstrate capsules, most commonly of *Cryptococcus neoformans* in cerebrospinal fluid.

Histology

This is a very important procedure for the detection of fungi for the following reasons.

- A fungal etiology for the disease might not have be suspected, so that appropriate cultures were not performed.
- Fungus might be seen on a histologic preparation, but does not grow in culture.

Culture

Fungi can be cultured, as bacteria, on agar-based media. A variety of media can be used, some of which are permissive for many different fungi, and some of which are selective for a limited range of fungi, or prevent the growth of bacteria in specimens from non-sterile sources. Several of the more commonly used media and their functions are detailed in Table 18.2.

Table 18.2 Commonly used mycology media.

Media	Function
Brain heart infusion agar (BHIA)	Culture of fastidious dimorphic fungi; all fungi and yeast
BHIA + 5% sheep blood	Improves growth of fastidious, slow growers (e.g. *Histoplasma capsulatum*). Note: addition of blood to media inhibits conidiation
Sabouraud dextrose agar (SAB)	Cultivation of all fungi; 4% glucose, pH 5.6; may contain additives: chloramphenicol, cycloheximide (this may inhibit fungi*), or penicillin, gentamicin
SAB-Emmons	As above but 2% glucose, pH 6.9–7.0; improves sporulation
Mycosel/Mycobiotic	For recovery of fungi from sites containing normal flora; contains chloramphenicol and cycloheximide (see above)
Potato dextrose agar (PDA)	Enhances sporulation; enhances pigment; used in slide cultures
Cornmeal agar	Enhances sporulation; used to observe yeast morphology
Dermatophyte test medium	Identification of dermatophytes by red color production
SAB+ heart infusion (SABHI)	Cultivation of all fungi, yeasts
Inhibitory mold agar	Cultivation of dimorphic fungi; chloramphenicol, gentamicin inhibits organisms from non-sterile sites
Yeast extract phosphate	Ammonia added to surface; inhibition of bacteria, yeasts, many fungi; enhances growth of slow-growing dimorphic fungi
CHROMagar®	Allows differentiation of yeasts based on color; most useful to detect mixed cultures of yeast

*Cycloheximide may inhibit *Pseudallescheria boydii*, *Cryptococcus neoformans*, *Trichosporon asahii*, some species of candida and most zygomycetes (mucorales).

Note that culturing **blood** requires special processing for **molds**, not **yeast**. The use of a biphasic bottle (containing both broth and agar) or the Isolator System (Wampole Labs, Cranbury, NJ) is standard. Once the blood is collected in the proprietary tube, cells are lyzed in the medium and the tube is centrifuged. The sediment is plated directly on fungal media.

Primary incubation

All media are routinely incubated aerobically at 30°C. Exceptions to this are when testing for temperature tolerance (e.g. *Aspergillus fumigatus*, maximum temperature 48°C) and conversion of dimorphic fungi from hyphal to yeast phase (or spherule in the case of *Coccidioides immitis*) at 37°C. In addition, recovery of some organisms requires the addition of special components directly to the agar (e.g. olive oil or palmitic acid for *Malassezia furfur*). **Therefore, the clinician should always communicate with the laboratory personnel when special handling is required, as in the above cases, and whenever a dimorphic fungus is suspected.**

Identification

The rate of growth, colony morphology, characteristics of the top side of the culture (texture and color), color of the reverse side, and the type and mode of sporulation provide the basis for phenotypic identification of mycelial fungi (Figs 18.4 & 18.5).

Once the fungus is detected on agar plates, a "tease" preparation is performed. In this procedure, a drop of lactophenol cotton blue or lactofuchsin (LPCB/F) reagent is placed on a microscope slide. Portions of the colony from the periphery and from near the center are placed in the medium and gently pulled apart with inoculating needles. Note that inflexible straight needles are used instead of loops (used in conventional bacteriology).

Fig. 18.4 Culture of *Trichophyton mentagrophytes*, showing the front side of the petri dish. Courtesy of PHIL, CDC.

Fig. 18.5 Culture of *Trichophyton mentagrophytes*, showing the reverse side of the petri dish. Courtesy of PHIL, CDC.

The slide is then covered with a cover slip and visualized for the type and formation of the conidia (spores). They are examined microscopically at 500× magnification; 1000× magnification may be required for critical evaluation of conidial ontogeny (Fig. 18.6). Alternatively, a scotch tape preparation may be performed by gently placing a small piece of clear scotch tape sticky side down on to the colony and placing it directly on to the slide containing LPCB/F.

If no conidia are demonstrated on a tease preparation, a slide culture is performed. In this procedure, a small 1 cm agar block is cut from potato flake agar (or other conidia-inducing medium) and placed on a microscope slide. The medium is then inoculated on each side with a portion of the mycelial growth. A cover slip is applied and the slide is placed on a V-shaped glass rod in a petri dish. Sterile water is added (5 mL) to the dish to keep the preparation hydrated. Incubation is at 30°C and once growth is detected, the cover slip is removed and placed in LPCB/F on a microscope slide and examined for characteristics (type and mode of conidial production) unique for each fungus. Slide cultures should be limited to organisms of low virulence. Communication with the laboratory personnel is key.

Additional tests used to identify fungi include:

- nutritional studies and hair penetration for dermatophytes
- urea hydrolysis for yeast and dermatophytes
- chemical detection: detection of fungal-specific metabolites
- molecular methods: polymerase chain reaction (PCR) and genetic sequence analysis
- MALDI-TOF MS: detects fungal proteins, creating a fingerprint for comparison with fungal database for identification.

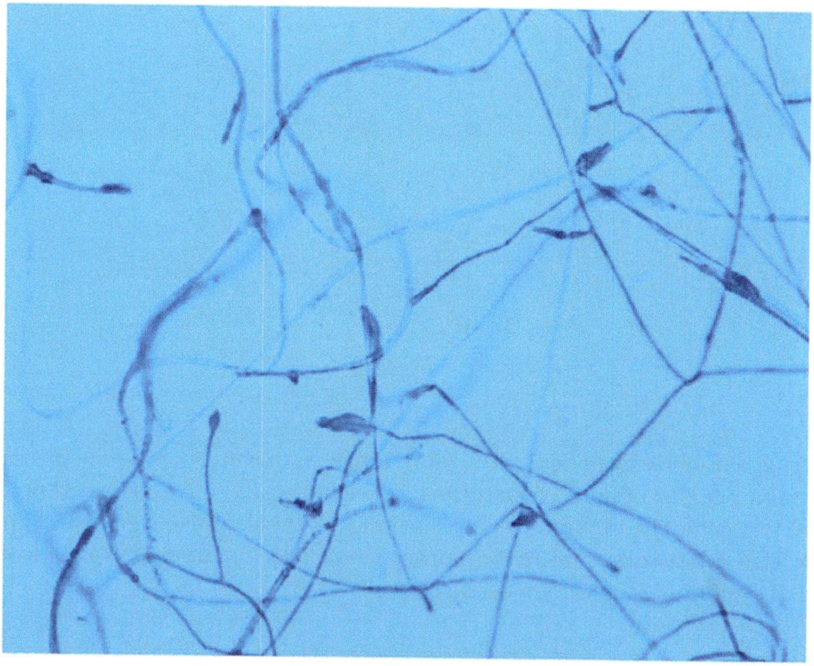

Fig. 18.6 Lactophenol cotton blue preparation showing conidia of *Bipolaris* spp.

Detection of fungal components in body fluids

Fungal antigen detection is used for the diagnosis of several fungal infections, namely Cryptococcus (cerebrospinal fluid, serum), Histoplasma (urine, serum), and Blastomyces (urine, serum).

The detection of fungal components in serum or bronchoalveolar lavage fluid is also sometimes used for the diagnosis of some fungal infections, namely β-glucan (several yeasts) and galactomannan (aspergillus). However, their utility in clinical decision making is controversial, due to their variability in sensitivity and specificity.

Molecular tests performed on serum or bronchoalveolar lavage fluid are promising but not yet in general use.

Serology is of value in diagnosing some endemic fungal infections (histoplasmosis, blastomycosis, and coccidioidomycosis).

Further reading

Arvanitis M, Anagnostou T, Fuchs BB, et al. (2014) Molecular and nonmolecular diagnostic methods for invasive fungal infections. Clin Microbiol Rev 27; 490–526.

Guarner J, Brandt ME (2011) Histopathologic diagnosis of fungal infections in the 21st century. Clin Microbiol Rev 24: 247–80.

Peman J, Canton E, Espinel-Ingroff A (2009) Antifungal drug resistance and mechanisms. Expert Rev Anti Infect Ther 7: 453–60.

Pfaller MA (2012) Antifungal drug resistance: mechanisms, epidemiology, and consequences for treatment. Am J Med 125(suppl 1): S3–S13.

Vandeputte P, Ferrari S, Coste AT (2012) Antifungal resistance and new strategies to control fungal infections. Int J Microbiol article ID 713687; doi:10.1155/2012/713687.

CHAPTER 19

Yeasts

These fungi reproduce by budding. The buds, called blastoconidia, can remain attached to one another, forming a chain called a pseudohypha (pleural pseudohyphae) (Fig. 19.1). Medically important yeasts belong to the following genera: Candida, Cryptococcus, Malassezia, and Trichosporon. In addition, Pneumocystis is a yeast-like fungus, of significant medical importance.

Candida

Several species of candida (this means white in Latin) can cause disease in humans. The most important are *C. albicans* (also meaning white, and by far the most important), *C. parapsilosis, C. tropicalis, C. krusei,* and *C. glabrata* (formerly called *Torulopsis glabrata*). Rare causes of human disease include *C. guillermondii* and *C. lusitaniae.*

Candida albicans forms part of the normal mucosal and skin flora. In overgrowth, it can cause mucocutaneous disease such as oral candidiasis, which is very common in young infants (Fig. 19.2), vaginal candidiasis, intertrigo, and angular stomatitis. In individuals with impaired cell-mediated immunity, it can cause esophagitis and chronic nail infections, and in premature infants and individuals with neutropenia, it causes disseminated infection affecting any viscus. Candida species are very important causes of healthcare-associated bloodstream infection, primarily due to infection of intravascular catheters, on which they form a biofilm (Figs 19.1 & 19.3).

Yeasts can be demonstrated by Gram stain of fluid and pus (Figs 19.1 & 19.4).

Candida albicans can be differentiated from the other species of candida by formation of foot processes at the base of the colonies on blood agar, and the production of a germ tube when grown for a few hours in pooled serum (Figs 19.5 & 19.6). (*Candida dubliniensis* also produces a germ tube.) The others can be differentiated from one another by their biochemical profile, and their growth and color production on CHROMagar® (Fig. 19.7).

Treatment of superficial infections can be accomplished with a topical azole (e.g. clotrimazole) or polyene (e.g. nystatin). Patients with systemic disease require treatment

Practical Medical Microbiology for Clinicians, First Edition. Frank E. Berkowitz and Robert C. Jerris.

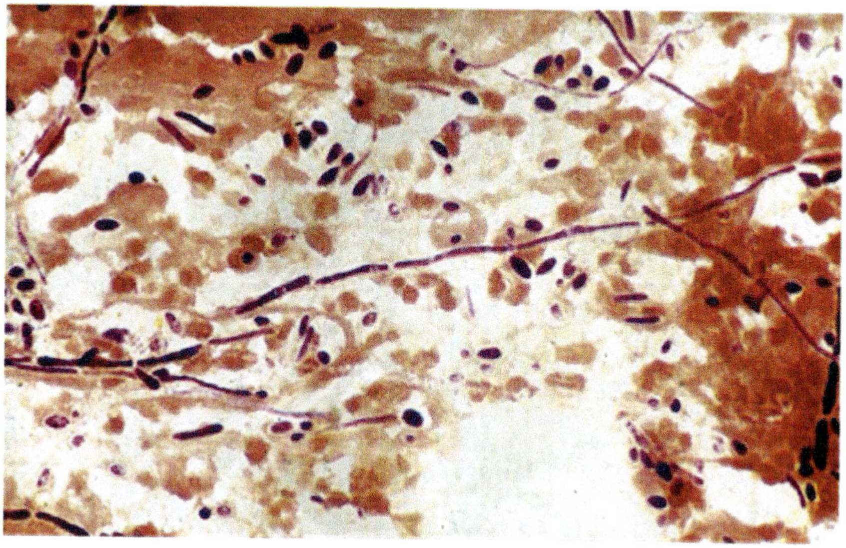

Fig. 19.1 Pseudohyphae of *Candida albicans* in pus from a perivenous abscess caused by an intravenous catheter (see Fig. 19.3). Copyright © Frank E. Berkowitz. Reprinted with the permission of Cambridge University Press.

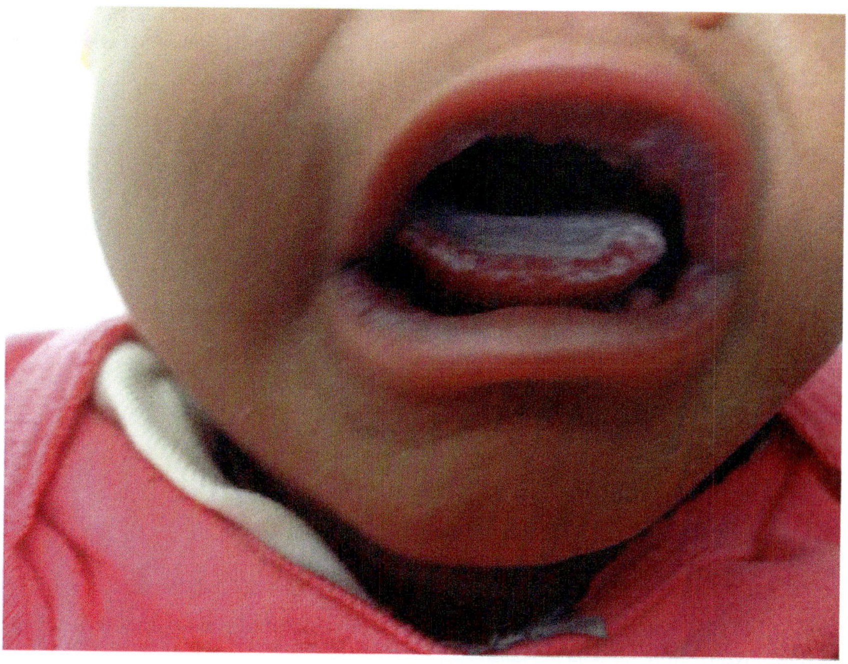

Fig. 19.2 Oral candidiasis (thrush) in a normal newborn infant.

with a systemic azole such as fluconazole, an echinocandin, or occasionally with amphotericin B. However, *C. krusei* is predictably resistant to fluconazole, while *C. glabrata* is relatively resistant to this drug. The rarer species of candida have variable antifungal susceptilities.

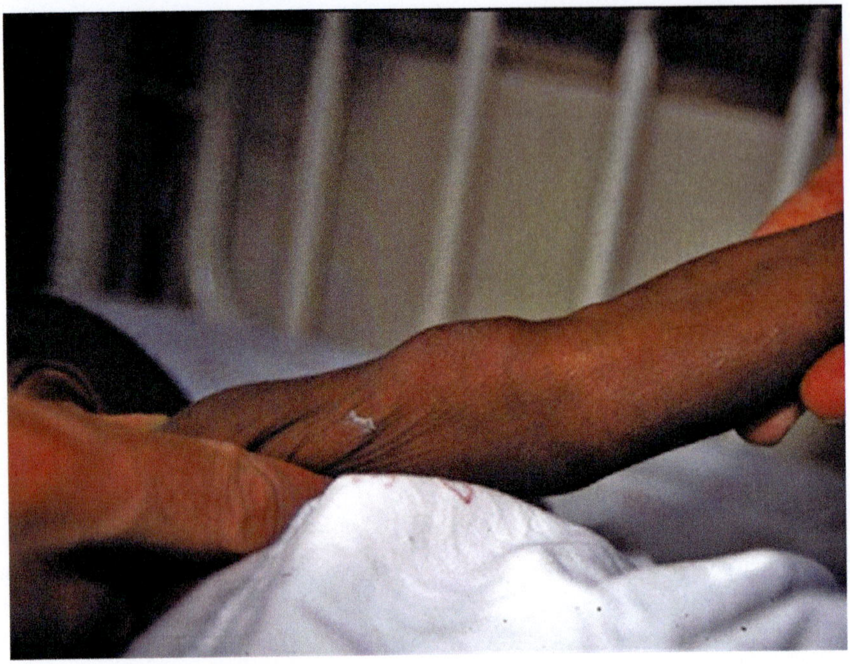

Fig. 19.3 Perivenous abscess at an intravascular catheter site, caused by *Candida albicans*. The Gram stain of an aspirate is shown in Fig. 19.1. Copyright © Frank E. Berkowitz. Reprinted with the permission of Cambridge University Press.

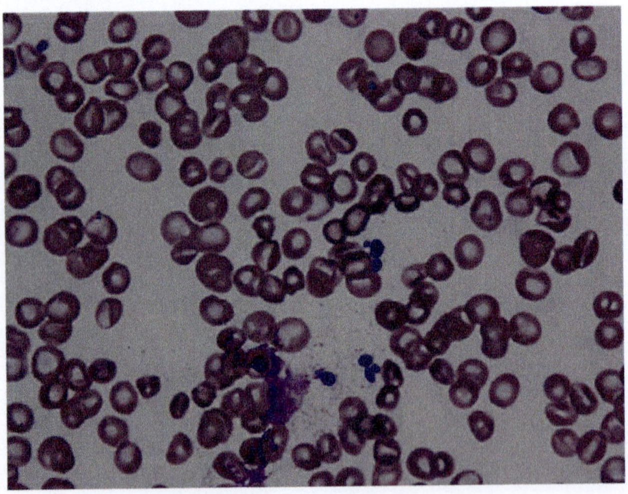

Fig. 19.4 Gram stain of a buffy coat preparation obtained from a central venous catheter, showing budding yeasts, which were *Candida tropicalis*.

Cryptococcus

These are environmental yeasts that grow on organic material such as bird or bat guano. The two main species are *C.neoformans* and *C. gatti*.

Cryptococcus neoformans is a well-recognized cause of chronic meningitis in individuals with deficiencies in cell-mediated immunity. It is a very important opportunistic

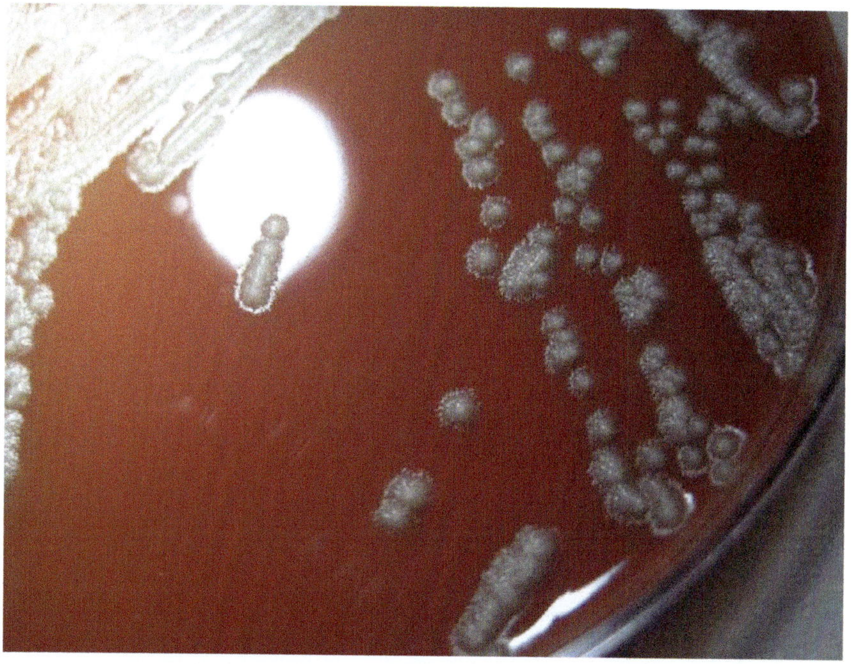

Fig. 19.5 "Foot processes" on colonies of *Candida albicans*.

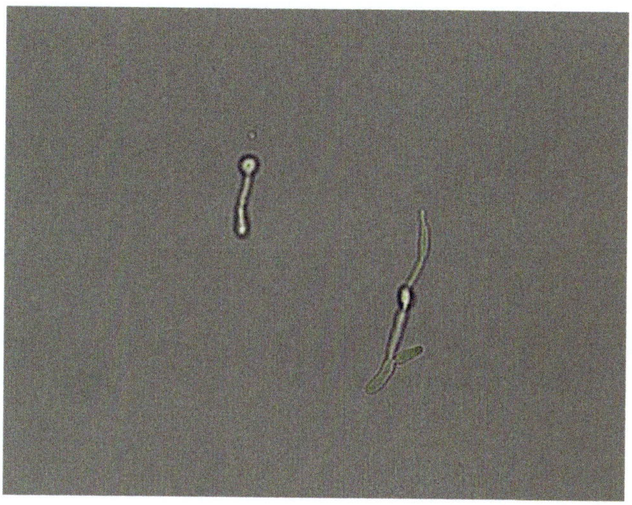

Fig. 19.6 Germ tube production by *Candida albicans*.

infection in individuals with AIDS. The organism enters the body via the lungs, where it can cause pneumonia, and it spreads to the meninges via the blood. The meningitis can have a very insidious progression over months. The organism can be detected in cerebrospinal fluid by Gram stain (Fig. 19.8), but it is best seen by India ink preparation, in which its large capsule can be seen (Fig. 19.9). It can be

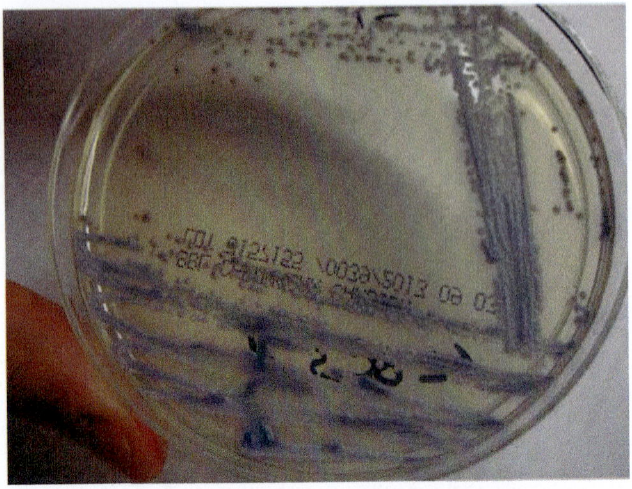

Fig. 19.7 Characteristic blue colonies of *Candida tropicalis* on CHROMagar®.

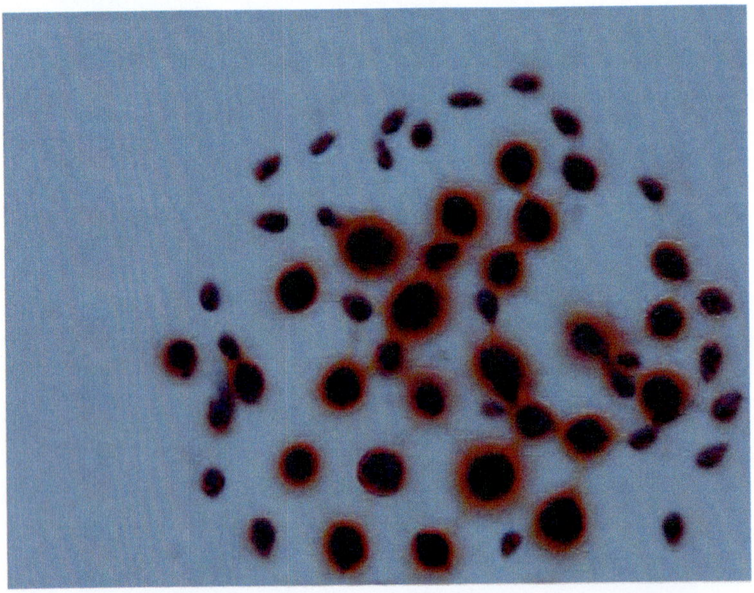

Fig. 19.8 Gram stain of cerebrospinal fluid showing *Cryptococcus neoformans*. Note budding.

cultured on blood and other agars (Fig. 19.10) but the most sensitive test for detecting the organism is the antigen test, which can be performed on cerebrospinal fluid and on serum.

Cryptococcus gatti is much less common than *C. neoformans*. Recently, many cases of this infection in normal hosts have been recognized in the northwestern USA and British Columbia.

Treatment of patients with cryptococcal meningitis consists initially of amphotericin B plus 5-flucytosine, followed by long-term fluconazole.

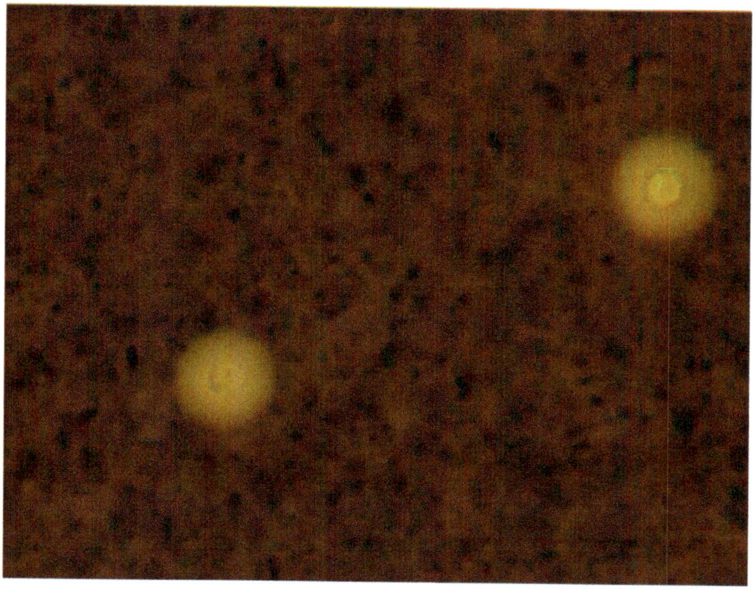

Fig. 19.9 India ink preparation of cerebrospinal fluid, showing *Cryptococcus neoformans* with a large capsule.

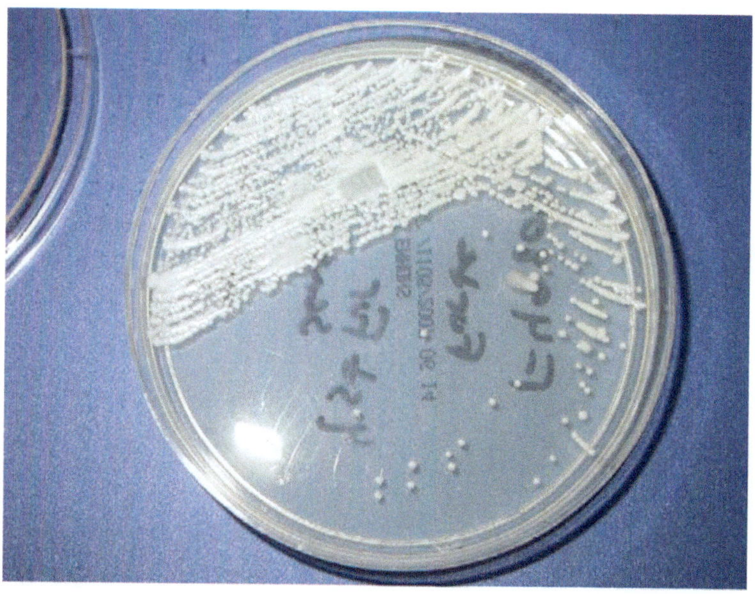

Fig. 19.10 *Cryptococcus neoformans* growing on agar.

Malassezia

These are dimorphic (both yeast and hyphal forms being present in the host), lipophilic fungi. There are several species, including *M. furfur* (the best known), *M. globosa*, *M. restricta*, and *M. sympodialis*. They are normal inhabitants of the superficial layers of human skin. *M. pachydermatis* is an animal pathogen, occasionally infecting humans. They cause a very common skin condition called pityriasis

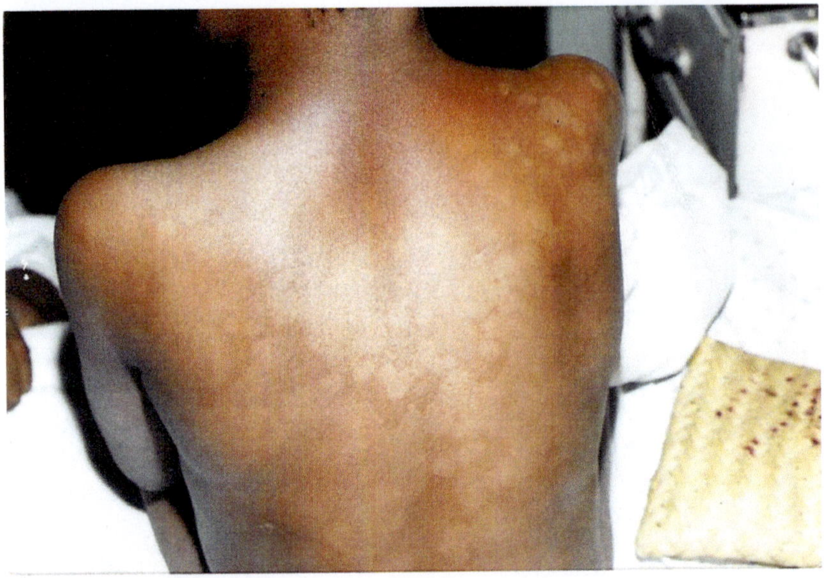

Fig. 19.11 A teenager with pityriasis versicolor due to *Malassezia* spp. Copyright © Frank E. Berkowitz. Reprinted with the permission of Cambridge University Press.

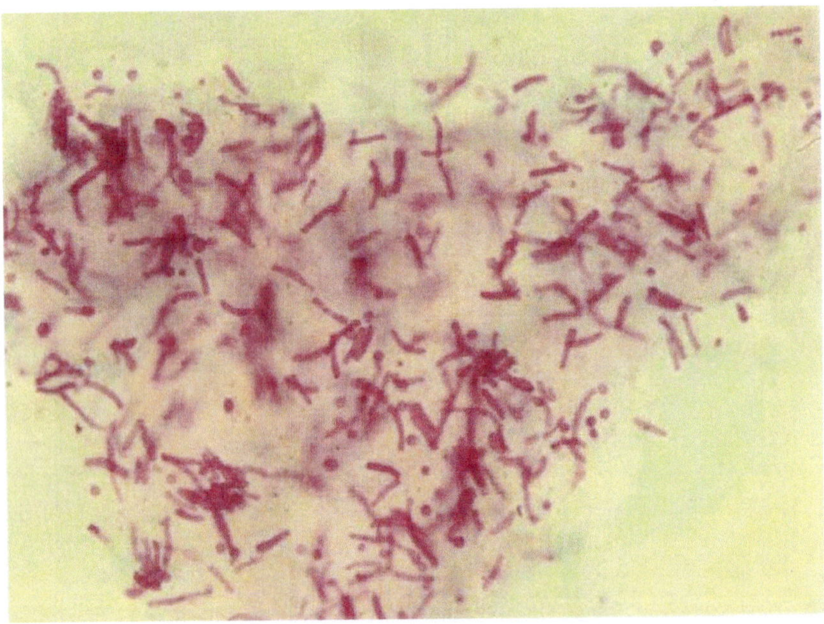

Fig. 19.12 *Malassezia furfur*, showing yeast and hyphal forms ("spaghetti and meatballs"). Courtesy of PHIL, CDC.

versicolor, which is characterized by slightly scaly areas of hypopigmentation on the neck and upper trunk, especially in adolescents (Fig. 19.11). The organism, which is lipophilic, can be seen on a skin scraping, stained with methylene blue, or cleared with KOH, as filaments and balls ("spaghettii and meatballs") (Fig. 19.12).

Systemic infection can occur in individuals receiving total parenteral nutrition that includes a lipid preparation. The presence of the central venous catheter and the lipid provides an ideal medium for the organism to grow. On a Gram stain, it has the appearance of a footprint, and it grows on media, such as blood agar, supplemented with fatty acid. Although olive oil can be used, the optimal fatty acid is palmitic acid.

Other yeasts

Blastoschizomyces capitatus (formerly *Geotrichium capitatum*), *Rhodotorula* species (*R. glutini, R. mucilaginosa,* formerly *R. rubra, and R. minuta*), and *Saccharomyces cerevisiae* (bakers' and brewers' yeast) are occasional causes of invasive disease, usually in individuals who are immunocompromised and/or have intravascular catheters.

Trichosporon spp. cause a hair infection called white piedra but can, occasionally, cause invasive disease in immunocompromised hosts. The former *T. beigelii* has been divided into several species, including *T. asahii, T. asteroides, T. cutaneum, T. inkin, T. mucoides,* and *T. ovoides.*

Pneumocystis

This genus was considered to belong to the protozoan parasites for many years, but is now classified as a fungus, related to yeasts. Originally, the organism infecting humans was called *P. carinii.* It is now known to be the species infecting rats. The species infecting humans is called *P. jiroveci.* It exists as a trophozoite, a precyst, and a cyst. It is transmitted by inhalation from other infected individuals. Almost all humans become infected within the first few years of life, and disease is largely confined to premature and malnourished children, and individuals of any age with severe impairment of cell-mediated immunity. It is a very important opportunistic infection in individuals with AIDS. It causes diffuse pneumonia that can be insidious or fairly acute in onset. Occasionally it spreads to organs other than the lung.

The diagnosis is made by demonstration of the organism in material from the lung, such as sputum (in heavily infected patients), bronchoalveolar lavage fluid, or lung biopsy. When stained by the Gomori silver impregnation stain, the cyst wall stains black (Fig. 19.13). When stained with Giemsa stain, the cyst wall is not seen, but the intracystic bodies can be seen. Immunofluorescent stains are available. Histologically the lung tissue shows interstitial infiltration of mononuclear cells and the alveoli are filled with foamy material (Fig. 19.14). The organism is not seen with hematoxylin and eosin stain.

Therapy consists of respiratory support and antimicrobial therapy. The mainstay of therapy is trimethoprim/sulfamethoxazole. Alternative drugs are atovaquone/proguanil, pentamidine, or primaquine plus clindamycin. Chemoprophylaxis should be used for individuals at risk or those previously infected.

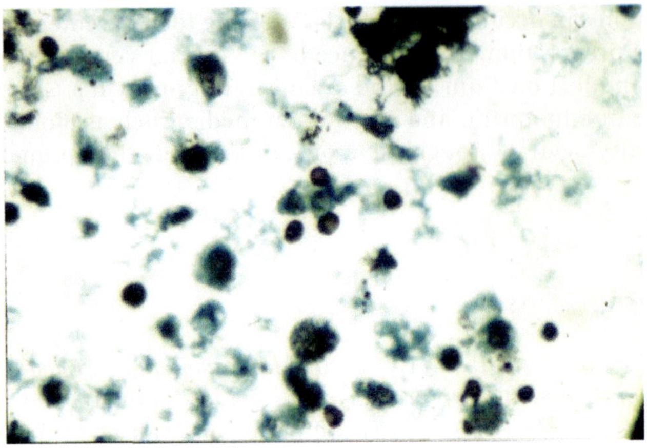

Fig. 19.13 The cysts of *Pneumocystis jiroveci*, in lung tissue from a child with AIDS, stained with Gomorri silver impregnation stain. Copyright © Frank E. Berkowitz. Reprinted with the permission of Cambridge University Press.

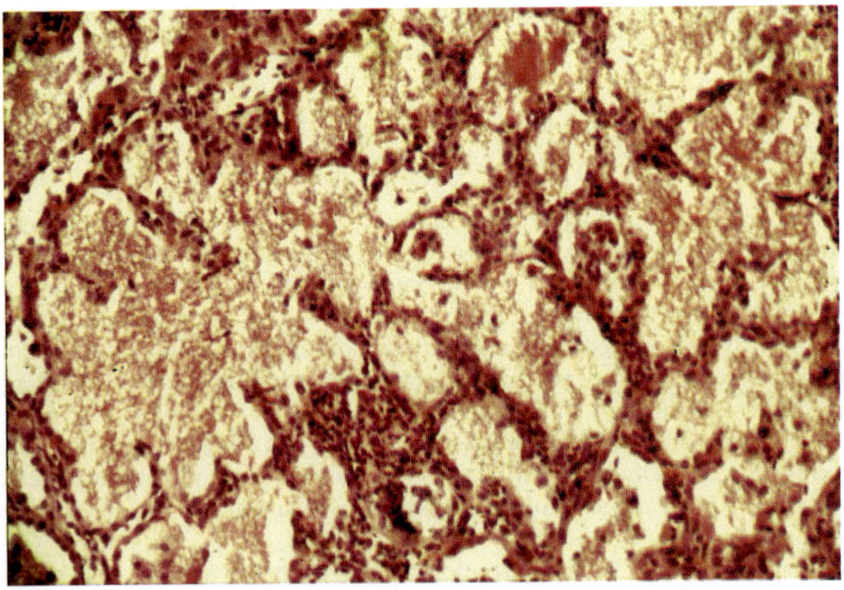

Fig. 19.14 The histologic appearance of the lung, stained with hematoxylin and eosin, caused by *Pneumocystis jiroveci* in a child with AIDS. Copyright © Frank E. Berkowitz. Reprinted with the permission of Cambridge University Press.

Further reading

Gilroy SA, Bennett NJ (2011) Pneumocystis pneumonia. Semin Respir Crit Care Med 32: 775–82.

Miceli MH, Lee SA (2011) Emerging opportunistic yeast infections. Lancet Infect Dis 11: 142–51.

Pappas PG, Kauffman CA, Andes D, et al. (2009) Clinical practice guidelines for the management of candidiasis: 2009 update by the Infectious Diseases Society of America. Clin Infect Dis 48: 503–35.

Perfect JR, Dismukes WE, Dromer F, et al. (2010) Practice guidelines for the management of cryptococcal disease: 2010 update by the Infectious Diseases Society of America. Clin Infect Dis 50: 291–322.

CHAPTER 20
Dimorphic endemic fungi

These are *Histoplasma capsulatum*, *Blastomyces dermatitidis*, *Coccidioides immitis*, *Paracoccidioides brasiliensis*, *Sporothrix schenkii*, *Penicillium marneffei*, and *Emmonsia* spp.

Histoplasma capsulatum

This organism, which causes histoplasmosis, lives as a mold in organic soil, such as that fertilized by bat or bird guano. It occurs in certain endemic areas of the USA, such as the Ohio and Mississippi River basins (Fig. 20.1), Mexico, and in other continents. When inhaled by a human, conidia of the mycelial form convert to yeasts, which are taken up by macrophages and can spread within the lung. A wide spectrum of clinical illness can occur, depending on the size of the inoculum and the immunocompetence of the host. This varies from asymptomatic disease to severe acute pneumonia (Fig. 20.2). Pulmonary infection can become subacute or chronic in individuals with underlying chronic obstructive airways disease. The organism can spread to mediastinal lymph nodes, and a non-infectious complication, mediastinal fibrosis, can occur. Non-infective inflammatory complications, such as pericarditis, erythema nodosum, and arthritis, can also occur. Hematogenous spread can occur to any organ, especially to the oral mucosa and intestine. Progressive disseminated histoplasmosis is a life-threatening condition.

The diagnosis is made by culture of material from potentially infective sites, and visualization of the yeast form in cytologic or histologic preparations of such material, stained with silver or PAS stain (Fig. 20.3). Culture must be performed in a laboratory in which appropriate precautions can be taken, because the organism can be aerosolized and cause disease in laboratory workers who inhale it. Although serologic tests, such as immunodiffusion, can be performed, they are not sensitive, nor do they determine when the infection occurred. However, antigen detection in urine or pulmonary fluid is useful.

Practical Medical Microbiology for Clinicians, First Edition. Frank E. Berkowitz and Robert C. Jerris.

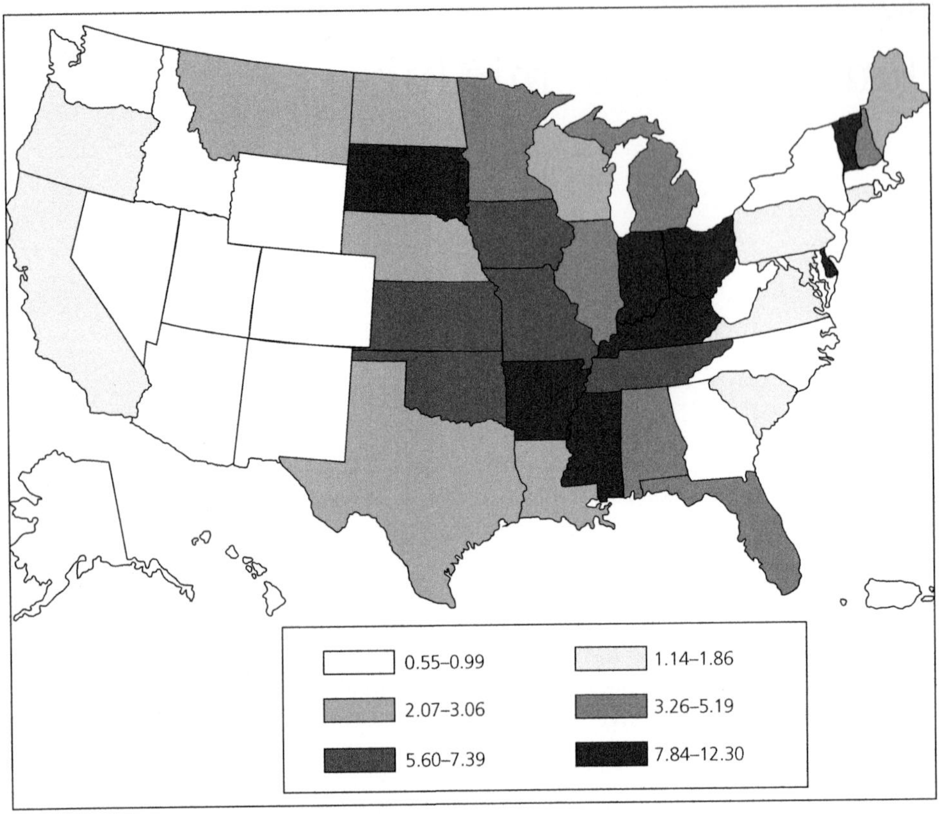

Fig. 20.1 Distribution of cases of histoplasmosis in individuals >65 years in the USA, 1999–2008. Numbers are cases per 100,000. Source: Baddley et al., 2011.

Treatment is not indicated in mild disease; in most cases, itraconazole is the usual treatment, but in severe and progressive disease, amphotericin B liposomal should be used.

A rare variety of *Histoplasma capsulatum, H. capsulatum var duboisii* (as opposed to *H. capsulatum var capsulatum*), occurs in Africa.

Blastomyces dermatitidis

This organism, which causes blastomycosis, grows as a mold in warm, moist soil, and is endemic mainly in the northern USA and Canada, although it has been reported from other continents (Fig. 20.4).

Humans inhale conidia, which infect pulmonary macrophages, where they develop into the yeast form, which causes pneumonia (Fig. 20.5). This can be acute, but it can become chronic. In many cases the organism spreads hematogenously, most commonly to the skin (hence the species name), and to bone.

The diagnosis is made by microscopy of infected material stained by Papanicolaou stain, Calcofluor white, PAS, or Gomori silver stain, or cleared with KOH. These reveal characteristic yeast forms (Fig. 20.6). Culture takes 2–4 weeks.

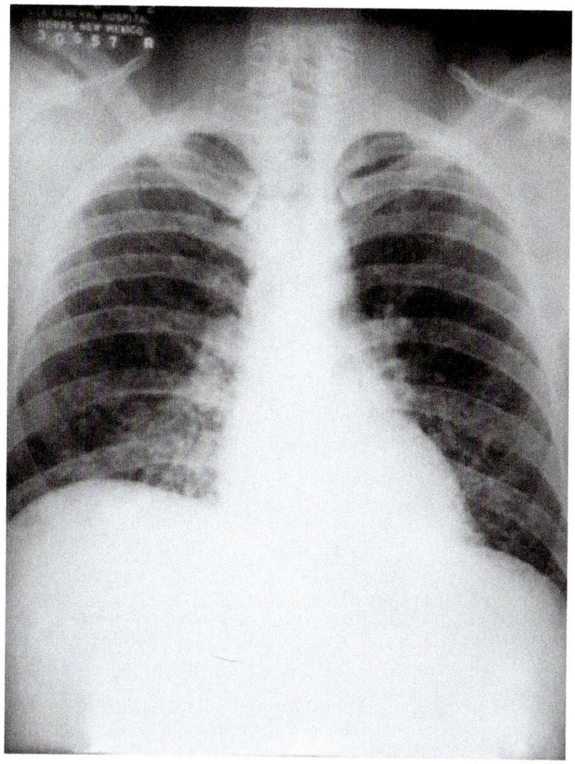

Fig. 20.2 Chest radiograph of a patient with acute pulmonary histoplasmosis. Courtesy of PHIL, CDC.

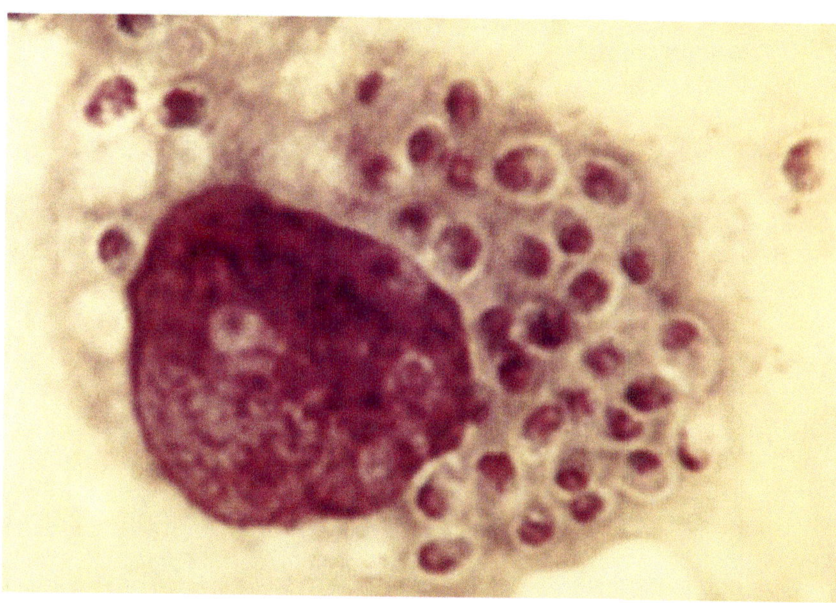

Fig. 20.3 *Histoplasma capsulatum* in a histiocyte. Courtesy of PHIL, CDC.

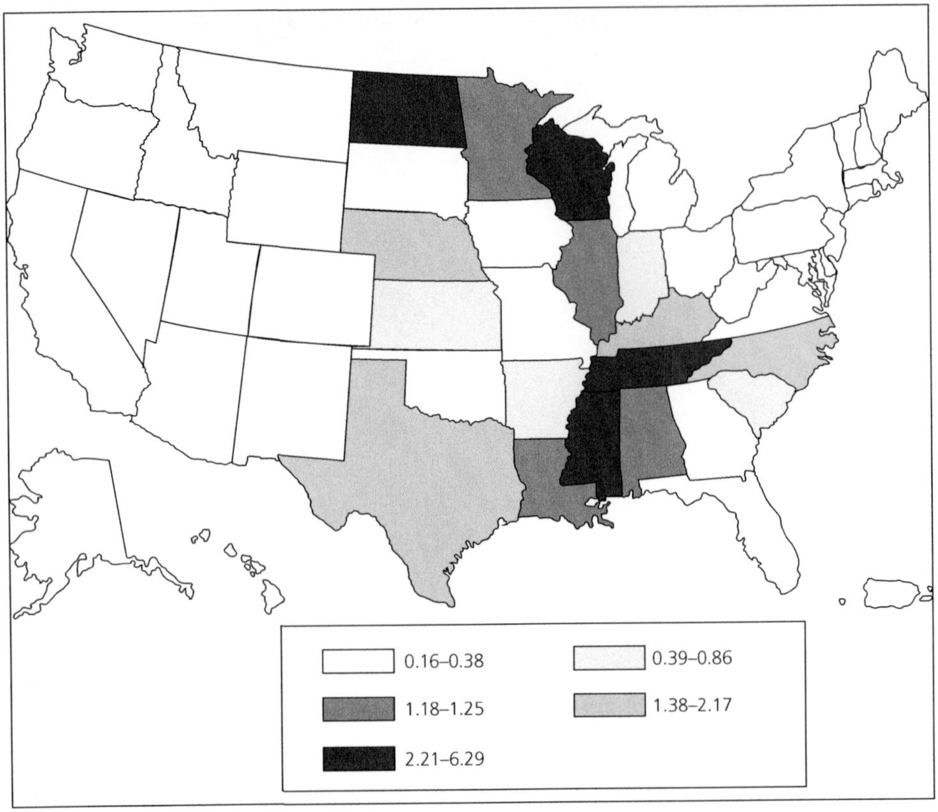

Fig. 20.4 Distribution of cases of blastomycosis in individuals >65 years old in the USA. Numbers are cases per100,000. Source: Baddley et al., 2011.

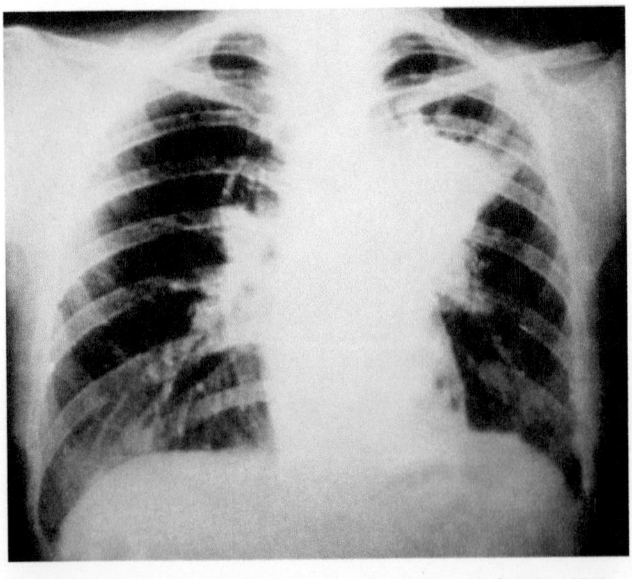

Fig. 20.5 Chest radiograph of a patient with blastomycosis. Courtesy of PHIL, CDC.

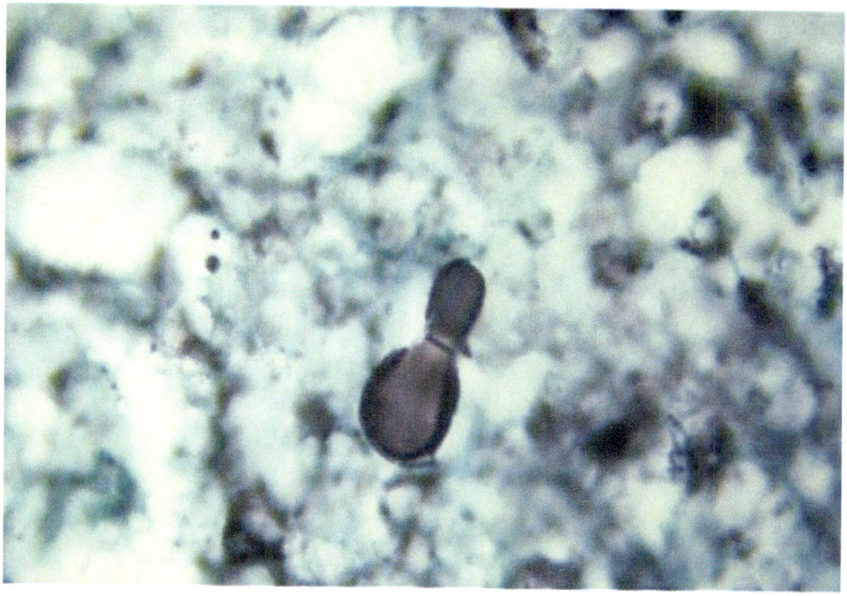

Fig. 20.6 Silver-stained section showing a budding yeast form of *Blastomyces dermatitidis*. Courtesy of PHIL, CDC.

Treatment, which all diagnosed patients should receive, consists mainly of itraconazole. In severe cases, liposomal amphotericin B should be used initially.

Coccidioides

These organisms, which cause coccidioidomycosis, are prevalent in the western USA (Fig. 20.7). There are two species: *C. immitis* and *C. posadasii*. They are present in the soil, in the mold form. They grow during the wet season, and when the environment is dry, the hyphae die, leaving arthrospores (also called arthroconidia). These are spread in dust, and can be inhaled by humans. In the lung, they form spherules. Within the spherules endospores form, which can spread within the lung (Figs 20.8 & 20.9).

The clinical illness is characterized mainly by pneumonia, which, if the patient is untreated, can persist for many weeks. The organism can spread to other organs, in particular the skin, the bone, and the meninges. Individuals who are at particular risk for severe or complicated disease are immunocompromised individuals, the elderly, and pregnant women.

The diagnosis is made by:

- visualization of spherules in sputum, bronchoalveolar lavage fluid or lung biopsy specimens. Although endospores can be seen, they are smaller and not as distinctive in appearance as spherules
- culture: this is hazardous in the laboratory and appropriate precautions should be taken (see Histoplasma)
- serology: the tests used are complement fixation and immunodiffusion tests. The presence of IgM is highly suggestive of the diagnosis, and the presence of antibody in cerebrospinal fluid is highly suggestive of meningeal infection.

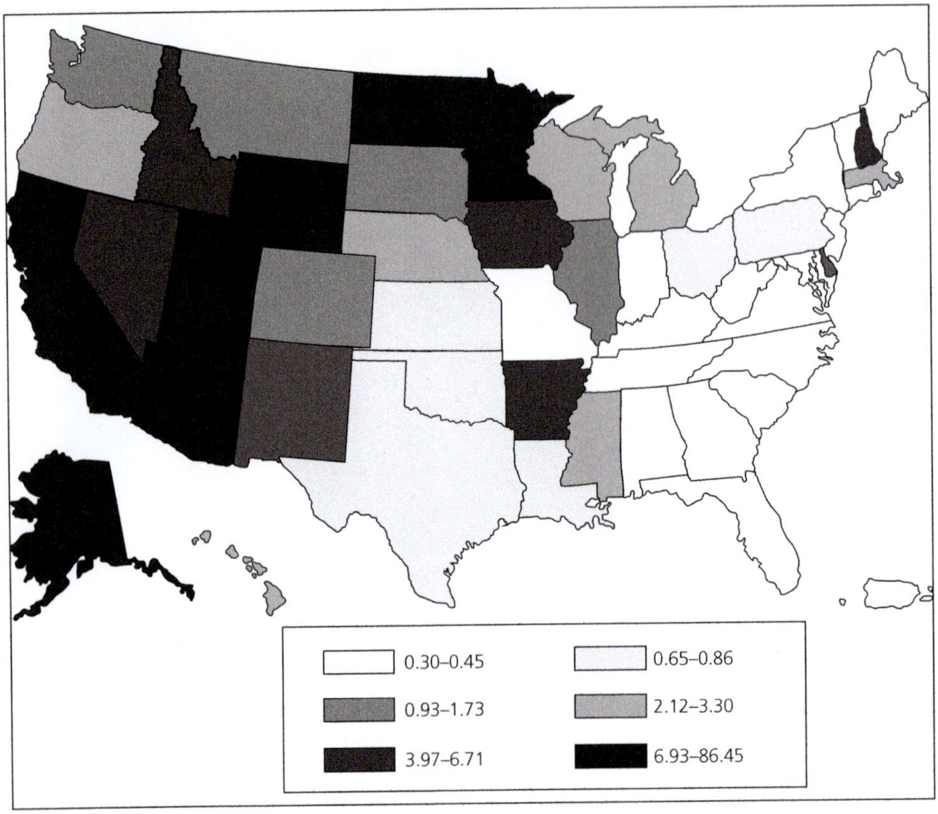

Fig. 20.7 Distribution of coccidioidomycosis in individuals >65 years old in USA. Numbers are cases per 100,000. Source: Baddley et al., 2011.

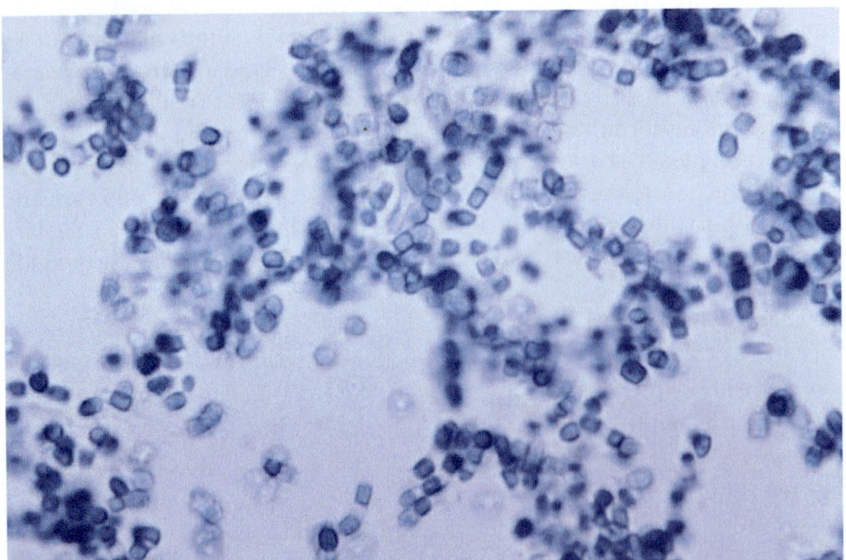

Fig. 20.8 Arthroconidia, the infectious form of Coccidioides. Courtesy of PHIL, CDC.

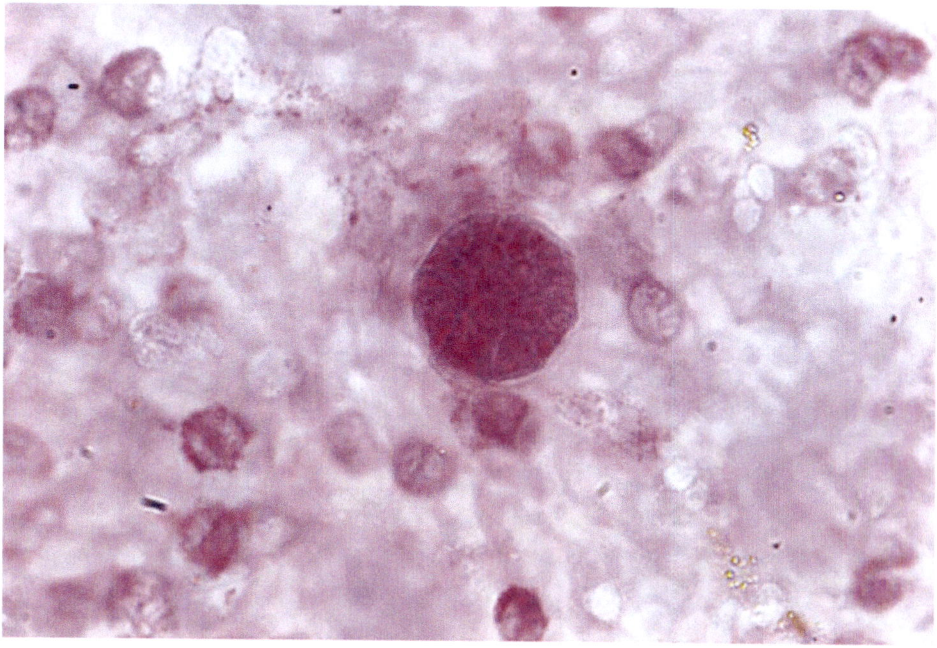

Fig. 20.9 Spherule of *Coccidioides immitis*, containing endospores. Courtesy of PHIL, CDC.

Treatment

Most cases do not require treatment. Although amphotericin B is used for severe cases, triazoles, namely fluconazole and itraconazole, can be used for most cases in whom therapy is indicated.

Paracoccidioides brasiliensis

This fungus, which causes South American blastomycosis, is prevalent in parts of South America, especially Brazil. It is present in the soil and causes infection by inhalation. There are two main forms of illness:

- the acute juvenile form, which has an equal sex ratio, and accounts for a small proportion of cases. The organism spreads to the reticuloendothelial system, and skin, and does not usually cause lung or mucous membrane disease (Fig. 20.10)
- the chronic adult form, which has a latency period of many years, causes slowly progressive pneumonia, and oral disease in many patients.

The differential diagnosis includes other diseases that cause chronic infection, such as tuberculosis, leishmaniasis, leprosy, and cancer.

The diagnosis is made by visualization of the budding yeast form in material from lesions (Fig. 20.11). Culture takes up to 1 month.

Treatment entails a prolonged course of itraconazole. Amphotericin B should be used for severe infection.

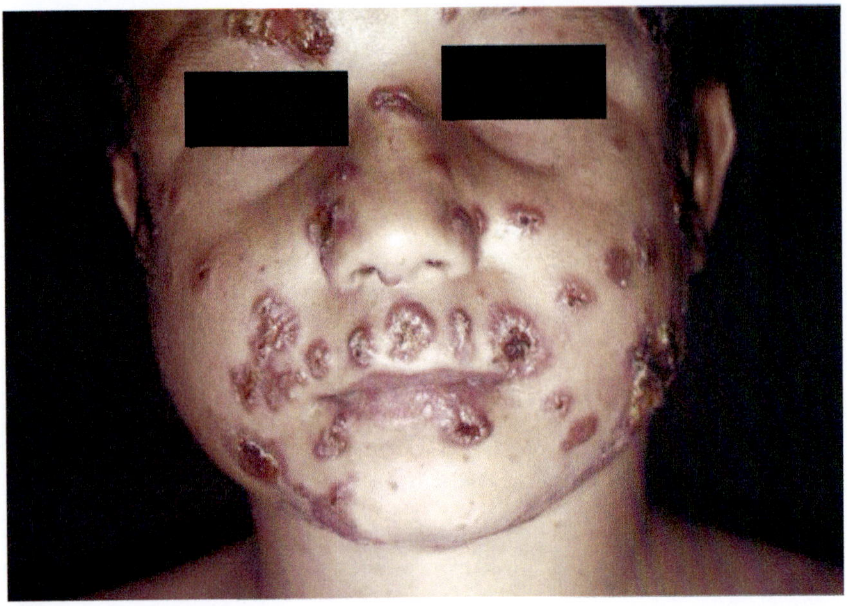

Fig. 20.10 Face of a Brazilian child with *Paracoccidioides brasiliensis* infection. Courtesy of PHIL, CDC.

Fig. 20.11 Budding yeast of *Paracoccidioides brasiliensis* in tissue. Courtesy of PHIL, CDC.

Sporothrix schenkii

This fungus, which causes sporotrichosis, is present on vegetation, especially thorns and sphagnum moss, and in soil worldwide (Fig. 20.12). Most infections occur as a result of cutaneous inoculation, but it can be inhaled. It causes a local skin lesion,

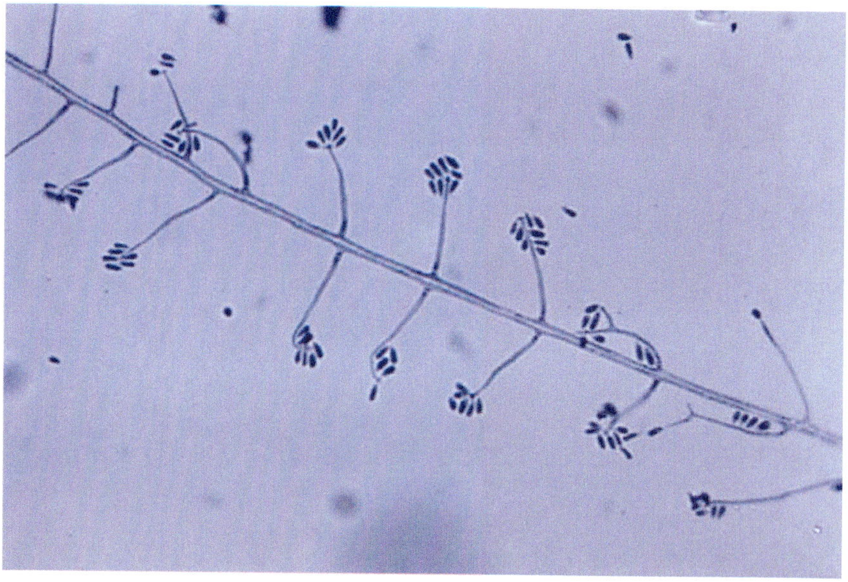

Fig. 20.12 Conidiophores with conidia of *Sporothrix schenkii* cultured from peat moss. Courtesy of PHIL/CDC.

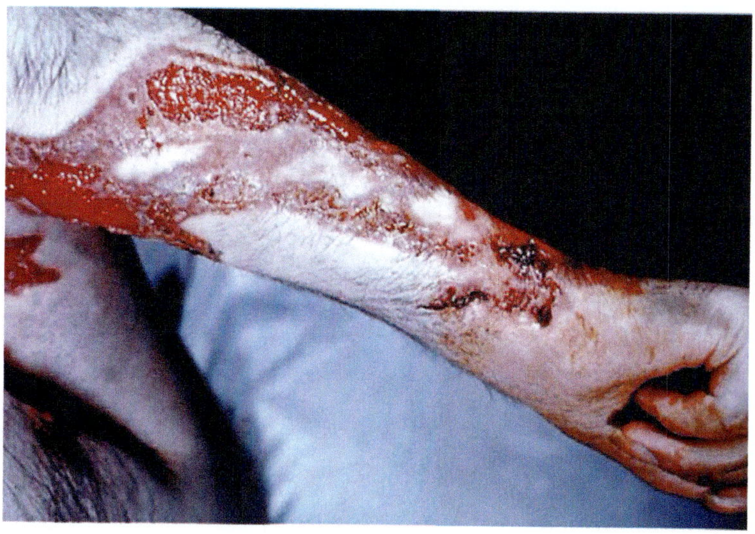

Fig. 20.13 Cutaneous infection caused by *Sporothrix schenkii*. Courtesy of PHIL/CDC.

and it can spread along the lymphatics (Fig. 20.13) and, occasionally, systemically. The diagnosis is made by visualization of the fungus and by culture (Fig. 20.14). Treatment consists of itraconazole. This has superseded treatment with a supersaturated solution of potassium iodide (SSKI), which was the mainstay of therapy for many years.

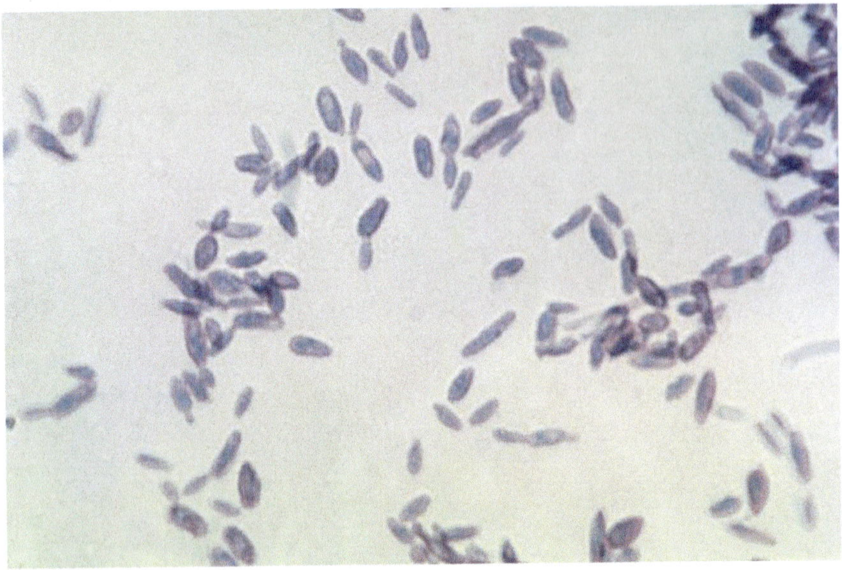

Fig. 20.14 Yeast form of *Sporothrix schenkii* in culture. Courtesy of PHIL/CDC.

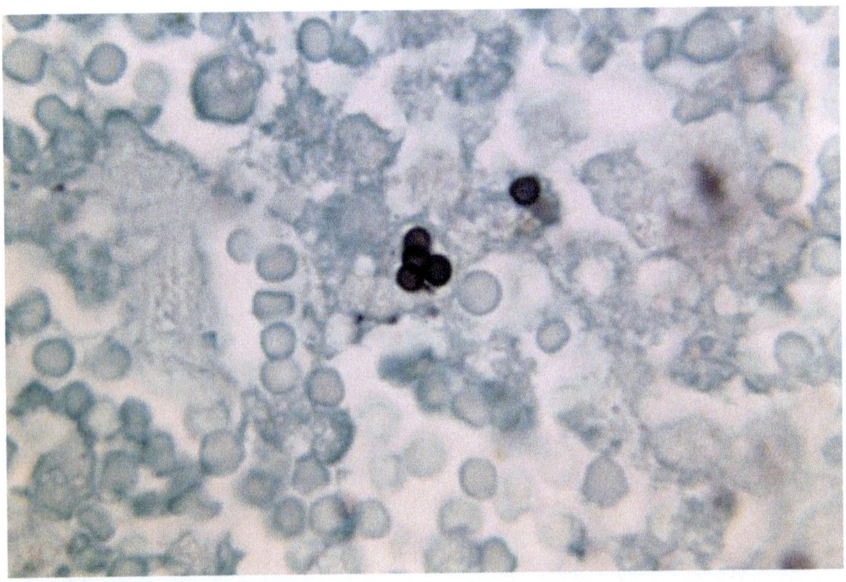

Fig. 20.15 *Penicillium marneffei* in human spleen, stained with Gomori methenamine silver stain. Courtesy of PHIL, CDC.

Penicillium marneffei (penicilliosis)

During the past few decades, this has become recognized as a very important pathogen in immunocompromised patients, in particular those with AIDS. It occurs mainly in South East Asia, especially in Thailand and Vietnam, but as far west as India. The source of the fungus is unclear, but it is known to infect bamboo rats. It causes a systemic infection, involving lungs, blood, bone marrow, liver, spleen, and skin. It can be demonstrated microscopically and cultured from material from any infected site, including blood (Figs 20.15, 20.16, & 20.17). Treatment consists of amphotericin B in severe cases, or itraconazole.

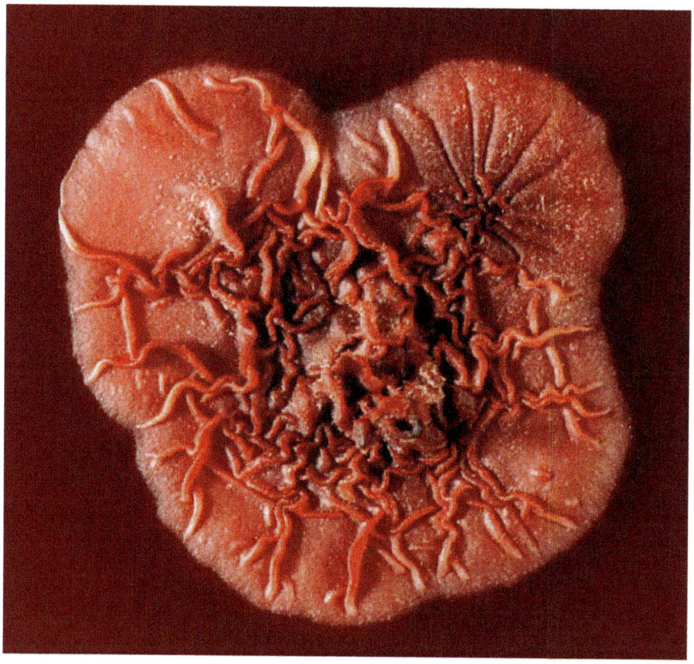

Fig. 20.16 Culture of *Penicillium marneffei*. Courtesy of PHIL, CDC.

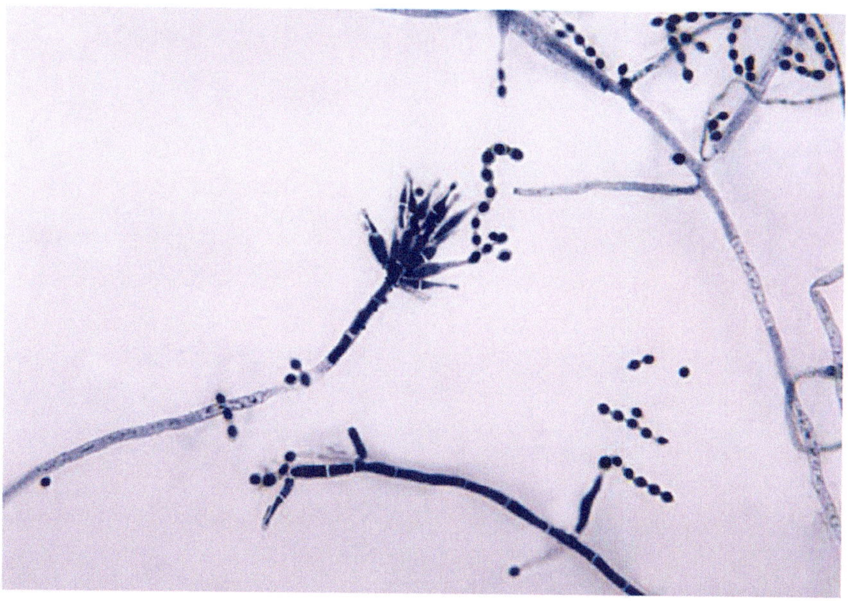

Fig. 20.17 Conidiophores laden with conidia of *Penicillium marneffei*. Courtesy of PHIL, CDC.

Other dimorphic endemic fungi

Recently, a series of cases of infections with a dimorphic fungus closely related to *Emmonsia pasteuriana* has been reported in patients with AIDS in Cape Town, South Africa.

Further reading

Ameen M, Talhari C, Talhari S (2009) Advances in paracoccidioidomycosis. Clin Exp Dermatol 35: 576–80.

Baddley JW, Winthrop KL, Patkar NM, et al. (2011) Geographic distribution of endemic fungal infections among older persons, United States. Emerg Infect Dis 17: 1664–9.

Chapman SW, Dismukes WE, Proia LA, et al. (for the Infectious Diseases Society of America) (2008) Clinical practice guidelines for the management of blastomycosis. Clin Infect Dis 46: 1801–12.

Galgiani JN, Ampel NM, Blair JE, et al. (for the Infectious Diseases Society of America) (2005) Practice guidelines for the treatment of coccidioidomycosis. Clin Infect Dis 41: 1217–23.

Kenyon C, Bonorchis K, Corcoran C, et al. (2013) A dimorphic fungus causing disseminated infection in South Africa. N Engl J Med 369; 1416–24.

Ustianowski AP, Sieu TPM, Day JN (2008) *Penicillin marneffei* infection in HIV. Curr Opin Infect Dis 21: 31–6.

Vyas KS, Bariola JR, Bradsher RW Jr (2010) Treatment of endemic mycoses. Expert Rev Respir Med 4: 85–95.

Wheat LJ (2006) Histoplasmosis: a review for clinicians from non-endemic areas. Mycoses 49: 274–82.

Wheat LJ, Freifeld AG, Kleiman MB, et al. (2007) Clinical practice guidelines for the management of patients with histoplasmosis: 2007 update by the Infectious Diseases Society of America. Clin Infect Dis 45: 807–25.

CHAPTER 21
Molds

Molds are filamentous fungi with cylindrical cells, called hyphae (singular – hypha) that often branch. A mass of hyphae is called a mycelium. They reproduce asexually by producing spores, called conidia, in addition to reproducing sexually. The morphology of the conidia and the structures that bear them (conidiophores) are frequently used to identify different molds. However, the diagnosis is often made histologically.

Aspergillus

This genus, named for the aspergillum used for sprinkling in Catholic church rituals (*aspergere* = to sprinkle, in Latin), contains several species, which are environmental agents. They form spores which enable them to spread in the environment through the air. They are 3–6 microns in diameter, with parallel walls, are septate, and they branch into two at about a 30° angle. The main species infecting humans are, in decreasing order of frequency, *A. fumigatus*, *A. flavus*, *A. terreus*, *A. niger*, and *A. nidulans*.

As human pathogens, they can cause several categories of disease.

- Colonization of anatomic spaces, such as paranasal sinuses which are chronically filled with fluid due to allergy. Their role in chronic rhinosinusitis is unclear.
- Colonization of a lung cavity caused by another disease, such as tuberculosis. In this situation they can form a fungus ball, which can result in hemorrhage.
- Allergic pneumonia, called allergic bronchopulmonary aspergillosis (ABPA), which is characterized by wheezing, flitting pulmonary infiltrates, and eosinophilia.
- Invasive aspergillosis, which generally occurs in immunocompromised individuals, such as those who are neutropenic as a result of cancer chemotherapy. This is a very severe infection, carrying a high fatality rate. The organism enters the body through the lung (it can also invade the paranasal sinuses), and spreads by invading blood vessels (angio-invasion). It can spread systemically to any organ.

Diagnosis

The diagnosis of aspergillus infection depends on obtaining specimens from the purported site of infection. In the case of invasive disease, this often requires the

Practical Medical Microbiology for Clinicians, First Edition. Frank E. Berkowitz and Robert C. Jerris.

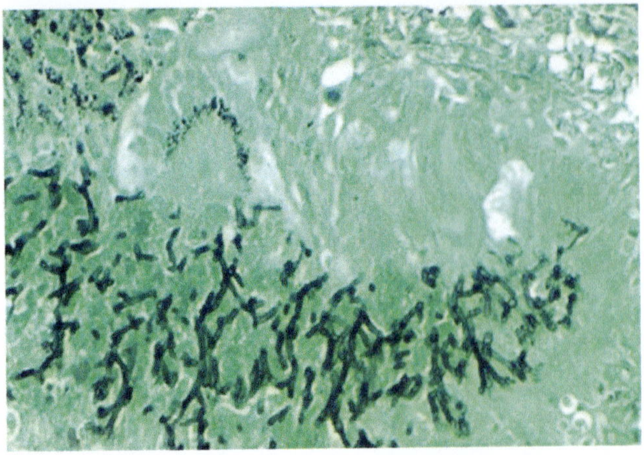

Fig. 21.1 Aspergillus in lung tissue stained with Gomori silver impregnation stain.

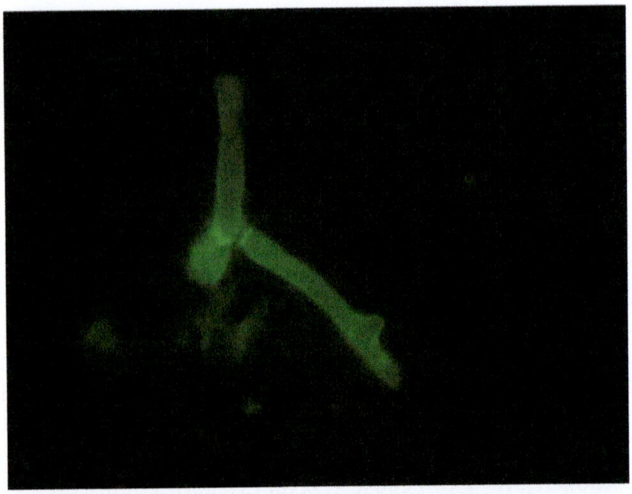

Fig. 21.2 Aspergillus cultured from a lung biopsy and stained with Calcofluor white.

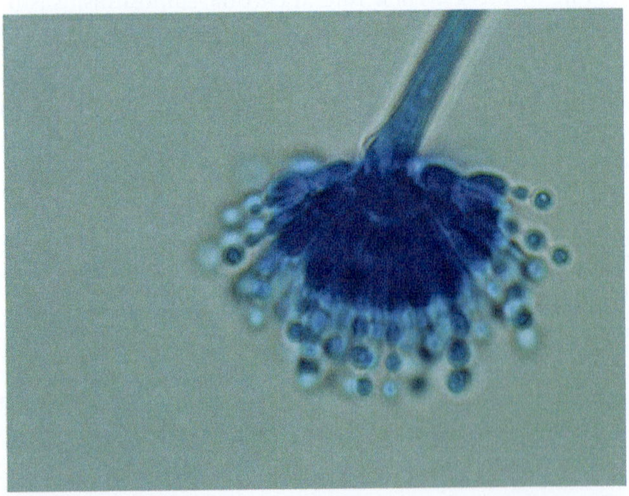

Fig. 21.3 Aspergillus hypha, cultured from a lung biopsy, and stained with lactophenol cotton blue. The preparation was made by applying tape to the mycelia of a culture. The round structures at the tips of the "aspergillum" are the conidia.

performance of a biopsy. The organism can be visualized by a variety of stains, such as Gomori silver impregnation stain (Fig. 21.1), periodic acid-Schiff (PAS), and Calcofluor white (Fig. 21.2) but **not** by Gram stain. It grows readily on many media, including blood agar. Because it is widespread in the environment, it is an important cause of laboratory contamination of agar plates and viral culture medium. The different species are identified by their growth and morphology in culture (Fig. 21.3). Detection of aspergillus antigen (galactomannan) in serum of high-risk individuals, such as stem cell transplant patients, can be useful in anticipating clinical evidence of infection (see Chapter 18).

Treatment of patients with aspergillus infections often entails surgery, to debride infected tissue when possible, in addition to antifungal therapy. Formerly, antifungal therapy was limited to amphotericin B deoxycholate. Nowadays, there are several agents active against most species of the organism. Voriconazole, which is fungicidal, is preferable for most patients with invasive disease. Although the echinocandins are active, they are only fungistatic. Amphotericin B remains a useful drug. Its dosage can be increased by the use of liposomal forms. However, it is not active against *A. terreus*.

Mucorales

This family includes fungi that have hyphae with wide and irregular diameters (6–15 microns), that branch at a right angle, and that are non- or sparsely septate (Fig. 21.4). Their asexual reproductive structures are sporangiospores, which are borne on sporangiophores. The sporangiospores can be used for identification (Fig. 21.5). This family contains several genera: Rhizopus (of which *R. orynae* is the most important), Mucor, Lichtheimia (formerly called Absidia), Rhizomucor, Cunninghamella, and Apophysomyces. They are saprobes in the environment. They cause invasive disease in immunocompromised individuals, including those with neutropenia, diabetic ketoacidosis, premature infants, and those with iron overload

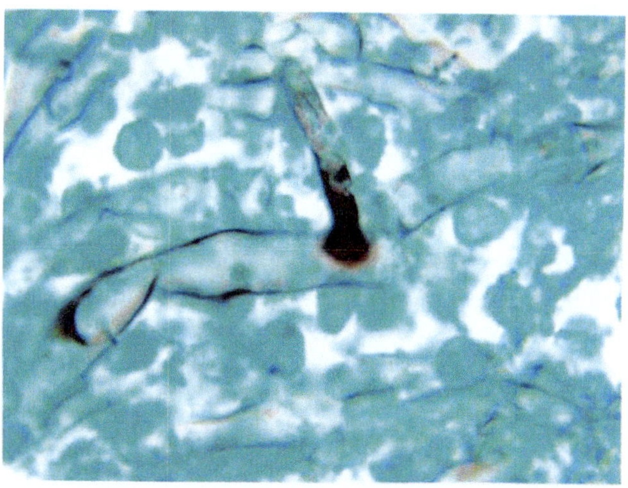

Fig. 21.4 *Mucor* spp. showing wide angle branching and absence of septa. Courtesy of PHIL, CDC.

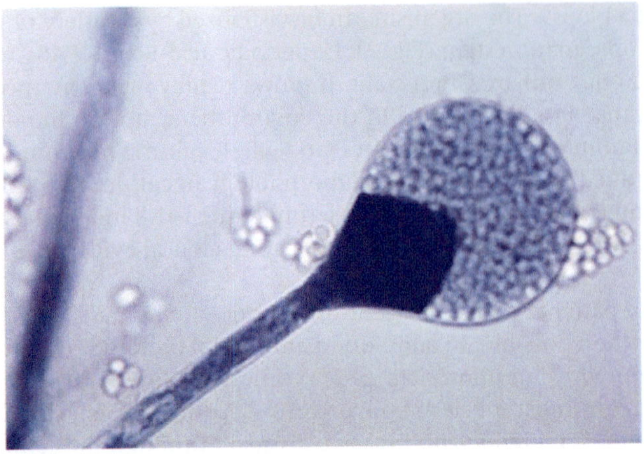

Fig. 21.5 Sporangium of Mucor. Courtesy of PHIL, CDC.

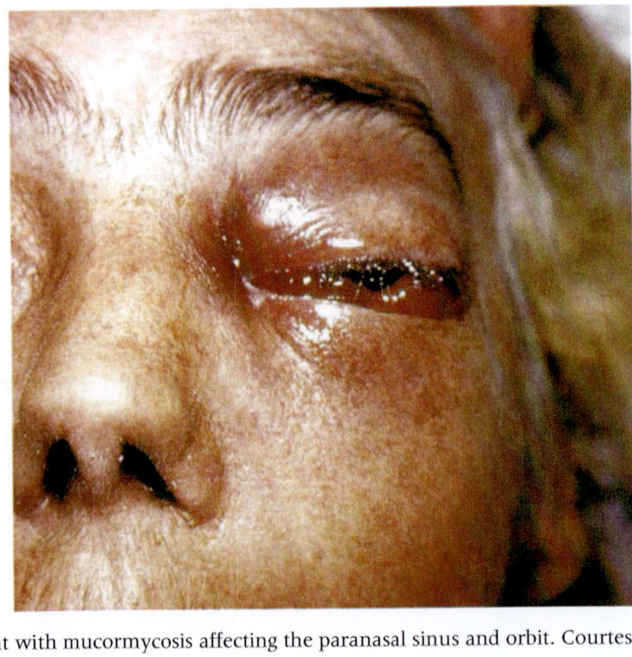

Fig. 21.6 Patient with mucormycosis affecting the paranasal sinus and orbit. Courtesy of PHIL, CDC.

being treated with desferoxamine. They invade the lung and paranasal sinuses, and can be inoculated into the skin, causing necrotizing cellulitis (Fig. 21.6). They can spread hematogenously, by angio-invasion. They grow rapidly, so disease can spread very quickly.

Aspergillus, Mucorales, and *Pseudomonas aeruginosa* cause black spreading lesions in the skin (ecthyma gangrenosum), because they cause ischemic necrosis (Fig. 21.7).

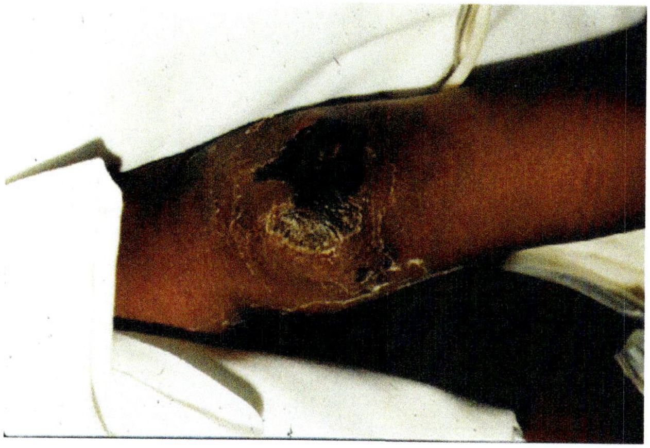

Fig. 21.7 Skin lesion (ecthyma gangrenosum) caused by *Mucor* spp. in a child with AIDS. Copyright © Frank E. Berkowitz. Reprinted with the permission of Cambridge University Press.

Patients with such lesions should be managed aggressively. The lesion should be biopsied as an **emergency**, so that smears and tissue sections can be stained for fungi and submitted for culture. Serologic and molecular assays are being investigated but their value is not yet established.

Treatment for patients suspected of having invasive mucormycosis should consist of surgical debridement and antifungal therapy. Voriconazole is active against aspergillus but **not** against mucorales. Posaconazole is active against both these fungal groups. It can be given orally or intravenously. If the patient cannot take this drug then treatment should entail amphotericin B.

Fusarium

This mold is widespread in the environment. It can cause localized infections in normal hosts, especially keratitis in those wearing contact lenses, and skin infections following injuries. However, it can become invasive in immunocompromised patients, entering the host through the lung and disseminating to viscera as well as the skin. Like Aspergillus and Mucorales, it invades blood vessels (Figs. 21.8 & 21.9). There are many species of Fusarium, the most important in human disease being *F. solani, F. oxysporum, F. verticillioides,* and *F. moniliforme*.

In immunocompromised hosts, two features suggest the diagnosis of this infection rather than aspergillosis: (a) isolation of the organism from blood culture (due to the production of yeast-like structures, a process called adventitious sporulation); and (b) metastatic skin lesions (Fig. 21.10). In addition, in histologic sections, both hyphae and yeast-like structures can be seen, which is not the case in aspergillosis. It is identified by the appearance of its macroconidia in culture.

Treatment of patients with fusariosis is very challenging because the organism is not very susceptible to most antifungal agents. For keratitis, topical natamycin

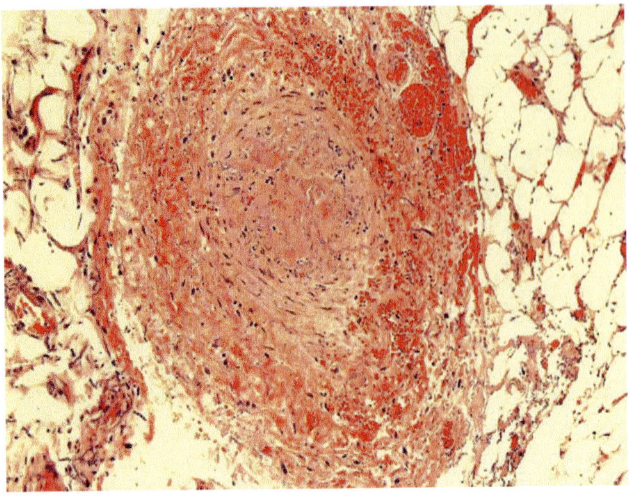

Fig. 21.8 Fusarium in a skin biopsy of the patient shown in Fig. 21.10. Although visible in this hematoxylin and eosin-stained section, the fungi are difficult to see well. Compare this with Fig. 21.9.

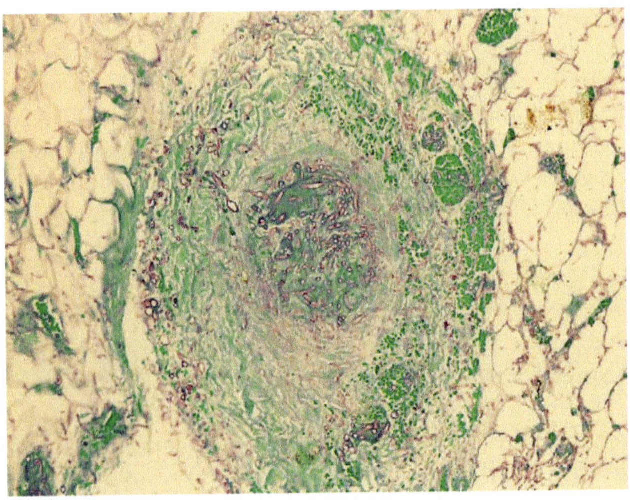

Fig. 21.9 The same biopsy material as shown in Fig. 21.8, but stained with Gomori–Grocott silver impregnation stain. Note how easily the fungi can be seen.

can be used. For systemic disease, amphotericin B, voriconazole, or posaconazole should be used. However, the success rate depends largely on host factors.

Scedosporium

Fungi of this genus, *S. prolificans*, which is often resistant to antifungal therapy, and *S. apiospermum*, also cause invasive mycoses in immunocompromised hosts.

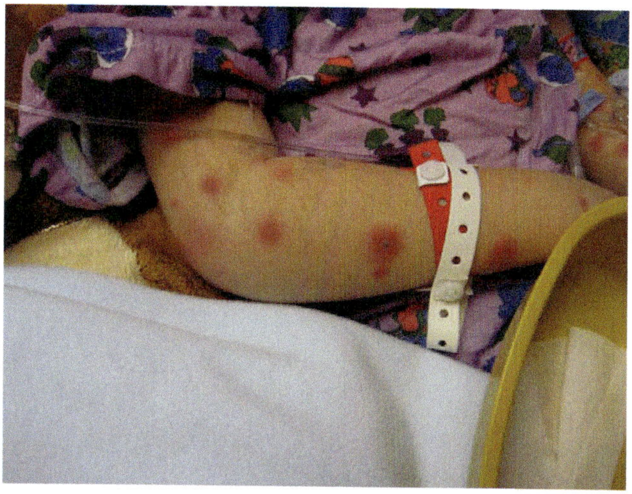

Fig. 21.10 Two-year-old girl with relapsed acute myeloid leukemia and neutropenia, illustrating skin lesions due to disseminated fusarium infection.

Fig. 21.11 Child with tinea capitis.

Dermatophytes

This is a group of filamentous fungi that cause common superficial skin, hair, and nail infections by infecting keratin. These are called Tinea (which means worm in Latin), followed by the Latin name for the anatomic site of the infection, e.g. *T. capitis* (head), *T. corporis* (body), *T. pedis* (foot), and *T. barbae* (beard) (Figs 21.11, 21.12, & 21.13). They belong to the genera Trichophyton, Epidermophyton, and Microsporum. Some are present in the environment, while some are present only on animals and/or humans.

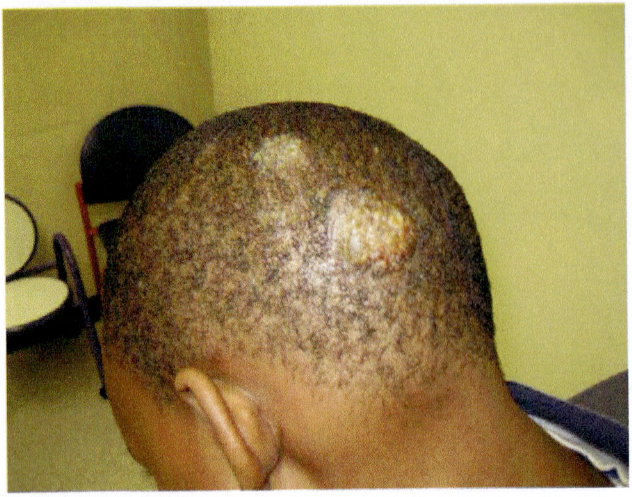

Fig. 21.12 Child with kerion, a hypersensitivity reaction to tinea capitis.

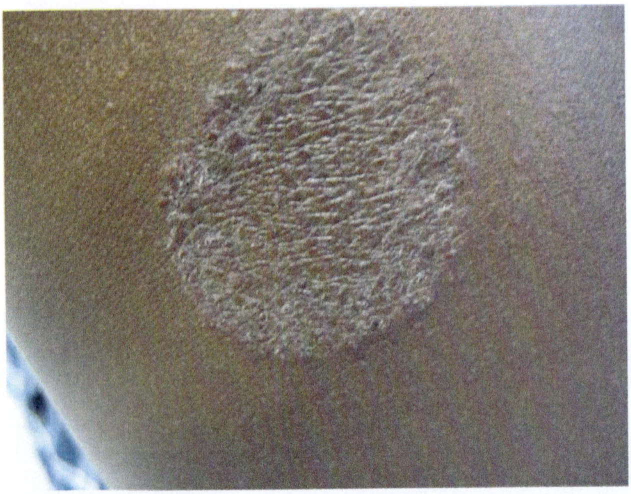

Fig. 21.13 Child with tinea corporis.

Diagnosis

- Specimen collection: specimens of hair can be obtained with tweezers, by scraping with a blade, or rubbing the affected area with a wet swab or gauze.
- Material can be cleared with heated KOH and visualized directly under the microscope, or stained with Congo red or Calcofluor white (which requires visualization under UV light) in order for hyphae to be seen.
- Culture can be performed on Sabouraud's or other selective media, and the fungi can be identified by the appearance of their macro- and microconidia. In most clinical circumstances, the diagnosis can be made without culture confirmation, and when a culture is positive, specific identification is not necessary, unless treatment fails (Figs. 21.14 & 21.15).

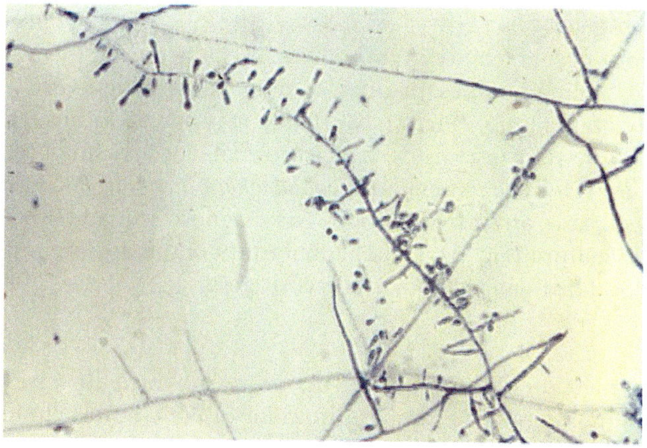

Fig. 21.14 *Trichphyton tonsurans* hyphae and microconidia. Courtesy of PHIL, CDC.

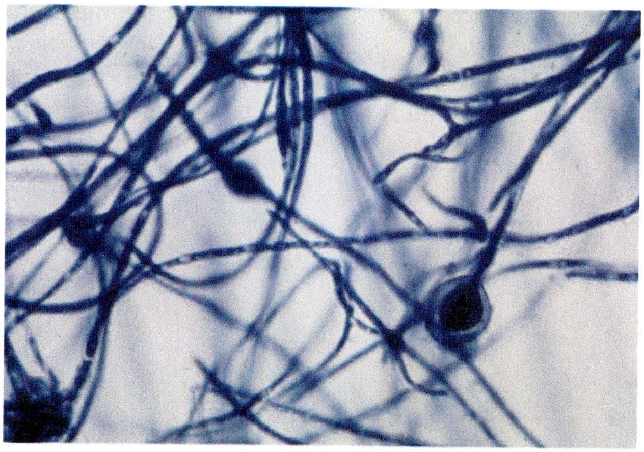

Fig. 21.15 Apiculate terminal chlamydospores of *Microsporum audouini*. Courtesy of PHIL, CDC.

Treatment

If this is on glabrous skin, the patient can usually be treated topically, with azoles or terbenafine, or, if on the toes (athlete's foot), with salicylic acid preparations. If the scalp is infected, treatment consists of systemic therapy with griseofulvin, terbenafine, or fluconazole.

Subcutaneous mycoses

There are five main types of subcutaneous mycoses, which can be caused by a large number of fungi, mostly molds: sporotrichosis (discussed above under dimorphic fungi), chromoblastomycosis, mycetoma, phaeohyphomycosis, and lobomycosis (lacaziosis). These infections, which are caused by fungi present in the environment, e.g. on vegetation, result from inoculation of the organism into the skin as a result of trauma (which might be minor).

Chromoblastomycosis

This disease, which occurs mainly in the tropics, is characterized by a variety of types of skin lesions, including nodules, plaques, tumor-like swellings, verrucae, and scars, occurring mainly on the limbs (Fig. 21.16). These may spread locally, causing lymphedema. The main causative organisms, which are dematiaceous fungi (producing melanin pigment), are *Fonsecaea pedrosoi, Cladophialophora carrionii, Phialophora verrucosa, Rhinocladiella aquaspersa,* and *Exophiala* spp. The diagnosis is made by visualization of the organism and culture (Fig. 21.17). The mainstays of therapy are oral itraconazole or terbenafine, local heat application, and cryosurgery.

Mycetoma

This is a chronic infection of the subcutaneous tissue in which abscesses and draining sinuses form. The pus contains granules, the color of which can help in determining the causative organism. The infection can extend into deeper structures such as bone. It occurs mainly in dry areas within the tropics. There are two types of mycetoma: those caused by fungi – eumycetoma – and those caused by actinomycetes – actinomycetoma. Eumycetoma, which occurs mostly in Africa and India, is caused mainly by *Madurella somaliensis, Pseudallescheria boydii, Madurella grisea, Leptosphaeria senegalensis, and Scedosporum apiospermum*. Actinomycetoma, which occurs mostly in Mexico and Central America, is caused mainly by *Nocardia brasiliensis, Streptomyces somaliensis, Actinomadura madurae,* and *Actinomadura pelletierii*. The microbiologic diagnosis is made

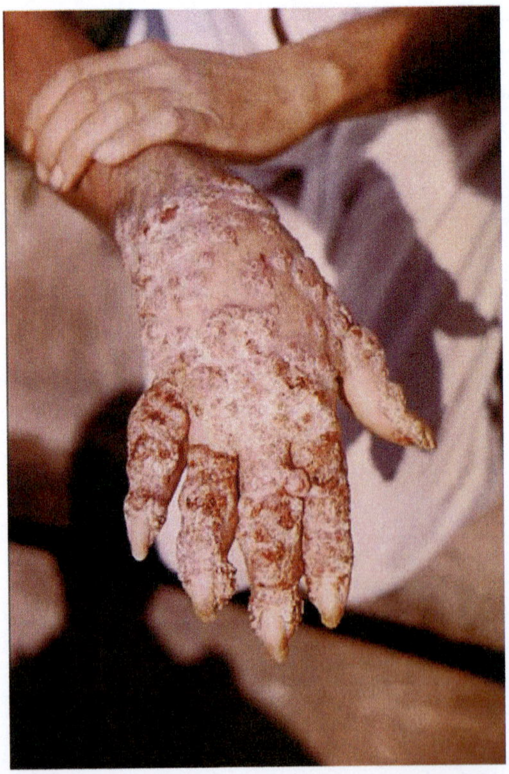

Fig. 21.16 Chromoblastomycosis caused by *Phialophora verrucosa*. Courtesy of PHIL, CDC.

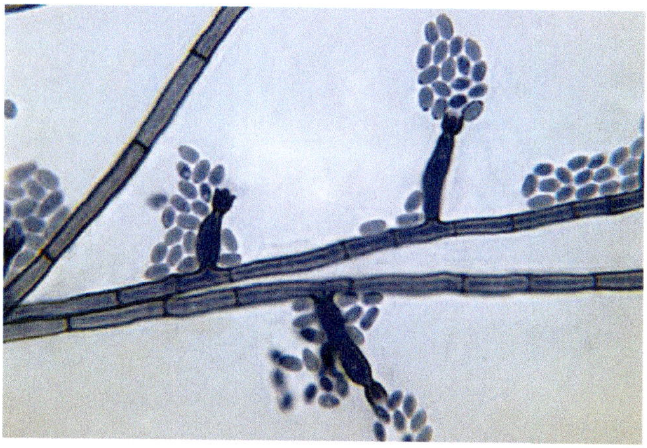

Fig. 21.17 *Phialophora verrucosa* in culture. Courtesy of PHIL, CDC.

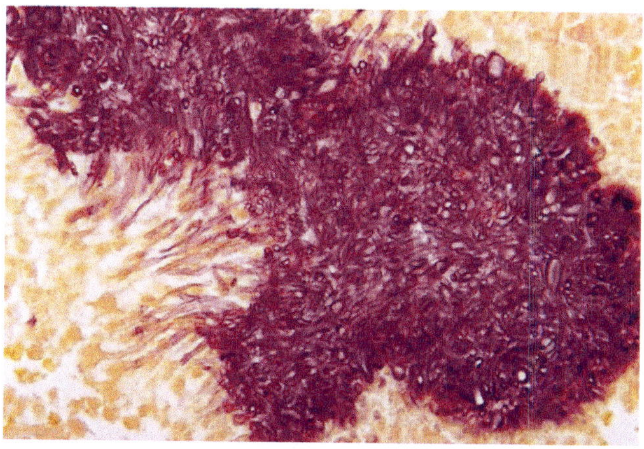

Fig. 21.18 Histology of "black grain" eumycetoma due to *Madurella mycetomatis*, stained with Gridley stain. Courtesy of PHIL, CDC.

by the color of the granule and by culture (Fig. 21.18). Eumycetoma does not respond well to antifungal therapy. Actinomycetoma can be treated with dapsone and streptomycin. Other antibacterial agents can also be tried, including rifampin, amikacin, fusidic acid, and imipenem.

Phaeohyphomycosis

This is a subcutaneous cyst caused by dematiaceous fungi, the most important being *Exophiala jeanselmei*, *Wangiella dermatitidis*, and *Bipolaris* spp (Figs 21.19 & 21.20). The diagnosis is made by staining and culture of biopsy material. Treatment consists of itraconazole and surgery.

Lobomycosis (lacaziosis)

This mycosis, which occurs in Central and South America, is caused by *Lacazia (*formerly *Loboa) loboi*. It is characterized by verrucae and plaques. It is diagnosed by characteristic yeast-like structures in a biopsy, and the treatment is surgical (Figs 21.21 & 21.22).

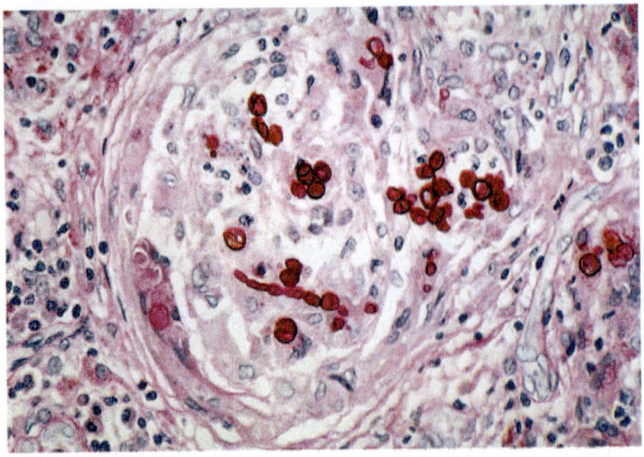

Fig. 21.19 Histology of phaeohyphomycosis due to *Wangiella dermatitidis*, stained with PAS. Courtesy of PHIL, CDC.

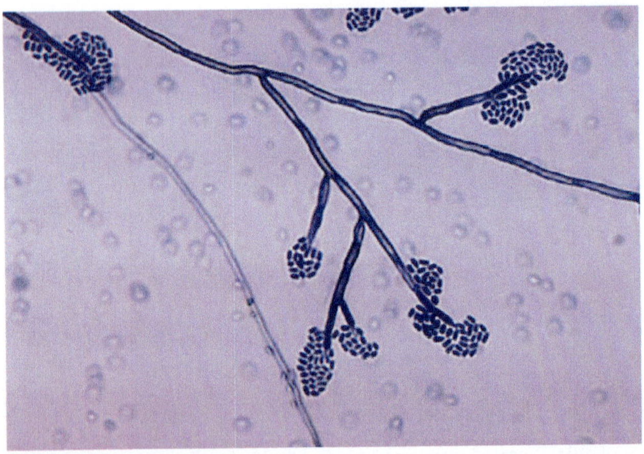

Fig. 21.20 Appearance of *Exophiala jeanselmei* in culture. Courtesy of PHIL, CDC.

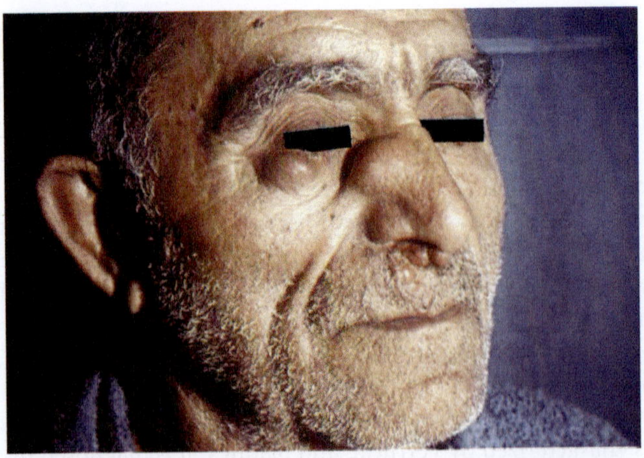

Fig. 21.21 Brazilian man with lobomycosis. Courtesy of PHIL, CDC.

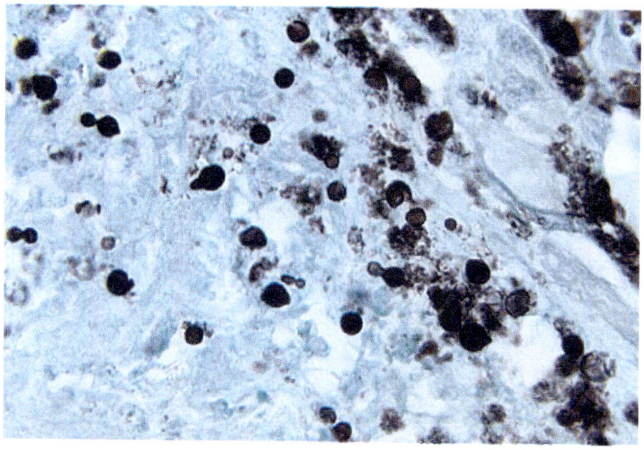

Fig. 21.22 Appearance of *Lacazia loboi* in tissue stained with Gomori silver impregnation stain. Courtesy of PHIL, CDC.

Further reading

Baddley JW, Winthrop KL, Patkar NM, et al. (2011) Geographic distribution of endemic fungal infections among older persons, United States. Emerg Infect Dis 17: 1664–9.

Castro LGM, de Andrade JS (2010) Chromoblastomycosis: still a therapeutic challenge. Expert Rev Dermatol 5: 433–43.

Garnica M, Nucci M, Queiroz-Telles F (2009) Difficult mycoses of the skin: advances in the epidemiology and management of eumycetoma, phaeohyphomycosis and chromoblastomycosis. Curr Opin Infect Dis 22: 559–68.

Guarner J, Brandt ME (2011) Histopathologic diagnosis of fungal infections in the 21st century. Clin Microbiol Rev 24: 247–80.

Lackner M, Caramalho R, Lass-Florl C (2014) Laboratory diagnosis of mucormycosis: current status and future perspectives. Future Microbiol 9: 683–95.

Nucci M, Anaissie E (2007) *Fusarium* infections in immunocompromised patients. Clin Microbiol Rev 20: 695–704.

Ostrosky-Zeichner L (2012) Invasive mycoses: diagnostic challenges. Am J Med 125: S14–S24.

Pang KR, Wu JJ, Huang DB, Tyring S (2004) Subcutaneous fungal infection. Dermatol Ther 17: 523–31.

Patterson TF (2005) Advances and challenges in management of invasive mycoses. Lancet 366: 1013–25.

Prabhu RM, Patel R (2004) Mucormycosis and entomophthoramycosis: a review of the clinical manifestations, diagnosis and treatment. Clin Microbiol Infect 10(suppl 1): 31–47.

Reiss E, Shadomy HJ, Lyon GM III (2012) *Fundamental Medical Mycology*. New Jersey: Wiley-Blackwell, 2012.

Walsh TJ, Anaissie EJ, Denning DW, et al. (2008) Treatment of aspergillosis: clinical practice guidelines of the Infectious Diseases Society of America. Clin Infect Dis 46: 327–60.

SECTION V
Parasitology

CHAPTER 22
Parasitology

General description of parasites

The microorganisms that are, by tradition, referred to as "parasites" are eukaryotes. They are classified, for convenience, into protozoa (single-celled organisms), worms, and ectoparasites. As a general principle, **protozoa multiply within the host** (as do viruses and bacteria). In contrast, with a few exceptions, e.g. *Strongyloides stercoralis*, **worms do not multiply within the host**. The "worm burden" is the number of worms entering the host. The protozoa are classified into four groups: flagellates, ciliates (of which only one infects humans), amebae, and apicomplexa (sporozoa). The worms are classified into three groups: roundworms (nematodes), flukes (trematodes), and tapeworms (cestodes). From a clinical standpoint, it is also convenient to classify the internal parasites into those infecting primarily the gut (even if they pass through the blood and tissues to reach there, in the case of some worms) and those infecting primarily the blood or tissues other than the gut.

Life cycle

This describes the pathway that the parasite takes from the time of entering the host, its pathway through the host, its exit and spread to another host, until its progeny reach the same point in another host. An understanding of the life cycle is essential for understanding the pathogenesis of the infection, its epidemiology, and its clinical manifestations. It thus provides the information on which methods for preventing transmission can be based. Many parasites that affect humans, in particular worms, have an animal host as part of their life cycle in the following ways.

- The parasite is an animal parasite and humans are incidental hosts, e.g. Babesia, Toxocara, Echinococcus.
- An animal is one of the hosts in a cycle that normally involves humans, e.g. Schistosoma spp., in which a snail is the intermediate host.

Practical Medical Microbiology for Clinicians, First Edition. Frank E. Berkowitz and Robert C. Jerris.

- An animal, usually an arthropod, is a vector that transmits the parasite to humans, e.g. *Plasmodium* spp., which is transmitted by a mosquito.

The **definitive host** is the one in which the adult lives, e.g. human for *Taenia solium*, or in which the sexual part of the cycle occurs, e.g. mosquito for *Plasmodium* spp.

The **intermediate host** is the animal in which the larval stage lives, e.g. pig for *Taenia solium*, or in which the asexual part of the cycle takes place, e.g. vertebrate for *Plasmodium* spp.

Diagnostic tests

The traditional tests depend on visualization of the parasites, either under the microscope, in the case of protozoa or worm eggs, or with the naked eye, in the case of worms. Various concentration methods have been devised to increase the sensitivity of stool tests for eggs, and of blood for trypanosomes. Because most parasitic diseases are rare in the USA, few microbiology laboratories have individuals skilled in the identification of parasites. Newer methods entail antigen detection and genome detection by PCR. The Division of Parasitic Disease of the Centers for Disease Control and Prevention, USA, provides assistance in diagnosing parasitic diseases (http://dpd.cdc.gov/dpdx).

A simple taxonomy of parasites is shown in Figure 22.1.

Taxonomy of internal parasites

Protozoa

Gut

Flagellates:
Giardia intestinalis

Ciliate:
Balantidium coli

Amebae:
Entamoeba histolytica (and related amebae)
Entamoeba coli
Endolimax nana

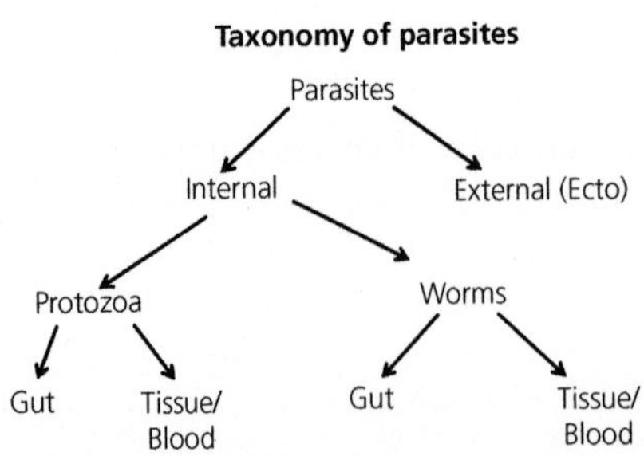

Fig. 22.1 Simple taxonomy of parasites.

Iodamoeba bütschlii
Dientamoeba fragilis

Apicomplexa:

Cystoisospora belli
Cryptosporidium spp.
Cyclospora cayetanensis

Tissue and blood

Flagellates:

Leishmania spp.
Trypanosoma brucei
Trypanosoma cruzi
Trichomonas vaginalis

Amebae:

Naegleria fowleri
Acanthamoeba spp.
Balamuthia mandrillaris
Sappinia spp.

Apicomplexa:

Plasmodium spp.
Babesia spp.
Toxoplasma gondii
Sarcocystis spp.

Worms

Gut

Nematodes (roundworms):

Ascaris lumbricoides
Strongyloides stercoralis
Ancyclostoma duodenale (hookworm)
Necator americanus (hookworm)
Trichuris trichiura (whipworm)
Enterobius vermicularis (pinworm)
Anisakis spp.
Capillaria philippinensis

Trematodes (flukes):

Fasciolopsis buski

Cestodes (tapeworms):

Taenia solium
Taenia saginata
Diphylobothrium lata
Hymenolepis nana
Dipylidium caninum

Tissue/blood

Nematodes (roundworms):

Toxocara cati, T. canis (dog and cat ascaris, respectively)
Baylisascaris procyonis (raccoon ascaris)

Ancyclostoma brasiliensis
Ancylostoma canis
Dracunculus medinensis (Guinea worm)
Angiostrongylus costaricensis
Gnathostoma spinegerum

Filaria:
Wuchereria bancrofti
Brugia malayi
Brugia timori
Onchocerca volvulus
Loa loa

Trematodes (flukes):
Schistosoma spp. – blood
Fasciola hepatica – liver
Clonorchis sinensis – liver
Opisthorchis viverrini - liver
Paragonimus spp. – lung

Cestodes (tapeworms):
Echinococcus granulosus
Multiceps multiceps
Coenuris coenuris
Cysticercus (see *Taenia solium*)

Antiparasitic drugs

Antimalarial drugs

The plasmodium parasites that cause malaria have a complicated life cycle, in two hosts (see Chapter 24). None of the antimalarial drugs are active against all stages of the parasite. Therapy of ill individuals is directed at the exoerythrocytic stage. Prevention of relapses in the cases of *Plasmodium vivax* and *P. ovale* infections is directed at the hypnozoites in the liver. Prevention of transmission of infection to the mosquito depends on elimination of gametocytes. Since there is no drug active against the sporozoite, chemoprophylaxis against infection depends on elimination of the hepatic or erythrocytic stages, once infection occurs.

Artemisinins

These sesquiterpene lactones (artemether and artesunate) are derivatives of the Chinese plant qinghaosu (*Artemisia annua*). They are the most rapidly schizonticidal antimalarials available currently, and they form the basis of therapy for individuals with *P. falciparum* infection in most parts of the world. They are also gametocidal. Their mechanism of action is unclear, but seems to be related to oxidative injury to the parasite.

Atovoquone

This hydroxynaphthaquinone is an analog of ubiquinone, which is necessary for parasite pyrimidine synthesis. The drug inhibits electron transport in the mitochondrial cytochrome bc_1 complex, which prevents regeneration of ubiquinone. For synergy, it is used together with proguanil, which is an inhibitor of the bifunctional enzyme dihydrofolate reductase-thymidylate synthase. This combination is active against the erythrocytic stage of the infection. Although it is active against the hepatic stage of *P. falciparum*, it is not active against the hypnozoites of *P.vivax*.

Folate antagonists (see Antibacterial agents, Chapter 9)

Pyrimethamine and sulfadoxine were previously widely used, mainly for prophylaxis. However, this is no longer the case, due to adverse effects and to resistance, which is the result of multiple mutations in the dihydrofolate synthase–thymidylate synthase complex and dihydropteroate synthase respectively.

Quinolines

This group of drugs has been the mainstay of antimalarial therapy for several decades, until the advent of the artimisinins.

Chloroquine is a 4-aminoquinoline. It was previously used for treating patients with infections caused by all species of plasmodium. However, widespread resistance in *P. falciparum*, the species that causes most cases of serious disease, has restricted its role in this infection. In susceptible strains, it is active against only the erythrocytic stage. The mechanism of action is thought to be as follows: within an acidic vacuole, the parasite digests hemoglobin. This generates free radicals and ferroprotoporphyrin IX, which are toxic to the parasite. The parasite therefore sequesters the ferroprotoporphyrin as an inert compound called hemazoin. Chloroquine binds to the heme, thus preventing this sequestration and leading to oxidative damage to the parasite. Chloroquine resistance is due to a reduction in the amount of chloroquine entering the vacuole where it should exert its effect. Other quinolines, including amodiaquine, quinine, quinidine, mefloquine, halofantrine and lumefantrine, are thought to have a similar mode of action. Primaquine is the only antimalarial active against the hepatic stage of plasmodia, but it is not active against the erythrocytic stage. It is therefore useful for eliminating the hypnozoite of *P. vivax* and *P.ovale*. It causes hemolysis in individuals with glucose-6-phosphate dehydrogenase deficiency, and should not be used in such individuals nor during pregnancy.

Antiprotozoal drugs (excluding antimalarials)

Antimonials

Sodium stibogluconate and meglumine antimonate are pentavalent antimonial compounds used for the treatment of patients with leishmaniasis. They are prodrugs, being converted inside the macrophage to trivalent antimonials, which interfere with the parasite's trypanothione redox system.

Eflornithine

This is an inhibitor of ornithine decarboxylase, an enzyme necessary for the synthesis of polyamines such as spermidine. The polyamines are necessary for cell division and for the synthesis of trypanothione, which prevents oxidative damage to the organism. It is used for patients with *Trypanosoma brucei gambiense* infection.

Melarsoprol

This is an arsenical prodrug which is converted to melarsen oxide. This reacts with sulfhydryl groups, affecting several enzymes. It is used for the treatment of neurologic disease caused by *Trypanosoma brucei rhodesiense*. Although it has a high rate of causing encephalopathy, which carries a significant fatality rate, it is the only drug available for this otherwise fatal stage of the infection. Resistance is due, at least partially, to a mutation in a transporter. This also causes cross-resistance to pentamidine.

Metronidazole and tinidazole

These are used for the treatment of individuals with flagellate (*Giardia intestinalis, Trichomonas vaginalis*) and amebic (*Entamoeba histolytica*) infections. Metronidazole is also used for many anaerobic bacterial infections (see Chapter 14). These organisms utilize ferrodoxin as a transporter of electrons, which it receives as a result of the activity of the enzyme pyruvate: ferrodoxin oxidoreductase. Ferrodoxin donates electrons to metronidazole, which forms a reactive nitro radical that kills the organism.

Miltefosine

This is an alkylphosphocholine, which is used for the treatment of individuals with leishmaniasis. Its mode of action is unknown.

Nifurtimox and benznidazole

These are a nitrofuran and a nitrobenzimidazole respectively. They are used for treatment of individuals with *Trypanosoma cruzi* and, in the case of nifurtimox, *Trypanosoma brucei gambiense* infections. They interfere with NADH-dependent mitochondrial nitroreductase, resulting in the formation of nitro radicals, which are trypanocidal.

Nitazoxanide

This is used for the treatment of several protozoal infections, and is the only drug available for the treatment of individuals with cryptosporidiosis. It interferes with pyruvate:ferrodoxin oxidoreductase-dependent electron transfer (see Metronidazole earlier).

Paromomycin

This is a non-absorbable aminoglycoside, used enterally for treating individuals with the luminal form of *Entamoeba histolytica*, and locally in individuals with cutaneous leishmaniasis.

Pentamidine

This is an aromatic diamidine, used for the treatment of individuals with *Trypanosoma brucei gambiense* and *Pneumocystis jiroveci* (formerly considered a protozoan but now considered a fungus) infections. Its mode of action is unclear but it binds to DNA and inhibits mitochondrial topoisomerase, which might explain its activity.

Suramin

This is a polysulfonated naphthalene derivative of urea used for patients with *Trypanosoma brucei rhodesiense* infection. Its mechanism of action is unclear but it inhibits endocytosis and interferes with parasite metabolism.

Antihelminthic drugs

Benzimidazoles (albendazole, mebendazole, and thiabendazole)

These are widely used antinematode drugs. Their mode of action is by binding to parasite β-tubulin, causing inhibition of microtubule polymerization. They may also affect some metabolic pathways. Resistance is due to mutations in the β-tubulin. Mebendazole is no longer available in the USA.

Diethylcarbamazine

This drug is used for treatment of infections caused by several different filaria, namely *Wuchereria bancrofti, Brugia malayi, Brugia timori,* and *Loa loa*. It causes organelle damage in microfilaria, but its mechanism of action on adult worms is unknown.

Ivermectin

This is a macrocyclic lactone derived from *Streptomyces avermitilis*. It activates ligand-gated chloride channels, causing tonic paralysis of the worm. It is the treatment for infections caused by *Onchocerca volvulus* and *Strongyloides stercoralis*, and can be used for patients with scabies and pediculosis.

Praziquantel

This is a pyrazinoisoquinoline that is the drug of choice for the treatment of schistosomiasis and most trematode infections. Its mode of action is unclear but includes the following possibilities: affecting the β subunit of voltage-gated calcium channels, causing an influx of calcium; inhibition of adenosine uptake; and causing alterations in the parasite integument, rendering the parasite more exposed to immune attack.

Pyrantel pamoate

This is a neuromuscular blocking agent and inhibitor of cholinesterase. It causes muscular paralysis of worms.

Further reading

Bogitsh BJ, Carter CE, Oeltmann TN (2013) *Human Parasitology*, 4th edn. Amsterdan: Elsevier.

Doenhoff MJ, Cioli D, Utzinger J (2008) Praziquantel: mechanisms of action, resistance and new derivatives for schistosomiasis. Curr Opin Infect Dis 21: 659–67.

John DT, Petri WA Jr (2006) *Markel and Voge's Medical Parasitology*, 9th edn. St Louis: Saunders Elsevier 2006.

Kappagoda S, Singh U, Blackburn BG (2011) Antiparasitic therapy. Mayo Clin Proc 86: 561–83.

McCarthy J, Loukas A, Hotez PJ (2011) Chemotherapy of helminth infections. In: Brunton LL, Lazo JS, Parker KL (eds) *Goodman and Gilman's The Pharmacological Basis of Therapeutics*, 12th edn. New York: McGraw-Hill.

Muller IB, Hyde JE (2010) Antimalarial drugs: modes of action and mechanisms of resistance. Future Microbiol 5: 1857–73.

Phillips MA, Stanley SL Jr (2011) Chemotherapy of protozoal infections: amebiasis, giardiasis, trichomoniasis, trypanosomiasis, leishmaniasis, and other protozoal infections. In: Brunton LL, Lazo JS, Parker KL (eds) *Goodman and Gilman's The Pharmacological Basis of Therapeutics*, 12th edn. New York: McGraw-Hill.

Vinetz JM, Clain J, Bounkeua V, et al. (2011) Chemotherapy of malaria. In: Brunton LL, Lazo JS, Parker KL (eds) *Goodman and Gilman's The Pharmacological Basis of Therapeutics*, 12th edn. New York: McGraw-Hill.

Wilkinson SR, Kelly JM (2009) Trypanosomal drugs: mechanisms, resistance and new targets. Expert Rev Mol Med 11: e31.

CHAPTER 23
Intestinal protozoa

Flagellates

Giardia intestinalis

This is a common cause of diarrhea in many parts of the world, due to the presence of the organism in bodies of water in which animals or humans have defecated. It is an important pathogen spread in day care centers, and individuals with IgA deficiency are at particular risk for this infection. The trophozoite form of the parasite infects the proximal small bowel, causing non-bloody diarrhea. This can be acute in onset, but the disease can become chronic if the patient is untreated, leading to malabsorption and malnutrition. The parasite forms cysts, which are passed in the stool, and which enable it to survive in the environment and to infect other hosts. The infective dose is only a few organisms.

The diagnosis depends on visualizing the trophozoite (Fig. 23.1) or cyst (Fig. 23.2) in the stool, or in material obtained from the duodenum. The latter can be obtained by passing a string with a weight attached into the duodenum. On removal, the duodenal fluid on the string is examined under the microscope, directly or using immunofluorescent antibody staining. More recently, antigen and molecular tests performed on stool have become available.

Several drugs can be used for treating patients with giardiasis, including metronidazole, tinidazole, nitazoxanide, and quinacrine.

Other flagellates

Chilomastix mesnili and *Pentatrichomonas hominis* are non-pathogenic flagellates found in humans, while *Dientamoeba fragilis* may be pathogenic.

Amebae

Several amebae infect the intestine, namely *Entamoeba histolytica*, *Entamoeba dispar* and *E. moshkovskii*, which are morphologically indistinguishable from one another, as well as *Entamoeba coli*, *Entamoeba hartmanni*, *Entamoeba polecki*, *Iodamoeba bütshlii*, and

Practical Medical Microbiology for Clinicians, First Edition. Frank E. Berkowitz and Robert C. Jerris.

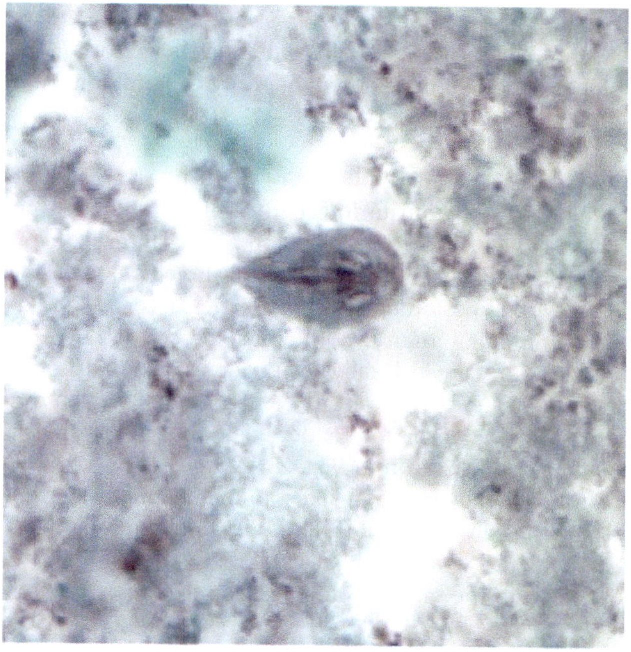

Fig. 23.1 *Giardia intestinalis* trophozoite. Courtesy of DPDx, CDC.

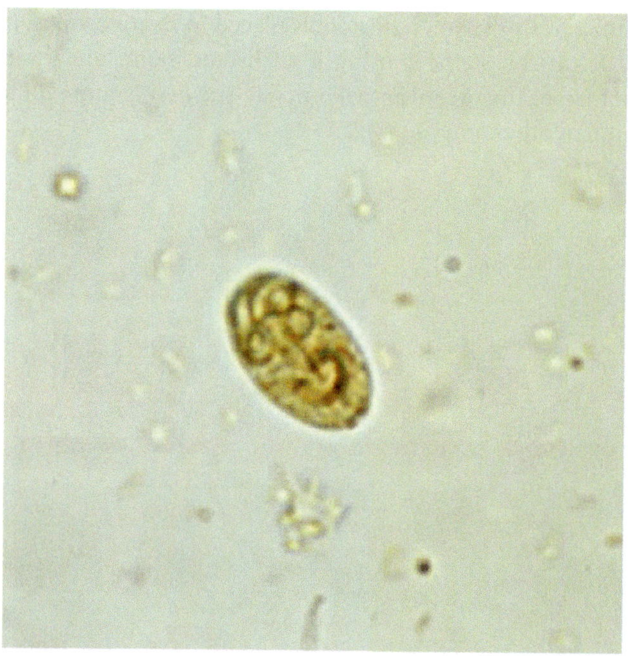

Fig. 23.2 *Giardia intestinalis* cyst. Courtesy of DPDx, CDC.

Endolimax nana. Among these, only *E. histolytica* is pathogenic. The others may be found in the stool, and their significance is in differentiating them from *E. histolytica*, and in indicating that the patient might be exposed to fecally contaminated food or water.

Entamoeba histolytica

This is a parasite of worldwide distribution, prevalent in areas with poor sanitation. The trophozoite stage of the organism infects the colon, where it causes ulceration, resulting in diarrhea, which might become bloody (dysentery). It can cause flask-shaped ulcers as it progresses through each layer of the bowel wall (Fig. 23.3). These can ultimately cause bowel perforation, which carries a very high fatality rate. The organism can cause inflammation around the bowel circumference, causing narrowing (ameboma), and it can spread locally to invade the genital tract. An important complication is spread via the portal vein to the liver, where it can cause an abscess. This manifests with prolonged fever and tender hepatomegaly. It can spread further to the pleural space, lung, and pericardium (Fig. 23.4). In cases of colitis, rectal examination can reveal a mucosa that is rough like sandpaper or cobblestones. When the organism is excreted, or even within the colon, where it can remain without causing disease, it forms cysts, which enable it to survive in the environment. When the cysts are ingested, they excyst in the intestine and the trophozoites are released.

Diagnosis

Colitis

- Wet preparation: fresh stool is mixed with warm saline on a microscope slide and examined immediately. The amebae, which are slightly larger than a leukocyte, are seen to put out pseudopodia; ingested red cells confirm that these are *E. histolytica* (Fig. 23.5).
- Fixed specimen: a fresh specimen of stool is placed in fixative and sent to the laboratory, where it can be stained, using a trichrome stain, and examined. If the specimen is not fresh, the organism will encyst. The cyst is more difficult to distinguish from those of other amebae (Fig. 23.6).

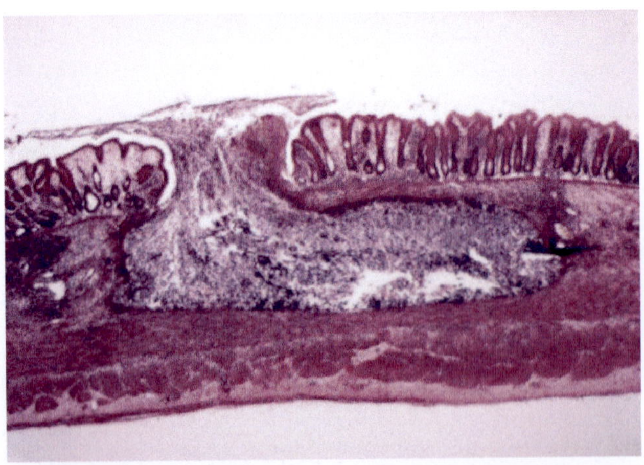

Fig. 23.3 Amebic ulcer. Note the undermining of the mucosa. Courtesy of PHIL, CDC.

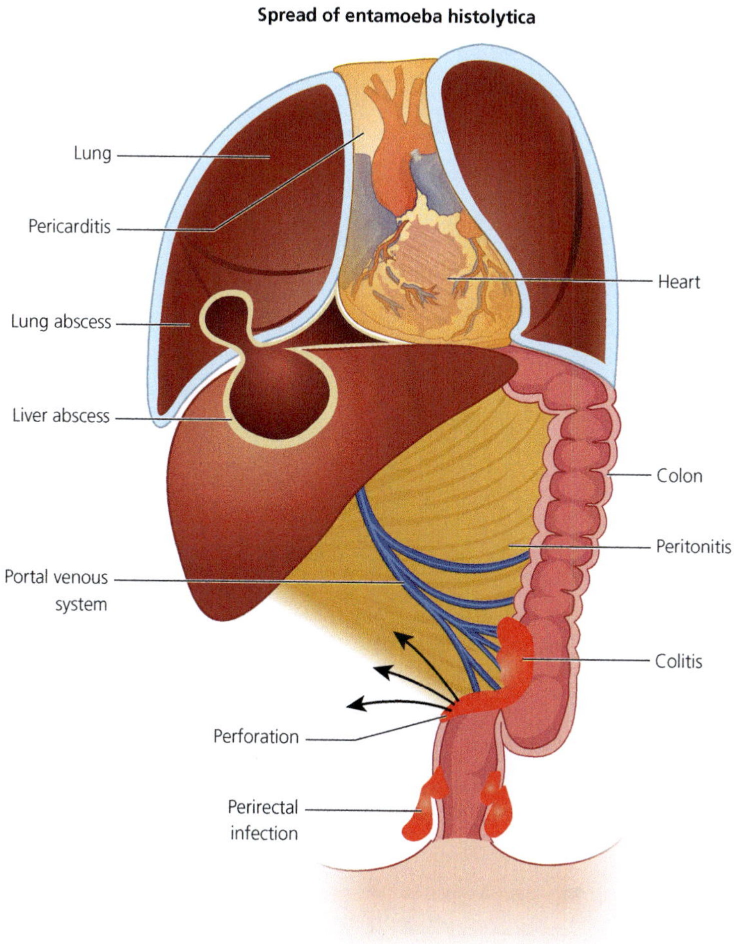

Fig. 23.4 Pathogenesis of amebiasis. Copyright © Frank E. Berkowitz. Reprinted with the permission of Cambridge University Press.

- Antigen and molecular methods of detection: nucleic acid methods are necessary to distinguish *E. dispar* and *E. moshkovskii* from *E. histolytica*.

Liver abscess

- Imaging: this can confirm the presence of an abscess, although it cannot determine whether this is amebic or bacterial. That is determined by the clinical scenario.
- Serology: this is very sensitive in patients with invasive disease.
- Examination of material from the abscess is **not** useful for the following reasons:
  - if an abscess is thought to be caused by *E. histolytica*, the patient should be treated medically, unless imminent rupture is thought likely
  - the organism is present only in the edge of the abscess, which is not likely to be sampled, not in the center.

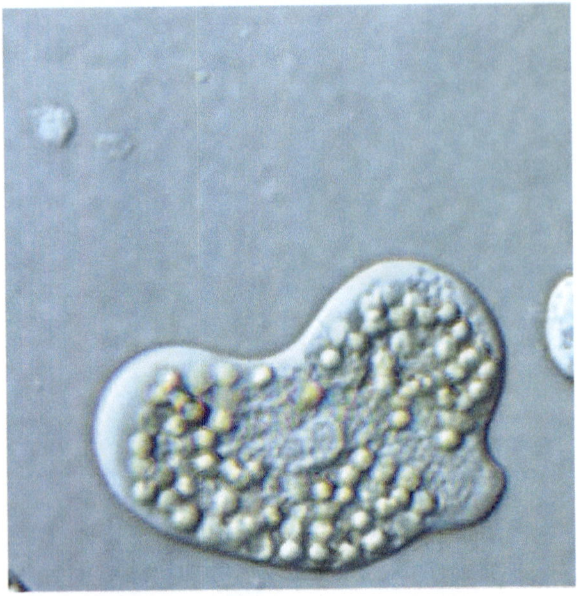

Fig. 23.5 *E. histolytica* in a fresh stool specimen, examined by phase contrast microscopy, showing numerous ingested erythrocytes. Courtesy of DPDx, CDC.

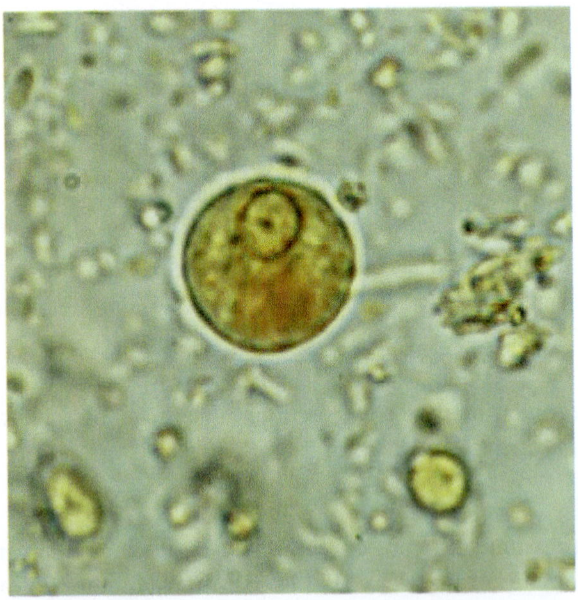

Fig. 23.6 Cyst of *E. histolytica*, stained with iodine. Courtesy of DPDx, CDC.

Treatment

The treatment of patients with amebic colitis or abscess consists of metronidazole or tinidazole. Luminal organism can be eliminated with iodoquinol or paromomycin.

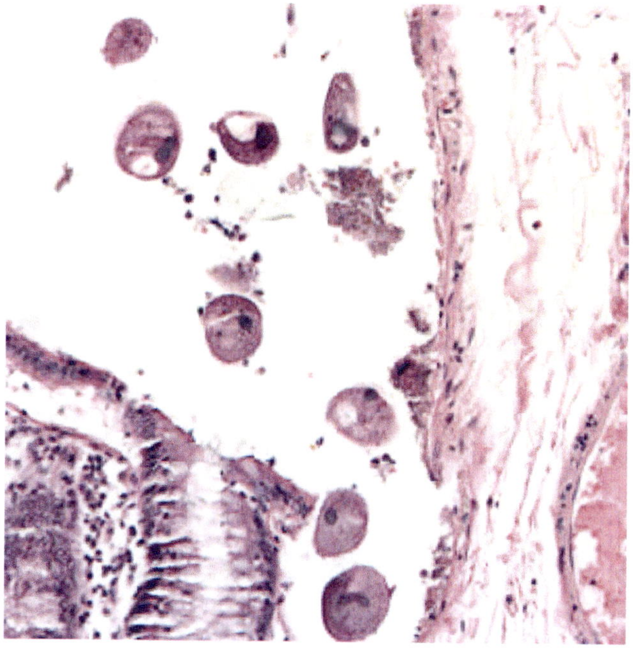

Fig. 23.7 *Balantidium coli* in colonic epithelium. Courtesy of DPDx, CDC.

Ciliates

Balantidium coli

This large protozoan is the only ciliated parasite affecting humans. It is spread by the fecal–oral route, via food or water; pig feces is an important source of infection. It is particularly common in eastern Asia and Bolivia. It causes a disease similar to that caused by *Entamoeba histolytica*. The trophozoite form invades colonic epithelium, causing diarrhea and dysentery, and it can cause colonic perforation (Fig. 23.7).

It forms a cyst, which enables it to survive in the environment and to spread to other hosts. The diagnosis is made by visualization of the organism in stool (Fig. 23.8). In a fresh wet preparation, it can be seen to have a rotatory motion.

Treatment consists of tetracycline, metronidazole, iodoquinol, or nitazoxanide.

Apicomplexa (sporozoa)

These organisms are characterized by having an "apical complex" at their anterior end. There are several different intestinal and tissue apicomplexa, whose life cycles share a common pattern. Their parallels are shown in Table 23.1.

Cryptosporidium

There are many species of this organism, the most important in humans being *C. hominis* and *C. parvum*. Several others can infect humans, namely *C. meleagridis*, *C. felis*, and *C. canis*, and perhaps others. The disease is both anthroponotic and zoonotic.

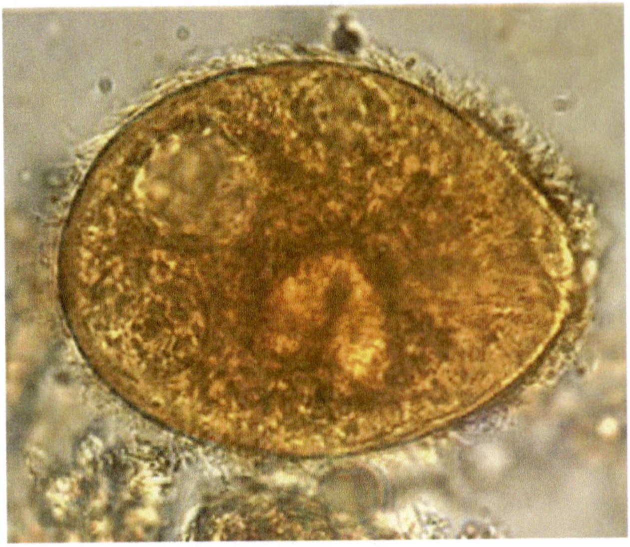

Fig. 23.8 Trophozoite of *Balantidium coli* in stool. Note the cilia on the outside. Courtesy of DPDx, CDC.

Although it has both sexual and asexual cycles, these both occur in the same host. The infectious stage is the oocyst, which measures 4–6 μ in diameter. This sporulates within the host, so that it is immediately infectious (Fig. 23.9). This is in contrast to *Cyclospora cayetanensis*, whose oocyst requires several days in the external environment for sporulation to occur (see later in this chapter). Thick-walled oocysts spread to other hosts, while thin-walled oocysts cause autoinfection of the host. The infective dose is very small, ranging from 9 to a few thousand oocysts. The sporozoites infect intestinal epithelial cells, lying in a parasitophorous vacuole between the cell membrane and the cytoplasm. Each sporozoite forms several merozoites, which can then infect other cells. Some merozoites form macro- and microgamonts, which form a zygote and initiate the sexual cycle. The disease is spread in food or water, and by close personal contact. This organism caused the largest water-borne disease outbreak in USA history, affecting approximately 400,000 individuals in Milwaukee, Wisconsin, in 1993.

The main clinical manifestation of infection is diarrhea, which is acute in onset, after an incubation period of approximately 1 week. In normal hosts, it is self-limiting. In individuals with cell-mediated immune deficiencies, such as those with AIDS, it can cause chronic diarrhea, which can be severe enough to resemble cholera. In such hosts, it can also affect the respiratory tract and the biliary tree, causing sclerosing cholangitis. The diagnosis is based on detection of the oocysts in the stool (Fig. 23.10). This can be done by modified acid-fast staining, which has a low sensitivity and specificity, immunofluorescence staining (Fig. 23.11), and, optimally, by antigen detection. PCR methods are now commercially available.

Therapy is mainly supportive. The only antimicrobial drug that has been effective in reducing the duration of symptoms is nitazoxanide. This has not been effective in severe cases in AIDS patients. Antiretroviral proteases seem to have some anticryptosporidial activity, but their main value is in correcting the immune deficiency.

Table 23.1 Parallels between the life cycles of different apicomplexan parasites.

Stage	Plasmodium		Babesia		Toxoplasma		Cryptosporidium		Sarcocystis	
Asexual stages										
Infective stage	Sporozoite	Mosquito ↓ Human	Sporozoite	Tick ↓ Vertebrate	Oocyst	Cat ↓ Human/ other animal	Oocyst	Human	Sporocyst	Human/ carnivore
	Merozoite	Human	Trophozoite	Vertebrate	Tachyzoite	Human/ other animal	Sporozoite	Human	Sporozoite	Cow/pig
	Trophozoite	Human	Merozoite	Vertebrate			Trophozoite	Human	Merozoite	Cow/pig
	Schizont	Human			Bradyzoite	Human/ other animal	Merozoite	Human	Bradyzoite	Cow/pig
	Gametocytes	Human ↓ Mosquito	Isogamete	Tick	Ingestion by cat/ human/ other animal		Gamont	Human		
Sexual stages		Mosquito		Tick		Cat		Human	Gametocytes	Human/ carnivore
Fertilization	Zygote	Mosquito	Zygote	Tick		Cat	Zygote	Human		Human/ carnivore
	Oocyst	Mosquito	Ookinete	Tick	Oocyst	Cat	Oocyst	Human	Oocyst	Human/ carnivore
	Sporozoite	Mosquito	Sporozoite	Tick	Oocyst	Cat	Oocyst	Soil	Sporocyst	Human/ carnivore

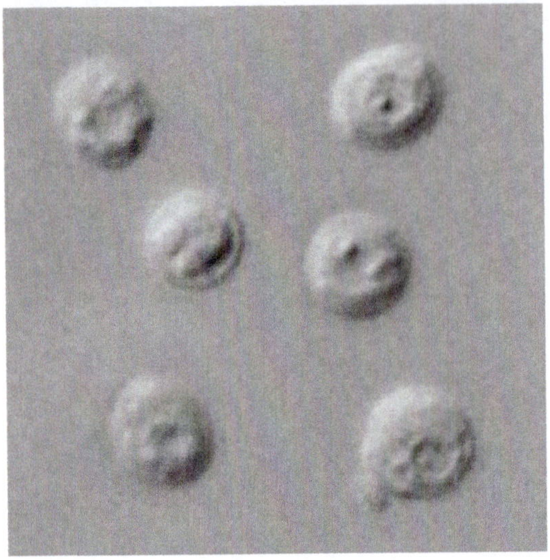

Fig. 23.9 Wet preparation of stool showing cryptosporidium oocysts containing sporozoites. Courtesy of DPDx, CDC.

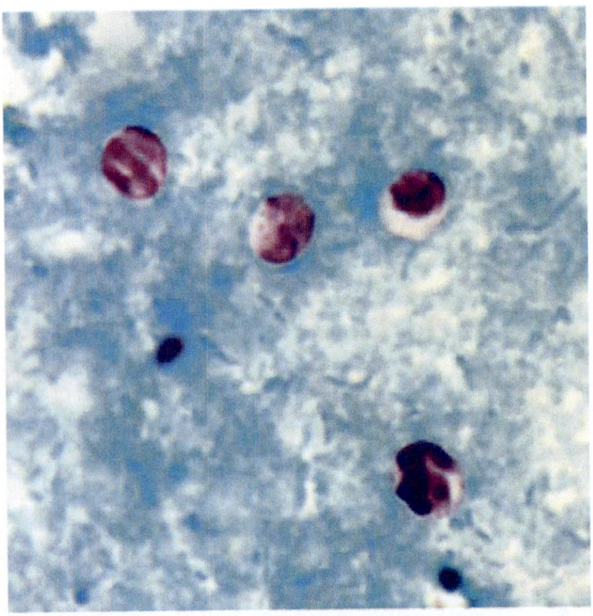

Fig. 23.10 Cryptosporidium oocysts stained with a modified acid-fast stain. Courtesy of DPDx, CDC.

Cyclospora cayetanensis

This is a sporozoan parasite named for for the University of Cayetano Heredia, in Lima, Peru. It has a worldwide distribution and is spread by contamination of food or water by human feces. Whether it can be transmitted by animals is unclear. Several large outbreaks have occurred in the USA. Raspberries, to which the oocysts can adhere particularly well, have been incriminated in several of these. Cyclospora

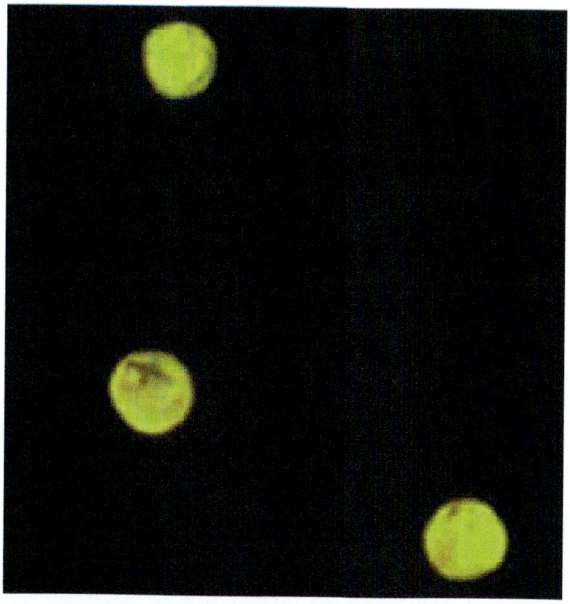

Fig. 23.11 Cryptosporidium oocysts stained with auramine-rhodamine fluorescent stain. Courtesy of DPDx, CDC.

oocysts differ from those of cryptosporidium in two important ways: they are larger than those of cryptosporidium (8–10 μ, compared with 4–6 μ) and they do not sporulate within the host, requiring a 1–2-week period in the environment to sporulate. Therefore, they are not immediately infectious after being passed in the stool, and spread by close personal contact does not occur. The life cycle is very similar to that of cryptosporidium. The oocyst is ingested and the sporozoites infect epithelial cells of the proximal small bowel. The sporozoites become type I meronts containing merozoites. These infect other cells, constituting the asexual cycle. Type II meronts develop into type II merozoites, which become gametocytes, which begin the sexual cycle. The microgamete fertilizes the macrogamete, forming a zygote, which develops into the oocyst, which is passed in the stool.

The clinical manifestations of cyclosporiasis are those of an acute infectious diarrhea, with anorexia, vomiting, abdominal cramps, and diarrhea.

The diagnosis is based on demonstrating the presence of the organism in stool. This can be done by phase-contrast microscopy, autofluorescence (Fig. 23.12), acid-fast staining, safranin staining (Fig. 23.13), and by molecular methods.

Treatment consists of trimethoprim/sulfamethoxazole. Ciprofloxacin has been successfully used in some patients.

Cystoisospora (Isospora) belli

This sporozoan parasite causes diarrhea, mainly in immunocompromised individuals such as those with HIV infection. The organism's life cycle is similar to that of cryptosporidium. The cyst, containing sporocysts, is ingested and the sporozoites infect intestinal cells. Inside the cell, schizogony occurs, with the formation of merozoites, which infect other cells. A sexual cycle is initiated by the development of micro- and macrogametocytes. Fertilization results in the formation of an oocyst, within which is a sporoblast. After excretion of the oocyst in the feces, the sporoblast divides into two

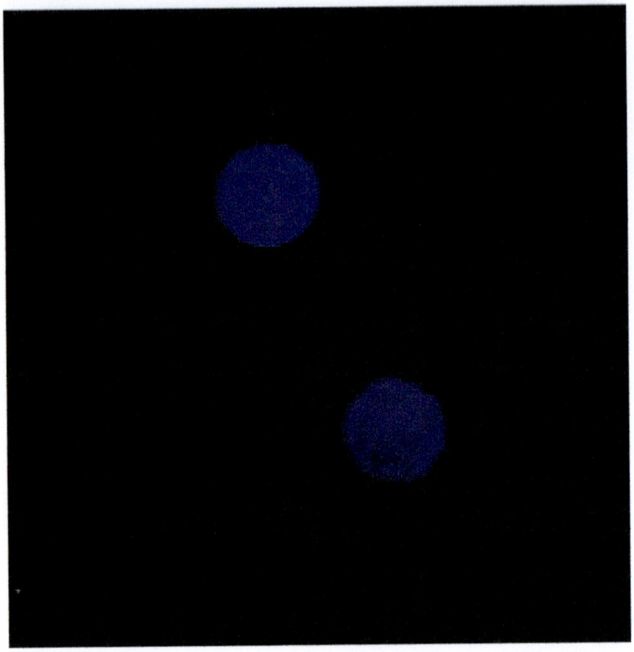

Fig. 23.12 Autofluorescence of *Cyclospora cayetanensis* under ultraviolet light. Courtesy of DPDx, CDC.

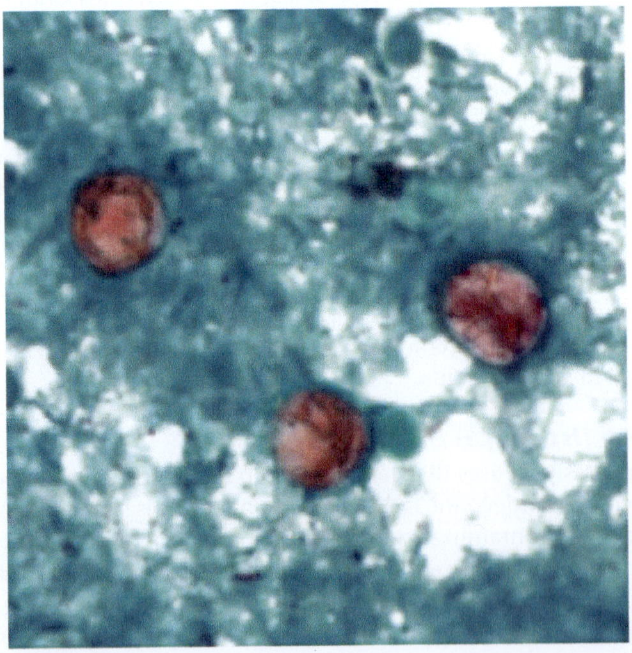

Fig. 23.13 Oocysts of *Cyclospora cayetanensis* stained with safranin. Courtesy of DPDx, CDC.

and each of the sporoblasts forms a sporocyst. Four sporozoites develop within each sporocyst. This environmental stage takes about 1 week.

Diagnosis of this infection is accomplished by examination of stool for oval oocysts. These autofluoresce under UV light and can be stained with a modified acid-fast stain (Figs 23.14 & 23.15). The treatment is trimethoprim/sulfamethoxazole.

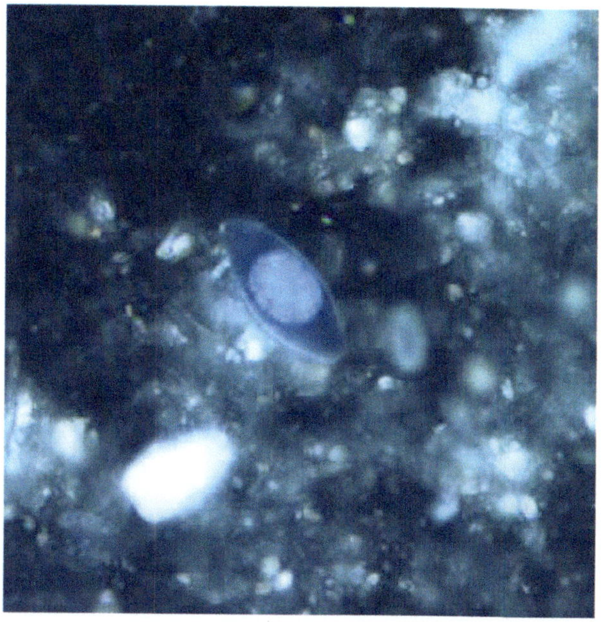

Fig. 23.14 *Cystoisospora belli* oocyst fluorescing under UV light. Courtesy of DPDx, CDC.

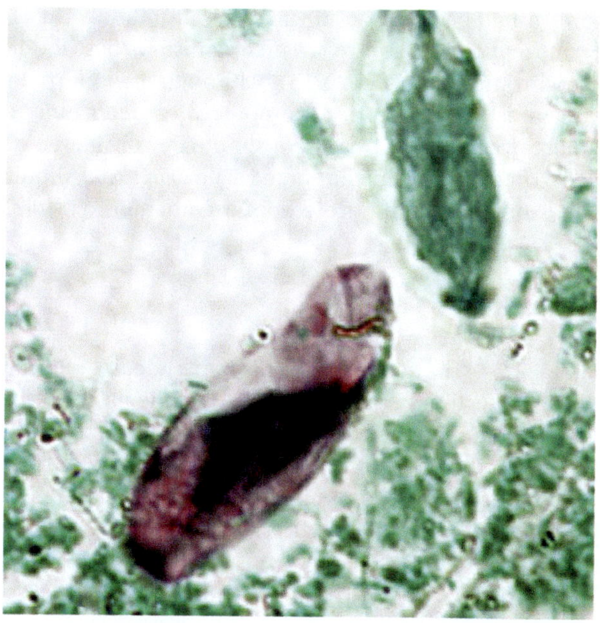

Fig. 23.15 *Cystoisospora belli* oocyst stained with an acid-fast stain. Courtesy of DPDx, CDC.

Sarcocystis

These are rare parasites of humans. There are about 120 species, two of which infect humans, namely *S. hominis*, which is acquired from cattle, and *S. suihominis*, acquired from pigs. The life cycle is as follows.

The sexual cycle occurs in the gut of a carnivore:

bradyzoites (within muscle metrocysts) → macro- and microgametocytes → fertilization → oocysts → sporocysts → feces → environment

The asexual cycle occurs in a herbivore:

sporocyst in carnivore feces → intestine → blood → endothelial cell→ schizogony → merozoites → blood → myocytes → cysts (called metrocysts, containing bradyzoites) → ingestion by carnivore

Humans can act as the definitive host (carnivore), the stage affecting the intestine, in which case intestinal symptoms may occur, or as the intermediate host (herbivore), in which case there may be systemic symptoms, evidence of myositis, subcutaneous nodules, and lymphadenopathy.

The diagnosis can be made by visualization of the sporocysts in centrifuged stool or, in the case of systemic disease, by muscle biopsy. The cysts must be differentiated from those of toxoplasmosis. Serology can also be performed. The treatment is albendazole.

Blastocystis hominis

This is an organism that is often considered among the protozoal parasites, but it does not belong to the protozoa. It belongs to the class Blastocystea within the kingdom Chromista. It varies considerably in size and form. Whether it causes intestinal symptoms, in particular diarrhea, is controversial. In some studies of patients with diarrhea, in whom the organism had been found in the stool, more individuals receiving antimicrobial therapy improved than controls. The organism is best found by staining of fixed stool with a trichrome stain (Fig. 23.16). Although it is referred to as *Blastocystis hominis*, routine diagnostic laboratories cannot really determine the species, so *Blastocystis* spp. is more appropriate. Treatment should be considered only when other possible causes of diarrhea have been eliminated. The drugs that should be considered are nitazoxanide, metronidazole, and trimethoprim/sulfamethoxazole.

Microsporidium

These are small (1–3 × 1.5–4 μ), ovoid organisms that are currently classified as fungi. They are included in this section because they have, in the past, been considered protozoa, and protozoa are often included in the differential diagnosis of the diseases they cause. They are present in the environment and colonize many different animals. The organism has the form of a spore, containing sporoplasm. It has a polar tube through which it injects sporoplasm to infect a new host cell (Fig. 23.17). There are about 150 genera of microsporidium with 1300 species, but members of only 14 species, within eight genera, are known to have infected humans: *Enterocytozoon bieneusi*, *Encephalitozoon intestinalis*, *Vittaforma* spp., *Pleistophora* spp., *Trachipleistophora* spp., *Nosema* spp., *Brachiola* spp., and *Microsporidium* spp. In humans, they cause primarily opportunistic

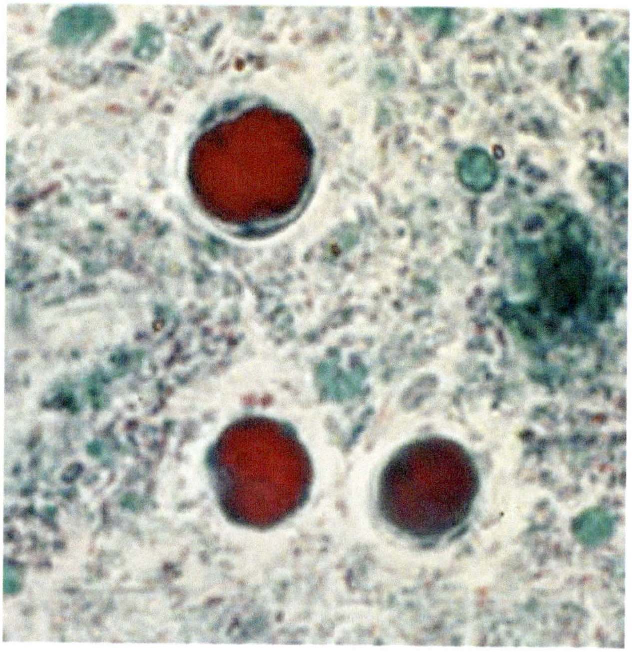

Fig. 23.16 *Blastocystis* spp. in stool, stained with trichrome stain. Courtesy of DPDx, CDC.

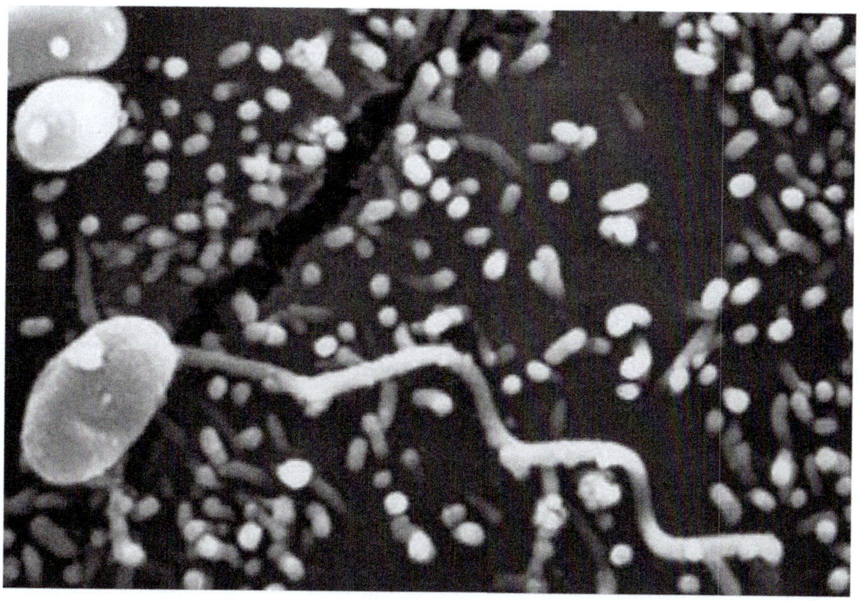

Fig. 23.17 Polar tube of a microsporidium inserted into a host cell. Courtesy of DPDx, CDC.

infections, particularly of the intestine, but they can also affect the biliary and respiratory tracts. They can also cause keratitis.

They are identified by visualization by light microscopy after staining with a modified trichrome stain (chromotrope 2R) (Fig. 23.18), modified acid-fast stain, and

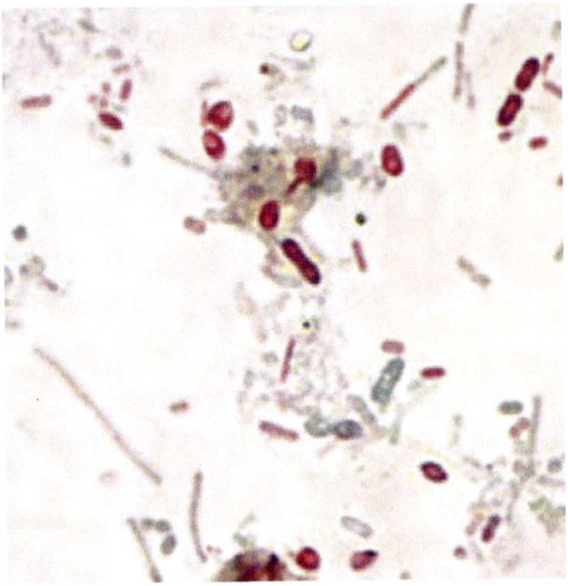

Fig. 23.18 Microsporidium stained with Chromotrope 2R. Courtesy of DPDx, CDC.

Table 23.2 Diagnostic tests used for intestinal protozoa.

Organism	Diagnostic tests
Giardia	Stool: Micro, Ag, Mol
E. histolytica	Stool: Micro, Ag, Mol; serum: antibody; Histo
Balantidium	Stool: Micro; Histo
Cryptosporidium	Stool: AFB stain, Ag, immunofluorescence, Mol
Cyclospora	Stool: AFB stain, autofluorescence, Mol
Cystoisospora	Stool: AFB stain, autofluorescence
Sporocystis	Stool: Micro; serum: antibody; Histo
Blastocystis	Stool: Micro; trichrome stain
Microsporidium	Stool, corneal scraping, BAL: trichrome, AFB, fluorochrome stains, Histo; Mol

AFB, modified acid-fast stain; Ag, antigen; BAL, bronchoalveolar lavage; Histo, histology; Micro, microscopy; Mol, molecular methods, e.g. PCR.

under ultraviolet light after being stained with a fluorochrome stain such as Calcofluor white. Fluorescent antibody stains and molecular methods are also available.

Treatment of patients with intestinal or systemic infection consists of albendazole, while patients with keratitis should be treated topically with fumagillin.

Diagnostic tests

The diagnostic tests used for detecting intestinal protozoal infections are summarized in Table 23.2.

Further reading

Anuradha D (2013) Current laboratory diagnosis of opportunistic enteric parasites in human immunodeficiency virus-infected patients. Trop Parasitol 3: 7–16.

Chacin-Bonilla L (2010) Epidemiology of *Cyclospora cayetanensis*: a review focusing in endemic areas. Acta Tropica 115: 181–93.

Chalmers RM, Davies AP (2010) Minireview: clinical cryptosporidiosis. Exp Parasitol 124: 138–46.

Choudhuri G, Rangan M (20121) Amebic infection in humans. Indian J Gastroenterol 31: 153–62.

Collinet-Adler S, Ward HD (2010) Cryptosporidiosis: environmental, therapeutic, and preventive challenges. Eur J Clin Microbiol Infect Dis 29: 927–35.

Esposito DH, Stich A, Epelboin L, et al. (2014) Acute muscular sarcoscystosis: an international investigation among ill travelers returning from Tioman Island, Malaysia, 2011–2012. Clin Infect Dis 59: 1401–10.

Fayer R (2004) *Sarcocystis* spp. in human infections. Clin Microbiol Rev 17: 894–902.

Gray T, Kwan Y, Phan T, et al. (2013) *Dientamoeba fragilis*: a family cluster of disease associated with marked peripheral eosinophilia. Clin Infect Dis 57: 845–8.

Ortega YR, Sanchez R (2010) Update on *Cyclospora cayetanensis*, a food-borne and waterborne parasite. Clin Microbiol Rev 23: 218–34.

Schuster FL, Ramirez-Avila L (2008) Current world status of *Balantidium coli*. Clin Microbiol Rev 21: 626–38.

Sharma S, Das S, Joseph J, et al. (2011) Microsporidial keratitis: need for increased awareness. Surv Ophthalmol 56: 1–22.

Stanley SL (2003) Amoebiasis. Lancet 361: 1025–34.

Tan KSW (2008) New insights on classification, identification, and clinical relevance of *Blastocystis* spp. Clin Microbiol Rev 21: 639–65.

CHAPTER 24
Tissue and blood protozoa

The tissue and blood protozoa include the apicomplexa (formerly called sporozoa) (*Plasmodium* spp., *Babesia* spp., and *Toxoplasma gondii*), the flagellates (*Leishmania* spp., *Trypanosoma* spp., and *Trichomonas vaginalis*), and the free-living amebae (*Naegleria fowleri, Acanthamoeba* spp., *Balamuthia madrillaris,* and *Sappinia* spp.). *Sarcocystis* spp., which can invade tissue in humans, are considered with the gut protozoa.

Apicomplexa

Plasmodium

This genus of parasite causes malaria, a name that is derived from the Latin for "bad air." It is one of the most important diseases worldwide, and by far the most important parasitic disease affecting humans. There are five species that infect humans: *P. falciparum* (the most important species), *P. vivax, P. ovale, P. malariae,* and *P. knowlesi. P. knowlesi,* a well-known monkey parasite, has only recently been recognized as a human parasite. Several other species affect animals. Plasmodia are endemic in many tropical areas and some temperate areas (Fig. 24.1). They are transmitted by the bite of a female mosquito of the genus Anopheles.

The life cycle follows the pattern of apicomplexan life cycles, described in the section on gut protozoa (see Chapter 23, Table 23.1). The sexual stage occurs in the gut of the mosquito, while the asexual stage occurs in the liver and blood of the vertebrate host. When a mosquito bites its host, it injects sporozoites into the blood. These travel to the liver and infect the hepatocyte, where they undergo schizogony (a type of merogony), resulting in many merozoites. This is called the extraerythrocytic stage of the infection. These merozoites leave the liver and infect erythrocytes, initiating the erythrocytic stage. Intraerythrocytic schizogony results in 8–24 merozoites, depending on the species. These then infect more erythrocytes, amplifying the parasite load. Some merozoites become micro- and macrogametocytes, the stage of the parasite infectious to the mosquito. They are taken up by the mosquito during feeding and

Practical Medical Microbiology for Clinicians, First Edition. Frank E. Berkowitz and Robert C. Jerris.

Fig. 24.1A Distribution of malaria in the Western Hemisphere. Courtesy of Centers for Disease Control and Prevention. CDC Health Information for International Travel 2014. New York: Oxford University Press, 2014. For a more detailed, interactive map, see www.cdc.gov, and search the alphabetic index for malaria.

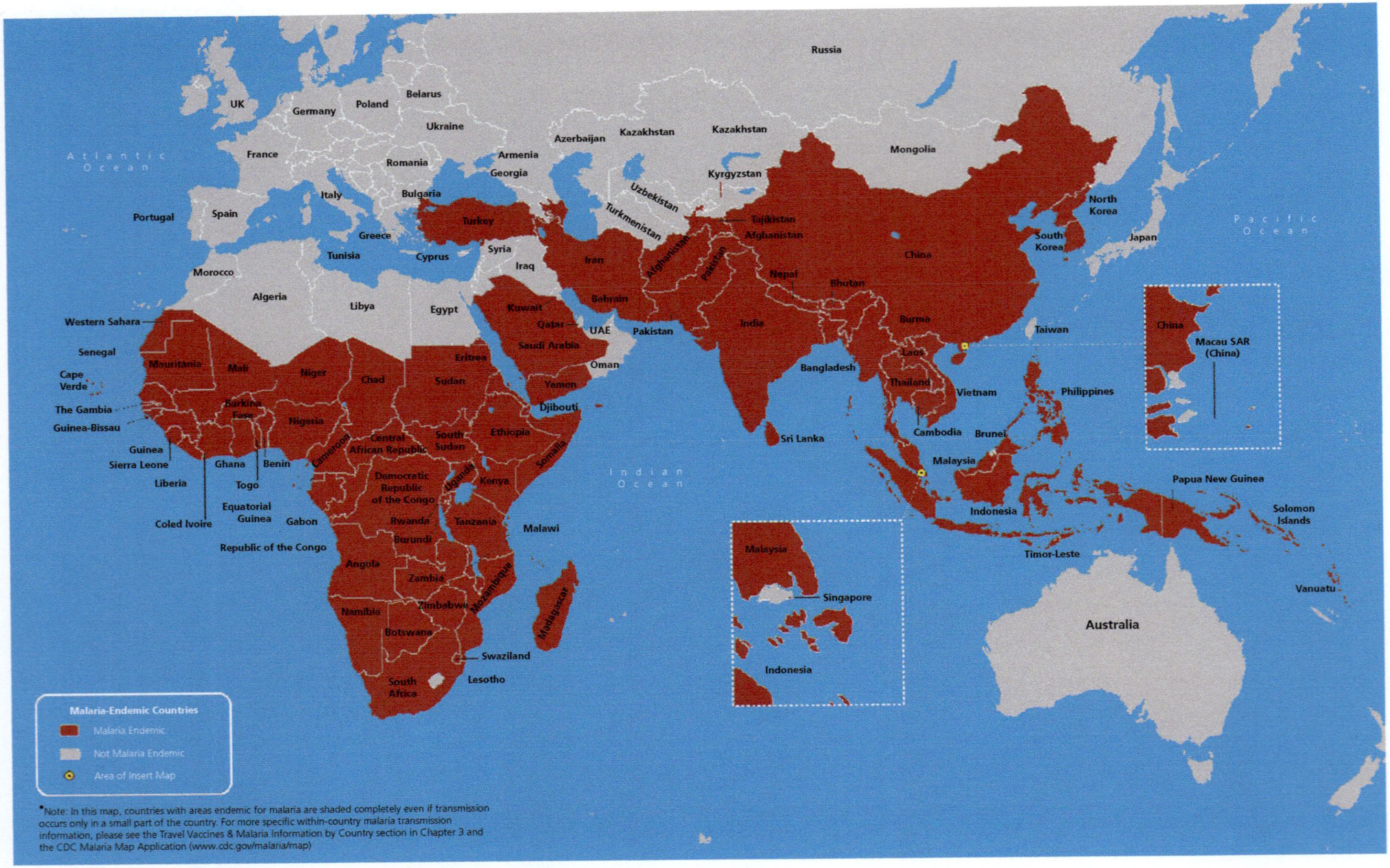

Fig. 24.1B Distribution of malaria in the Eastern Hemisphere. Courtesy of Centers for Disease Control and Prevention. CDC Health Information for International Travel 2014. New York: Oxford University Press, 2014. For a more detailed, interactive map, see www.cdc.gov, and search the alphabetic index for malaria.

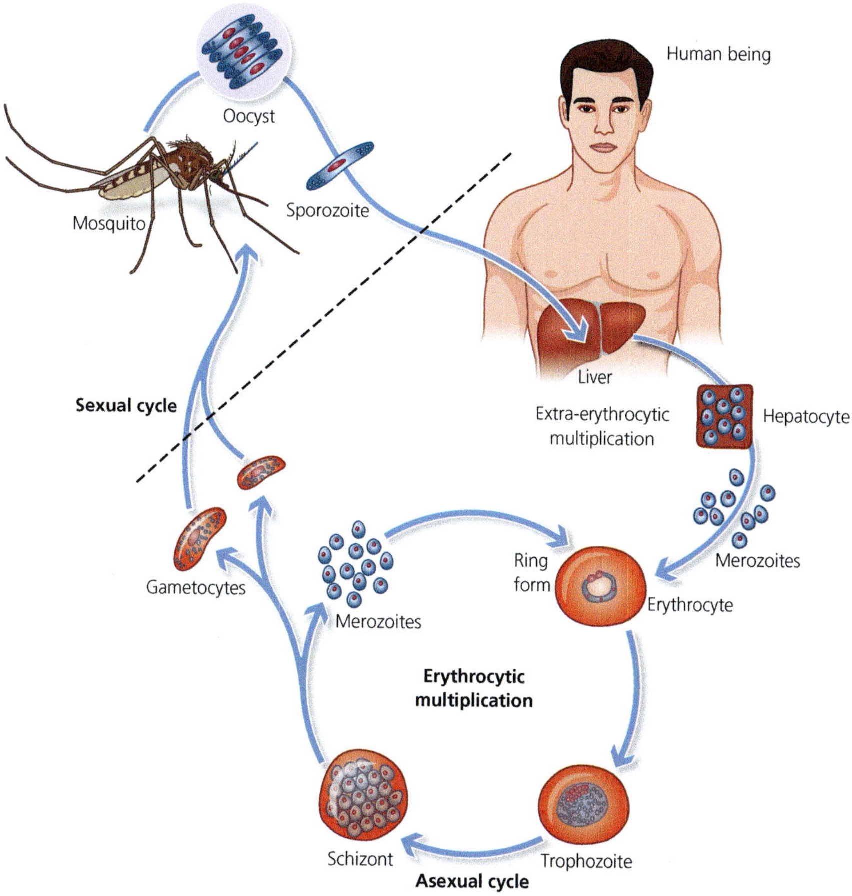

Fig. 24.2 Life cycle of Plasmodium. Copyright © Frank E. Berkowitz. Reprinted with the permission of Cambridge University Press.

form a zygote, which develops into an oocyst in the mosquito's gut lining. The sporozoites, formed within the oocyst, complete the cycle (Fig. 24.2).

In the cases of *P. vivax* and *P.ovale* infection, some parasites remain dormant in the liver as hypnozoites. These may emerge from the liver subsequently and initiate a new erythrocytic stage, causing a clinical relapse.

Malaria can be spread by contaminated blood, such as blood transfusion or needle sharing by intravenous drug addicts. It can also be transmitted transplacentally, although this is relatively uncommon. In these latter circumstances, since the infection is initiated in the erythrocytic stage and the liver stage does not occur, hypnozoites are not present and relapses do not occur.

The clinical manifestations of malaria are related to three main phenomena: hemolysis; cytokine production by the host; and in the case of P. *falciparum* infection, vascular obstruction. The main and sometimes the only clinical feature is fever. This may be associated with severe shaking chills. As the infection progresses, anemia and jaundice may develop. In the case of falciparum malaria, multiorgan disease may occur. Therefore this accounts for most malaria deaths.

Table 24.1 Differences between the different species of plasmodium.

	P. falciparum	*P. vivax*	*P. ovale*	*P. malariae*	*P. knowlesi*
Distribution	All areas	Asia, C. America, Africa	Africa	All areas	SE Asia
Hypnozoite stage	No	Yes	Yes	No	No
Erythrocyte	All	Young	Young	Old	All
Life cycle (days)	2	2	2	3	1
Stages seen	Ring	All	All	All, band	All, band
Merozoites per schizont	8–24	12–24	6–14	6–12	Up to 16
Gametocyte shape	Sickle	Round	Round	Round	Round
Chloroquine resistance	Widespread	Limited	None	None	None

This is due, at least in part, to the expression of a particular protein on the erythrocyte (*Plasmodium falciparum* erythrocyte membrane protein) which causes adhesion of the erythrocyte to the endothelium, resulting in capillary obstruction. Death is due to severe anemia or organ failure. Of particular importance is brain involvement (cerebral malaria). The main differences between the different plasmodial species are shown in Table 24.1.

Diagnosis

As with any disease, the diagnosis depends on initial clinical suspicion. This depends, in turn, on knowing that the patient has been exposed, in most cases as a result of living in or travel to an endemic area.

Because malaria can be fatal, making the diagnosis is a matter of urgency. The diagnosis consists of three parts: is there malaria? what species of plasmodium is the cause? what is the degree of parasitemia? The standard method for making the diagnosis of malaria is visualization of the erythrocytic stage of the parasite in a blood smear. Two kinds of smears are made.

- Thin smear: in this smear, the tail and edges of the smear are only one cell thick. The smear is fixed with a fixative and stained with a Romanowsky stain such as Giemsa. The slide is examined using a high-power (oil immersion) lens (×1000). The parasites are seen within the erythrocyte, whose morphology is well preserved. The appearance of the parasite, as well as the morphology of the erythrocyte, is used to determine the species of parasite. A thin smear also enables one to determine the percentage of erythrocytes parasitized (degree of parasitemia), which is a factor in determining the severity of the infection. This is determined by counting the number of parasitized erythrocytes and the total number of erythrocytes in the same field. Another method involves counting the number of parasitized erythrocytes and the number of leukocytes in the same field, and using the ratio of total leukocyte count and total erythrocyte count (measured independently in mm^3) to determine the number of parasitized erythrocytes per mm^3. The disadvantage of the thin smear is that only a very small amount of blood is examined within each field.

- Thick smear: the smear is made thicker (so that newsprint can just be read through it); it is stained without being fixed, so that the erythrocytes are not preserved and form a mush. The advantage of this is that each field contains more blood and the sensitivity of the examination is thus increased (Figs 24.3, 24.4 & 24.5). The disadvantage of a thick smear is that the erythrocyte morphology cannot be seen.

The main disadvantage of examining a blood smear is that it requires significant expertise, as well as a microscope and stains. This might not be available in areas where malaria is prevalent (e.g. a platelet, lying on an erythrocyte, might be called a parasite) (Fig. 24.6). Therefore, alternative rapid tests have been developed.

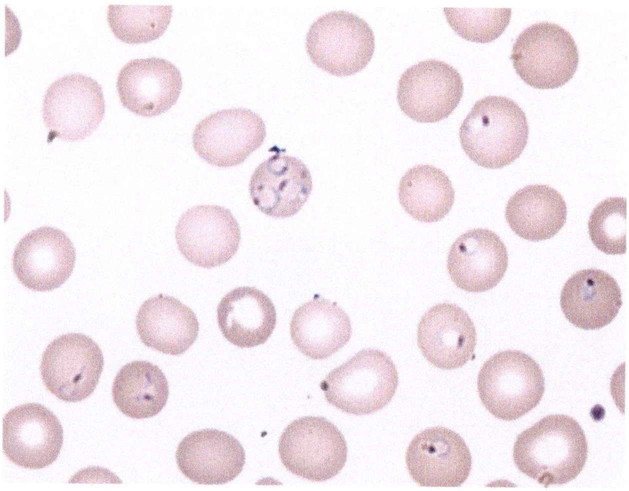

Fig. 24.3 Thin blood smear with many ring forms of *Plasmodium falciparum*. This is a very high-grade parasitemia.

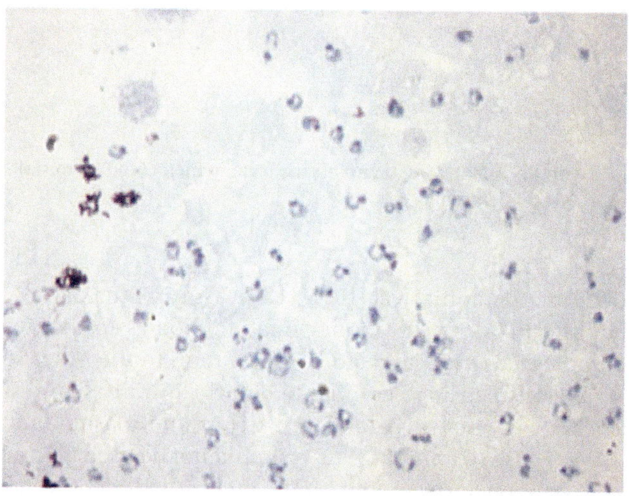

Fig. 24.4 Thick blood smear with multiple ring forms of *Plasmodium falciparum*. Note the difference in overall appearance from the thin smear.

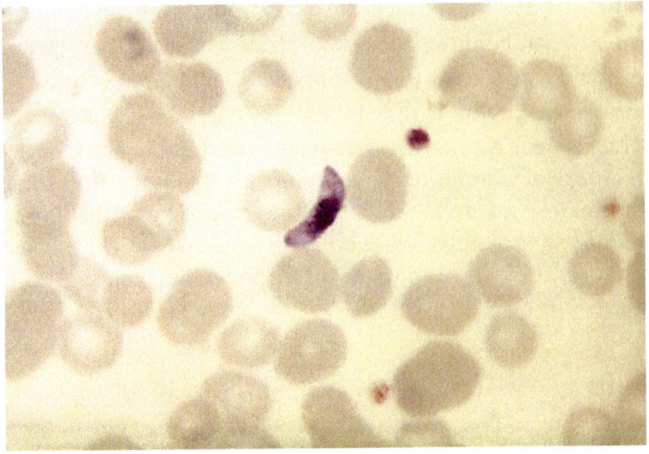

Fig. 24.5 Thin blood smear showing a gametocyte of *Plasmodium falciparum*. Courtesy of PHIL, CDC.

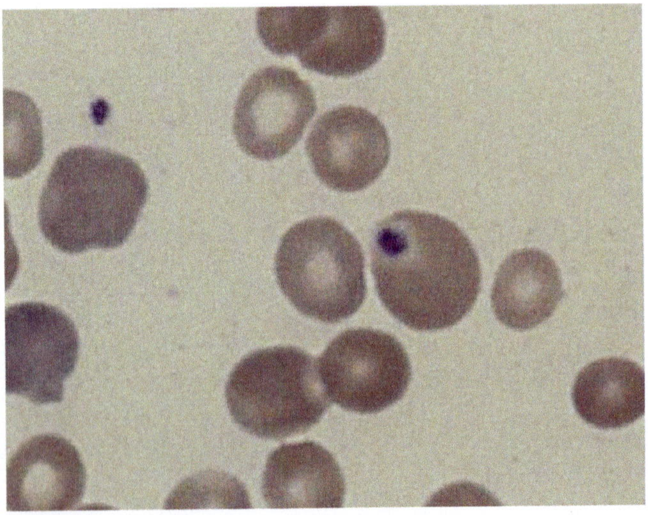

Fig. 24.6 Blood smear with a platelet lying on an erythrocyte, which could be mistaken for a malarial parasite by an inexperienced microscopist.

An immunochromatographic test (ICT) card assay (BinaxNOW Malaria) has been approved by the FDA. It uses monoclonal antibodies to detect histidine-rich protein 2 antigen of *P. falciparum*, and aldolase common to the other species of plasmodium. This is generally sensitive enough to detect clinically significant infections. PCR tests, which are very sensitive and can determine plasmodium species, are available at the CDC. These ICT and PCR tests should not be used to monitor response to therapy, because antigen and genome can be present in blood even after parasites have been killed.

Serologic tests can be used to determine whether an individual has had malaria.

Treatment

Supportive: ensuring adequate hemoglobinization, hydration, and blood glucose.

Antiparasitic therapy: antimalarial drugs are discussed in Chapter 22. In the case of falciparum malaria, the main therapeutic goal is to reduce the burden of erythrocytic parasites as quickly as possible. For many years, chloroquine was the mainstay of therapy for all types of malaria. However, the emergence of chloroquine-resistant *P. falciparum*, and its current prevalence in almost all endemic areas, have rendered this drug largely ineffective for this infection. Chloroquine resistance has also emerged in *P. vivax* in South East Asia. Currently, the most effective drugs for patients with *P. falciparum* infection are the artimisinin derivatives (artemether and artesunate). These should never be used alone, but should be used together with a longer-acting drug, e.g. lumefantrine. Other drugs that can be used are quinine, atovaquone/proguanil, and mefloquine. In the USA, quinine for intravenous use (which would be used for severe cases) is not available. Its alternative is quinadine. In cases of infection by species that have hypnozoites, radical cure is necessary to prevent relapses. The only drug available to accomplish this is primaquine. Its disadvantage is that it causes severe hemolysis in individuals with glucose-6-phosphate dehydrogenase deficiency, and it is unsafe in pregnancy. Therefore these conditions should be excluded before it is administered.

- A case of malaria in an area that is non-endemic but where the vector is present poses a risk for spread of the infection. Therefore, this is a public health issue. In the USA, malaria is a reportable disease. If the infection has occurred in a traveler, co-travelers should be evaluated for possible malaria.
- Prevention: in malaria-endemic areas antimalarial prophylaxis should be used for individuals at high risk for severe disease (infants and pregnant women) as well as travelers to such areas. Insecticide-impregnated bednets are also very important.

Babesia

This is a parasite of animals, and is a zoonosis in humans, who are accidental hosts. There are about 100 different species of babesia, the most important in humans being the *B. microti* group, *B. divergens, B. bovis, B. canis, B. duncani, B. venatorum*, and *B. divergens*-like (B. K01). The organism occurs worldwide and is transmitted by *Ixodes* ticks of all stages. In the USA, most cases occur in the north east, midwest, and north west, and most are due to *B. microti*, for which the main reservoir host is the white-footed mouse (*Peromyscus leucopus*). In the tick, gametocytes ingested with the blood meal fuse to form an ookinete (equivalent of an oocyst), which develops sporoblasts. When the next stage of the tick takes a blood meal, the sporoblast develops into many sporozoites, which are injected into the host. On entering the host, the sporozoites attack erythrocytes and develop into trophozoites which multiply by budding. The resulting merozoites enter other erythrocytes. Some of these merozoites become gametocytes, which are taken up by a tick, completing the cycle (Fig. 24.7). The infection can also be transmitted by blood transfusion and transplacentally.

The effects of the infection are hemolysis and an inflammatory response. The clinical manifestations in humans range from asymptomatic infection, prolonged febrile illness

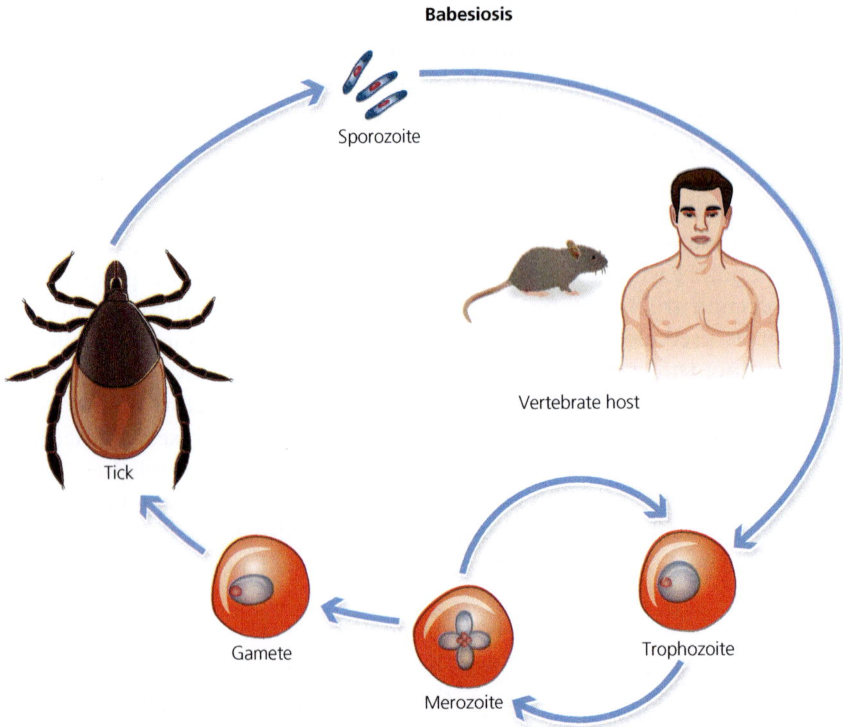

Fig. 24.7 Life cycle of *Babesia* spp.

with non-specific features of myalgia, chills, and cough to severe, life-threatening disease with acute respiratory failure, coagulopathy, and multiorgan disease. This occurs primarily in immunocompromised individuals (asplenia, immunoglobulin deficiency, and cell-mediated immunodeficiency) and the elderly.

Laboratory diagnosis

The main method of parasite detection is the Giemsa-stained blood smear (as for malaria). Different forms can be seen, namely rings, bands, tetrads (Maltese cross), and pyriforms (Figs 24.8, 24.9 & 24.10). The degree of parasitemia, which is of value in management, can be determined as described above for malaria. Although babesia parasites resemble malaria parasites, they differ in that they are pleomorphic, and they do not produce pigment.

Other methods include the following: animal inoculation, which is not readily available and takes several weeks; PCR, using 18s ribosomal DNA; *in vitro* culture, which takes 2–4 weeks; and serology for *B. microti*, using a direct immunofluorescent test. This may take weeks or months to become positive and is therefore not sensitive.

Management

Asymptomatic individuals do not require treatment unless the cause is *B. divergens*. Treatment of symptomatic individuals consists of the combinations of quinine and clindamycin, or of atovaquone plus azithromycin.

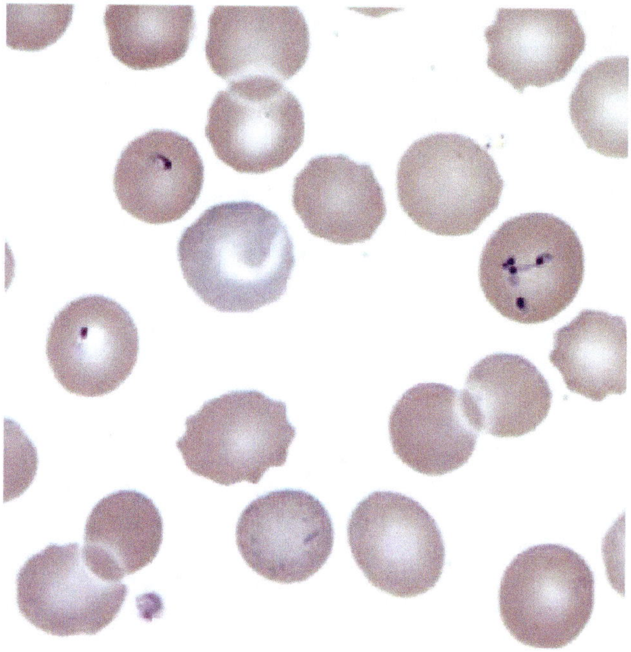

Fig. 24.8 Blood smear showing *Babesia* spp. Courtesy of DPDx, CDC.

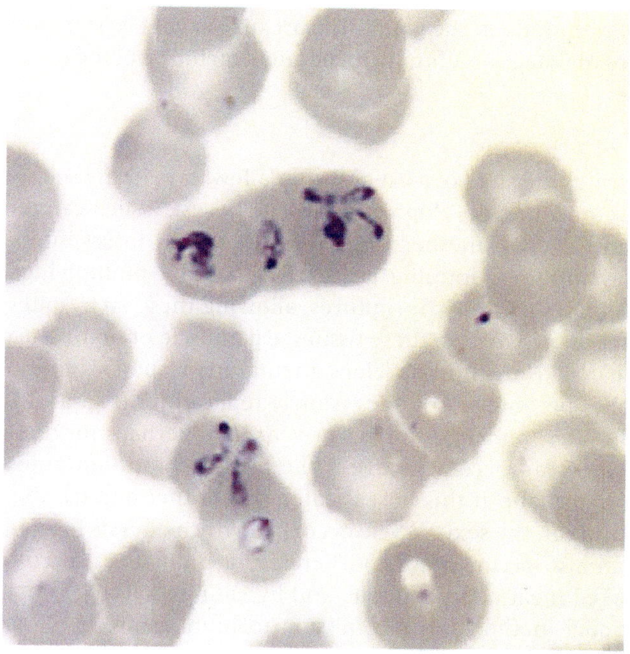

Fig. 24.9 Blood smear showing tetrad forms of *Babesia* spp. Courtesy of DPDx, CDC.

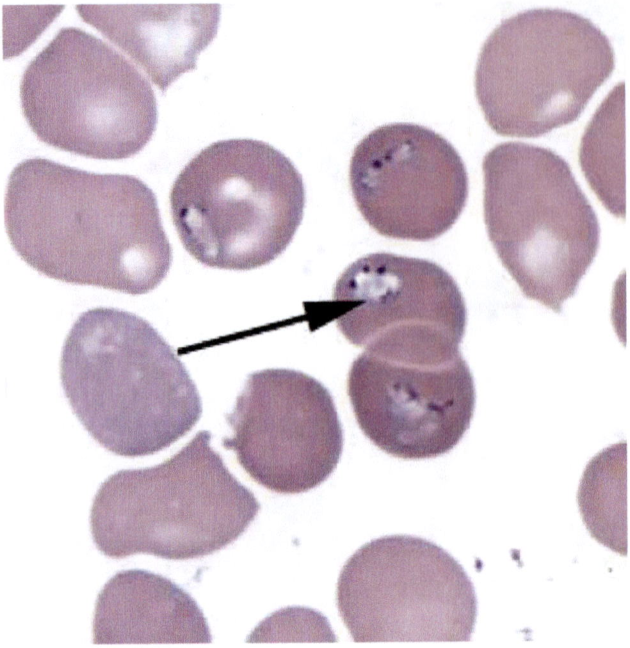

Fig. 24.10 Blood smear showing vacuolation in *Babesia* spp. parasites. Courtesy of DPDx, CDC.

Toxoplasma gondii

This apicomplexan parasite is named for its bow shape (*toxo* is the Greek word for bow) and the North African rodent, the gondi. The asexual stages occur in herbivores or carnivores, but the sexual stage occurs only in the gut of members of the cat family (Felidae).

Life cycle (Fig. 24.11)

Oocysts are passed in the stool of a cat and after a few days in the soil, two sporocysts, each with four sporozoites, develop within it. The oocyst, which can remain infective for months, is ingested by another host. The sporozoites infect intestinal epithelial cells, where they multiply, forming tachyzoites (rapidly multiplying) (Fig. 24.12). When the cell is full of parasites, it ruptures, allowing these tachyzoites to spread via the blood to other tissues. Within these tissues, e.g. muscle or brain, the parasites form cysts. Within the cyst, as the host develops an immune response, multiplication slows and the tachyzoites become bradyzoites (slowly replicating) (Fig. 24.13). These cysts, containing bradyzoites, remain dormant for years and perhaps for the lifespan of the host. However, they can become active if the host becomes immunosuppressed, as with AIDS or a transplant. If this host is eaten by another animal, the asexual cycle recurs. If eaten by a cat, the sexual cycle occurs, in which oocysts form in the gut. (The asexual cycle can also occur in a cat.) Therefore a human can become infected by eating undercooked meat of any animal, and by eating anything contaminated with cat feces. Additional methods of transmission to humans are by blood transfusion, organ transplantation, and transplacentally.

The hosts that are particularly vulnerable to disease are individuals with depressed cell-mediated immunity (see earlier), and the fetus.

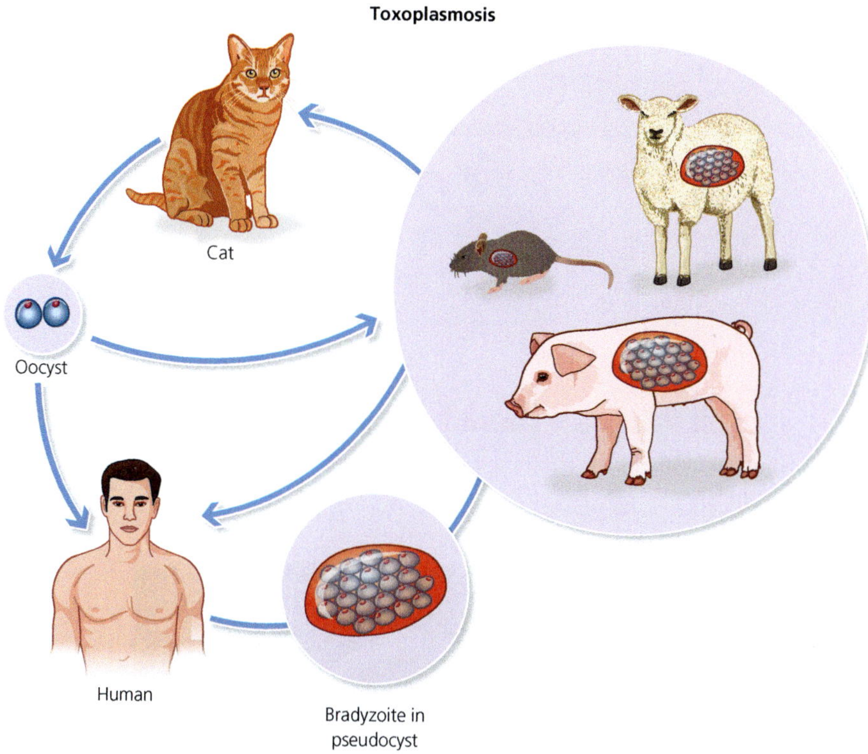

Fig. 24.11 Life cycle of *Toxoplasma gondii*.

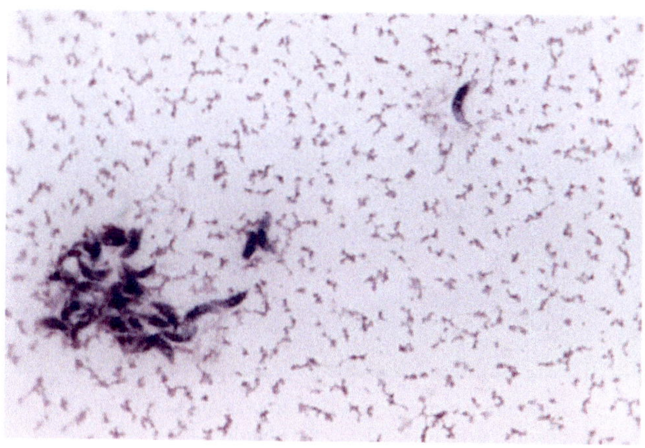

Fig. 24.12 Tachyzoites of *Toxoplasma gondii*. Courtesy of PHIL, CDC.

Clinical features

In normal hosts, the infection may be asymptomatic or characterized by an infectious mononucleosis-like syndrome (see Cytomegalovirus and Epstein–Barr virus, Chapter 5), with fever, sore throat, and generalized lymphadenopathy. It is usually self-limited.

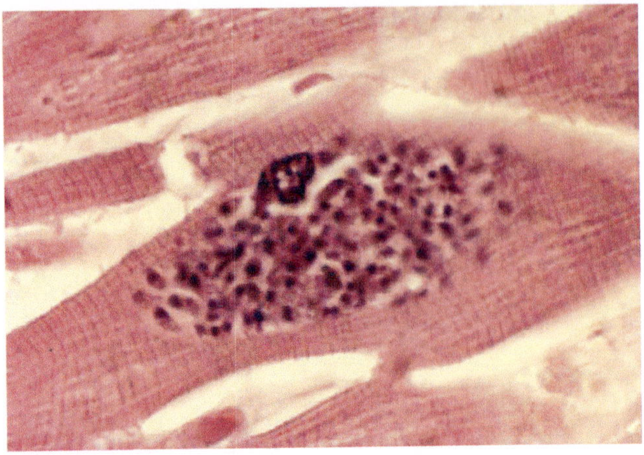

Fig. 24.13 Bradyzoites of *Toxoplasma gondii* inside a pseudocyst in cardiac muscle. Courtesy of PHIL, CDC.

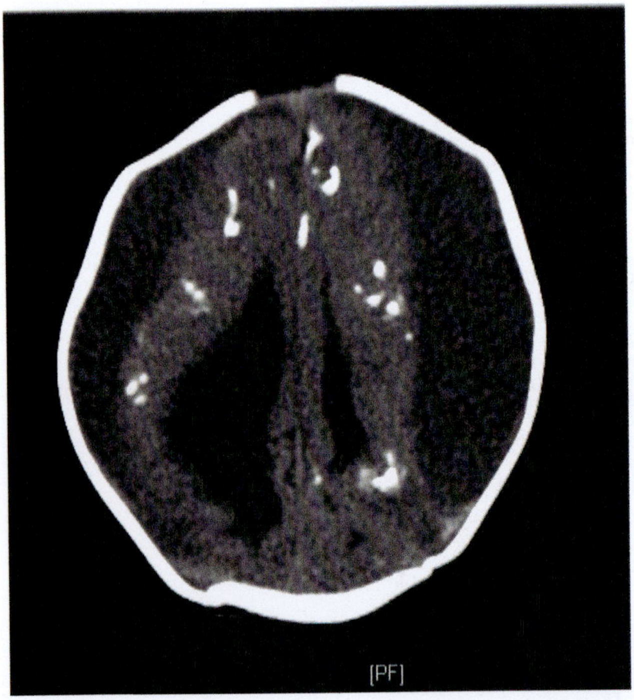

Fig. 24.14 Brain CT scan of an infant with congenital toxoplasmosis, showing hydrocephalus, and diffuse parenchymal calcification. Courtesy of Dr Craig Shapiro.

In immunosuppressed individuals, it can cause focal brain lesions and retinitis, but also pneumonia, myocarditis, and lesions in other organs. In the fetus, a disseminated infection may occur, with evidence of skin, brain, liver, and retinal disease. The brain disease can be very destructive, and the retinitis is particularly damaging because it characteristically affects the posterior pole of the eye (Figs 24.14 & 24.15).

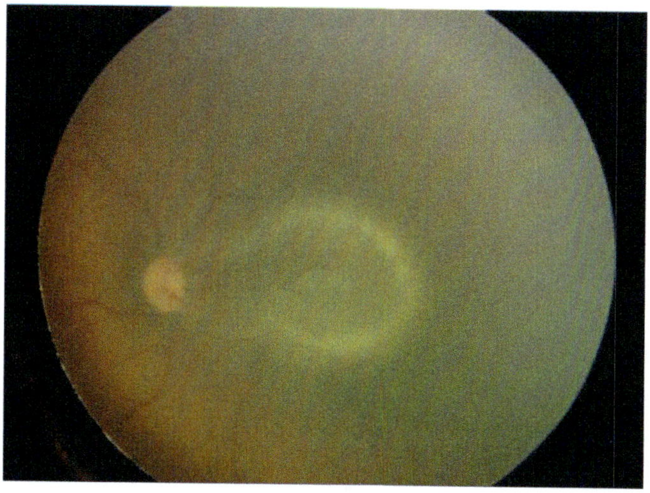

Fig. 24.15 Child whose CT scan is shown in Fig. 24.14, showing retinitis affecting the posterior pole of the eye. Courtesy of Dr Craig Shapiro.

Diagnosis

The diagnosis of toxoplasmosis differs from that of other apicomplexan parasites in that the parasite can seldom be seen in tissue fluids. The diagnosis is made serologically in most cases, by detection of the parasite histologically (which requires tissue) or by PCR of body fluids, such as blood, cerebrospinal fluid, amniotic fluid, ocular fluid, or bronchoalveolar fluid. There are several serologic tests available. It is important to use tests that can differentiate between past infection and current infection.

Treatment

The mainstay of treatment entails use of the folate antagonist pyrimethamine and a sulfonamide. Folinic acid is also given to prevent folate deficiency in the patient.

Hemoflagellates

Hemoflagellates, which infect blood and tissues other than the intestine, belong to the order Kinetoplastida. Two genera of hemoflagellates, belonging to the family Trypanosomatidae, infect humans, namely Leishmania and Trypanosoma. They are transmitted by arthropod vectors, and they have similar life cycles.

Trypanosomes

Two species infect humans: *T. brucei*, which causes African trypanosomiasis (sleeping sickness), and *T. cruzi*, which causes American trypanosomiasis (Chagas disease).

African trypanosomiasis

This is one of the major health problems in tropical Africa. It is caused by two subspecies of *T. brucei*, namely *T. brucei gambiense*, which occurs in West and Central Africa, and causes most cases of the disease, and *T. brucei rhodesiense*, which occurs in East and

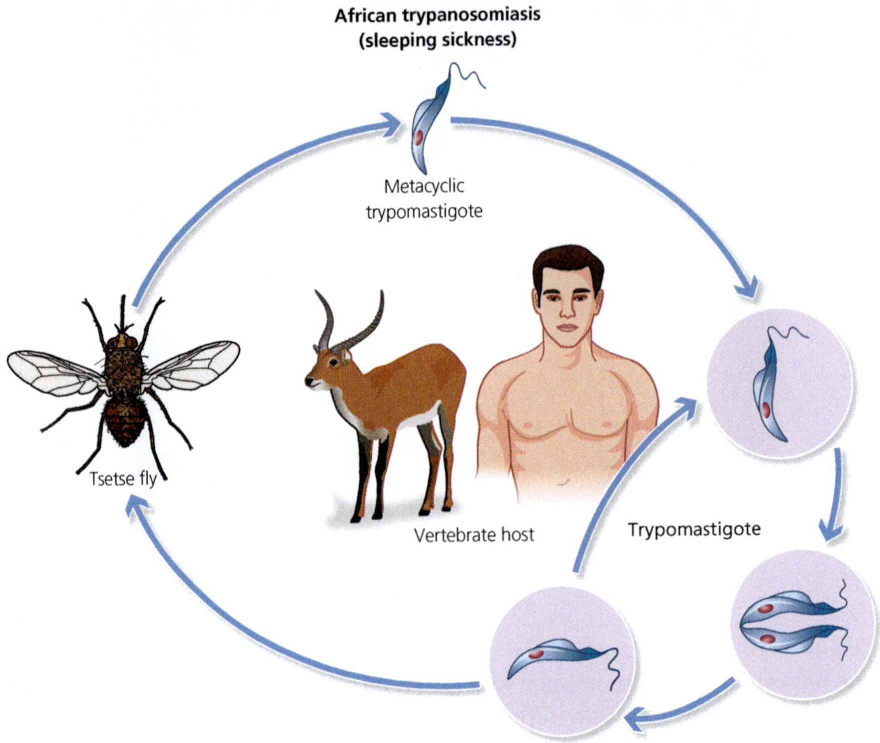

Fig. 24.16 Life cycle of *Trypanosoma brucei.*

South-Central Africa. The former is anthroponotic. The latter is zoonotic, the main reservoirs being cattle and wild animals. Although the latter subspecies accounts for only about 5% of all cases, it is the subspecies more likely to infect tourists, because they visit game parks, particularly in East Africa. The infection is transmitted by the bite of either sex of the tsetse fly (*Glossina* spp.), which is painful. The life cycle is shown in Fig. 24.16.

The fly injects metacyclic trypomastigotes in its saliva into the mammalian host. These spread from the lymphatics to the blood, where they become trypomastigotes, which multiply by binary fission. They do not enter cells but remain in extracellular fluids. After a variable period (weeks to months), they enter the central nervous system. When a fly bites an infected human, it takes up the trypomastigotes. These become procyclic trypomastigotes, which multiply in the midgut. These, in turn, become epimastigotes, which enter the salivary gland, where they become metacyclic trypomastigotes, thus completing the cycle.

The clinical features include fever, rash, and lymphadenopathy. Some cases of *T. b. rhodesiense* infection have an eschar at the site of the fly bite. The fever is relapsing due to antigenic variation of the variant surface glycoprotein (see borrelial relapsing fever, Chapter 16). This is due to the large number of genes that the organism possesses for this glycoprotein. As the host develops antibodies to the surface glycoprotein and the number of organisms decreases, the glycoprotein changes. Many organs can be affected, including the liver, spleen, heart, and endocrine system. Once the brain is involved, there is a progressive deterioration, with encephalopathy, disturbance of

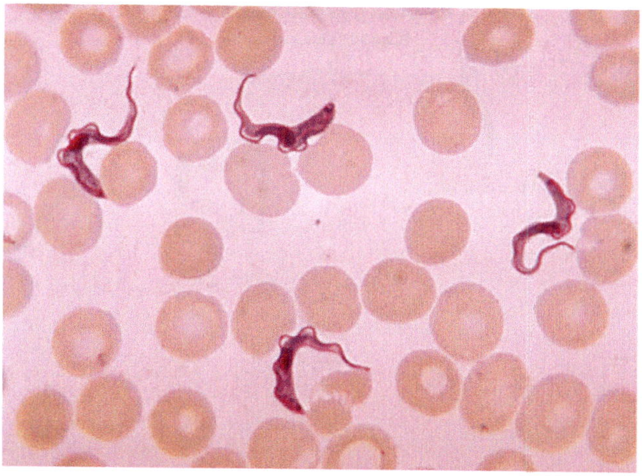

Fig. 24.17 Giemsa-stained thin blood smear showing trypomastigotes of *Trypansoma brucei*. Courtesy of PHIL, CDC.

sleep-wake cycle (hence the name sleeping sickness), and eventually death. The course of the infection caused by *T. b. rhodesiense* is much more rapidly progressive (weeks to months) than that caused by *T. b. gambiense* (months to years).

The diagnosis can be considered at three different levels.

- In the field screening of at-risk populations for *T. b. gambiense*. This involves adding a drop of patient blood to a card containing trypanosomal antigen, and observing agglutination. (Card agglutination test for trypanosomiasis – CATT.)
- Diagnosis of a suspect case by various forms of microscopy: wet preparations of lymph node aspirates, Giemsa-stained thin and thick blood smears (Fig. 24.17), as are done for diagnosing malaria, examination of unstained buffy coat preparations from centrifuged microhematocrit tubes for motile parasites, staining with acridine orange and visualization under UV light, and an anion-exchange centrifugation method.
- Determination of whether the central nervous system is involved by examination of cerebrospinal fluid. This is very important, because therapy is significantly influenced by the presence or absence of nervous system involvement.

The geographic location of acquisition of the infection is also very important, because the subspecies cannot readily be differentiated in a routine laboratory, but therapy is determined by the likely species. The drugs used include pentamidine, eflornithine, nifurtimox, and melarsoprol for *T. b. gambiense* infection, and suramin and melarsoprol for *T. b. rhodesiense* infection. Melarsoprol is very toxic, so the accurate diagnosis of neurologic disease in individuals with *T. b. rhodesiense* infection is very important, as there is currently no alternative to this drug.

American trypanosomiasis

This infection, caused by *Trypanosoma cruzi*, occurs in South and Central America. It is transmitted by reduviid bugs ("kissing bug," "assassin bug") of the genera Triatoma, Panstrongylus, and Rhodnius. In sylvatic transmission, the vector lives on plants, and various animals serve as reservoirs. In urban transmission, the vector lives in cracks

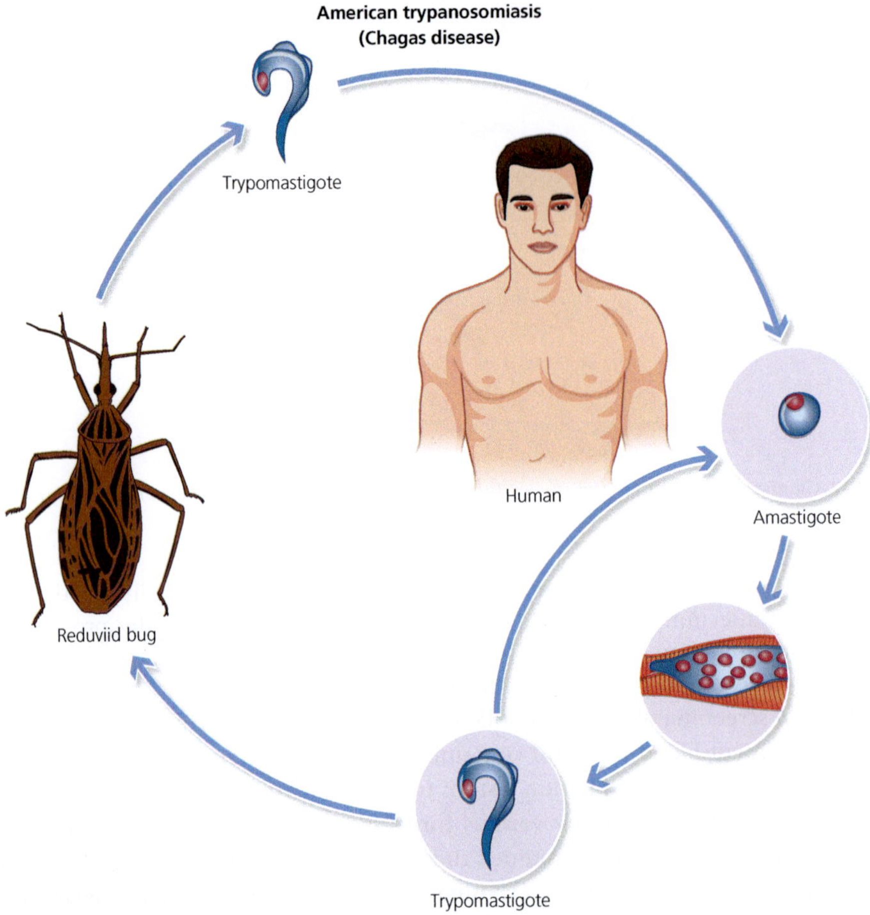

Fig. 24.18 Life cycle of *Trypanosoma cruzi.*

in the walls of houses. It is therefore a problem associated with poverty and poor housing. In addition to vector transmission, the organism can be transmitted by blood transfusion, needle sharing, transplacentally, and by ingestion of food contaminated by bug feces. In recent years there has been a very successful effort to reduce vector-borne and blood transfusion transmission. Therefore congenital cases constitute a greater proportion of cases than previously.

The life cycle (Fig. 24.18) is as follows. While taking a blood meal, the triatomid bug passes metacyclic trypomastigotes in its feces. These enter the wound and infect cells, where they become amastigotes. These divide by binary fission and form a pseudocyst, full of amastigotes. After about 5 days, the amastigotes become trypomastigotes, which can enter the blood and infect other cells. There are two forms of trypomastigotes: a slender form, which infects other cells, and a broad form, which is taken up by a biting triatome bug. In the bug, the trypomastigotes become epimastigotes, which multiply in the midgut. In the hindgut they become metacyclic trypomastigotes, which completes the cycle.

The clinical manifestations can be considered as acute and chronic. The acute stage is characterized initially by a sore at the site of inoculation, localized edema, and

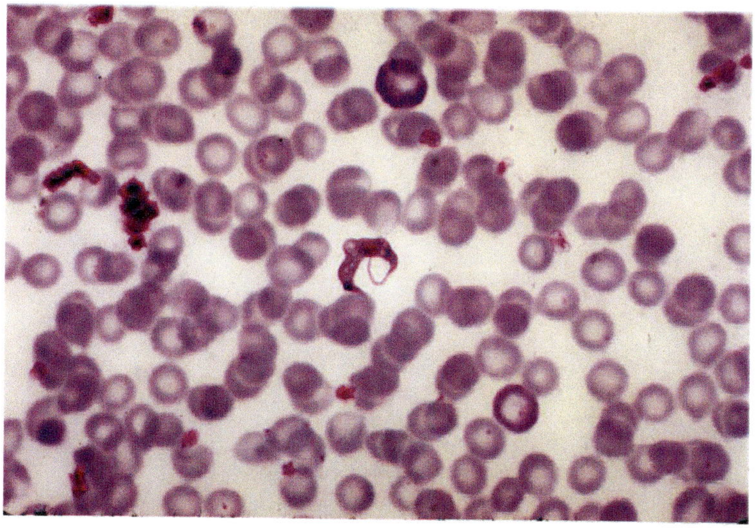

Fig. 24.19 Trypomastigote of *T. cruzi* in a Giemsa-stained blood smear. Courtesy of PHIL, CDC.

localized lymphadenopathy. After a few days to weeks, systemic features may occur. These include fever, rash, generalized lymphadenopathy, enlargement of liver and spleen, myocarditis, and meningoencephalitis. The chronic stage manifests years later and is the result of infection of the heart and the nerves of the gastrointestinal tract. Cardiac manifestations include cardiomyopathy and conduction defects. The gastrointestinal manifestations are megaesophagus and megacolon.

Diagnosis

In the acute stage, attempts should be made to detect the organism visually in the blood in stained smears, and by using concentration methods (see African trypanosomiasis) (Fig. 24.19). Molecular methods can be used on tissue fluids when a high parasite concentration is likely, as in acute infection and in transplant recipients. Animal inoculation and xenodiagnosis can also be used. The latter entails allowing a triatome bug that has previously fed on only birds (which are not susceptible to the infection) to bite the patient. After 30 days its feces is examined for the parasites.

In the chronic or the indeterminate stage, the diagnosis is made serologically. Several different tests can be used, and the diagnosis should depend on two different tests.

Treatment

Two drugs can be used for treating patients with American trypanosomiasis, benznidazole and nifurtimox, both of which have significant adverse effects. All cases of acute and congenital infection, and younger adults (<50 years) with cardiac disease, should receive treatment.

Leishmaniasis

This is a group of chronic infections caused by flagellates of the genus Leishmania, transmitted by certain flies in five main geographic areas. In the "Old World" (Sudan/East Africa, Mediterranean basin and Middle East, Central Asia, and India), it is

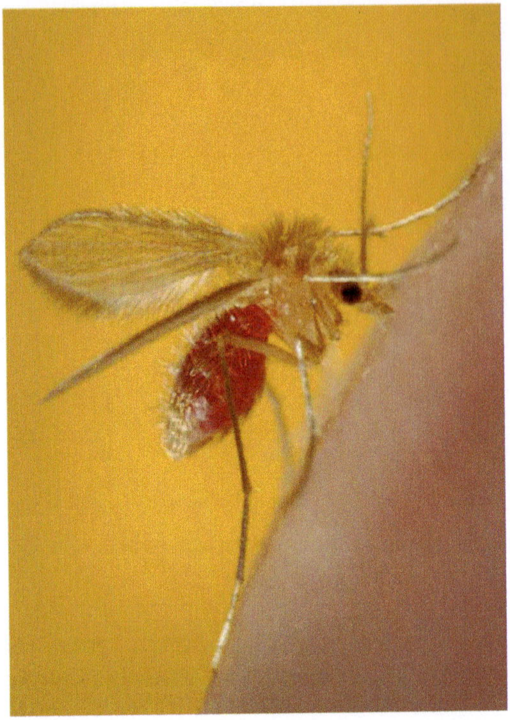

Fig. 24.20 *Phlebotomus papatasi*. Courtesy of DPDx, CDC.

transmitted by flies of the genus Phlebotomus, while in the "New World" (Central and South America), it is transmitted by flies of the genus Lutzomyia (Figs 24.20, 24.21 & 24.22). There are many different species of Leishmania, which cannot be differentiated from one another morphologically. They include *L. donovani* complex, *L. mexicana* complex, *L. tropica, L. major, L. aethiopica,* and the genus Viannia. Many different animals, such as rodents, canines, and hyraxes, but also humans, serve as reservoirs for the organism.

The life cycle of the parasite is shown in Fig. 24.23.

The organism exists as a flagellated promastigote in the insect vector and a non-flagellated amastigote in the mammalian host. When a fly feeds on the mammal, it injects promastigotes into the blood. These enter macrophages, where they become amastigotes. These spread to other cells, including macrophages. When a fly feeds on the host, these macrophages, containing amastigotes, are taken up. The amastigotes change into promastigotes and migrate to the proboscis, ready for injection into another host.

There are three forms of the disease: cutaneous, visceral (also called kala-azar), and mucocutaneous. Cutaneous disease, which accounts for the majority of cases of leishmaniasis, can take several different forms, including chronic ulcers or nodules (Fig. 24.24). These may be single or multiple. Most cases resolve spontaneously after many months. Visceral disease is characterized by significant systemic symptoms, including fever and weight loss. The main clinical signs are anemia and marked enlargement of liver and spleen. Darkened skin, the source of the name (from "black fever" in Hindi), is unusual. Other organs such as lung, intestine, and kidney can be involved. Mucocutaneous disease occurs only in South America. It may develop after

Status of endemicity of cutaneous leishmaniasis, worldwide, 2012

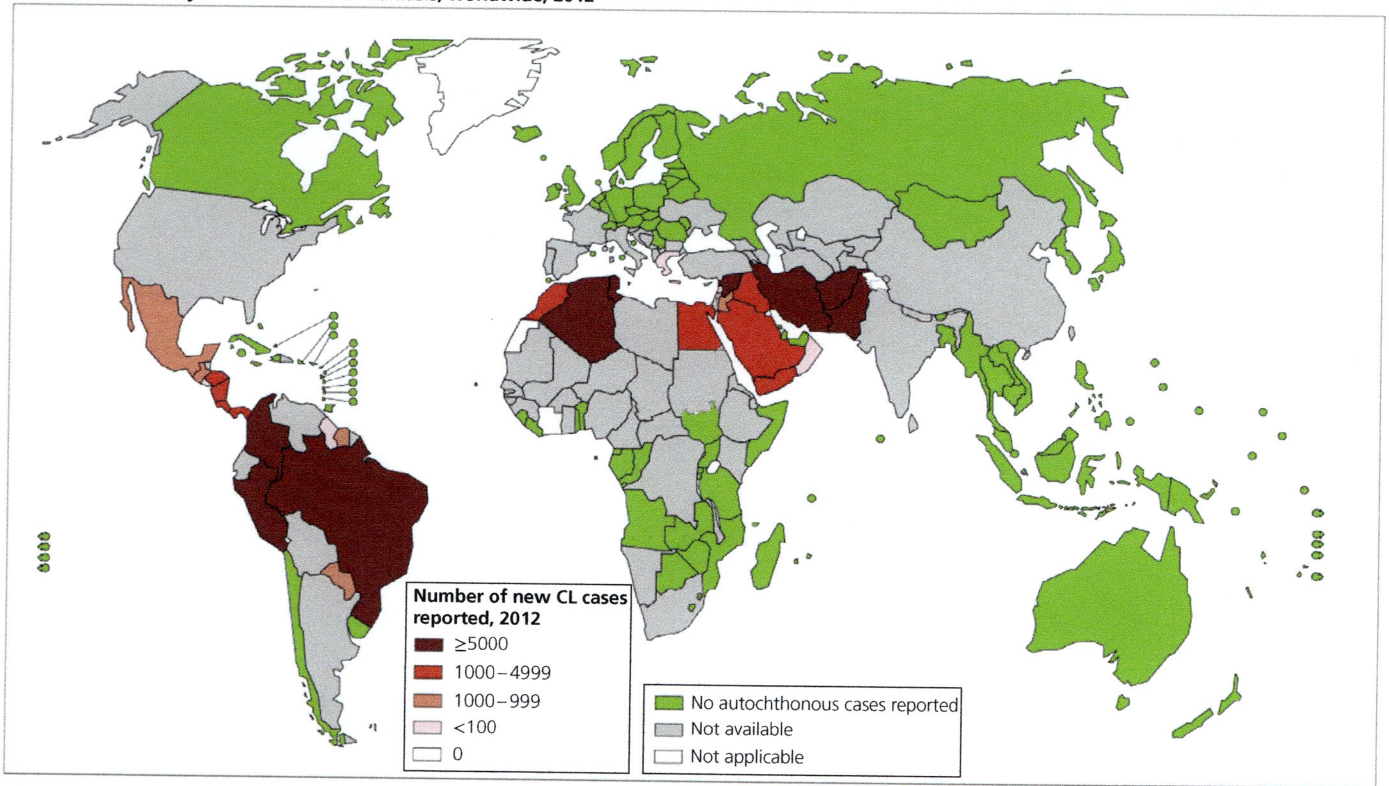

The boundaries and names shown and the designations used on this map do not imply the expression of any opinion whatsoever on the part of the World Health Organization concerning the legal status of any country, territory, city or area or of its authorities, or concerning the delimitation of its frontiers or boundaries. Dotted lines on maps represent approximate border lines for which there may not yet be full agreement.

Data Source: World Health Organization
Map Production: Control of Neglected Tropical Diseases (NTD)
World Health Organization

World Health Organization

Fig. 24.21 Worldwide distribution of cutaneous leishmaniasis. With permission from the World Health Organization.

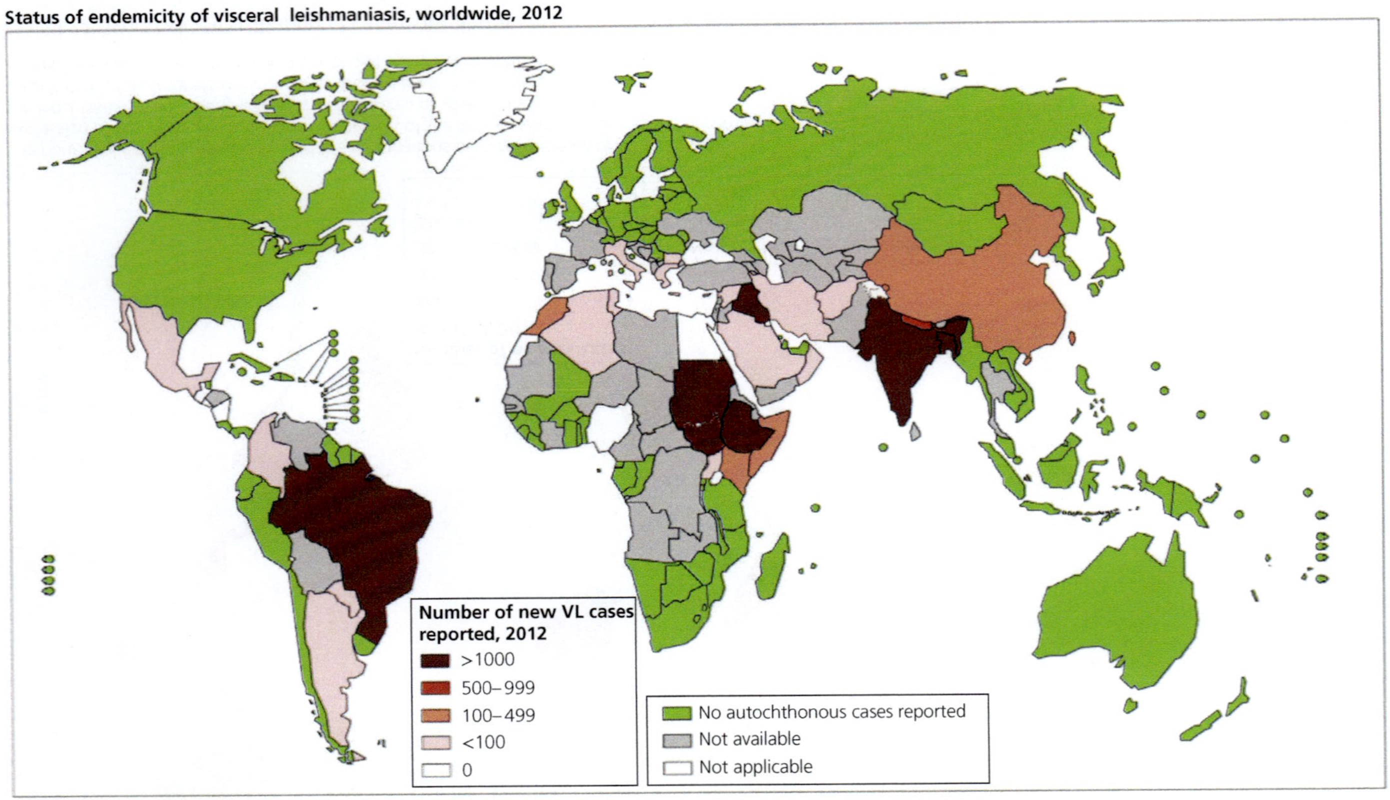

The boundaries and names shown and the designations used on this map do not imply the expression of any opinion whatsoever on the part of the World Health Organization concerning the legal status of any country, territory, city or area or of its authorities, or concerning the delimitation of its frontiers or boundaries. Dotted lines on maps represent approximate border lines for which there may not yet be full agreement.

Data Source: World Health Organization
Map Production: Control of Neglected Tropical Diseases (NTD)
World Health Organization

World Health Organization

Fig. 24.22 Worldwide distribution of visceral leishmaniasis. With permission from the World Health Organization.

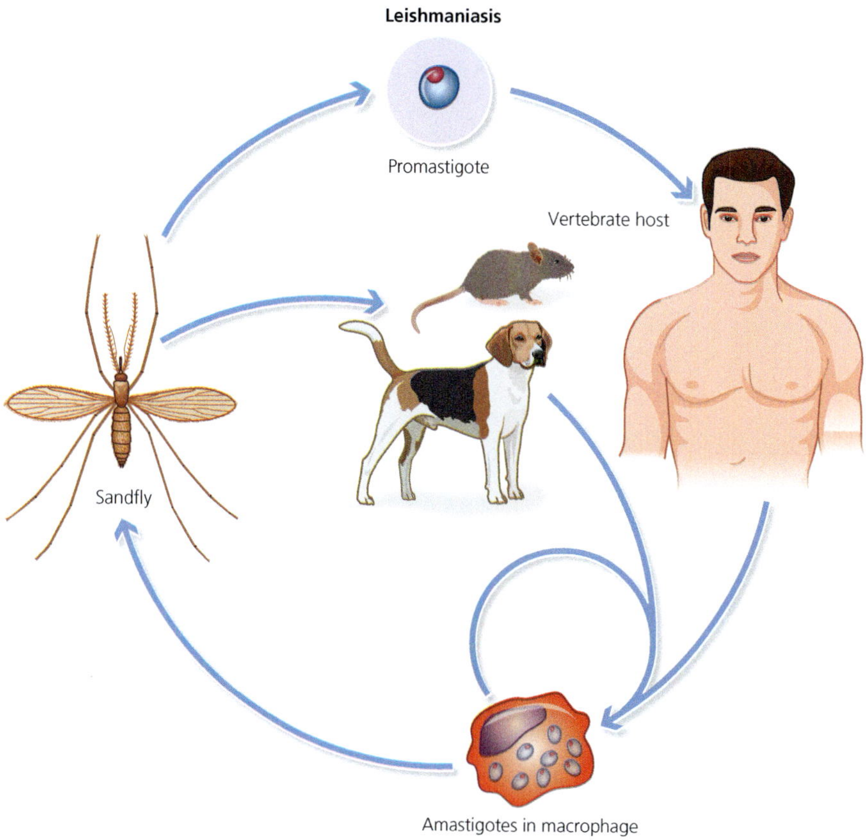

Fig. 24.23 Life cycle of *Leishmania* spp.

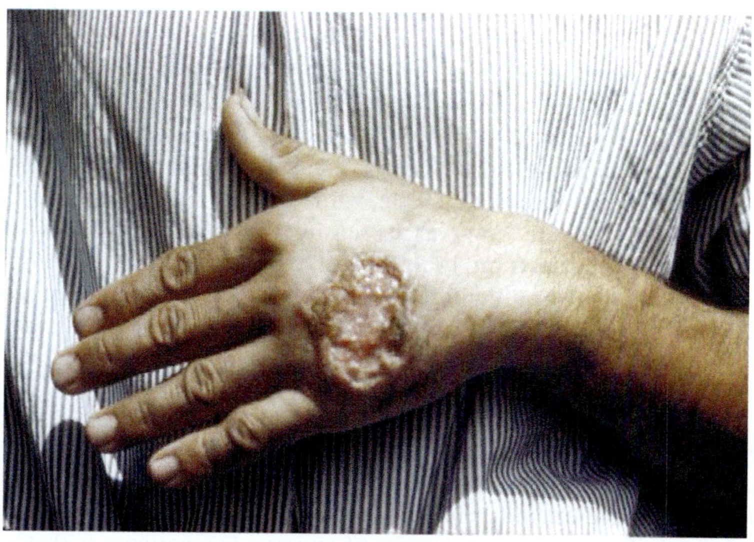

Fig. 24.24 Individual with cutaneous leishmaniasis. Courtesy of PHIL, CDC.

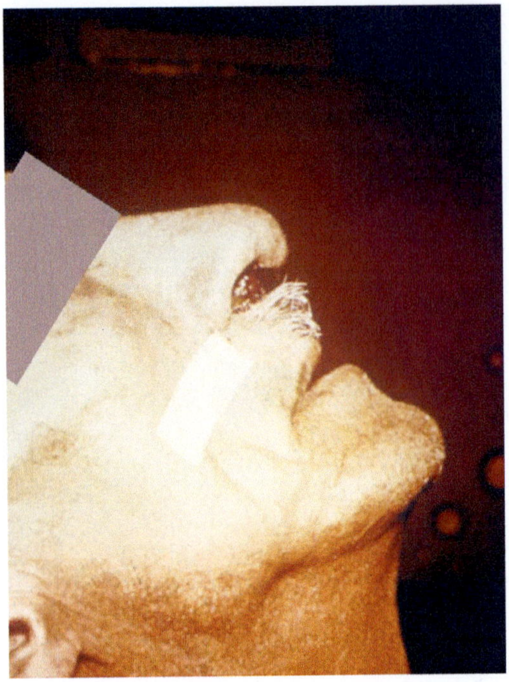

Fig. 24.25 individual with mucocutaneous leishmaniasis causing destruction of the nasal septum and consequent nasal deformity. Courtesy of PHIL, CDC.

cutaneous disease caused by *L. braziliensis* and *L. panamanensis*. It is a very destructive disease affecting the nose, cheeks, mouth, and larynx (Fig. 24.25).

Diagnosis

The main method is visualization of the organism in Giemsa- or immunoperoxidase-stained scrapings of skin lesions, tissue aspirates, e.g. of bone marrow or spleen, or biopsies (Fig. 24.26). The organism is seen within macrophages. An important differentiating feature is the presence of the kinetoplast, a dot near to but smaller than the nucleus of the organism. This is absent in histoplasma-infected cells, the main cytologic or pathologic differential diagnosis (see Fig. 20.3, Chapter 20).

Many different serologic tests have been evaluated for visceral leishmaniasis, with variable sensitivities and specificities. That with the best performance is the rk39 immunochromatographic test. However, its sensitivity is lower for East African cases than for Indian cases. PCR tests can be performed on blood, and the organism can be cultured *in vitro*, which enables identification of the species to be made, by isoenzyme analysis. Because cases of cutaneous disease caused by *L. braziliensis* and *L. panamanensis* can progress to mucocutaneous disease, and should therefore be treated, it is important that these species be specifically identified.

Treatment

Cutaneous disease is usually self-resolving, and treatment has many adverse effects so is often not indicated. Many different forms of therapy have been used, both antimicrobial and physical, with varying degrees of success. For patients with visceral or mucocutaneous disease, antimicrobial therapy should always be given. For many

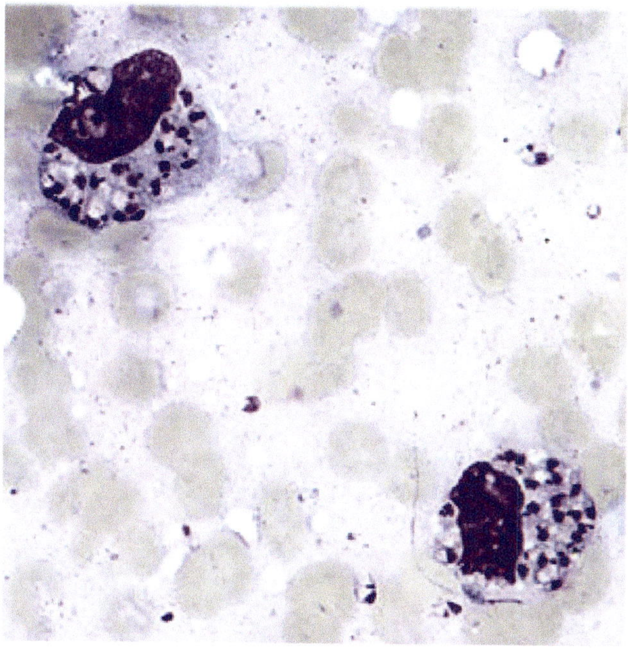

Fig. 24.26 Bone marrow smear showing Leishmania amastigotes inside macrophages. The kinetoplasts are the small dots adjacent to the nucleus. Courtesy of DPDx, CDC.

years, antimonial compounds, namely sodium stibogluconate and meglumine, were the only drugs available. However, they are being surpassed by newer drugs, where possible. The main drug for use, where available, is liposomal amphotericin B. Other available drugs are amphotericin B deoxycholate (regular amphotericin B), miltifosine, and paromomycin.

The different hemoflagellates are compared in Table 24.2.

Tissue flagellate

Trichomonas vaginalis

This is a flagellate that causes trichomoniasis, a common sexually transmitted infection causing symptoms mainly in females. It does not have a cyst form, so it cannot survive in the external environment. It multiplies by binary fission. It adheres to vaginal mucosa, causing injury to the epithelium. This results in inflammation and vaginitis. Most cases are asymptomatic but in symptomatic cases, the main symptom is a vaginal discharge and vaginal odor. Urinary symptoms may also occur. This organism can play a role in the pathogenesis of pelvic inflammatory disease, and the vaginal inflammation can predispose the host to other sexually transmitted infections such as herpes simplex virus and HIV infections. The main differential diagnosis is bacterial vaginosis.

The most commonly used method of detecting this organism is the "wet-prep" (or wet mount) in which a sample of vaginal discharge is mixed with normal saline and viewed under the microscope (Fig. 24.27). The presence of motile cells about the size of leukocytes is diagnostic of the infection. Although this method is very specific, it has low sensitivity. Its advantage is that it can be performed in the clinic.

Table 24.2 Comparison of the hemoflagellates.

Disease	African trypanosomiasis	American trypanosomiasis	Leishmaniasis
Alternative name	African sleeping sickness	Chagas disease	"Kala-azar" for visceral form
Causative organism	*Trypanosoma brucei*	*Trypanosoma cruzi*	*Leishmania* spp. (many)
Vector	Tsetse fly	Triatome bug	Sandfly
Geography	Tropical Africa	South and Central America, Mexico	Mediterranean basin, Middle East, Central Asia, India (Bihar state), Sudan, East Africa, South and Central America
Vector stage	Procyclic trypomastigote ↓ Epimastigote ↓ Metacyclic trypomastigote	Trypomastogote ↓ Epimastigote ↓ Metacyclic trypomastigote	Amastigote ↓ Promastigote
Mammalian stage	Trypomastigote	Metacyclic trypomastigote ↓ Amastigote ↓ Trypomastigote	Promastigote ↓ Amastigote
Infectious fluid (vector)	Saliva	Feces	Saliva
Other modes of transmission	Congenital - rare	Blood, congenital, ingestion	
Target cell in host	Extracellular	Muscle	Macrophage
Target organs	Systemic, brain	Systemic, heart, gut	Skin, mucocutaneous tissue, reticuloendothelial system
Diagnosis visualization	Blood, lymph node juice, cerebrospinal fluid	Blood	Tissue: scraping, smear, aspirate, biopsy
Serology	*T. gambiense*	Two different tests	Not very useful
Prognosis untreated	Fatal *T. b. gambiense* years *T. b. rhodesiense* months	30-40% chronic heart or intestinal disease	Cutaneous – good Mucocutaneous – poor Visceral - poor

Other diagnostic methods include the following.

- Culture: this is the gold standard. Its disadvantage is that it takes about 5 days.
- Nucleic acid detection (nucleic acid amplification test – NAAT) and transcription-mediated amplification (TMA); this latter has very high performance characteristics.
- Rapid antigen detection tests.

Treatment consists of metronidazole or tinidazole.

Tissue amebae

There are four known genera of free-living amebae that can cause invasive disease in animals and humans: Acanthamoeba, Balamuthia, Naegleria, and Sappinia.

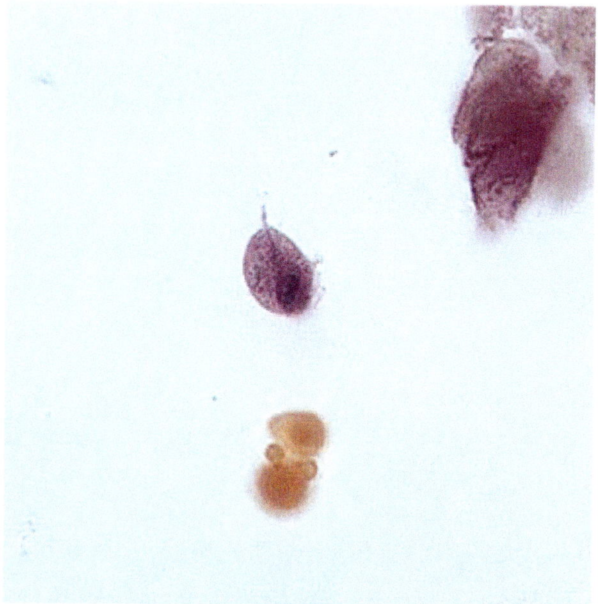

Fig. 24.27 *Trichomonas vaginalis* trophozoite. Courtesy of DPDx, CDC.

Acanthamoeba

There are at least 24 species of this organism, which lives in the environment, including soil and water. They exist as trophozoites, which feed on bacteria, and as double-walled cysts. The trophozoite has thorn-like projections, which give it its name (*acanthos* = thorn in Greek) (Fig. 24.28).

The organism can be grown in the laboratory on plates with *E. coli* or other bacteria, and once isolated, can be grown in axenic culture (no other organisms present).

Clinically, it causes chronic infections of the brain and other organs, including the skin, in immunocompromised individuals, and keratitis in individuals wearing contact lenses. Although the route by which the organism enters the host is unclear, it probably enters via the respiratory tract, from which it spreads hematogenously. It causes granulomatous amebic encephalitis (GAE), which has an insidious progression over weeks to months, and if untreated is fatal. Making the diagnosis is difficult because the organism is not seen in the cerebrospinal fluid, although the fluid is abnormal (usually a lymphocytic pleocytosis). Most cases are diagnosed pathologically, unfortunately at autopsy. The treatment that has been used includes pentamidine + sulfadiazine+ flucytosine + fluconazole.

Amebic keratitis can be diagnosed using visualization and culture of material from corneal scrapings (Fig. 24.29). Molecular technology and matrix-assisted laser desorption ionization time of flight (MALDI-TOF – see Chapter 2) can be used to identify the organism.

The main forms of treatment of keratitis are chlorhexidine gluconate, polyhexamethylene biguanide or propamidine isethionate + dibromopropamidine isethionate. Voriconazole has been shown to have activity against this organism.

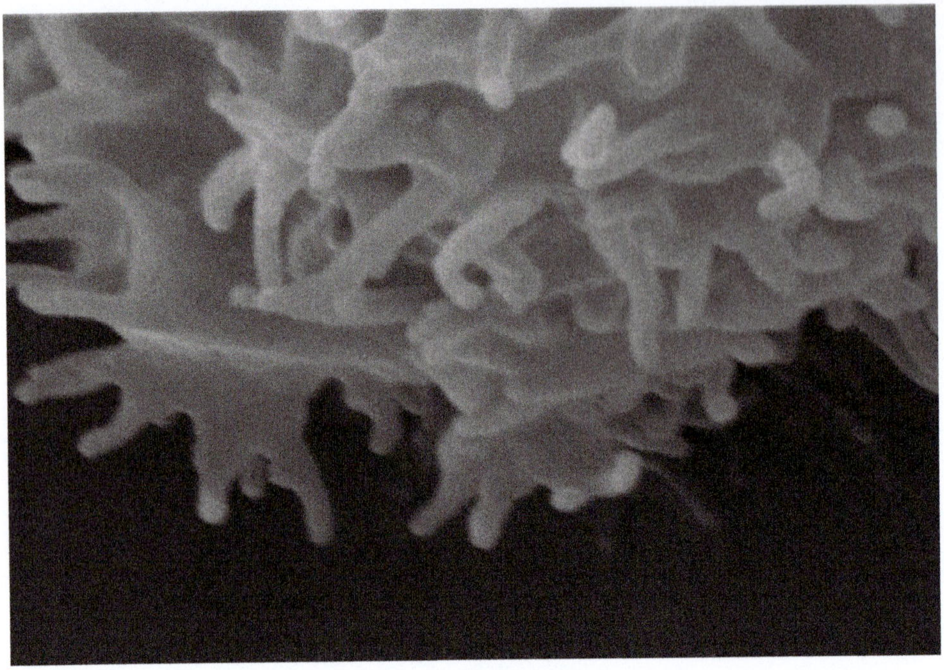

Fig. 24.28 *Acanthmoeba polyphaga* viewed by scanning electron microscopy. Courtesy of PHIL, CDC.

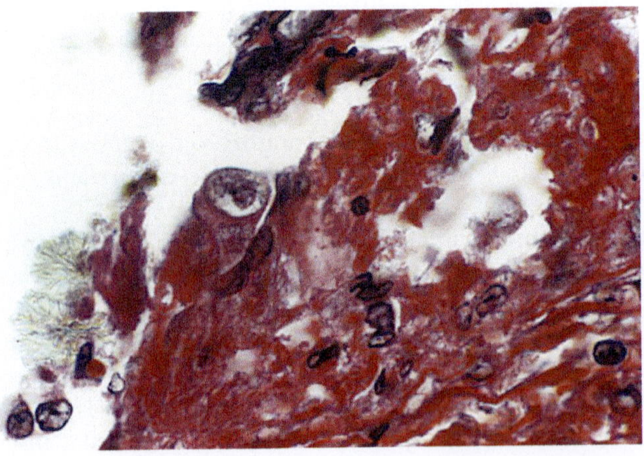

Fig. 24.29 Corneal scraping showing *Acanthamoeba polyphaga*. Courtesy of PHIL, CDC.

Balamuthia mandrillaris

This is the only species of this genus so far identified. It was first found in the brain of a mandrill baboon. It is present in soil and causes the same type of disease (skin lesions and granulomatous amebic encephalitis) as does *Acanthamoeba* spp., but it does not cause keratitis (Fig. 24.30). The CDC offers a PCR test for diagnosing this organism.

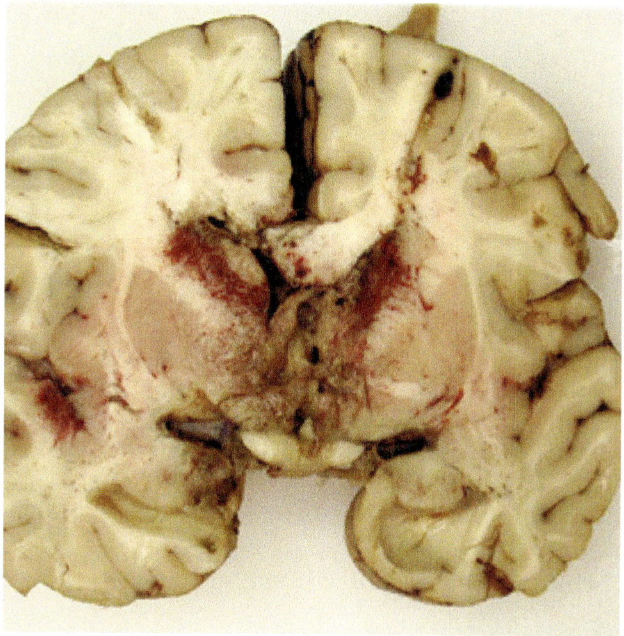

Fig. 24.30 Brain showing the damage caused by *Balamuthia mandrillaris*. Courtesy of DPDx, CDC.

The optimal treatment of patients is unclear. Two survivors were treated with a combination of pentamidine isethionate, sulfadiazine, clarithromycin, flucytosine, and fluconazole. Miltefosine, which is very useful for the treatment of individuals with leishmaniasis, also has activity against this organism.

Naegleria fowleri

This is a thermophilic ameboflagellate. It exists as an ameboid trophozoite, which divides by binary fission, as a non-dividing flagellate, and as a cyst. It is present in bodies of water, such as ponds, and causes invasive disease in individuals exposed to the water by invading the olfactory nerves through the cribriform plate and entering the brain. There it causes a rapidly progressive, destructive meningoencephalitis (primary amebic meningoencephalitis), that clinically cannot be distinguished from acute bacterial meningitis, and which is almost always fatal (Fig. 24.31). The organism can be detected in cerebrospinal fluid (as opposed to cases of acanthamoeba and balamuthia cerebral infection) by the following methods: wet preparation, in which motile trophozoites are seen, preparations stained with Giemsa and other stains (the specimen should not be heat-fixed), immunofluorescent staining, PCR, and culture on a lawn of *E. coli* (see CDC website: www.cdc.gov).

In vitro, it is susceptible to amphotericin B, azithromycin, voriconazole, chlorpromazine (a phenothiazine), and miltefosine. The optimal form of therapy should include amphotericin B.

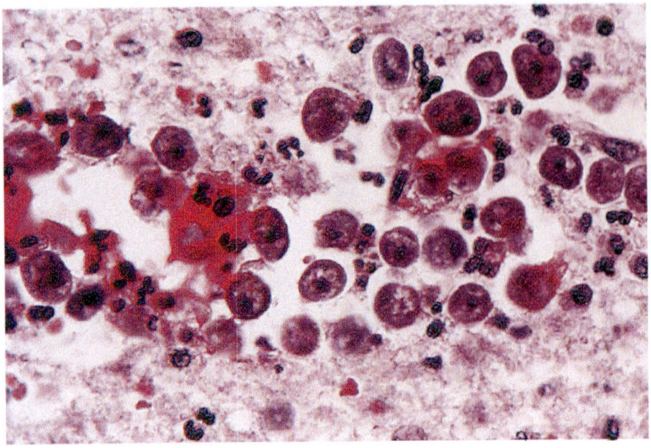

Fig. 24.31 *Naegleria fowleri* in brain tissue. Courtesy of PHIL, CDC.

Sappinia diploidea

Only a single case of brain infection due to this organism has been reported, in a normal host.

Further reading

Bern C (2011) Antitrypanosomal therapy for chronic Chagas' disease. N Engl J Med 364: 2527–34.

Boelaert M, Verdonck K, Meaten J, et al. (2014) Rapid tests for the diagnosis of visceral leishmaniasis in patients with suspected disease. Cochrane Database Syst Rev 6: CD009135.

Brun R, Blum J, Chappuis F, et al. (2010) Human African trypanosomiasis. Lancet 375: 148–59.

Chappuis F, Loutan L, Simarro P, et al. (2005) Options for field diagnosis of human African trypanosomiasis. Clin Microbiol Rev 18: 133–46.

Crawley J, Chu C, Mtove G, et al. (2010) Malaria in children. Lancet 375: 1468–81.

Elmahallawy EK, Martinez AS, Rodriquez-Granger J, et al. (2014) Diagnosis of leishmaniasis. J Infect Dev Ctries 8: 961–72.

Fox MC, Lakdawala N, Miller AL, et al. (2013) A patient with syncope. N Engl J Med 369: 966–72.

Hunfeld KP, Hildebrand A, Gray JS (2008) Babesiosis: recent insights into an ancient disease. Int J Parasitol 38: 1219–37.

Huppert JS (2009) Trichomoniasis in teens: an update. Curr Opin Obstet Gynecol 21: 371–8.

Kantele A, Jokiranta TS (2011) Review of cases with the emerging fifth human malaria parasite, *Plasmodium knowlesi*. Clin Infect Dis 52: 1356–62.

Lescure FX, LeLoup G, Freilij H, et al. (2010) Chagas disease: changes in knowledge and management. Lancet Infect Dis 10: 556–70.

McAuley JB (2008) Toxoplasmosis in children. Pediatr Infect Dis J 27: 161–2.

Montoya JG, Liesenfeld O (2004) Toxoplasmosis. Lancet 363: 1965–76.

Murray HW, Berman JD, Davies CR, et al. (2005) Advances in leishmaniasis. Lancet 366: 1561–77.

Opsteegh M, Kortbeek TM, Havelaar AH, et al. (2015) Intervention strategies to reduce human *Toxoplasma gondii* disease burden. Clin Infect Dis 60: 101–7.

Petersen E (2007) Toxoplasmosis. Semin Fetal Neonat Med 12: 214–23.

Rassi A Jr, Rassi A, Marin-Neto JA (2010) Chagas disease. Lancet 375: 1388–402.

Rosenblatt JE (2009) Laboratory diagnosis of infections due to blood and tissue parasites. Clin Infect Dis 49: 1103–8.

Travassos MA, Laufer MK (2009) Resistance to antimalarial drugs: molecular, pharmacologic, and clinical considerations. Pediatr Res 65(5 pt 2): 64R–70R.

Vannier E, Krause PJ (2012) Human babesiosis. N Engl J Med 366: 2397–407.

Vannier E, Gewurz BE, Krause PJ (2008) Human babesiosis. Infect Dis Clin North Am 22: 469–88.

Visvesvara GS (2010) Amebic meningoencephalitis and keratitis: challenges in diagnosis and treatment. Curr Opin Infect Dis 23: 590–4.

Visvesvara GS, Moura H, Schuster FL (2007) Pathogenic and opportunistic free-living amoebae: Acanthamoeba spp., Balamuthia mandrillaris, Naegleria fowleri, and Sappinia diploidea. FEMS Immunol Med Microbiol 50: 1–26.

White NJ, Pukrittayakamee S, Hien TT, et al. (2014) Malaria. Lancet 383: 723–35.

Wilson ML (2013) Laboratory diagnosis of malaria. Arch Pathol Lab Med 137: 805–11.

Yager PH, Luginbuhl LM, Dekker JP (2014) Case 6-2014: a 35-day-old boy with fever, vomiting, mottled skin, and severe anemia. N Engl J Med 370: 753–62.

CHAPTER 25

Helminths

Introduction

Helminths (worms) are common parasites of animals and humans worldwide. The hosts of the adult worm are considered the definitive hosts, while those harboring intermediate stages are termed the intermediate hosts. The infectious stage is usually the egg, laid by the adult, but in some cases it can be a larva. In general, with few exceptions, worms do not multiply within the definitive host. The worms that enter the definitive host comprise the host's worm burden. This is very different from the case of protozoa, which multiply within the host. The worm burden is an important concept in that a small worm burden might not cause any clinical problem, while a large burden will.

Helminths are categorized into three main groups:

- roundworms (round cross-section) (nematodes)
- flatworms or flukes (flat cross-section) (trematodes)
- tapeworms (cestodes).

From a clinical viewpoint, as with protozoa, it is convenient to consider worms in terms of gut worms (the gut being the site where the adult lives) and tissue worms. (Many worms have a life cycle in which different stages pass through non-enteric organs, but in which the adult lives in the gut.)

Intestinal nematodes

These are prevalent where hygiene and sanitation are poor. The main drugs for treatment are the benzimidazoles, namely albendazole and mebendazole (which is no longer available in the USA). These kill the parasites by binding to and inhibiting the activity of β-tubulin, which is necessary for microtubule polymerization.

These parasites have four basic types of life cycles.

1 The egg, excreted by another individual of the same species of host, is ingested, and it develops into an adult worm within the intestine of the host (e.g. *Enterobius vermicularis, Trichuris trichiura*).

Practical Medical Microbiology for Clinicians, First Edition. Frank E. Berkowitz and Robert C. Jerris.

2 The egg (which has been excreted in the stool of another host, and is present in the soil) is ingested by the host. The larvae, after hatching in the intestine, pass through the systemic venous blood, through the right side of the heart to the pulmonary capillaries, emerge into the alveoli, ascend into the upper respiratory tract, and are swallowed. Thus the adults reach the intestine (e.g. *Ascaris lumbricoides*).
3 The egg, passed by another individual of the same species, hatches in the soil. The larvae enter the skin of the new host and follow the same pathway as described in (2) above (e.g. hookworm).
4 The egg hatches within the intestine of the host, and the larva is passed in the stool. The subsequent course is the same as in (3) (e.g. *Strongyloides stercoralis*).

With the exception of Strongyloides, these are diagnosed by visualization of the adult worm, passed per rectum or in stool, or by microscopic examination of the stool for eggs. Various methods are used to concentrate the stool.

Enterobius vermicularis (pinworm)

This is a common infection in young children, whose poor hygiene facilitates its spread. Its life cycle follows the pattern described in (1) above. The adult worm lives in the large bowel and rectum. At night the female emerges from the anus and lays eggs. These can cause pruritus ani. The infection is not associated with any significant morbidity. The child (usually), by scratching the perianal area and sucking his/her fingers, can self-inoculate. The diagnosis is made by examination of the stool for adult worms, which look like threads of white cotton about 1.5 cm long; microscopy of the stool for eggs; and the "tape test." The goal of the tape test is to obtain eggs which the female has laid on the perianal skin, for microscopy. It involves placing clear tape over the perianal area of the patient in the morning, and then placing it on a microscope slide for examination (Figs 25.1 & 25.2).

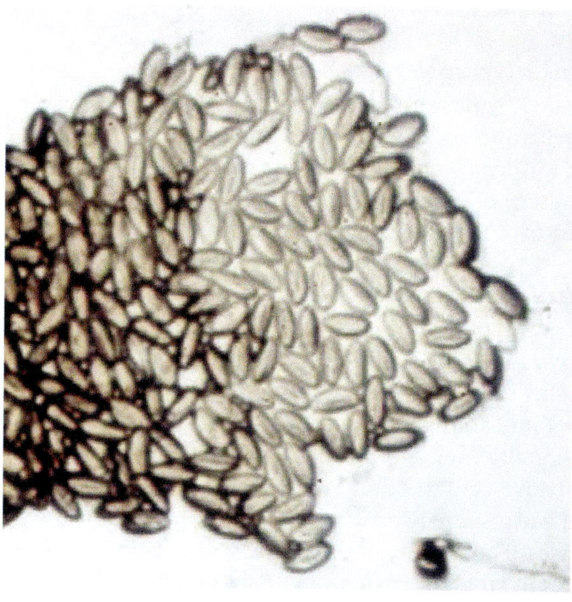

Fig. 25.1 Eggs *of Enterobius vermicularis* obtained by tape. Courtesy of DPDx, CDC.

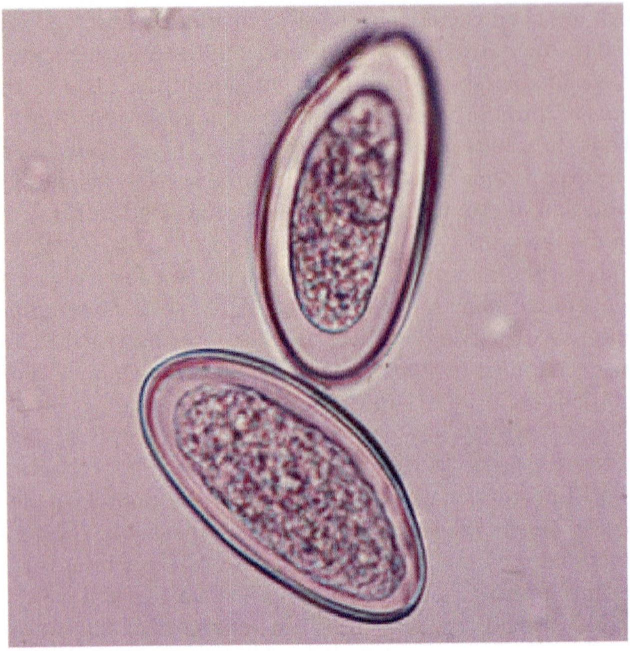

Fig. 25.2 Eggs of *Enterobius vermicularis*. Courtesy of DPDx, CDC.

Treatment consists of albendazole, mebendazole (no longer available in the USA), or pyrantel pamoate.

Soil-transmitted helminths are common worldwide. They remain in the gut for many years. The females lay thousands to hundreds of thousands of eggs per day. Their main adverse effects are on physical and intellectual development of children.

Trichuris trichiura (whipworm)

This organism (Fig. 25.3) has the life cycle described in (1). The adult worm lives in the large bowel, with the head in the mucosa. It causes inflammation which, in a heavy infestation, can lead to colitis, dysentery, and rectal prolapse. The diagnosis is made by demonstration of eggs in the stool (Fig. 25.4).

Treatment consists of mebendazole or albendazole.

Ascaris lumbricoides

This is one of the most common infections of mankind. The life cycle is described in (2) above and shown in Fig. 25.5. When larvae are migrating through the lungs, some respiratory symptoms can occur, such as cough and wheezing. This is called visceral larva migrans (see Toxocariasis later in this chapter). The adult worms may be passed rectally, and occasionally they ascend the esophagus and emerge from the mouth or nose. The main adverse clinical effect is intestinal obstruction due to a large bolus of worms. Individual worms can also ascend the bile or pancreatic duct, causing obstruction or liver abscesses. The diagnosis is made by visualization of the adult worm, or by visualization of eggs in stool by microscopy (Figs 25.6 & 25.7).

The treatment consists of albendazole, mebendazole, or ivermectin.

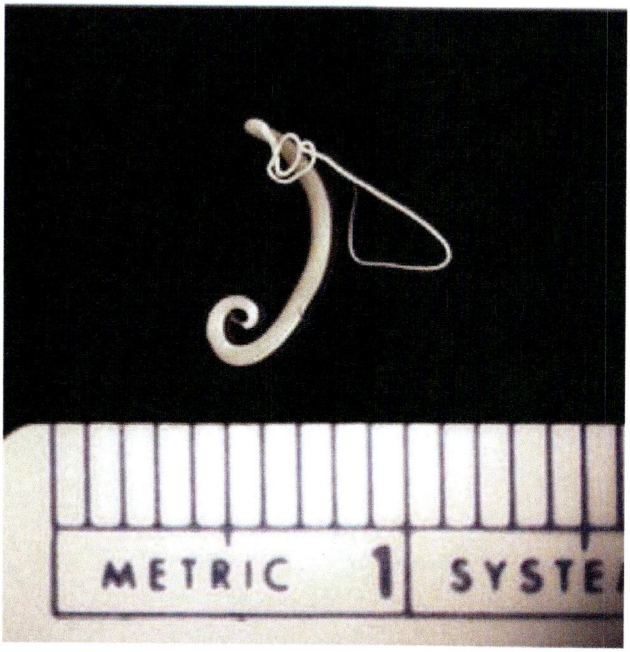

Fig. 25.3 Adult *Trichuris trichiura.* Courtesy of DPDx, CDC.

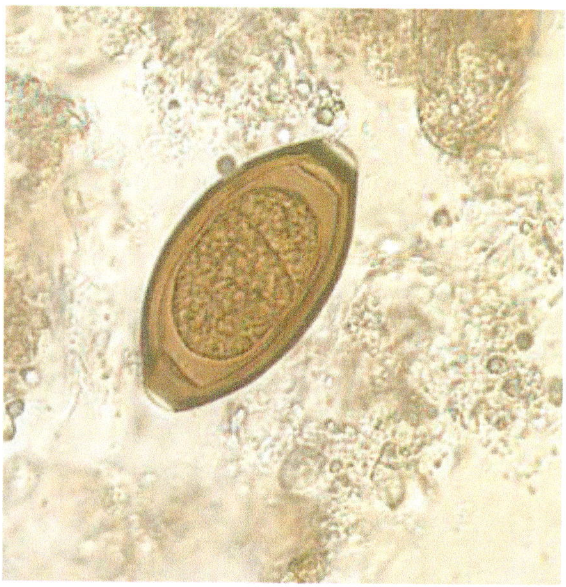

Fig. 25.4 Egg of *Trichuris trichiura* in feces. Note the opercula at both ends. Courtesy of DPDx, CDC.

Visceral larva migrans

Animal ascarids cannot complete their maturation within the human host. They migrate within tissues causing inflammation and, often, significant eosinophilia. The following syndromes are caused by *Toxocara cati* and *Toxocara canis*: pneumonia,

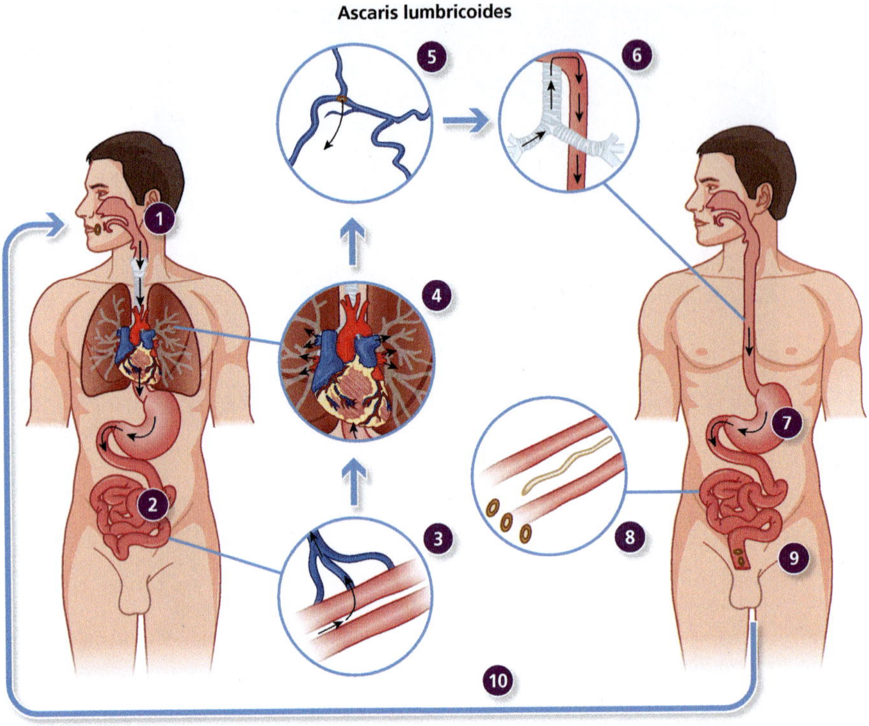

Fig. 25.5 Life cycle of *Ascaris lumbricoides*. 1. Eggs are ingested in soil; 2. The eggs hatch in the intestine; 3. The larvae penetrate the intestine and enter the mesenteric venous system; 4. The larvae pass through the right side of heart and into the lung; 5. The larvae pass from the pulmonary capillary into the alveolus; 6. The larvae ascend the respiratory tree and are swallowed; 7. The larvae reenter the intestine; 8. The larvae mature into adults; 9. The female lays eggs, which are passed out in feces; 10. The eggs mature in soil, in which they are ingested, completing the cycle.

Fig. 25.6 Adult *Ascaris lumbricoides* worm.

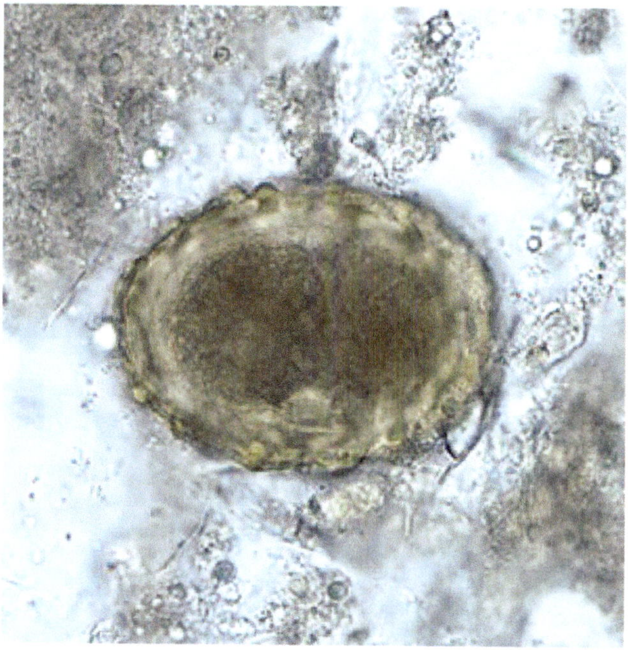

Fig. 25.7 *Ascaris lumbricoides* egg showing an embryo within. Courtesy of DPDx, CDC.

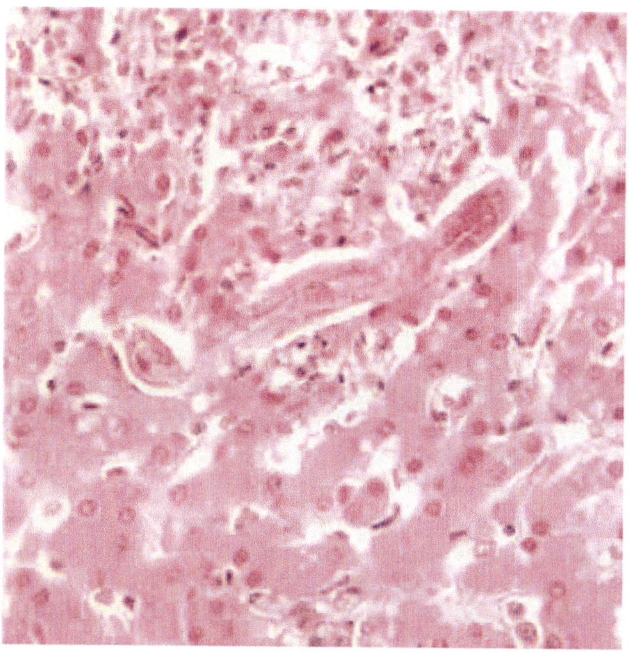

Fig. 25.8 *Toxocara* in liver tissue. Courtesy of DPDx, CDC.

hepatomegaly, and retinitis (Fig. 25.8). *Baylisascaris procyonis*, the raccoon ascaris, causes severe encephalitis that is often fatal (Fig. 25.9). Because the life cycle is not completed in the human host, eggs cannot be detected. The diagnosis is usually made serologically, and occasionally by the visualization of larva in tissue.

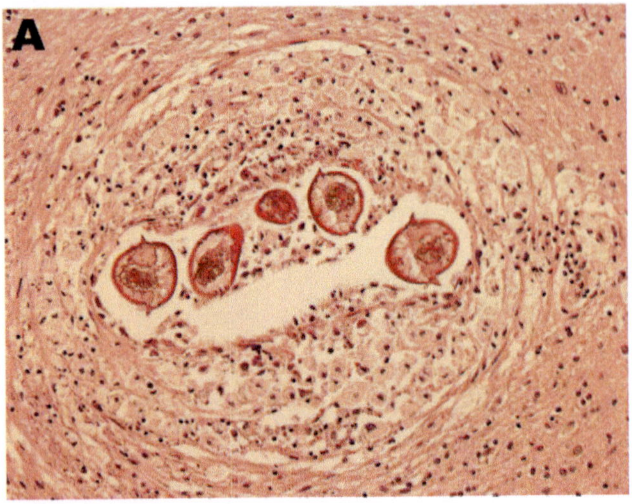

Fig. 25.9 *Baylisascaris procyonis* in brain tissue. Source: Hung et al. (2012).

Hookworms: *Necator americanus* and *Ancylostoma duodenale*

These parasites have the life cycle described in (3). The adults attach to the microvilli of the small intestine, by virtue of teeth (*A. duodenale*) or cutting plates (*N. americanum*) (Fig. 25.10), and drink the host's blood. Heavy infestations result in iron deficiency anemia and hypoalbuminemia. The diagnosis is made by visualization of the eggs in feces (Fig. 25.11).

Treatment consists of albendazole, mebendazole, or pyrantel pamoate.

The hookworm of dogs (*Ancylostoma braziliense*) cannot complete its migration in the human host, and the filiform larva wanders around in the skin, causing an itchy rash along its tract. This is called cutaneous larva migrans (Fig. 25.12).

Strongyloides stercoralis

This parasite's life cycle is complex and is described partially by (4). The larval stage that is passed in the stool is the rhabditiform larva. This undergoes two molts in the soil to develop into the filariform larva, which is infectious, and which enters the host's skin. This parasite can multiply within the host as a result of rhabditiform larvae within the gut becoming filariform larvae, and invading the blood through the gut wall or the perianal skin. These parasites can remain in the gut for many years. A particularly severe form of infection can occur if such a host becomes immunosuppressed as a result of an organ transplant or HTLV-1 infection (see Retroviruses, Chapter 6). The parasite can multiply and spread systemically, causing a disease called hyperstrongyloidiasis. The parasites can carry Gram-negative rods systemically. This condition carries a very high fatality rate. Because the parasite can live in the intestine for long periods, this complication can occur many years after the initial infection.

Diagnosing strongyloidiasis can be difficult, and depends mostly on visualization of the worm in stool (Fig. 25.13) or duodenal aspirate, or culture on plates inoculated

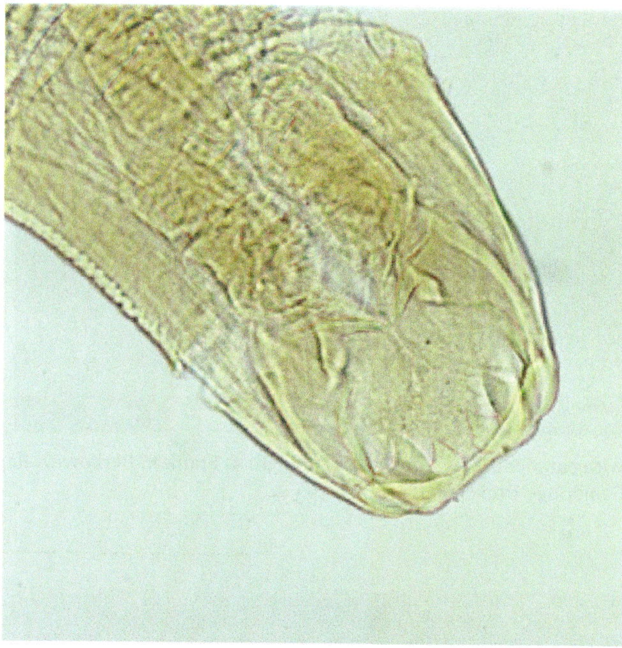

Fig. 25.10 Head of adult *Ancylostoma duodenale,* showing teeth. Courtesy of DPDx, CDC.

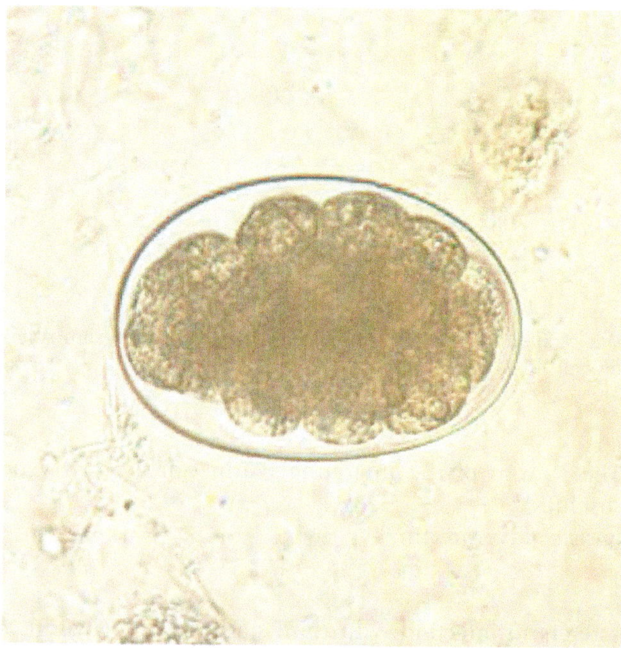

Fig. 25.11 Egg of hookworm in stool (wet preparation). Courtesy of DPDx, CDC.

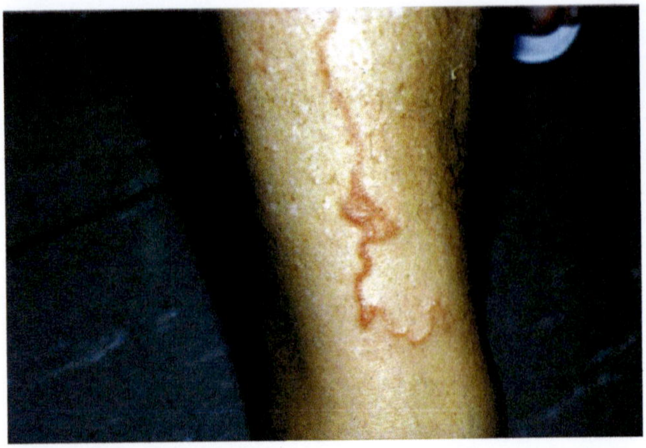

Fig. 25.12 Patient with cutaneous larva migrans. Copyright © Frank E. Berkowitz. Reprinted with the permission of Cambridge University Press.

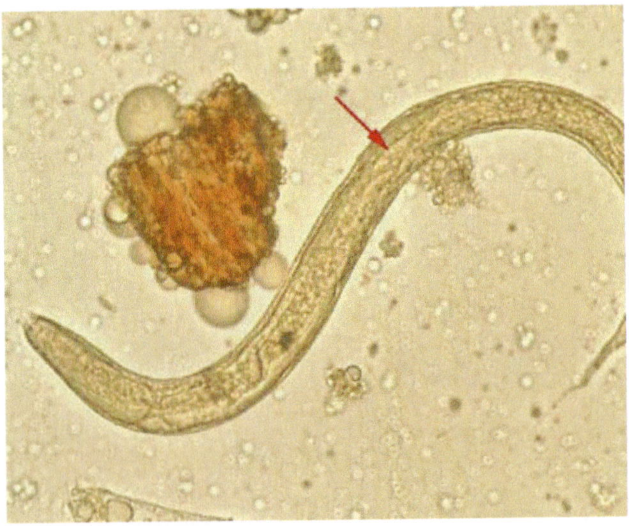

Fig. 25.13 *Strongyloides stercoralis* rhabditiform larva in stool. The red arrow shows the genital primordium. Courtesy of DPDx, CDC.

with bacteria. The pattern of the migrating worm can be observed on the agar. Serology is also useful.

Treatment consists of ivermectin.

Anisakiasis

In humans, this refers to infection with third-stage larvae of worms of the genera Anisakis and Pseudoterranova. These are parasites of marine mammals. These mammals excrete eggs, from which second-stage larvae enter crustaceans. These develop into third-stage larvae, which enter squid and fish. Their predators become infected with these larvae, as do humans. Adults develop from larvae only in marine mammals. In humans, the larvae cause throat irritation and gastritis. Penetration through the intestinal wall can occur. Treatment entails physical removal by endoscopy or surgery, which also provides the means of confirming the diagnosis.

Capillaria philippinensis

This is a zoonosis, which humans contract by eating raw fish containing larvae. The worms infect the small bowel. Heavy infections can cause significant intestinal symptoms and malnutrition. The diagnosis can be made by visualization of eggs or worms in the stool, and treatment consist of albendazole.

Tissue nematodes

Filaria

The female adults, in the vertebrate host, give birth to modified eggs called microfilariae. They are long, thin, and flexible, and have a sheath corresponding to the shell of an egg. These are the infective stage for the insect vector, the intermediate host, in which the larvae develop. When the insect feeds on the vertebrate host, it injects the larvae, which develop into adult worms of both sexes. The microfilariae of the different species can be differentiated from one another microscopically.

Lymphatic filariasis

This is caused by three parasites: *Wuchereria bancrofti, Brugia malayi,* and *Brugia timori*. The vectors of these parasites are mosquitoes. The life cycle of *W. bancrofti* is shown in Fig. 25.14. The larvae enter lymph nodes and lymphatics, where they mature. They cause lymphangitis and lymphatic incompetence, resulting in lymphedema in the corresponding areas of drainage. Wolbachia, which are endosymbionts belonging to the Rickettsiales, play a role in the inflammation. The skin changes result in secondary bacterial infection, which aggravates the condition. These factors can lead to elephantiasis, marked swelling of the scrotum, vulva, breast, and other areas (Fig. 25.15). The microfilariae are detected in blood smears, a technique that can be enhanced by concentration methods (Fig. 25.16). In some strains, their presence in blood has a diurnal pattern. *W. bancrofti* has a widespread distribution within the tropics (Asia, Africa, and America), while *B. malayi* occurs in South East Asia, and *B. timori* occurs in Indonesia and Timor. Antigen detection can be used for detecting *W. bancrofti*.

The treatment of lymphatic filariasis consists of diethylcarbamazine with or without albendazole. The addition of doxycycline, which kills Wolbachia, improves the success of therapy.

Onchocerca volvulus

This causes "river blindness." It was present in parts of tropical Africa, the Middle East, and tropical America, but due to very successful public health activities by The Carter Center in Atlanta, its range has been markedly curtailed. The life cycle is shown in Fig. 25.17. It is transmitted by flies of the genus Simulium, which breed near rapidly flowing rivers and streams. They inject larvae into the human hosts. These mature and the adults form a clump within a nodule (one male and a few females). The females give birth to microfilariae, which wander through the skin causing dermatitis.

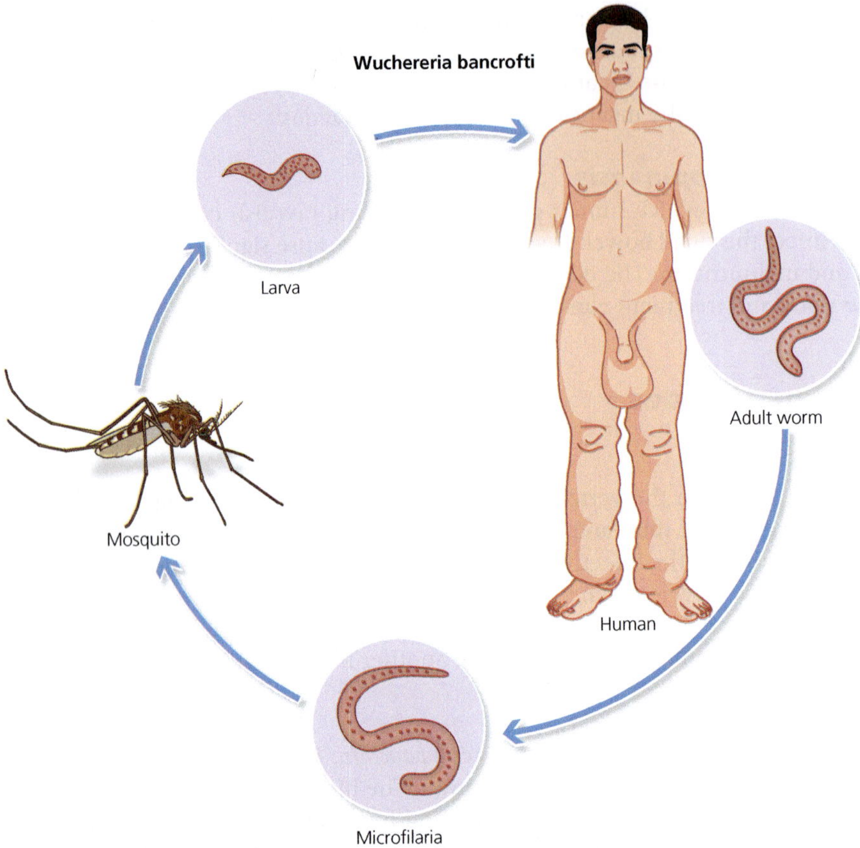

Fig. 25.14 Life cycle of *Wuchereria bancrofti*.

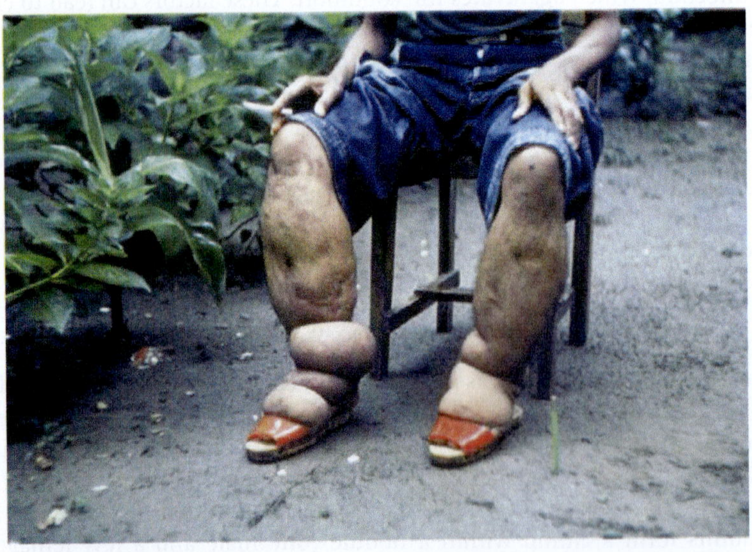

Fig. 25.15 Individual with elephantiasis due to lymphatic filariasis. Courtesy of PHIL, CDC.

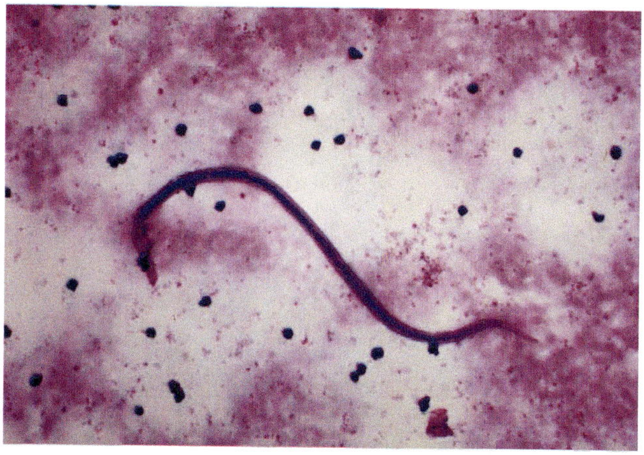

Fig. 25.16 Thick blood smear showing microfilaria of *W. bancrofti*. *Courtesy* of PHIL, CDC.

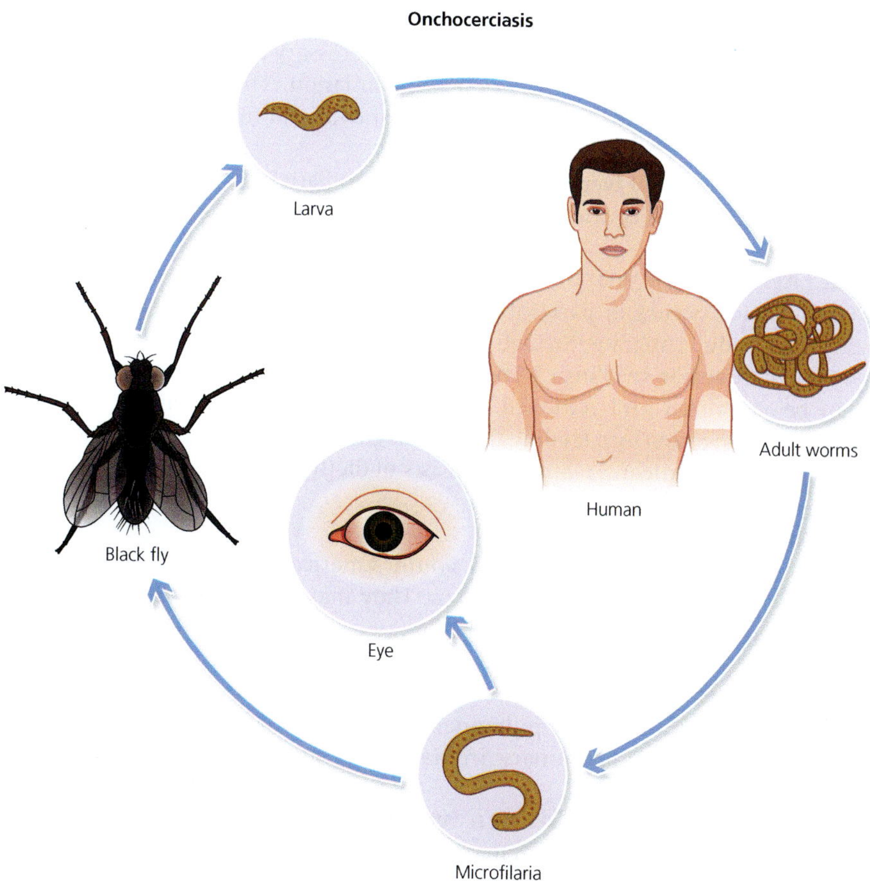

Fig. 25.17 Life cycle of *Onchocerca volvulus*.

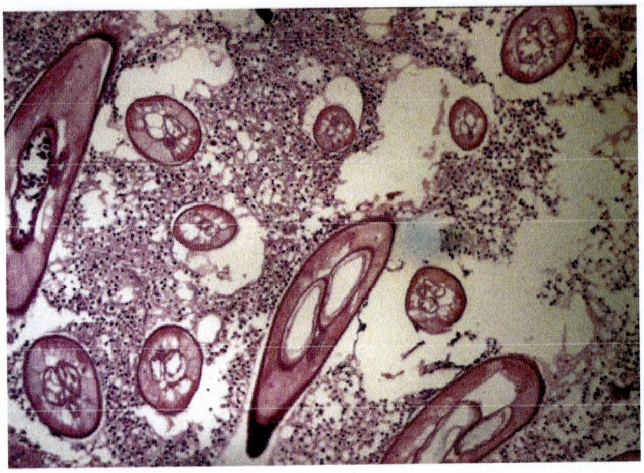

Fig. 25.18 Histological section of a nodule showing multiple sections of worms (*Onchocerca volvulus*). Copyright © Frank E. Berkowitz. Reprinted with the permission of Cambridge University Press.

They wander into the eye, causing inflammation of the anterior and posterior segments, leading to blindness. The microfilariae are taken up by biting flies, completing the cycle. Symbionts of the worms, Wolbachia, are necessary for fertility of the females, and probably play a role in causing ocular inflammation. The diagnosis is made by examination of skin snips for microfilariae, or removal and histology of a nodule, in which adult worms can be seen (Fig. 25.18).

Treatment, consisting of ivermectin, kills the microfilariae but not the adults. If this treatment is preceded by a course of doxycycline, the outcome of therapy is much better.

Loa loa (African eye worm)

This worm, which occurs in parts of tropical Africa, is transmitted by a mango fly (Chrysops). The adults migrate through the subcutaneous tissue, leaving inflammatory nodules called "Calabar swellings." They can migrate across the conjunctiva, where they can be visualized and removed. The microfilariae are present in the blood, whence they are ingested by a feeding fly. The diagnosis is made by visualization of the microfilaria on blood smear. Treatment consists of diethylcarbamazine.

Mansonella spp.

Mansonella ozzardi occurs in Central and South America, and the Caribbean, while *M. streptocerca* and *M. perstans* occur in Africa. They may cause skin disease, but most infections are asymptomatic.

Difilaria immitis

Difilaria immitis, which infects the pulmonary arteries of dogs, a disease called canine heartworm, can also cause pulmonary arterial disease in humans.

Dracunculus medinensis (Guinea worm)

This worm has been eliminated from many countries, especially in Asia, where it had previously been prevalent. It currently occurs in only a few African countries. Its life cycle is as follows: the adult female worm, which is up to 1 meter long, present in the

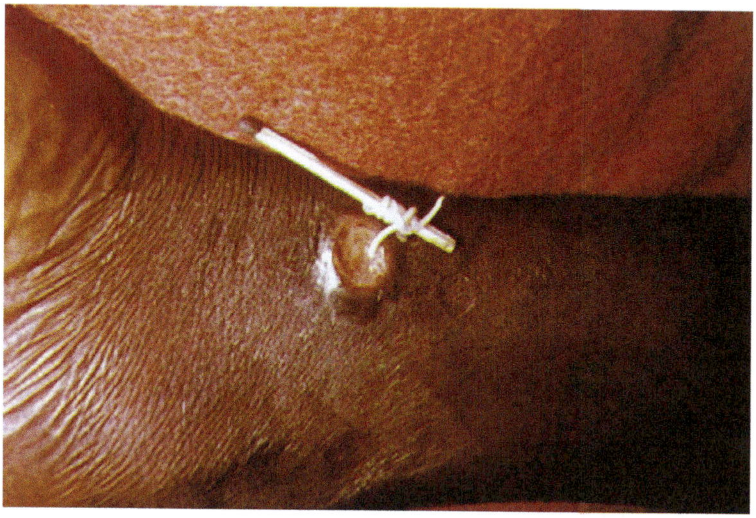

Fig. 25.19 Female Guinea worm being extracted from a patient's limb. Courtesy of PHIL, CDC.

subcutaneous tissue of the host, produces a blister on the skin surface, and when the skin is exposed to water, she emerges and emits the larvae. This occurs over several weeks. The larvae enter a copepod (e.g. *Cyclops* spp.). When the copepods are ingested by the mammalian host, the larvae burrow through the intestinal wall and migrate through deep tissues, as they mature, which takes about a year. Some reach the subcutaneous tissue. Most symptoms are due to the emergence of the adult worm, and bacterial infection that complicates this. The diagnosis is made by visualizing the worm. Because the copepods can be readily removed from water by simple filters, widespread distribution of filters has resulted in a marked decline in the prevalence of the disease, which could soon be eradicated. No antimicrobial therapy is available for treating infected patients, and treatment entails slow removal of the worm (Fig. 25.19).

Trichinella spiralis

This parasite, of worldwide distribution, is a zoonosis affecting mainly rats and pigs, but can affect any carnivore. The larvae are encysted, curled up in muscle. When the muscle is eaten, the larvae mature rapidly in the intestine of the new host and, after a few days, produce larvae. These migrate through the blood to various tissues, especially muscle (Fig. 25.20).

During the migratory phase, they can cause fever, periorbital edema, splinter hemorrhages, and myositis. This is usually accompanied by eosinophilia. The main causes of mortality and severe morbidity are myocarditis, encephalitis, and pneumonitis. The diagnosis is usually made by history or serology. Muscle biopsy can demonstrate the encysted larvae (Fig. 25.21).

The treatment consists of albendazole, with the addition of corticosteroids in severe cases.

Other species of trichinella include *T. nativa*, present in the Arctic, *T. nelsoni*, in tropical Africa, and *T. britovi*, in Europe.

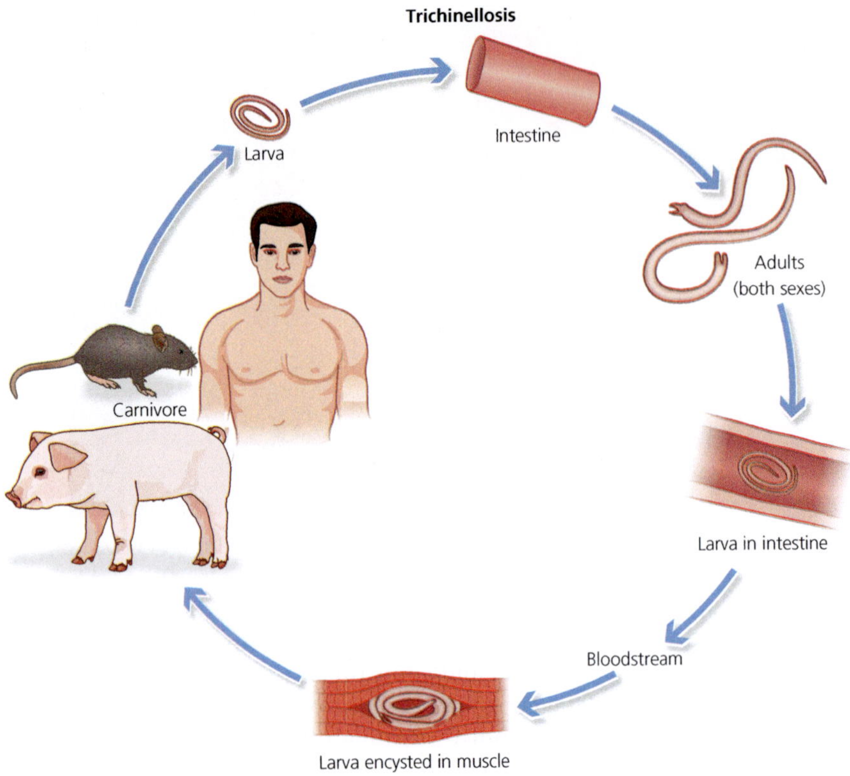

Fig. 25.20 Life cycle of *Trichinella spiralis*.

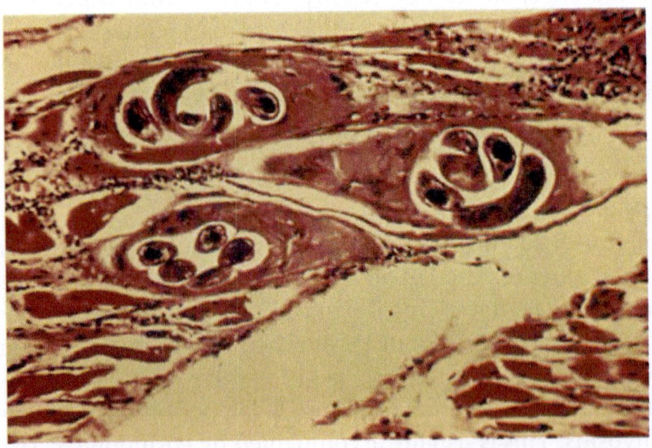

Fig. 25.21 *Trichinella spiralis* in human muscle. Courtesy of PHIL, CDC.

Angiostrongylus cantonensis

This is a parasite of rats, prevalent in Asia, the Pacific region, and the Caribbean. Larval stages live in snails and slugs, and can be present in their slime, which can contaminate vegetables. When the mollusk or its slime is ingested by a human, the larvae spread to the meninges, where they become adults. On dying, they cause inflammation. This is an important cause of eosinophilic meningitis. Larva can occasionally be found in cerebrospinal fluid. The CDC has a PCR for performance on cerebrospinal fluid specimens.

Angiostrongylus costaricensis

This is a parasite of Central America. The normal host is the cotton rat. Humans can be infected by ingestion of slugs. Larvae penetrate the intestinal mucosa and migrate to branches of the mesenteric vessels, causing eosinophilic granulomas adjacent to the bowel, especially the cecum.

Gnathostoma spinegerum (Fig. 25.22)

This parasite, present mainly in South East Asia, is a pathogen of cats and dogs. Larvae enter copepods, which can then be eaten by fish, frogs, snakes, or birds. When a mammal eats these, the larvae migrate through the intestine to other tissues, in particular the subcutaneous tissue, where they cause swellings. They can migrate to the brain.

Flukes (Order: Trematoda)

These are flat worms of variable size that are hermaphrodites, except for schistosomes. Their life cycles have the patterns shown in Fig. 25.23.

Treatment of patients with platyhelminth infections is summarized in Table 25.1.

The diagnosis of intestinal, liver and lung flukes has been traditionally based on demonstrating the presence of eggs in feces (and sputum in the case of lung flukes). This involves using various concentration methods, e.g. the Kato-Katz method. A recent method, using flotation in a "FLOTAC" apparatus, provides increased sensitivity. PCR performed on feces has an increased sensitivity as well as specificity, the latter largely because the morphologies of the eggs of different flukes are very similar. Immunoassays of blood and stool are also being developed. Although serologic tests are available for detecting some flukes, because these are very complex organisms, they may show cross-reactivity with one another.

Fig. 25.22 Head of *Gnathostoma spinegerum*. Courtesy of DPDx, CDC.

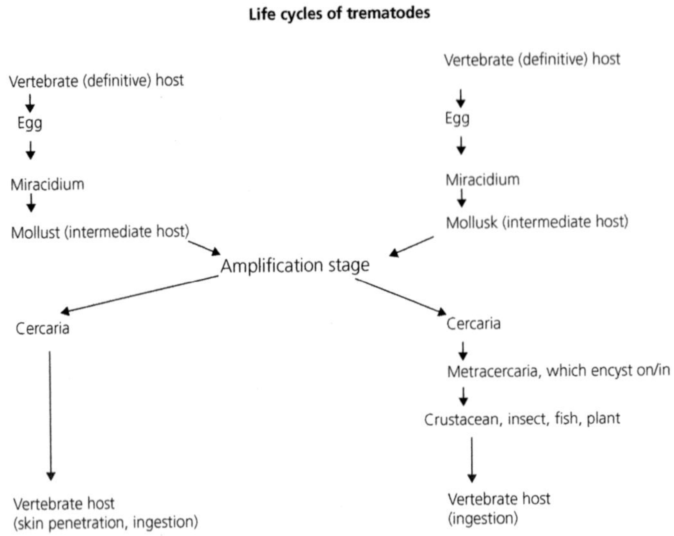

Fig. 25.23 Patterns of life cycles of flukes.

Table 25.1 Treatment for individuals with platyhelminth infections.

Parasite	Drug	Alternative
Intestinal flukes		
Fasciolopsis buski	Praziquantel	
Heterophyes heterophyes	Praziquantel	
Metagonimus yokogawai	Praziquantel	
Liver flukes		
Fasciola hepatica	Triclabendazole	Bithionol
Clonorchis sinensis	Praziquantel	Albendazole
Opisthorchis viverrini	Praziquantel	
Lung fluke		
Paragonimus spp.	Praziquantel	Triclabendazole Bithionol
Schistosomes		
S. haematobium	Praziquantel	Metrifonate
S. mansoni	Praziquantel	Oxamniquine
S. japonicum	Praziquantel	
S. mekongi	Praziquantel	
S. intercalatum	Praziquantel	
Tapeworms		
Taenia solium (intestinal)	Praziquantel	Niclosamide
T. solium (cysticercosis)	Albendazole	Praziquantel
T. saginata	Praziquantel	
Dipylidium caninum	Praziquantel	Niclosamide
Diphyllobothrium latum	Praziquantel	Niclosamide
Hymenolepis nana	Praziquantel	Niclosamide
Echinococcus granulosus	Albendazole	

Intestinal flukes

Fasciolopsis buski

This is the largest of the flukes, measuring up to 7 cm in length. It is prevalent in South and South East Asia. The life cycle is human/pig → egg, which matures in water → snail→ cercaria, which encysts as metacercaria on waterchestnut or bamboo → human/pig. It attaches to the bowel wall. Large infestations may cause abdominal pain, diarrhea, and malabsorption, leading to hypoproteinemia. The diagnosis is made by demonstration of eggs (which cannot readily be differentiated from those of *Fasciola hepatica*) or, occasionally, the adult fluke in stool.

Echinostoma spp.

These are small flukes which cause minimal disease.

Heterophyes heterophyes and *Metagonimus yokogawai*

These are small flukes prevalent in Asia and the Middle East. Metacercaria encyst under the scales of fish. If the adult worm burrows into the intestinal mucosa of the human host, the eggs that are laid can embolize.

Liver flukes

These all cause hyperplastic changes in the biliary tract epithelium where they live.

Fasciola hepatica (Fig. 25.24)

The usual mammalian host is the sheep, but humans can be infected. Metacercaria encyst on watercress or other water plants, which are then ingested by the definitive host. The worms burrow through the duodenal wall, cross the peritoneum and burrow through the liver capsule, to infect the bile ducts. Clinically this manifests with features of hepatitis. Eggs can be found in duodenal fluid or stool (Fig. 25.25), and serology can also be used for the diagnosis.

Fig. 25.24 *Fasciola hepatica* from a bile duct. Courtesy of DPDx, CDC.

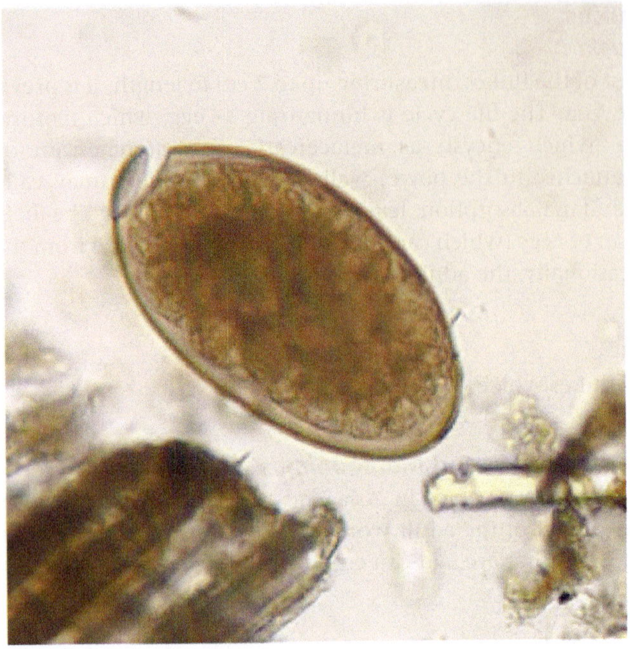

Fig. 25.25 Operculated egg of *Fasciola hepatica*. *Courtesy* of DPDx, CDC.

Clonorchis sinensis (Fig. 25.26) (prevalent in China and Vietnam), *Opisthorcis viverrini* (prevalent in South East Asia), and *Opisthorcis felineus* (prevalent in eastern Europe)

The metacercaria encyst in fish. After ingestion, they ascend the biliary ducts into the liver. Heavy infestations can cause cholangitis, and the chronic inflammation can lead to cholangiocarcinoma.

Dicrocoelium dentriticum

The metacercaria encyst in ants, so human infection, which is rare, requires that these insects are eaten.

Lung flukes

Paragonimus spp. (*P. westermani, P. ohirai, P. iloktsuenensis, P. kellicotti*)

These occur mostly in Asia, but also in Africa, South America and, rarely, North America (*P. kellicotti*). Many canines and felines are their natural definitive hosts. The metacercaria encyst in the tissues of crabs or crawfish. On ingestion, the worm penetrates the gut, migrates across the peritoneal cavity, through the diaphragm, and enters the lung, where it encysts. The adult lays eggs which are excreted via the bronchi or swallowed and excreted in the feces, enabling the diagnosis to be made (Fig. 25.27). During migration, they can cause liver disease. They cause chronic focal lung disease, and can occasionally spread to the brain.

Fig. 25.26 *Clonorchis sinensis*, showing its size. Courtesy of DPDx, CDC.

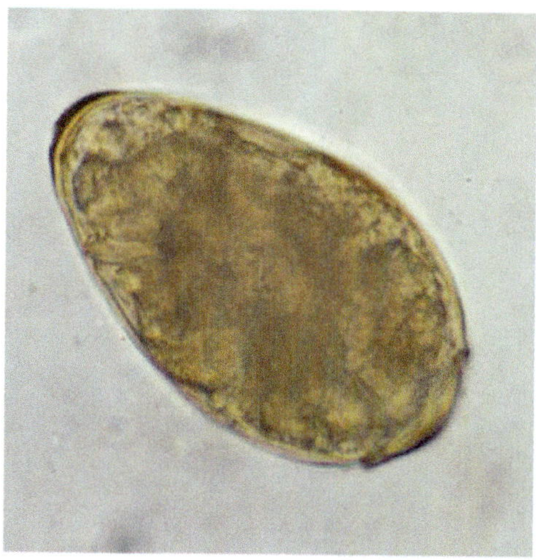

Fig. 25.27 Egg of *Paragominus westermani*. *Courtesy* of DPDx, CDC.

Blood flukes

Schistosoma spp. (schistosomiasis/bilharziasis)

These are very major causes of human disease. The distribution of schistosomiasis is shown in Fig. 25.28.

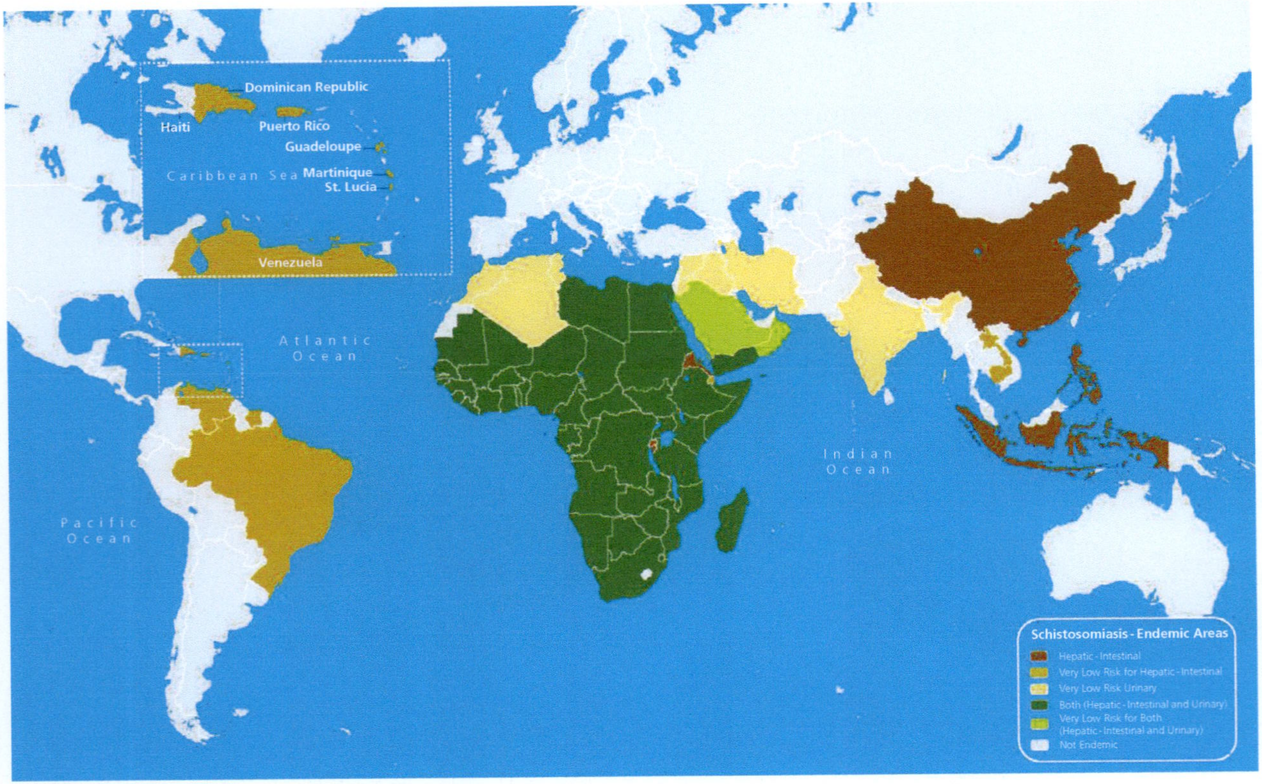

Fig. 25.28 World distribution of schistosomiasis. CDC, Yellow Book, 2014.

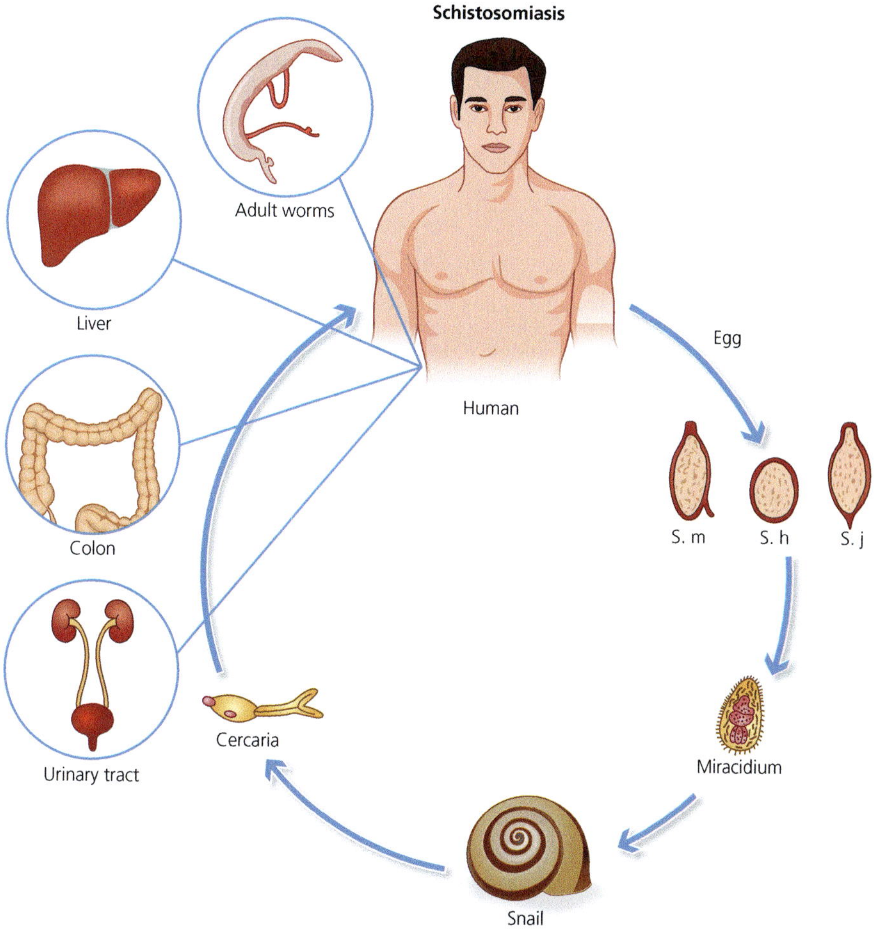

Fig. 25.29 Life cycle of *Schistosoma* spp. S. m, *Schistosoma mansoni*; S. h, *Schistosoma haematobium*; S. j, *Schistosoma japonicum*.

Six species affect humans, and many affect animals and birds. The latter can enter the human skin exposed to water and cause itching (swimmers' itch), but they do not spread in the human body.

The life cycle is shown in Fig. 25.29: eggs are passed in the human's feces or urine. On entering water, the miracidium hatches and enters a specific water snail. There it multiplies, and hundreds of cercaria leave the snail (Fig. 25.30). They must find a host within 24 hours. They enter through the host's skin, lose their tails and become schistosomulae. They enter the lymphatics and then the bloodstream. They settle as adults in the plexuses of the intestine (*S. mansoni*, *S. japonicum*, *S. mekongi*, *S. intercalatum*, *S. guineensis*) or the urinary tract (*S. haematobium*). The female attaches herself to the male in his gynecophoric canal, where her eggs are fertilized (Fig. 25.31). She leaves the male to lay eggs (hundreds per day) in venules. Some eggs embolize proximally,

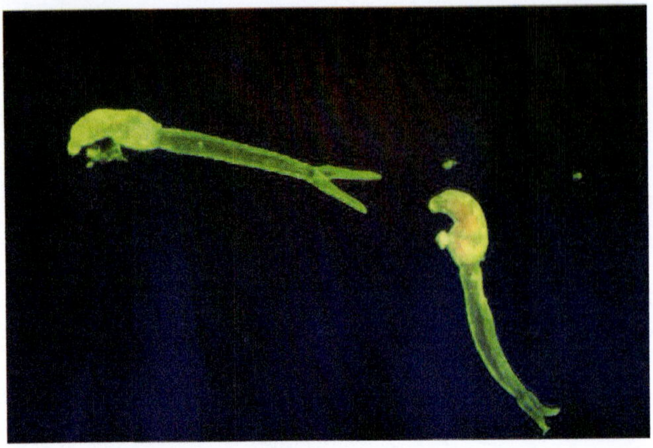

Fig. 25.30 Cercaria of *S. mansoni*. *Courtesy* of DPDx, CDC.

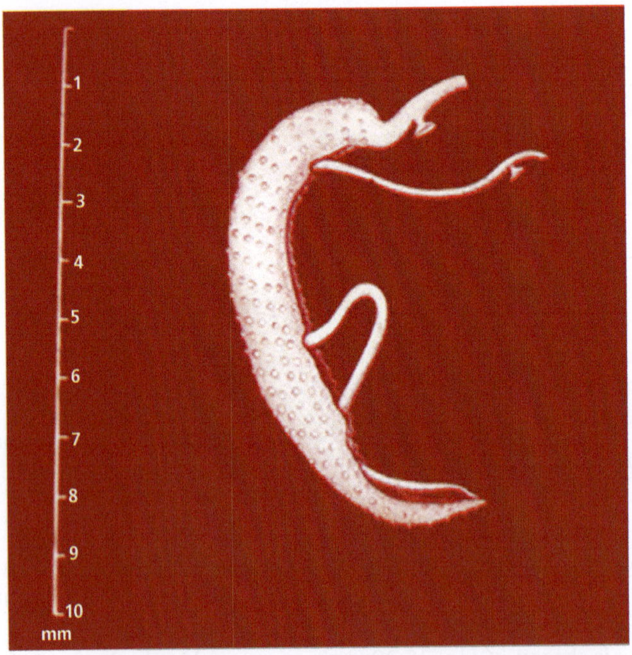

Fig. 25.31 Female within the gynecophoric canal of the male. Courtesy of DPDx, CDC.

e.g. to the liver, while others become embedded in the organ drained by the plexus (intestine or bladder), some of which are passed externally to complete the cycle. *S mansoni* and *S. haematobium* are parasites of only humans, while *S. japonicum* affects many mammals, and *S. mekongi* also affects pigs.

The eggs cause the pathology. Those that are wedged into the venules of the intestine cause focal inflammation, colitis, and polyps, and are excreted in the feces. Those that embolize to the liver stimulate the formation of granulomas, which leads to fibrosis (called Symmers' "pipestem" fibrosis), and which, in heavy infestations, leads to portal hypertension. They can also spread to other organs, particularly the brain and spinal cord. In the case of *S. haematobium*, the eggs are wedged into the venules of the bladder, where they cause inflammation and are excreted (Fig. 25.32). Bladder disease is characterized by hematuria, and can lead to urinary tract obstruction, bladder calcification, and cancer. Both male and female genital tracts can also be affected. Eggs can also embolize to the lung causing pulmonary hypertension.

Diagnosis

This is made optimally by demonstration of eggs: for intestinal plexus organisms, this is done by examination of stool or of a rectal mucosal biopsy. Granulomas around eggs can be shown in liver biopsies. Urinary schistosomiasis can be diagnosed by examination of urine. The eggs of *S. mansoni* have a lateral spine, while those of *S. haematobium* have a terminal spine (Figs 25.33 & 25.34).

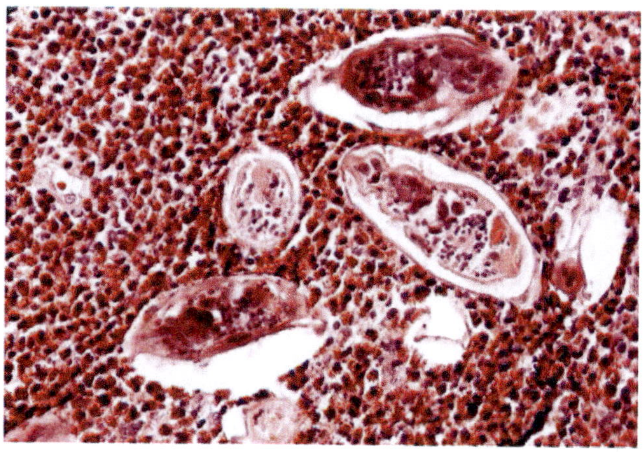

Fig. 25.32 Intense inflammation around eggs in the bladder, caused by *S. haematobium*. Courtesy of PHIL, CDC.

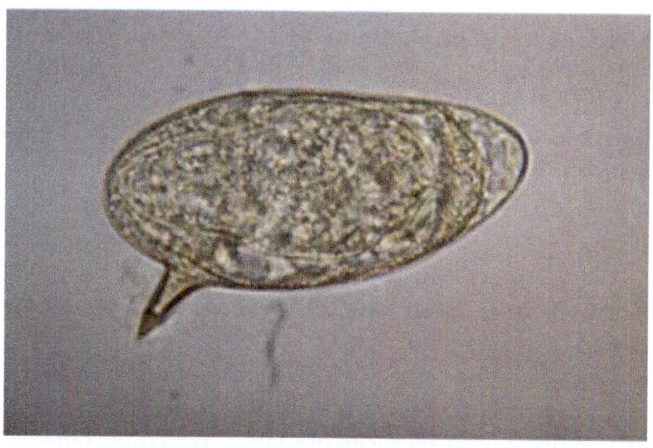

Fig. 25.33 Egg of *S. mansoni*; note the lateral spine. Courtesy of DPDx, CDC.

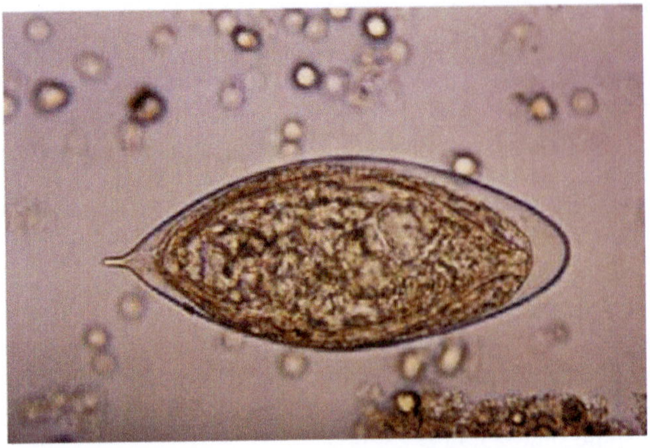

Fig. 25.34 Egg of *S. haematobium*; note the terminal spine. Courtesy of DPDx, CDC.

Viability of eggs can be demonstrated by allowing a few drops of water to trickle under a cover slip covering urine or feces. Miracidia can be seen to hatch. Serologic tests are available to determine whether an individual has been infected by a schistosome; antigen detection methods can be applied to serum and urine, and molecular methods for detection of eggs are promising.

Treatment

Praziquantel can be used for treating patients with all species of schistosomes, while oxamniquine can be used for patients with *S. mansoni*, and metrifonate for patients with *S. haematobium*.

Tapeworms (Order: Cestoda)

These are segmented worms. The scolex ("head") has four suckers for attachment to the bowel mucosa and, in some species, rows of hooklets. Growth occurs at the neck end, with the segments (proglottids) at the opposite end containing the eggs. They are hermaphrodites. They absorb nutrients through their integument. The diagnosis of infection by the adult worm is generally made by visualization of worm proglottids, passed in the stool, or of eggs in a microscopic examination of the stool. The life cycles have the basic pattern shown in Fig. 25.35.

Humans can serve as hosts in the following ways.

- Definitive host only (*Taenia saginata, Diphyllobothrium latum, Hymenolepis diminuta*)
- Definitive or intermediate host (*Taenia solium, Hymenolepis nana*)
- Intermediate host (*Echinococcus granulosus*)

Diphyllobothrium latum, D. nihonkaiense, and *D. pacificum* (fish tapeworms)

In these organisms, the egg hatches a corecidium embryo which is ingested by a copepod (the first intermediate host). When this is eaten by a fish (the second intermediate host), it becomes a plerocercoid larva, which becomes an adult when

Adult worm in intestine of definitive host

↓ Excretion

Egg

↓ Ingestion by intermediate host

↓

larva

↓ Ingestion by definitive host

↓

Adult worm

Fig. 25.35 Basic life cycle of cestodes.

the fish is eaten by a mammal (the definitive host). The diagnosis is made by demonstration of eggs in stool.

Sparganosis

This is an uncommon infection of muscle or subcutaneous tissue, caused by ingestion of water containing copepods infected with procercoid larvae of some diphyllobothrial worms within the genus Spirometra. It can also follow ingestion of uncooked snakes or frogs, containing plerocercoid larvae, or when these animals are placed on wounds.

Taenia solium (pork tapeworm)

This is the most important of the cestodes affecting humans, due to the larval stage, called cystercercosis. The life cycle is shown in Fig. 25.36. When the human becomes the intermediate host, as a result of eating eggs passed by another human (or possibly himself or herself), the larval stage enters the human's tissues. This is of particular importance in the nervous system (neurocysticercosis). It can cause a variety of neurologic syndromes, depending on the location of the cyst. The most common is epilepsy. Symptoms often occur when the worm dies. The circumstances necessary for cysticercosis to occur are humans eating undercooked pork (which allows for human infection with the adult worm), and poor hygiene (which allows for human fecal contamination of food for other humans).

The diagnosis of neurocysticercosis depends largely on imaging of the brain (Fig. 25.37). Definitive diagnosis is occasionally made if the cyst is removed (seldom necessary). Serologic tests are not sensitive, unless many cysts are present.

Whether to treat patients with neurocysticercosis with antiparasitic therapy has been controversial.

Taenia saginata (beef tapeworm)

This has a similar life cycle to that of *T. solium*, but humans do not serve as intermediate hosts.

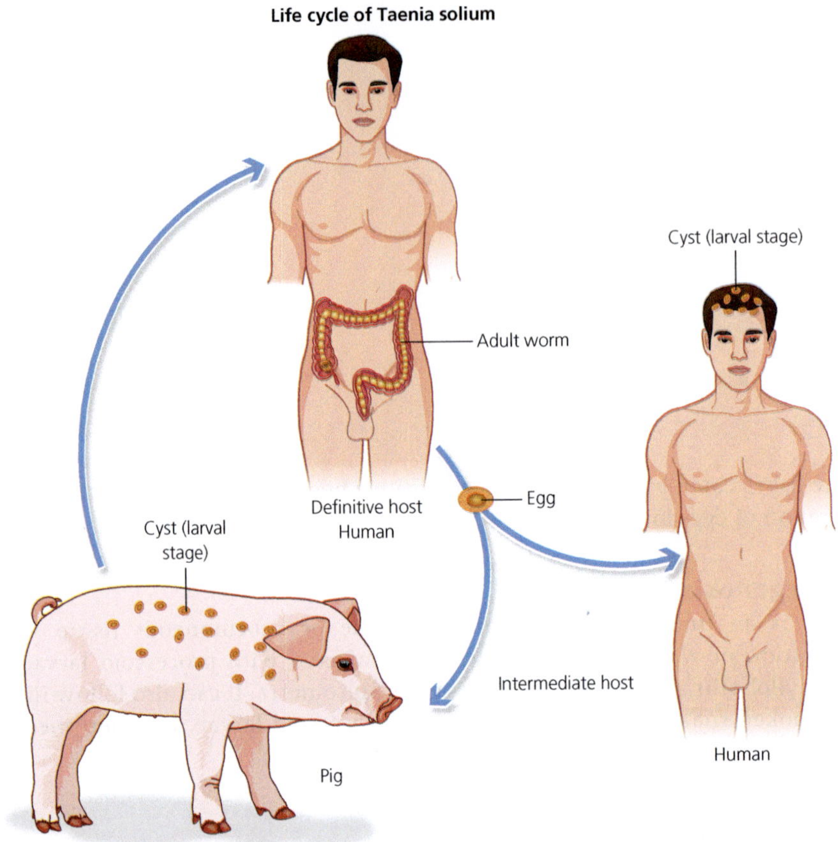

Fig. 25.36 Life cycle of *Taenia solium*. Copyright © Frank E. Berkowitz. Reprinted with the permission of Cambridge University Press.

Echinococcus granulosus

This is a tiny worm (0.6 cm long). The life cycle is shown in Fig. 25.38. Humans become infected by the larval stage, which produces large cysts, called hydatid cysts, mainly in the liver and lung, but also in other organs, such as bone. They can cause problems by space occupation and compression and, if they leak, by causing anaphylaxis. Diagnosis is made by imaging (Fig. 25.39). If fluid from the cyst is examined microscopically, protoscolices can be seen (Fig. 25.40). Serologic tests are also available.

Treatment of hydatid disease often involves surgery, which must be performed in a way that prevents spillage of the cyst contents. Various techniques have been developed for this. When possible, the patients should first be treated with albendazole, in an attempt to kill the larvae within the cyst.

Multiceps multiceps causes a similar disease to that caused by *Echinococcus granulosus*, affecting mainly the brain.

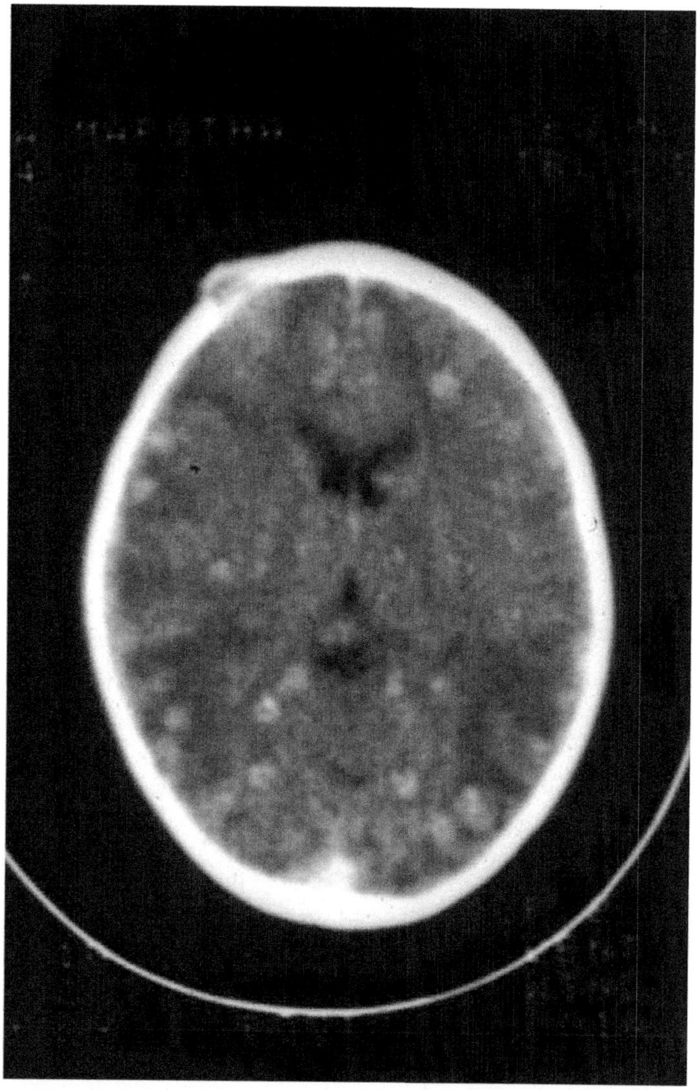

Fig. 25.37 Brain computer tomographic scan of an 8-year-old child from Honduras, who presented with a seizure. Multiple neurocysticerci are seen. Copyright © Frank E. Berkowitz. Reprinted with the permission of Cambridge University Press.

Dipylidium caninum

The adult of this worm, which grows up to 80 cm long, lives in the cat or dog. The intermediate host is the flea. Humans (usually young children) become infected by eating an infected flea.

Hymenolepis nana

The definitive host is the mouse, and humans can become infected both by eating eggs (passed in the stools of rodents or other humans), and by eating certain insects

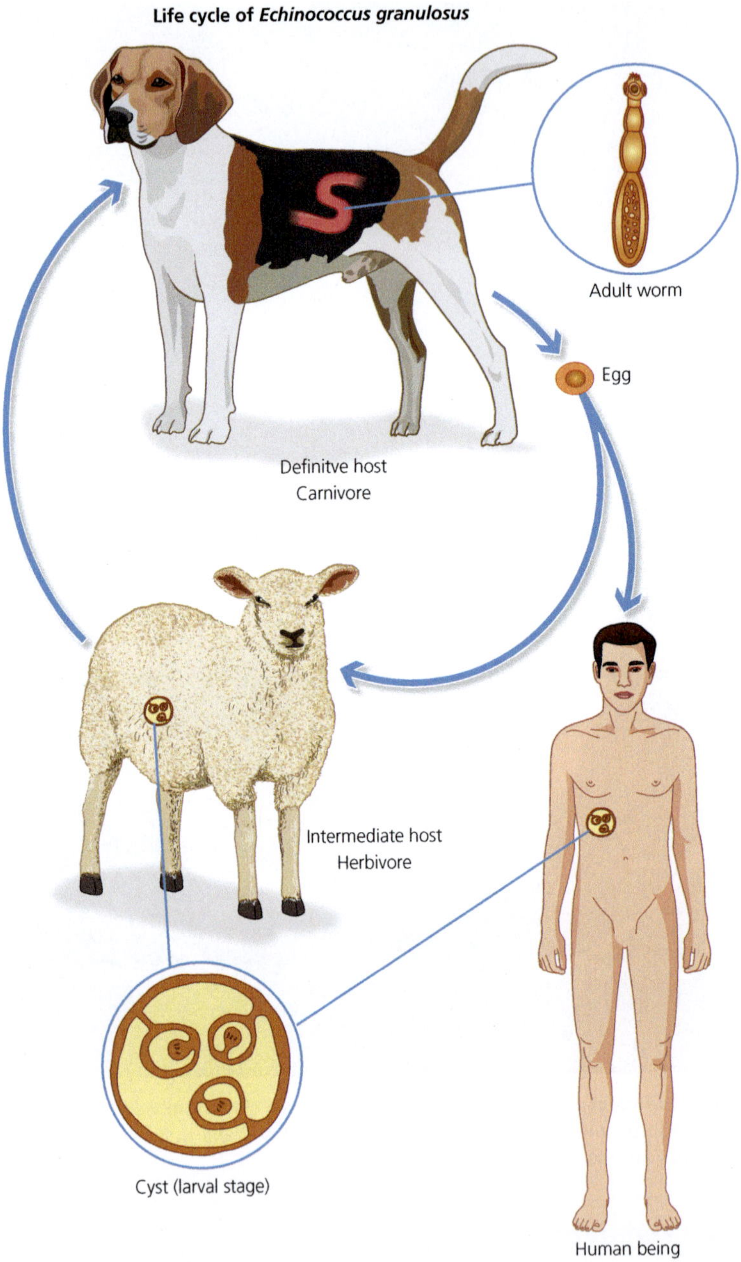

Fig. 25.38 Life cycle of *Echinococcus granulosus*. Copyright © Frank E. Berkowitz. Reprinted with the permission of Cambridge University Press.

containing cysticercoids (the intermediate stage). Autoinfection can also occur in humans, when the egg releases its embryo within the intestine.

Hymenolepis diminuta

The definitive host is the rat. Humans can become infected by eating an insect that contains the cysticercoid.

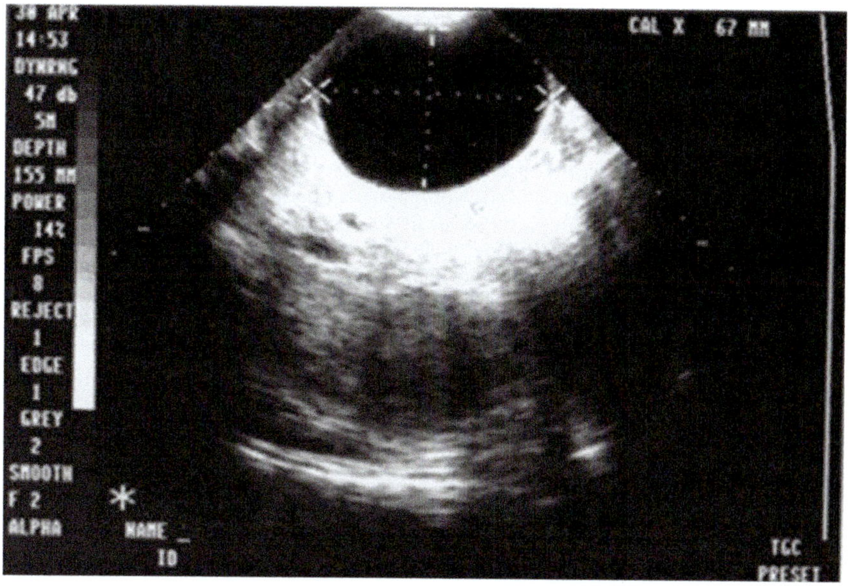

Fig. 25.39 Hydatid cyst in the liver of a child. Copyright © Frank E. Berkowitz. Reprinted with permission of Cambridge University Press.

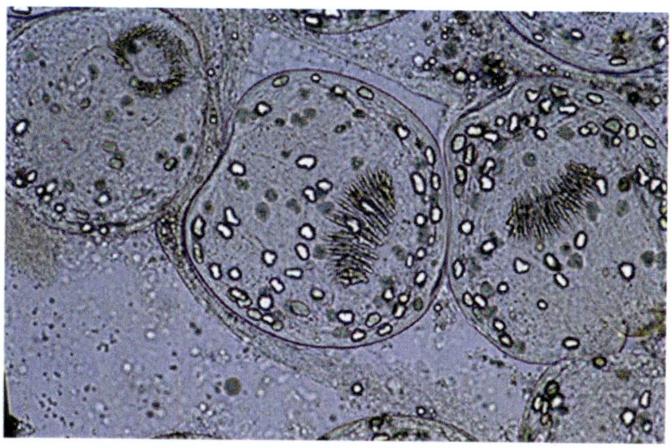

Fig. 25.40 Contents of the cyst shown above, revealing protoscolices. Copyright © Frank E. Berkowitz. Reprinted with permission of Cambridge University Press.

Further reading

Barry M (2007) The tail end of guinea worm – global eradication without a drug or a vaccine. N Engl J Med 356: 2561–4.

Bethany J, Brooker S, Albonico M, et al. (2006) Soil-transmitted helminth infections: ascariasis, trichuriasis, and hookworm. Lancet 367: 1521–32.

Cavalcanti MG, Silva LF, Peralta RHS, et al. (2013) Schistosomiasis in areas of low endemicity: a new era in diagnosis. Trends Parasitol 29: 75–82.

Colley DG, Bustinduy AL, Secor WE, et al. (2014) Human schistosomiasis. Lancet 383: 2253–64.

Furst T, Sayasone S, Odermatt P, et al. (2012) Manifestation, diagnosis, and management of foodborne trematodiasis. BMJ 344: 1–11.

Gryseels B, Polman K, Clerinx J, et al. (2006) Human schistosomiasis. Lancet 368: 1106–18.
Hung T, Neafie RC, Mackenzie IRA. (2012) Unusual Baylisascaris procyonis infection in elderly person, British Columbia, Canada. Emerg Infect Dis 18: 341.
John DT, Petri WA Jr (2006) *Markell and Voge's Medical Parasitology*. St Louis: Saunders Elsevier.
Kappgoda S, Singh U, Blackburn BG (2011) Antiparasitic therapy. Mayo Clin Proc 86: 561–83.
Keiser J, Duthaler U, Utzinger J (2010) Update on the diagnosis and treatment of food-borne trematode infections. Curr Opin Infect Dis 23: 513–20.
King CH (2009) Toward the elimination of schistosomiasis. N Engl J Med 360: 106–9.
Martin C, Gavotte L (2010) The bacteria *Wolbachia* in *Filaria*, a biological Russian doll's system: new trends in antifilarial treatments. Parasite 17: 79–89.
Olsen A, van Lieshout L, Marti H, et al. (2009) Strongyloidiasis – the most neglected of the neglected tropical diseases? Trans R Soc Trop Med Hyg 103: 967–72.
Pukkila-Worley R, Nardi V, Branda JA (2014) Case 28-2014: a 39-year-old man with a rash, headache, fever, nausea, and photophobia. N Engl J Med 371: 1051–60.
Roxby AC, Gottlieb GS, Limaye AP (2009) Strongyloidiasis in transplant patients. Clin Infect Dis 49: 1411–23.
Taylor MJ, Hoerauf A, Bockarie M (2010) Lymphatic filariasis and onchocerciasis. Lancet 376: 1175–85.

CHAPTER 26
Ectoparasites

These are arthropods that can live on the host, sometimes burrowing into the skin. Although many different arthropods (flies, fleas, mites, ticks, lice, mosquitoes) can transmit infectious agents, including viruses, bacteria, and parasites, in this section only arthropods that are considered a problem in themselves are discussed. The exclusion of ticks and mosquitoes is arbitrary.

Flies

Fly infestation is called myiasis. It may be considered in the following broad categories.

- The vertebrate host is a necessary element in the fly's life cycle. *Dermatobia hominis*: the adult female lays eggs on a mosquito. When the mosquito attacks a vertebrate host, the fly eggs are brushed on to the host's skin. The larvae (Fig. 26.1) hatch and burrow into the skin, producing a nodule. *Cordylobia anthropophaga* (Tumbu fly): the fly lays its eggs on soil or linen. The eggs hatch and burrow into the skin of a host in contact with the linen, producing a local area of inflammation. Others include bot-flies and screwworms.
- The fly usually lays its eggs in decaying organic matter, but can do so in wounds, resulting in an opportunistic infestation by the larvae. Various bot flies and blow flies are included in this group.

Fleas

The sand (Chigoe) flea (*Tunga penetrans*) causes tungiasis. It spread from America to Africa, and it is now endemic in many tropical areas. The female flea penetrates the host's epidermis, often on the nailbeds of the toes, where it sucks blood. It becomes rotund and lays thousands of eggs. It causes very painful lesions, which are susceptible to secondary bacterial infection (Fig. 26.2). Treatment is surgical removal of the flea.

Practical Medical Microbiology for Clinicians, First Edition. Frank E. Berkowitz and Robert C. Jerris.

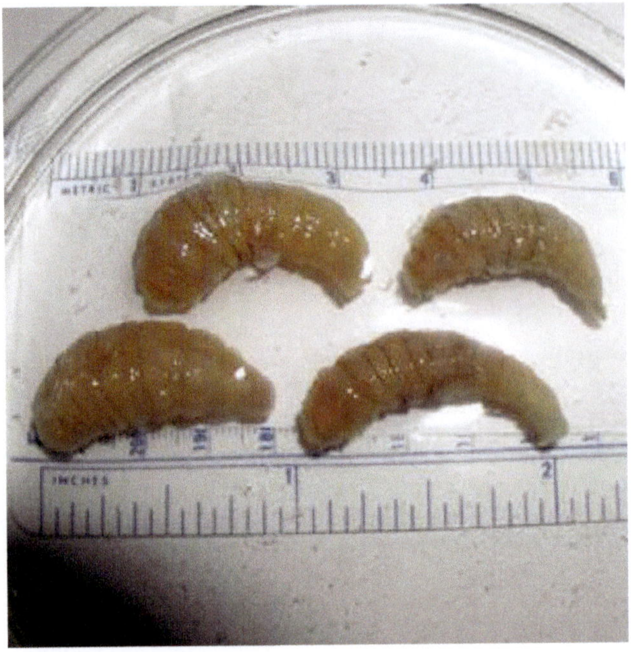

Fig. 26.1 Dermatobia larvae removed from the skin. Courtesy of DPDx, CDC.

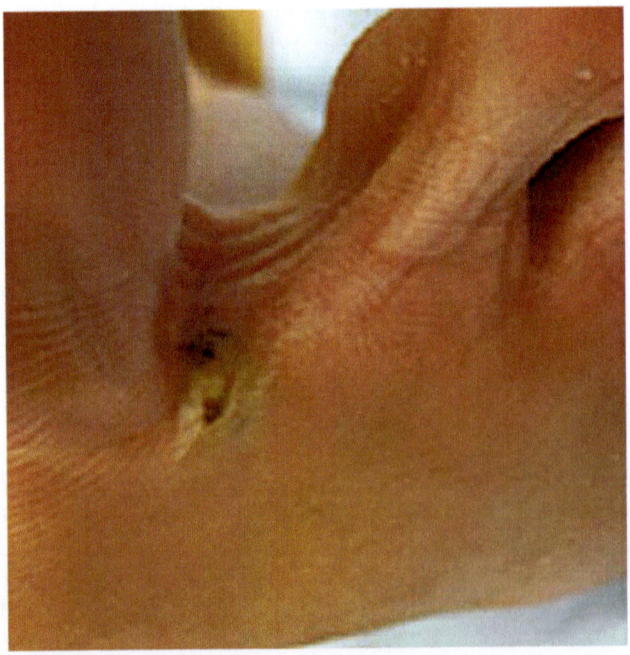

Fig. 26.2 Skin lesion caused by *Tunga penetrans*. Courtesy of DPDx, CDC.

Mites

Sarcoptes scabiei

This causes a common skin infection called scabies. It is spread from person to person by direct contact; fomites, such as clothes and bed linen, play a minor role in transmission. The female mite burrows into the stratum corneum of the skin, where she lays eggs. The tissue around the burrows becomes inflamed, causing itchy papules (Fig. 26.3). The areas usually affected are the web spaces and lateral edges of the hands, the ventral surfaces of the forearms, the axillary folds, and the breasts. In infants, the affected areas are less specific. A severe form (crusted or Norwegian scabies) is characterized by extensive areas of inflammation with eczematization of the skin, and by the presence of thousands of mites. Although the diagnosis is usually made clinically (which may not always be reliable), it can be confirmed by visualization of the female mite, or eggs, in scrapings of the stratum corneum. These are placed in mineral oil under a cover slip and examined microscopically under low power (Fig. 26.4).

Many topical forms of treatment are available. These include permethrin, malathione, lindane, crotomiton, sulfur in petrolatum, and benzyl benzoate. Oral ivermectin can also be used.

Other mites

Chigger mites (*Eutrombicula alfredduges* and *E. splendens*): these mites of animals occasionally attach to the skin of humans, from whom they suck tissue juice, and cause an itchy dermatitis.

Demodex folliculorum and *D. brevis* are mites that live, respectively, in hair follicles and sebaceous glands of the skin, particularly of the face, scalp, and upper trunk. Whether they cause disease is controversial.

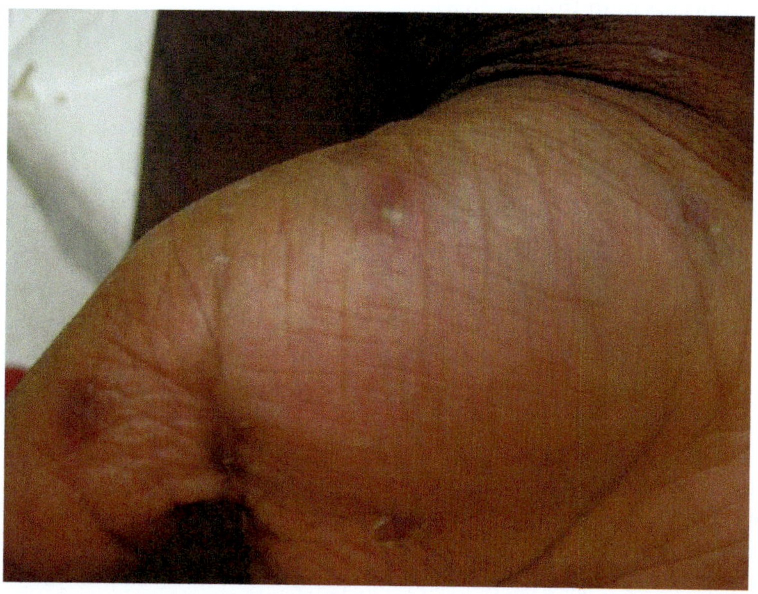

Fig. 26.3 Hand of an 8-year-old boy with scabies.

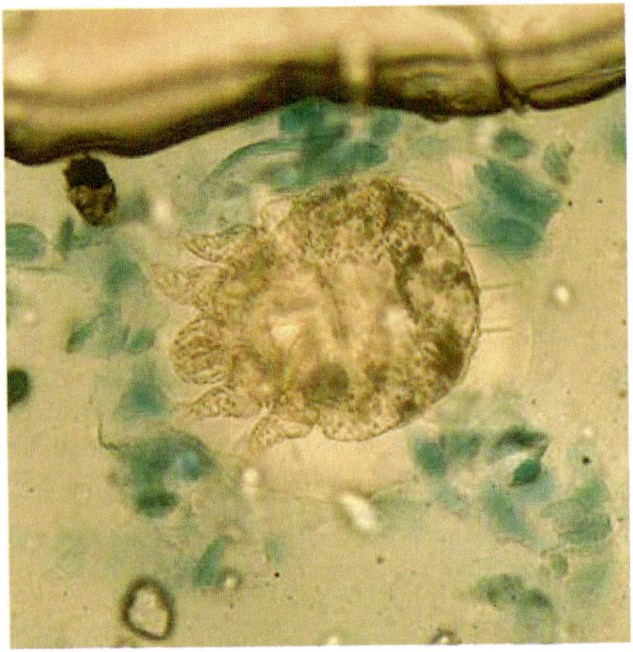

Fig. 26.4 *Sarcoptes scabiei* in a skin scraping. Courtesy of DPDx, CDC.

Bed bugs

Cimex lenticularius (Fig. 26.5) has a worldwide distribution, and has reemerged as a widespread nuisance in the USA. It lives in the crevices of beds and feeds on human blood. It causes skin lesions, often in groups, due to a reaction to its saliva (Fig. 26.6). *C. hemipterus* occurs in in tropical areas.

Lice

These are wingless insects that live on host blood. Three species affect humans: *Pediculus humanus capitis* (the head louse), *P. humanus corporis* (the body louse), and *Phthirus pubis* (the crab louse).

Pediculus humanus capitis (Fig. 26.7) is the most important louse in developed countries. It lives on the scalp, sucking blood and causing itching. It cannot survive without a meal for more than 3 days. It is transmitted by hair-to-hair contact and can be transmitted by fomites (e.g. shared hats). It lays its eggs (nits) on hair shafts. These hatch after 7–12 days and the emerging instars (nymphs) take 8–10 days to mature. These durations determine the optimal intervals between two treatments with topical agents that kill the adults but do not affect the eggs. The diagnosis is made by direct visualization of the adult or nits (which must be differentiated from hair casts). Treatment consists of topical application of shampoos containing malathione, lindane or pyrethroids. Oral ivermectin can also be used.

Pediculus humanus corporis (Fig. 26.8) is very important in circumstances of crowding and during wars. It lays its eggs in clothing. It is very important as a vector of three pathogens: *Rickettsia prowazeki*, the cause of epidemic typhus (see Chapter 15),

Fig. 26.5 *Cimex lenticularius*. Courtesy of PHIL, CDC.

Fig. 26.6 Skin lesion produced by a bed bug bite.

Bartonella quintana, the cause of trench fever (see Chapter 13), and *Borrelia recurrentis*, one of the causes of relapsing fever (see Chapter 16). Because they live in the clothing, not on the host, elimination should be directed at cleaning the clothing.

Phthirus pubis (Fig. 26.9) lives in pubic hair and sometimes eyelashes. It can therefore be a sexually transmitted infection. Its bite causes blue macules (maculae cerulae). Treatment consists of physical removal, shampoos used in the pubic hair, and, in the case of eyelid infection, occlusion with petroleum jelly.

Fig. 26.7 *Pediculus humanus capitis,* the human head louse. Courtesy of PHIL, CDC.

Fig. 26.8 *Pediculus humanus corporis,* the human body louse. Courtesy of PHIL, CDC.

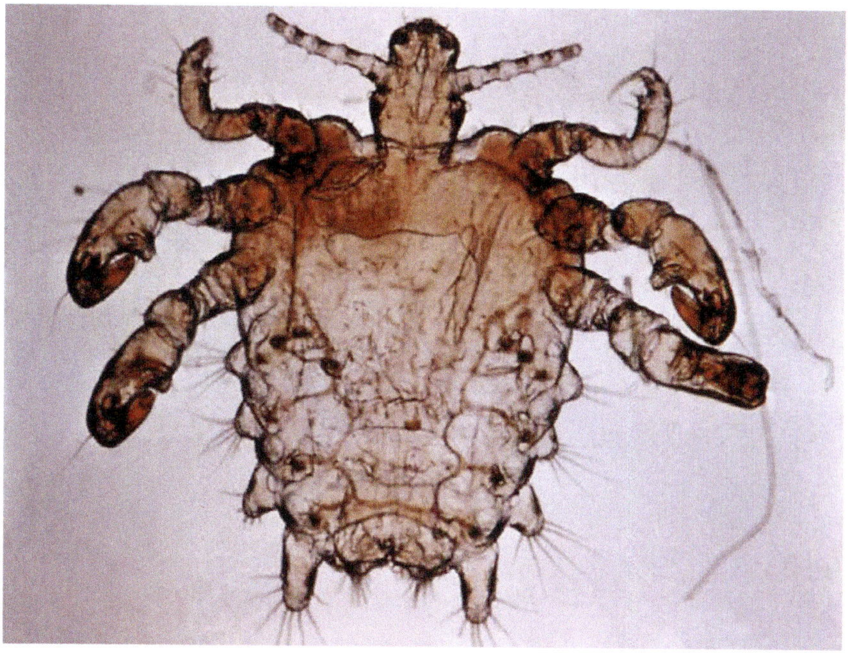

Fig. 26.9 *Phthirus pubis*, the pubic louse. Courtesy of PHIL, CDC.

Further reading

Hay R (2010) *Demodex* and skin infection: fact or fiction. Curr Opin Infect Dis 23: 103–5.
Nordlund JJ (2009) Cutaneous ectoparasites. Dermatol Ther 22: 503–17.
Shmidt E, Levitt J (2012) Dermatologic infestations. Int J Dermatol 512: 131–41.

SECTION VI
Clinical cases

CHAPTER 27

Cases

Case 1: A teenage boy with fever, neutropenia, and skin lesions.
Case 2: A 10-year-old Nigerian boy with fever.
Case 3: A 70-year-man with cough and fever.
Case 4: A 20-year-old woman with fever and back pain.
Case 5: A teenage boy with an arm abscess after trauma.
Case 6: A 4-year-old boy with meningitis after trauma.
Case 7: A newborn with fever and poor feeding.
Case 8: Evaluation of a new influenza diagnostic test.
Case 9: A 4-year-old boy with fever, weakness, and hallucinations.
Case 10: A teenage boy with fever and breathing difficulty.
Case 11: A 7-month-old girl with fever, cough, and red eyes.
Case 12: A 17-year-old girl with hemoptysis.
Case 13: A newborn with fever and seizures.
Case 14: A 6-week-old boy with runny nose and cough.
Case 15: A stabbed abattoir worker.
Case 16: A 2-year-old boy with fever and diarrhea.
Case 17: A teenage boy with an abscess after trauma in a river.
Case 18: A 1-year-old boy with fever and limp.
Case 19: A neutropenic boy with skin lesions.
Case 20: An infant with an itchy rash.
Case 21: A 4-year-old heart transplant recipient with pneumonia.
Case 22: A 2-year-old boy with a neck swelling.
Case 23: A 2-year-old girl with fever, headache, and seizure.
Case 24: A neonate with a rash and hepatosplenomegaly.
Case 25: A 4-month-old infant with a healthcare-associated infection.
Case 26: A 20-year-old girl with a Baclofen pump and meningitis.
Case 27: A 6-year-old girl with cellulitis after a cat bite.
Case 28: A 20-year-old man with jaundice and fever after travel.

Practical Medical Microbiology for Clinicians, First Edition. Frank E. Berkowitz and Robert C. Jerris.

Case 29: A 19-year-old man with fever and sore throat.
Case 30: A 16-year-old boy with cystic fibrosis and increased cough.
Case 31: A college student with a vaginal discharge.

The reader should contemplate the case before reading the discussion about diagnostic testing and management.

Case 1

A teenage boy, being treated for leukemia, developed fever and difficulty breathing. He was neutropenic. On examination, he was very ill-appearing and he had fever, poor perfusion, marked respiratory difficulty, and skin lesions (Fig. 27.1), but no other signs to indicate a focus of infection.

He was given a bolus of intravenous fluid and vasopressors, and placed on a ventilator.

Blood culture, urine culture, and a chest radiograph were performed, and broad-spectrum antimicrobial therapy was initiated with piperacillin/tazobactam.

The skin lesions probably represent metastatic infection, and provide an excellent opportunity to make an immediate microbiologic diagnosis. A lesion should be biopsied immediately, and the material submitted to the laboratory for histology (very important for detecting molds), smears, and culture. The material for culture should be minced rather than ground up, as mucorales are delicate and can be damaged by grinding. Smears should be made and stained immediately with Gram stain (to detect bacteria and yeasts), and PAS or a silver stain, to detect molds. While preparations for biopsy are being made, a lesion can be aspirated, and the aspirate submitted to the same procedures (other than histology).

Fig. 27.2 shows the Gram stain of an aspirate. This reveals multiple yeasts. Cultures of the aspirate as well as the blood subsequently grew out *Candida tropicalis*.

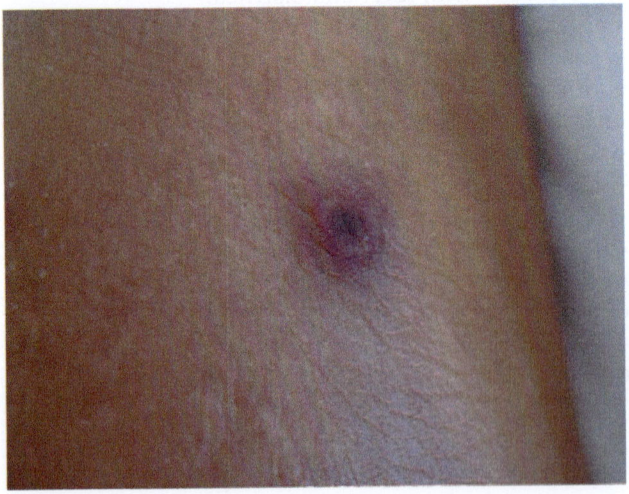

Fig. 27.1 Skin lesions in a febrile, neutropenic boy.

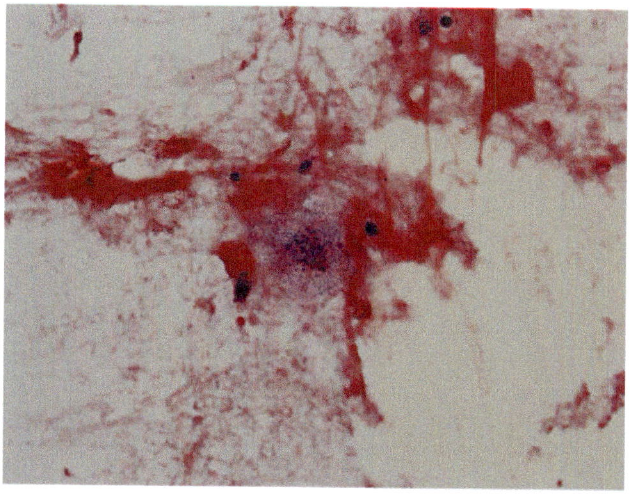

Fig. 27.2 Gram-stained smear from an aspirate of one of the skin lesions.

Case 2

A 10-year-old boy presented with fever for 3 days, starting 1 week after arriving in the USA from Nigeria. He had had some rigors with fever, but there was no cough, no urinary symptoms, nor diarrhea, but there was mild vague abdominal pain. On examination, he was ill-appearing, with a temperature of 39.5°C. He was alert and perfusing well. There was no pallor nor jaundice, and the rest of his examination was normal. In particular, the abdomen was soft and non-tender, and there was no enlargement of liver or spleen.

What is the differential diagnosis, and what laboratory tests would you use to arrive at a diagnosis?

Differential diagnosis

- Viral infections: influenza; hemorrhagic fever, including yellow fever and Lassa fever
- Bacterial infections: typhoid fever, meningococcal infection
- Parasitic infection: malaria

While tests are being performed, he should be placed in isolation. The greatest urgency is in diagnosing conditions that have the potential for causing severe morbidity or mortality, or that have significant public health importance due to their contagiousness.

Diagnostic tests

- Thick and thin smears for malaria
- Coagulation profile and liver transaminases to screen for hemorrhagic fevers
- Blood culture, which will detect most cases of typhoid fever and meningococcal infection

The thin blood smear is shown in Figs 27.3 and 27.4. What is the specific diagnosis?

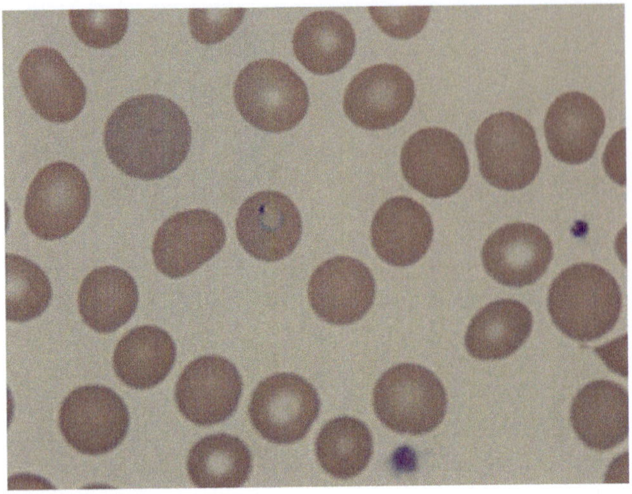

Fig. 27.3 Giemsa-stained thin blood smear.

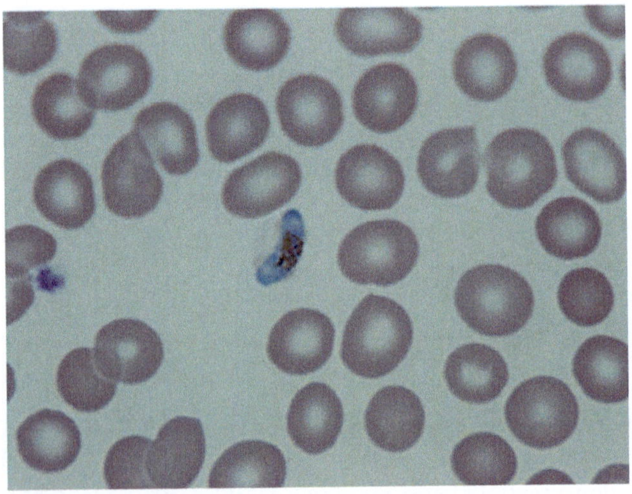

Fig. 27.4 Giemsa-stained thin blood smear.

Answer

Fig. 27.3 shows a single ring form within an erythrocyte. This is diagnostic of a plasmodium infection (malaria); however, this alone is not adequate for making a species diagnosis. Fig. 27.4 shows a sickle-shaped gametocyte. This is pathognomonic of *Plasmodium falciparum* and is the source of its name (*falx* is the Latin word for sickle).

Case 3 (hypothetical)

A 70-year-old man, who had been in good health, presented with cough and fever for 3 days. He sometimes looked after his 6-month-old great-grandson, who attended day care. On examination, he was ill-appearing, with a temperature of 39°C, a heart rate of 100/minute, and a respiratory rate of 40/minute. He had mild intercostal retractions, and there was dullness to percussion and decreased breath sounds, which were tubular, over his left lower lobe. He was coughing up blood-tinged, purulent sputum.

What is the likely diagnosis, and how would you make a microbiologic diagnosis?

He has evidence of lobar pneumonia, and he has a risk factor of exposure to *Streptococcus pneumoniae* from his great-grandson. However, one should consider other causes of community-acquired pneumonia, such as *Mycoplasma pneumoniae*, *Chlamydophila pneumoniae*, and *Legionella pneumophila*.

The sputum should be Gram stained, and cultured on blood and chocolate agar. If no bacteria are seen on the smear, the non-pneumococcal causes become more likely. For the diagnosis of mycoplasma, chlamydophila, and legionella infection, serologic tests can be used. A PCR test, performed on a respiratory specimen, can also be used for diagnosing mycoplasma infection. In addition, for the diagnosis of legionella infection, sputum can be cultured on charcoal yeast extract agar and submitted to direct immunofluorescence testing. A urinary antigen test can detect *Legionella pneumophila* type 1, the most common cause of legionellosis.

Case 4

A 20-year-old woman presented with fever and back pain for 2 days. On examination, she was ill-appearing, febrile, and had mild abdominal tenderness and severe costovertebral angle tenderness on the right. A clinical diagnosis of pyelonephritis was made.

How would you confirm this, and make a microbiologic diagnosis?

Urinalysis

On urine dipstick, the presence of leukocyte esterase and nitrites (produced by members of the Enterobacteriaceae) suggest the diagnosis of a urinary tract infection. This is further supported by the presence of leukocytes (more than 10/hpf) and bacteria on microscopy. A Gram stain of the urine showing bacteria also strongly supports this diagnosis.

The urine should be cultured on MacConkey agar and blood agar plates. The presence of pink colonies on the MacConkey agar indicates a lactose-fermenting organism, such as *Escherichia coli*, *Klebsiella* spp., or *Enterobacter* spp. Bacteria swarming all over the agar plate indicate *Proteus mirabilis* or *P. vulgaris* infection.

Case 5

A teenage boy injured his arm on a wooden post. Two days later he presented with arm pain and an abscess at the site of the injury, which was draining pus.

How would you determine the microbial cause of the infection?

Answer

The pus should be submitted to the laboratory for inoculation on to blood and chocolate agar, and for a smear to be Gram stained. A Gram stain of the pus is shown in Fig. 27.5. This shows Gram-positive cocci in chains, indicating streptococci. The most likely streptococcus in this situation is *Streptococcus pyogenes*.

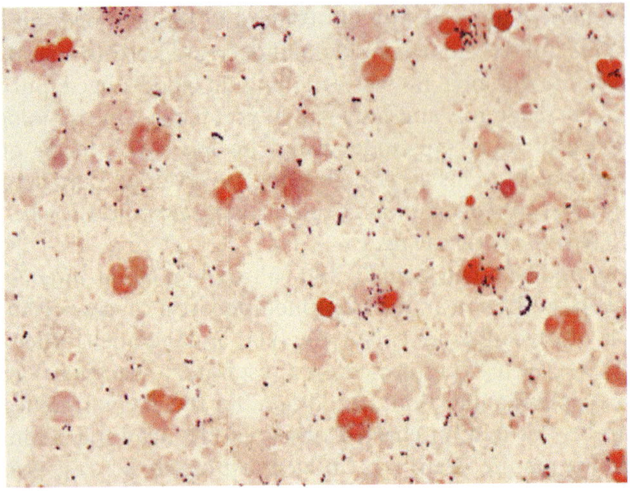

Fig. 27.5 Gram-stained smear of pus.

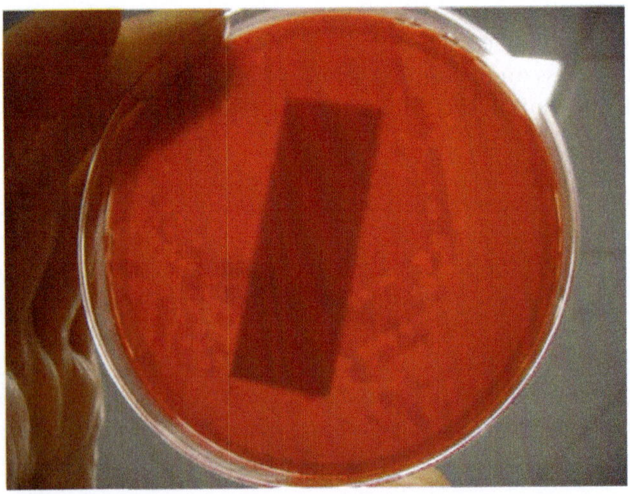

Fig. 27.6 Culture of the pus on blood agar.

The culture is shown in Fig. 27.6. This shows small gray colonies with a surrounding zone of complete (β) hemolysis.

The species of streptococcus can be determined by (i) a bacitracin test (to which *S. pyogenes* is susceptible, so that a zone of inhibition is seen) or (ii) agglutination with specific antiserum, which, in this case, confirmed the species as *S. pyogenes* (group A).

He was treated successfully with penicillin.

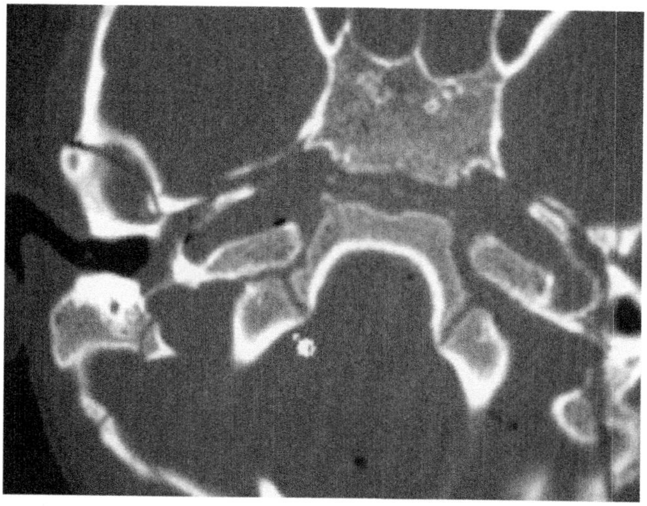

Fig. 27.7 CT scan showing basilar skull fractures.

Case 6

A 4-year-old child sustained a head injury with a basilar skull fracture (Fig. 27.7). At that time, his level of consciousness was normal, as was the rest of his neurologic examination. A few days later, he developed fever and a severe headache. Examination revealed neck stiffness and obtundation. A diagnosis of meningitis complicating the fracture was made and cerebrospinal fluid was obtained, revealing cloudy fluid.

What tests should be performed on the CSF?

Answer

- Cell count: more than 10 cells per mm^3 constitutes a pleocytosis.
- Measurement of protein and glucose concentrations: a protein concentration of greater than 30–40 mg/dl is abnormal; a glucose concentration less than 50% of the simultaneous blood glucose concentration is abnormal, unless hyperglycemia is present.
- An aliquot of the specimen should be centrifuged, and part of the pellet smeared on a slide and Gram-stained; the rest of the pellet should be cultured for bacteria on blood and chocolate agars.

The CSF showed the following: 62,000 leukocytes/mm^3, 90% neutrophils, a protein concentration of 199 mg/dL, and a glucose concentration of <20 mg/dL. The Gram stain is shown in Fig. 27.8.

The culture showed colonies growing on the chocolate agar but not the blood agar. Gram stain of these colonies revealed small Gram-negative rods. What is the likely identity of the organism?

Haemophilus influenzae.

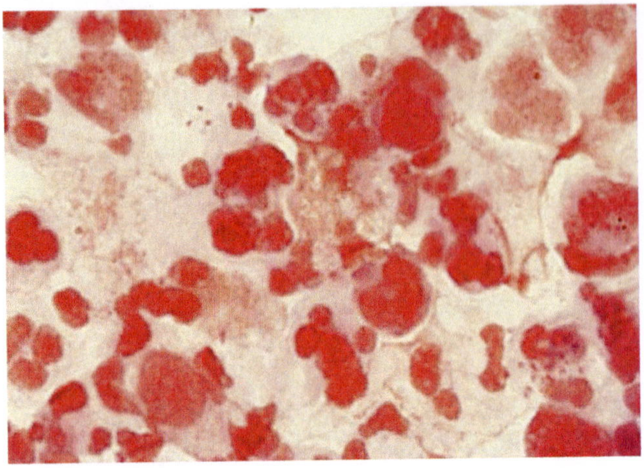

Fig. 27.8 Gram stain of CSF, showing small, pleomorphic Gram-negative rods.

Case 7

A 3-day-old boy presented with fever and poor feeding. Examination revealed a floppy infant, poorly responsive, and perfusing poorly. He was resuscitated with intravenous fluid, a blood culture was taken and antimicrobial therapy with cefotaxime and ampicillin was initiated. After 10 hours, the blood culture became positive, the Gram stain showing Gram-negative rods. How would you identify the organism?

Fluid from the blood culture bottle was inoculated on to blood agar, chocolate agar, and MacConkey agar. The following day, the blood and chocolate plates were growing large grayish colonies, while those on the MacConkey plate were pink. What can you conclude at this point?

This indicates a lactose-fermenting member of the Enterobacteriaceae, such as *E. coli, Klebsiella* spp. or *Enterobacter* spp. Biochemical tests can be used for further identification of the organism.

What is the next piece of information that you want?

Answer

The antimicrobial susceptibilities, which will take another day to obtain.

The identification was *E. coli*, and the antimicrobial susceptibility test results were as shown in Table 27.1.

What can you conclude about this organism regarding its susceptibilities to β-lactam drugs?

This is clearly resistant to multiple β-lactam antibiotics. The fact that it is, however, susceptible to ticarcillin/clavulanic acid, a β-lactam/β-lactam inhibitor combination, indicates that it is what is called an "extended-spectrum β-lactamase (ESBL) producer." Whether the ticarcillin/clavulanic acid is suitable for therapy is controversial. If meningitis is present, which could be the case in a newborn, this combination should not be used, because clavulanic acid does not enter the CSF well.

Meropenem would be the optimal treatment. Note that the laboratory can do additional tests to confirm the "ESBL" designation.

Table 27.1 Antibiotic susceptibilities of the *E. coli* isolate.

Antibiotic	Minimal inhibitory concentration (MIC) (μg/mL)
Ampicillin	>4
Ceftriaxone	>4
Ceftazidime	>8
Ticarcillin/clavulanic acid	<16
Meropenem	<2
Gentamicin	>8
Amikacin	<2
Trimethoprim/ sulfamethoxazole	<2/38
Ciprofloxacin	<2

Table 27.2 Data for the PCR test and rapid test.

		PCR test	
		Positive	**Negative**
Rapid test	Positive	35	20
	Negative	15	30

Case 8 (hypothetical)

There is a pandemic of a new influenza A virus, influenza A H3N2 Sakhalin. Your reference laboratory has established a PCR test that can be performed on pharyngeal swabs. It is considered the gold standard, but it cannot be performed rapidly. Your hospital laboratory has developed a rapid antigen test, which it is offering for patients, and which you would like to use. When the laboratory tested 100 specimens that had also been tested with the PCR method, the results were as shown in Table 27.2.

1 If your patient had a positive rapid test, what would be the probability that she actually had the infection? This is the positive predictive value.

2 If she had a negative rapid test, what would be the probability that she did not have the infection? This is the negative predictive value.

Answer

Table 27.3 shows how sensitivity, specificity, positive predictive value, and negative predictive value are calculated.

- Sensitivity = true positives/true positives + false negatives = a/a + c
- Specificity = true negatives/true negatives + false positives = d/d + b
- Positive predictive value = true positives/true positives+ false positives = a/a + b
- Negative predictive value = true negatives/true negatives + false negatives = d/d + c

1 35/35 + 20 = 35/55 = 0.64

2 30/30 + 15 = 30/45 = 0.67

This is not a good test!

Table 27.3 How sensitivity, specificity, positive predictive value, and negative predictive value are calculated.

		PCR test	
		Positive	**Negative**
Rapid test	Positive	a	b
	Negative	c	d

Cells: a, true positives; b, false positives; c, false negatives; d, true negatives.

Case 9

A 4-year-old boy presented in August 2007 with a history of fever, anorexia, headache, photophobia, and weakness for 2 days. On the day of presentation, he had developed an unsteady gait and visual hallucinations. One week earlier, he had had a sore throat, but this had resolved. He lived in Atlanta. He had not been in the woods, but he played outside in the dirt, and had been bitten by mosquitoes. There had been no exposure to cats, bats, or caves, and there had not been any travel. His father had filled up a plastic pool for paddling, which he had emptied after use.

On examination, he was ill-appearing. He was febrile, and the rest of his vital signs were normal. The main abnormalities were in the neurologic examination. He was conscious but confused. His tone and strength were normal, and he had normal deep tendon reflexes and down-going plantar responses. The cranial nerves were normal. He had an ataxic gait. During the examination, he exhibited myoclonic jerks.

What do you think is the problem and what would you do?

Clinically, he has evidence of an infection (fever) and a brain disorder. A clinical differential diagnosis of encephalitis, meningitis or meningoencephalitis was made.

A computer tomographic scan was normal.

Cerebrospinal fluid revealed the following: 62 leukocytes per mm^3, with 44% neutrophils and 50% lymphocytes and 6% monocytes; the glucose concentration was 80 mg/dL and the protein concentration was 46 mg/dL; the Gram stain was negative.

This confirmed a diagnosis of meningitis or meningoencephalitis.

Management should consist of:

- supportive care that is indicated (not discussed here)
- therapy for possible causes of this illness, for which treatment is available
- diagnostic testing to confirm or exclude, as far as possible, these diseases.

What is your microbial differential diagnosis, and how would you manage him further?

Answer

- Bacterial meningitis (due to *Streptococcus pneumoniae* or *Neisseria meningitidis*). Given the lack of a preponderance of neutrophils, a normal glucose concentration, and only a slightly elevated protein concentration in the CSF, and a negative Gram stain, acute bacterial meningitis is unlikely.
- Viral meningitis – due to enterovirus, arbovirus; the abnormality of brain function indicates that if this is viral in nature, it is meningoencephalitis or encephalitis rather than only meningitis.

Table 27.4 Results of arbovirus testing.

Virus	Result	Test type
Eastern equine encephalitis	<1:16	Titer
Western equine encephalitis	<1:16	Titer
California encephalitis	<1:16	Titer
St Louis encephalitis	<1:16	Titer
West Nile virus IgG	<1.30 (negative <0.9)	Optical density
West Nile virus IgM	4.13 (negative <0.90)	Optical density

- Encephalitis – due to *Herpes simplex* virus, enterovirus, arbovirus, *Rickettsia rickettsii*, or *Ehrlichia* spp.

He was treated initially with antibacterial agents (vancomycin, ceftriaxone, and doxycycline) and acyclovir, and he improved markedly over the following few days.

Bacterial cultures and HSV DNA PCR were negative. The negative bacterial cultures suggest strongly that this was not a case of acute bacterial meningitis, and the negative HSV PCR suggests that this was not a case of HSV encephalitis.

On discharge on the fifth day he was still ataxic. When he was seen 3 days later, his gait had returned to normal.

There are many viral causes of encephalitis, and the yields of diagnostic testing are low and disappointing. However, especially in the summer, it is worth pursuing the possibility of an arbovirus infection. This is usually accomplished by antibody testing of blood and, if available, of CSF.

On the day of discharge, antibody titers, obtained from blood, were as shown in Table 27.4.

A diagnosis of West Nile virus encephalitis was made.

Case 10

A 15-year-old boy presented with breathing difficulty, after having a sore throat and fever for 2 weeks. On presentation, he was febrile, in respiratory distress and hypoxic, and hypotensive. In addition, he had left-sided neck tenderness, but no palpable mass. Examination of his mouth, throat, and ears was normal. He was tachypneic but the lungs were otherwise normal; there was no skin disease, and the rest of the examination was normal.

1 What is your differential diagnosis?
2 What would you do?

Answers

1 Pneumonia with bacteremia, and septic shock
2 Lemierre's syndrome

Possible organisms include:

1 Pneumonia with septic shock: *Staphylococcus aureus*, *Streptococcus pneumoniae*
2 Lemierre's syndrome: oral anaerobes, in particular *Fusobacterium necrophorum*

Management

- Oxygen and fluid resuscitation.
- Blood culture in aerobic **and** anaerobic bottles.
- Antimicrobial therapy with a drug(s) active against the above organisms, e.g. vancomycin + metronidazole.
- Image his neck to look for a thrombus in his jugular vein or one of its tributaries.

He was treated as described above, and a neck CT was normal.

Blood culture grew out *Porphyromonas asaccharolytica*, an oral anaerobe, susceptible to metronidazole.

Case 11

A 7-month-old girl presented with a history of noisy breathing and was admitted to the hospital with the diagnosis of croup. The illness had started 7 days earlier with fever, cough, red, runny eyes, and runny nose. She was treated with antibiotics for otitis media. Two days later, a rash had developed, starting on her face and progressing to the rest of her body. Fever persisted for 5 more days, and the cough continued. She had traveled 5 days before the onset of the illness and returned home while ill. On examination, she was miserable but in no clinical distress. Her conjunctivae were mildly inflamed, her buccal mucosa was slightly red, the lungs were normal, and the skin is shown in Figs 27.9 and 27.10. Although the rest of her examination was normal, she was coughing frequently.

1 What is your differential diagnosis?
2 What would you do?

Answers

Differential diagnosis: measles, adenovirus infection, Kawasaki disease.

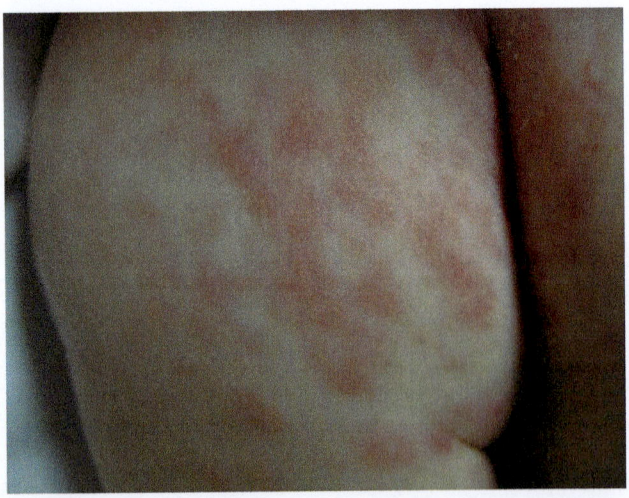

Fig. 27.9 Rash on the child's arm.

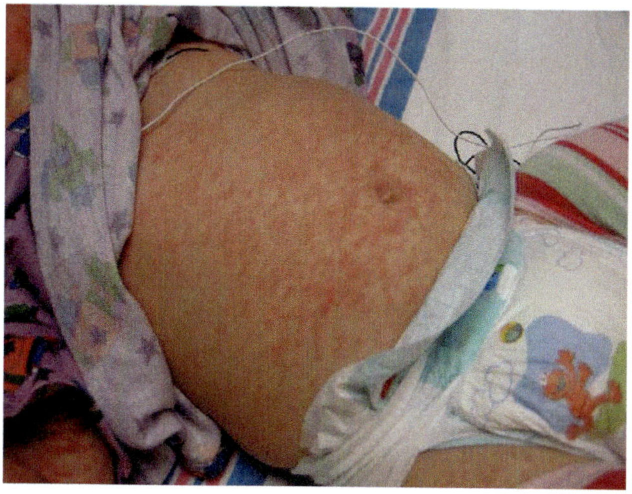

Fig. 27.10 Rash on the child's trunk.

Management

- Airborne isolation.
- Urgently confirm or exclude a diagnosis of measles, which is of great public health importance, by testing serum for IgM antibodies, or by immunofluorescent staining or PCR testing on a pharyngeal or urine sediment specimen.

The following morning, the diagnosis of measles was confirmed by a positive measles IgM test. The state health department was alerted and extensive contact tracing was initiated.

Case 12

A 17-year-old girl presented with a history of a single episode of hemoptysis. She had been cured of acute lymphoblastic leukemia 10 years earlier, and previously she had been treated with INH for latent tuberculosis infection (she had been a household contact of a proven case, and had had an ulcerating tuberculin skin test). She had not completed the INH therapy (she had taken this for only about 3 months). She had complained of right-sided chest pain for many months, and was treated for pneumonia with amoxicillin 10 days before this presentation. The chest X-ray at that time showed right upper lobe consolidation with a cavity (Fig. 27.11).

Examination was normal.

1 What is your differential diagnosis?
2 What would you do?

Answers

Differential diagnosis: infectious diseases: tuberculosis, histoplasmosis, acute bacterial pneumonia (but very unlikely due to chronic symptoms, and lack of fever); non-infectious diseases, such as granulomatosis with polyangiitis.

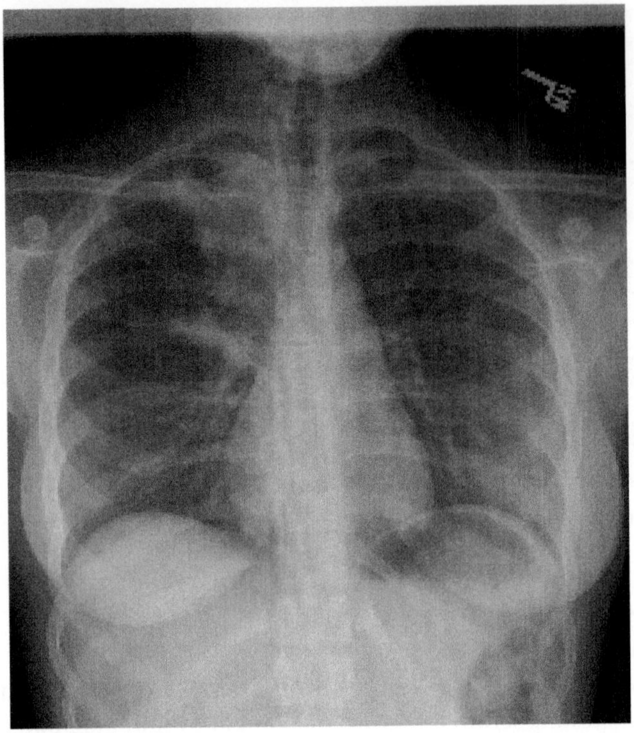

Fig. 27.11 Chest X-ray showing right upper lobe consolidation with cavitation.

Management

- Airborne isolation while in hospital.
- Collect sputum specimens for staining for pyogenic bacteria (Gram stain), for acid-fast bacteria (with fluorochrome stain), for yeast and other fungi (Calcofluor white stain), and culture for pyogenic bacteria, mycobacteria, and fungi.

Once three sputum specimens have been collected, considering the high suspicion for tuberculosis, initiate antituberculous therapy.

After three sputum specimens for mycobacterial stain and culture had been obtained, she was treated with INH, rifampin, pyrazinamide, and ethambutol, by directly observed therapy.

The culture grew out fully susceptible *M. tuberculosis*.

Case 13

A 2-week-old boy presented with a history of fever and seizures. He was born by vacuum extraction after a difficult labor, resulting in bilateral cephalhematomas. On examination, he was very ill-appearing and in shock, and his temperature was 38.8°C. His anterior fontanelle was bulging, and he had bilateral cephalhematomas. He was poorly responsive and hypotonic.

What is your initial assessment and what would you do?

Initial assessment

- Shock, probably due to:
  - sepsis
  - meningitis
- Possible metabolic disturbance, especially hypoglycemia

Initial management

- Fluid resuscitation
- Endotracheal intubation
- Check blood glucose
- Diagnostic tests: blood culture
- Initiate antimicrobial therapy, with drugs that are active against the most likely pathogens in this age group, namely *Streptococcus agalactiae* (group B streptococcus), *E. coli*, and *Listeria monocytogenes*. A combination of ampicillin and cefotaxime would be suitable.

Although CSF and urine specimens should, ideally, be obtained before the institution of antimicrobial therapy, this child's condition might be too unstable to permit this. CSF was, however, obtained in this patient, and the results were 1675 leukocytes per mm^3 with 90% neutrophils, and protein and glucose concentrations of 283 and <20 mg/dL respectively. The Gram stain is shown in Fig. 27.12.

- What does the CSF Gram stain reveal?
- What is the likely identity of the organism?
- How would you perform the identification?

The Gram stain shows many Gram-negative rods; the most likely Gram-negative rod causing neonatal meningitis is *Escherichia coli*. The identification is made by growing the organism on blood agar, and then performing biochemical tests on the colonies. Antimicrobial testing should also be performed on the isolate.

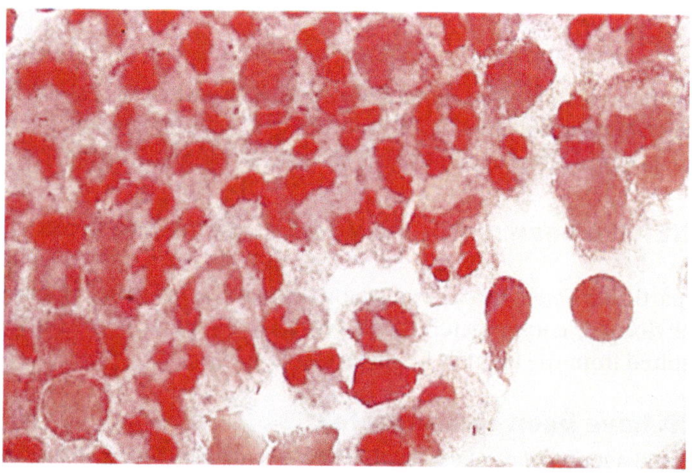

Fig. 27.12 Gram stain of the CSF.

Case 14

A 6-week-old boy presented with a history of running nose and cough for a few days; the cough was becoming severe, and associated with cyanosis. On examination, he was well-appearing and afebrile. Examination of his lungs and heart was normal. During the examination, he had an episode of coughing during which he became cyanosed and apneic for about 20 seconds. A chest X-ray showed mild peribronchial thickening, but was otherwise normal, largely excluding a pyogenic bacterial pneumonia.

What is the microbial differential diagnosis, and how would you investigate his illness?

This child likely has a respiratory tract infection that does not, according to the chest X-ray, appear to be a pyogenic pneumonia, making a bronchial or bronchiolar infection likely. Likely causes are respiratory viruses, in particular adenovirus, respiratory syncytial virus, and parainfluenza virus, and *Bordetella pertussis*.

To diagnose one of the viral pathogens, one should take a nasopharyngeal wash and submit it for respiratory viral culture in tissue culture, antigen testing, or PCR, which is the optimal method.

The value of knowing whether this is one of these viral infections lies primarily in withholding antibacterial therapy.

To diagnose *Bordetella pertussis* infection, a nasopharyngeal swab should be submitted for culture on selective medium (e.g. Regan Lowe or Bordet–Gengou medium), and PCR, which is the optimal method.

This child had a positive PCR for *Bordetella pertussis*. The diagnosis of pertussis is important for directing antimicrobial therapy (with a macrolide antibiotic) as a public health measure. At this stage of the child's illness, antibiotic will not ameliorate his condition but it will reduce his contagiousness. Chemoprophylaxis should be offered to the close contacts (family members, and healthcare staff, who might not have been protected while caring for the child).

Case 15 (hypothetical)

An abattoir worker was involved in an altercation with one of his co-workers, during which he was stabbed in the arm.

What were the infectious agents which he was at risk of acquiring?

What might have been on his skin that might cause wound infection?

Bacteria – in particular *Staphylococcus aureus* and *Streptococcus pyogenes*, but also coagulase-negative staphylococci, corynebacteria, and yeasts. In addition, he might have had an organism acquired from the animals he slaughtered.

What might have been on the knife?

- Environmental agents: *Clostridium tetani, Pseudomonas aeruginosa, Acinetobacter* spp.
- If the knife had been used to stab another person: hepatitis B and C; HIV (only if within the past few hours)

What might have been on/in the animals, which might be on the knife or on his skin?

- Bacteria: staphylococci, streptococci, including *S. suis*
- *Clostridium* spp., *Bacillus* spp., including *B. anthracis*
- *E. coli, Salmonella enterica, Campylobacter* spp.
- *Erysipelothrix rhusiopathiae*
- Protozoa: *Toxoplasma gondii*

Case 16

A few days after the Thanksgiving holiday, a 2-year-old boy presented with a 2-day history of abdominal pain, fever, vomiting, and diarrhea, which had become bloody. On examination, he was mildly ill-appearing, but not dehydrated. There was mild diffuse abdominal tenderness, and the rest of his examination was normal. There were no pets in the home, but the previous week he had visited a petting zoo, and there are tortoises at the day care he attends. His mother prepared chitterlings (pork intestines) and chicken for the Thanksgiving dinner.

1 Considering his exposure history, what might be the cause of his illness?
2 How would you identify such pathogens?

Answers

His exposures and their potential pathogens were:

- Petting zoo: *E. coli* (including enterohemorrhagic strains), *Salmonella* spp.
- Tortoises: *Salmonella* spp.
- Chicken: *Campylobacter* spp., *Salmonella* spp.
- Chitterlings: *Yersinia enterocolitica, Salmonella* spp.
- Day care: all most enteric pathogens, especially *Shigella* spp. (in addition to viral and protozoal pathogens)

The acute onset of illness, and the presence of blood in the stool, make a bacterial cause of the diarrhea very likely. This can be supported by the demonstration of leukocytes in the stool, which can be done as follows. Stool, preferably a part containing blood or pus, is smeared thinly on a microscope slide and allowed to dry. It is stained with methylene blue or a stain used for blood smears. A specimen containing many leukocytes is shown in Fig. 27.13.

Investigation of the bacterial causes of diarrhea consumes resources (both time and materials) far out of proportion to the number of cases identified, and to the value of the information, regarding influence on management, even when a bacterial pathogen is identified. This is because feces contains large numbers of bacteria that are not known to cause acute infectious diarrhea. Therefore, the bacteria of interest must be grown in such a manner that those of no interest do not grow. This requires steps using selective media (to select out those of interest), which is not required when culturing fluids that are normally sterile, such as blood. The selected bacteria can then be identified using phenotypic or molecular methods. This explains why stool cultures take longer than most other cultures.

Useful media and pathogen colonial morphology are as follows.

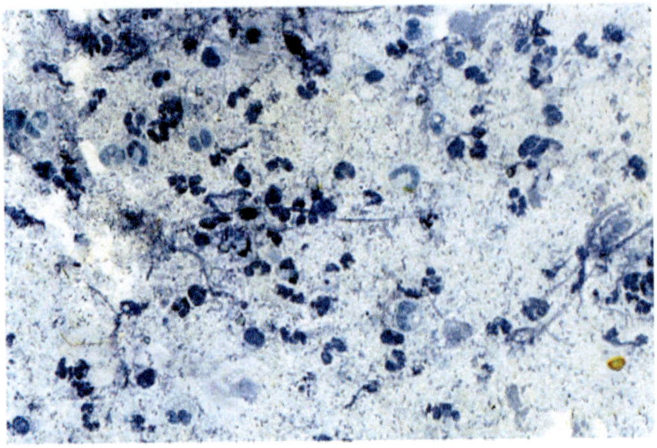

Fig. 27.13 Fecal smear, stained with methylene blue, demonstrating large numbers of leukocytes. Copyright © Frank E. Berkowitz. Reprinted with permission of Cambridge University Press.

- Blood (5% sheep) agar plate: *Aeromonas* spp./*Plesiomonas* spp. β-Hemolytic, oxidase-positive, Gram-negative rods.
- MacConkey agar: *Salmonella* spp. and *Shigella* spp. colonies appear colorless or transparent.
- Xylose-lysine-desoxycholate (XLD) agar: *Shigella* spp. colonies are colorless, *Salmonella* spp. colonies are black (due to the production of H_2S).
- Cefsulodin-irgastan-novobiocin (CIN) agar: *Yersinia* spp. colonies appear as "bullseye" colonies when incubated at room temperature.
- Campylobacter selective media: *Campylobacter* spp. grow optimally when incubated in a microaerobic environment (specialty gas or incubation system) at 42°C.
- Sorbitol MacConkey plate: *E. coli* O157:H7 colonies are colorless (non-sorbitol fermenting); follow-up testing should be performed with specific antiserum and shigatoxin detection.

Case 17

A 14-year-old boy was swinging on a tendril (Tarzan-like) over a river in north Georgia, USA, when the tendril broke, and he fell into the river and lacerated his arm on a rock. The wound developed purulent drainage. What types of organisms is he at risk of having in the wound?

Water organisms include non-fermenters, in particular *Pseudomonas aeruginosa, Acinetobacter* spp., and *Aeromonas hydrophila*, and members of the Enterobacteriaceae.

Cultures from his wound grew *Pseudomonas aeruginosa* and *Aeromonas hydrophila*.

Case 18

A 1-year-old boy presented with fever and a limp. On examination, he had swelling and limitation of movement of his left knee. A diagnosis of septic arthritis of the knee was made and a diagnostic aspirate was performed.

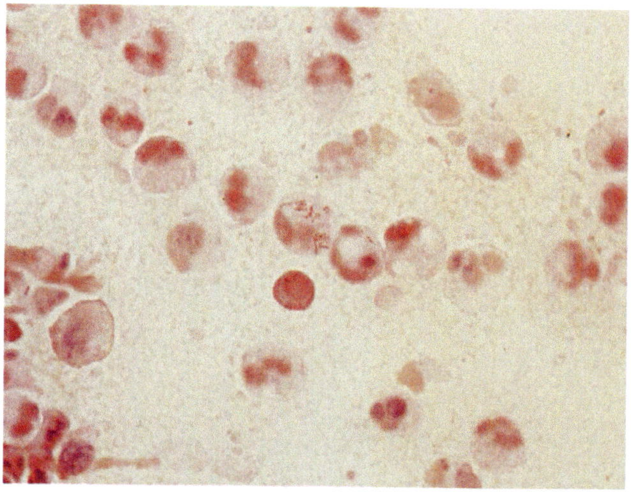

Fig. 27.14 Gram-stained smear of joint fluid.

1 What is your microbiological differential diagnosis?
2 What would you do with the fluid?

Answers

1 Microbiological differential diagnosis: *Staphylococcus aureus, Haemophilus influenzae* type b (in unimmunized populations), *Kingella kingae, Streptococcus pneumoniae*. Less likely in children of this age is *Streptococcus pyogenes*.
2 Processing the fluid: some fluid should be inoculated into a blood culture bottle (which increases the yield), and the rest should be used for a Gram-stained smear, and inoculation on to blood and chocolate agar plates.

The Gram-stained smear is shown in Fig. 27.14. This shows small, Gram-negative rods which reduces the microbial differential diagnosis to *Haemophilus influenzae* and *Kingella kingae*.

Culture of this fluid revealed *Kingella kingae*. Although this organism is often susceptible to amoxicillin, it can produce β-lactamase, which is inhibited by clavulanic acid. (The presence of this enzyme can be tested using nitrocefin – see Chapter 2.) Therefore he was treated with amoxicillin/clavulanic acid.

Case 19

A boy who was receiving chemotherapy for leukemia, which resulted in neutropenia, developed skin lesions on his back (Fig. 27.15).

1 What is the differential diagnosis?
2 What testing should be done?

Answers

Differential diagnosis

- Herpes virus infection
- Bacterial infection (metastatic)
- Fungal infection (metastatic)

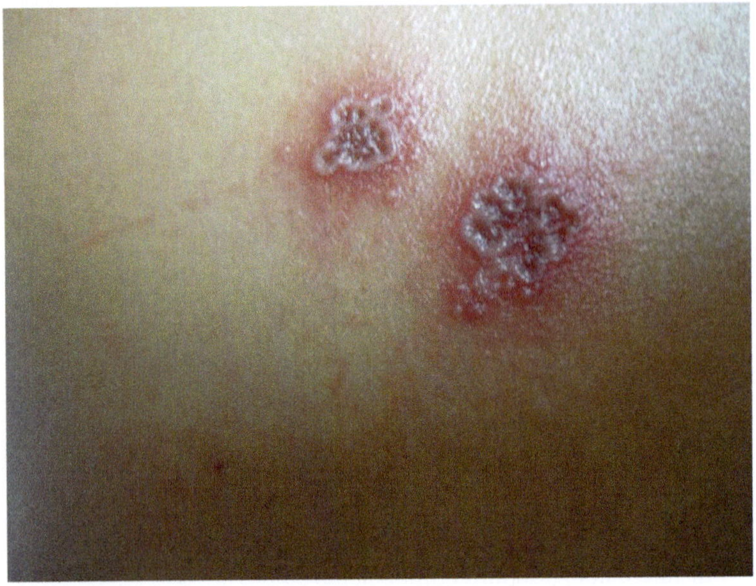

Fig. 27.15 Skin lesions.

Action (emergency)

Unroof the lesion and scrape its inside, and perform the following on the material obtained.

- For HSV: Tzanck preparation, HSV direct fluorescent antibody stain (DFA) (if available) and HSV culture (ELVIS rapid culture technique preferred).
- For bacteria: Gram stain and bacterial culture on blood and chocolate agar.
- For fungi: PAS, methylene blue stain or Calcofluor white (if available) and fungal culture.
- Biopsy the lesion (emergency): perform histology using hematoxylin and eosin (routine stain) as well as silver impregnation and PAS stains for detecting fungi, and culture for bacteria and fungi.

The studies revealed HSV infection. He was treated successfully with acyclovir.

Case 20

A 6-week-old infant was seen for an itchy rash on his limbs (Fig. 27.16). His mother had similar lesions (Fig. 27.17).

1 What is the likely diagnosis?
2 How could this be confirmed?

Answers

1 Diagnosis: scabies (*Sarcoptes scabiei*)
2 Confirmation: scrape lesion, and observe the mite or fecal balls.

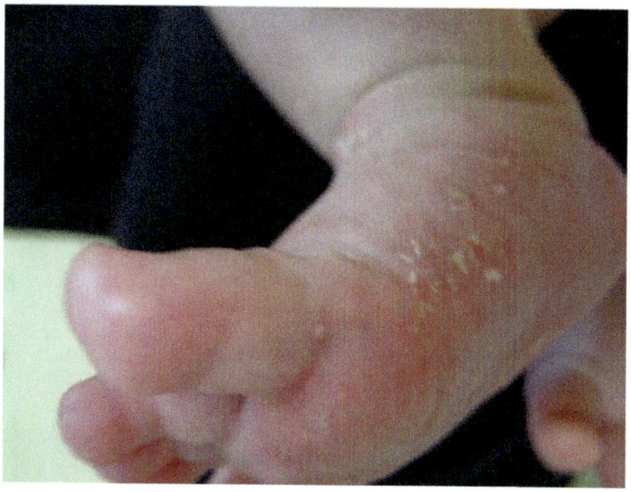

Fig. 27.16 Skin lesions on the infant's foot.

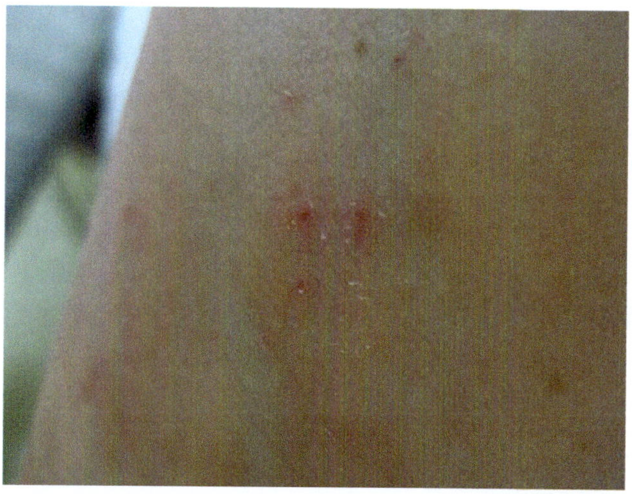

Fig. 27.17 Rash on mother's wrist.

Case 21

A 4-year-old boy who had undergone heart transplantation twice for restrictive cardiomyopathy, most recently 2 years earlier, presented with a 2-day history of fever and breathing difficulty. He was initially diagnosed with an RSV infection, but his condition deteriorated rapidly and he required mechanical ventilation. In addition to respiratory failure, he had scattered petechiae on his upper torso. His chest X-ray is shown in Fig. 27.18.

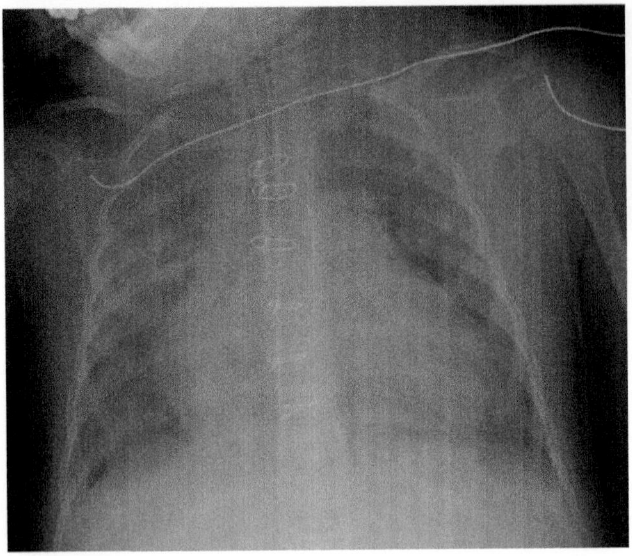

Fig. 27.18 Chest X-ray.

He was diagnosed with severe pneumonia, and despite treatment with very broad-spectrum antimicrobial therapy, his condition deteriorated.

Given his transplant status and immunosuppressive therapy:

1 what is your microbiologic differential diagnosis?
2 how would you make a definitive diagnosis?

Answers

Differential diagnosis

- Viral: respiratory viruses (RSV, parainfluenza viruses, adenovirus, influenza virus), CMV
- Bacteria: *Streptococcus pneumoniae, Staphylococcus aureus, Mycoplasma pneumoniae, Chlamydophila pneumoniae, Legionella pneumophila, Mycobacterium tuberculosis*
- Fungi: *Pneumocystis jiroveci*

Tests

- Nasopharyngeal swabs for respiratory virus antigen detection/PCR/culture
- CMV blood PCR
- Material must be obtained from the lungs for:
  - staining for bacteria (Gram stain), mycobacteria (acid-fast stain), and fungi (silver impregnation stain)
  - culture for bacteria, including *Legionella pneumophila*, mycobacteria, fungi, and viruses, including for respiratory viruses and CMV.

In this case a bronchoalveolar lavage was performed. Although the culture for CMV was negative, a blood PCR was positive with a high viral load. He was treated with ganciclovir, and improved.

Case 22

A 2-year-old boy presented with a 4-week history of left-sided neck swelling. He had not had fever. There was no known exposure to an individual with tuberculosis, nor exposure to cats or other animals. On examination, he was well-appearing. The only abnormality was a large lymph node (4.5 cm diameter) in the posterior triangle of the neck on the left side. It was mobile and non-tender, and the overlying skin was not erythematous (Fig. 27.19).

1 What is the differential diagnosis?

2 How would you make a diagnosis?

Answers

Differential diagnosis

Infectious:

- Acute bacterial infection – *Staphylococcus aureus* and *Streptococcus pyogenes* – unlikely
- Chronic bacterial infection:
  - Cat scratch disease – no exposure
  - Tuberculosis
  - Non-tuberculous mycobacterium

Non-infectious: lymphoma (unlikely, given focal nature and no other clinical findings)

Management

Place tuberculin skin test. The result is shown in Fig. 27.20. The diameter was 11 mm.

The differential diagnosis is largely narrowed to mycobacterial infection, whether tuberculosis or non-tuberculous. The chest X-ray was normal. Although this does not exclude tuberculosis, it makes it less likely.

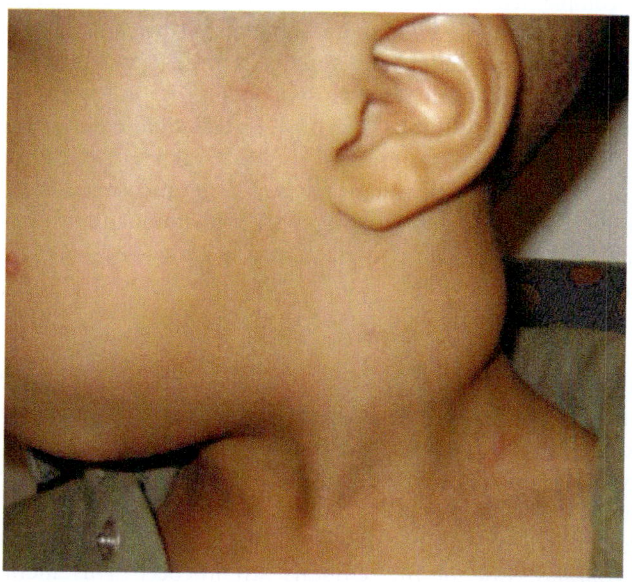

Fig. 27.19 Large lymph node.

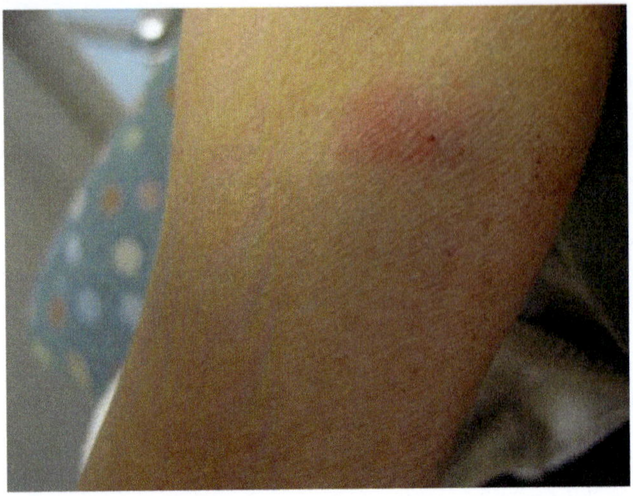

Fig. 27.20 Results of the tuberculin skin test.

The optimal diagnostic test is an excisional biopsy, with histology, and stains and culture for mycobacteria, as well as for fungi. This also constitutes optimal therapeutic management for non-tuberculous mycobacterial lymphadenopathy. When this cannot be performed (usually because important structures could be damaged in the course of the surgery), a needle aspiration to obtain material for staining and culture should be performed.

Case 23

A 2-year-old girl complained of fever and headache for 1 week. On the night of presentation, she had a focal seizure. Her examination revealed drowsiness, but was otherwise normal. A brain CT scan revealed a large abscess in the left parietal lobe (Fig. 27.21).

What is the microbial differential diagnosis?

Brain abscesses usually arise from infections in the middle ear, paranasal sinuses, teeth, or suppurative foci in a distant part of the body such as the lung (abscess, bronchiectasis), or bacteremia. Because the source is often a mucosal surface, the abscess is often polymicrobial, and anaerobes are often part of its flora. Other organisms include streptococci, such as *S. intermedius*, and oral Gram-negative rods. *Staphylococcus aureus* is relatively uncommon.

The abscess was drained surgically, and a Gram stain of the pus obtained is shown in Fig. 27.22.

What does the Gram stain show, and how would you further identify the organism?

The Gram stain shows numerous Gram-positive cocci, singly and in pairs, suggesting a streptococcus. The pus should be inoculated on to blood agar, chocolate agar, and on to media for anaerobic culture. Further identification should utilize the catalase test, to confirm that it is a streptococcus, and biochemical tests.

The culture of this organism had the odor of caramel, which is characteristic of *Streptococcus intermedius*, one of the viridans streptococci. This identification was confirmed by biochemical tests. Thorough evaluation of the heart, ears, and teeth

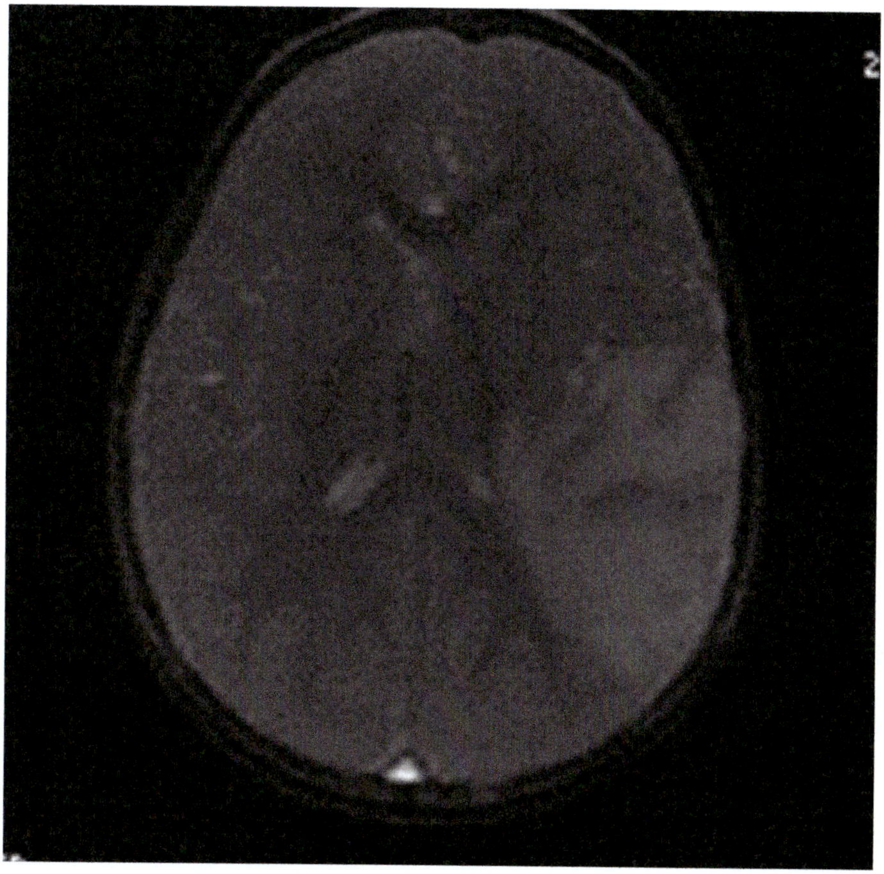

Fig. 27.21 CT scan showing a large left parietal lobe abscess.

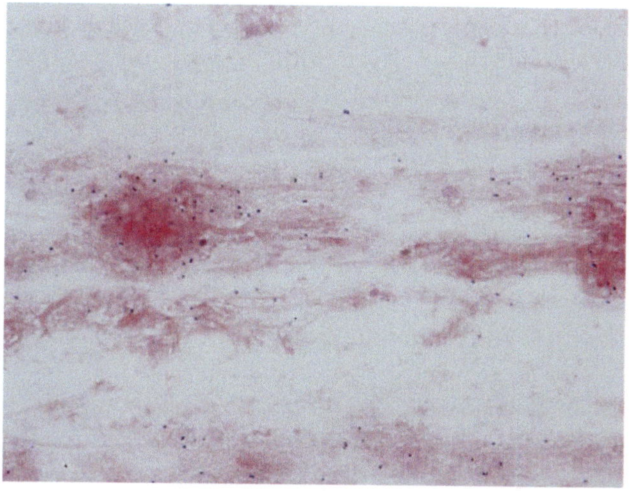

Fig. 27.22 Gram stain of the pus from the abscess.

failed to reveal a focus of origin of the abscess. She made a good recovery after surgery, and was treated with vancomycin and ceftriaxone.

Note that laboratory identification of viridans streptococci is difficult due to the constantly changing nomenclature and the scarcity of phenotypic and genotypic tests to accurately differentiate this group of organisms. For convenience's sake, related to clinical infections, the organisms have been classified into the following groups: *mitis*, *anginosis* (which includes *S. intermedius*), *mutans*, *salivarius*, *sanguinis*, and *bovis*. *Streptococcus pneumoniae* is genetically related to the *mitis* group but is differentiated from the viridans streptococci, for clinical purposes, by its sensitivity to optochin and its solubility in bile (see Chapter 10, Gram-positive cocci).

Case 24 (hypothetical)

A 1-week-old baby presents with a diffuse rash, and is found to have jaundice, and enlargement of liver and spleen. The mother had had a non-specific febrile illness during the second trimester of pregnancy.

1 What is the differential diagnosis?
2 How would you make the diagnosis?

Differential diagnosis

- Acute bacterial infection (sepsis)
- Congenital (intrauterine infection) due to:
  - CMV
  - Enterovirus
  - Herpes simplex (disseminated)
  - Rubella
  - Syphilis
  - Toxoplasmosis

The appearance of the rash could help to narrow the differential diagnosis.

- Petechiae, due to thrombocytopenia, could be due to all of the above.
- Macules, slightly scaly, suggest syphilis, especially if there is palmar or plantar peeling.
- Raised, purple lesions ("blueberry muffin" lesions) suggest CMV, rubella, or toxoplasmosis.
- Vesicles suggest HSV or VZV.

Making the diagnosis

- CMV: culture urine (tissue culture, using shell vial) for CMV.
- HSV: specimens for HSV culture from eye, mouth, rectum, and from vesicular skin lesions; PCR for HSV DNA on blood.
- Rubella: serology (IgM and IgG); if the mother had been rubella seropositive before this pregnancy, rubella would be extremely unlikely.
- Enterovirus: PCR on blood.
- Syphilis: obtain serology from mother (if RPR is positive, perform treponemal test) to confirm infection in her. Compare the infant's RPR titer with the mother's: if the infant's titer is fourfold of the mother's titer or greater, the diagnosis is confirmed.

Although dark-field microscopy of scrapings from lesions should be performed, this is rarely available.
- Toxoplasma: serology: IgM and IgG; toxoplasma PCR performed on blood.

Case 25

A 4-month-old girl, who had undergone surgery for hypoplastic left heart syndrome soon after birth, had a sudden cardiac deterioration. She had a cardiac catheterization during which she had a cardiac arrest. She was resuscitated. Two days later, she became febrile. A blood culture grew Gram-negative rods.

1 What is the microbial differential diagnosis of this healthcare-associated Gram-negative rod bacteremia?
2 What empiric antimicrobial therapy would you use?
3 How should this organism be further identified?

Answers

1 (i) Enterobacteriaceae, in particular *E. coli, Klebsiella pneumoniae,* and *Enterobacter* spp. (ii) *Pseudomonas aeruginosa*. One should be guided by the causes of infections in the particular unit (in this case, the cardiac surgery unit), as well as by prior antimicrobial therapy that the patient might have received.
2 Empiric therapy, that should be active against the above organisms, might consist of a carbapenem, e.g. meropenem. A combination of a β-lactam active against *Pseudomonas aeruginosa* (e.g. ceftazidime or piperacillin/tazobactam) in addition to an aminoglycoside (e.g. amikacin, to broaden the antimicrobial spectrum, pending identification and antimicrobial susceptibility test results) would also be suitable.
3 The blood culture broth should be inoculated on to chocolate, blood agar and MacConkey plates. Once there are colonies, an oxidase test should be performed. This will give an indication of the likelihood of *Pseudomonas aeruginosa,* which is oxidase positive, or a member of the Enterobacteriaceae, which are oxidase negative.

Further testing will involve biochemical tests. Antimicrobial susceptibility testing should be set up at the same time as the biochemical tests.

In this case, the organism was identified as *Klebsiella pneumoniae,* an important cause of healthcare-associated infections.

Case 26

A 20-year-old girl with spastic cerebral palsy had a Baclofen pump inserted into her spinal subarachnoid space to help control her spasticity. Two weeks later, she developed swelling and drainage at the surgical wound site (which communicates with the subarachnoid space). Cerebrospinal fluid showed 2323 leukocytes per mm^3, with protein and glucose concentrations of 446 and <20 mg/dL respectively. A Gram stain of the CSF was negative, but the cultures are shown in Fig. 27.23. Gram stain of the colonies showed Gram-negative rods.

What is the likely identity of this organism?

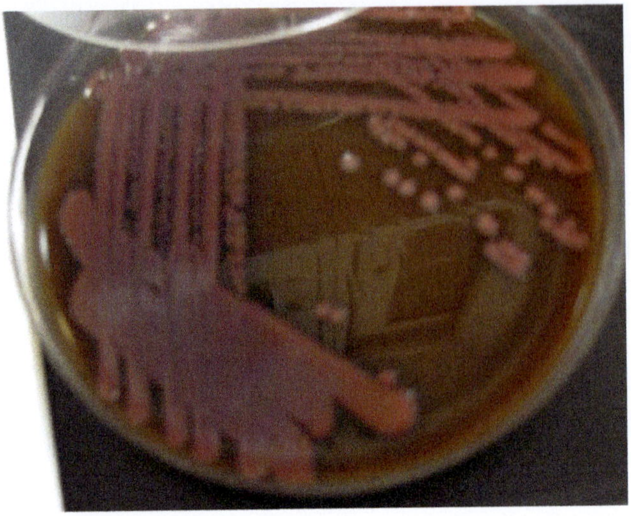

Fig. 27.23 Chocolate agar showing pink/red colonies.

Answer

Gram-negative rods producing pink/red pigment suggest *Serratia marcescens*. This is an important cause of healthcare-associated infections. Although susceptibility tests might indicate that it is susceptible to extended-spectrum cephalosporins, it often has inducible β-lactamases, which can render these antibiotics ineffective.

Case 27

Within 24 hours of being bitten by a cat, a 6-year-old child developed cellulitis at the bite site. Culture of the drainage grew colonies on both blood agar and chocolate agar, but not on MacConkey agar (Figs 27.24 & 27.25), the Gram stain of which is shown in Fig. 27.26. The oxidase and indole tests, performed on the colonies, are shown in Fig. 27.27.

What is the most likely identity of the causative organism?

Answer

The organism shows the following characteristics, illustrated in the above Figures: it grows on blood and chocolate agars but not on MacConkey agar; it is a small Gram-negative coccobacillus, and it is both indole and oxidase positive. These features identify it as *Pasteurella multocida*, an important cause of infections resulting from bites of cats and other animals.

Case 28 (hypothetical)

A 20-year-old man presented with a complaint of yellow eyes. He had had fever and generally felt ill for the past week. On examination, he appeared ill and had jaundice and fever. He had marked hepatomegaly, which was tender, but no splenomegaly. He

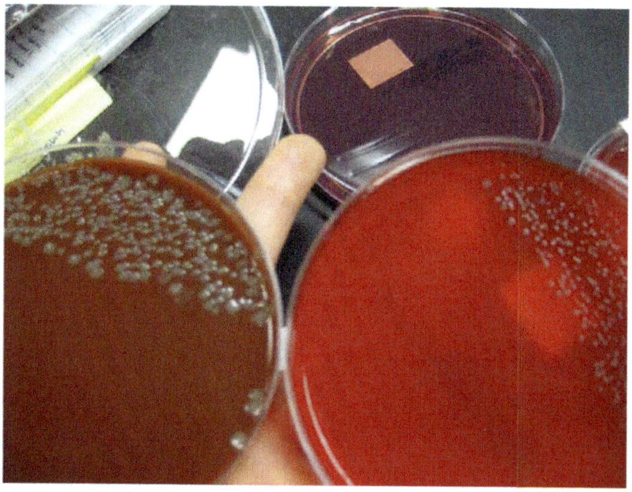

Fig. 27.24 Colonies growing on the blood and chocolate agar plates (*foreground*), but not on the MacConkey plate (*upper right*).

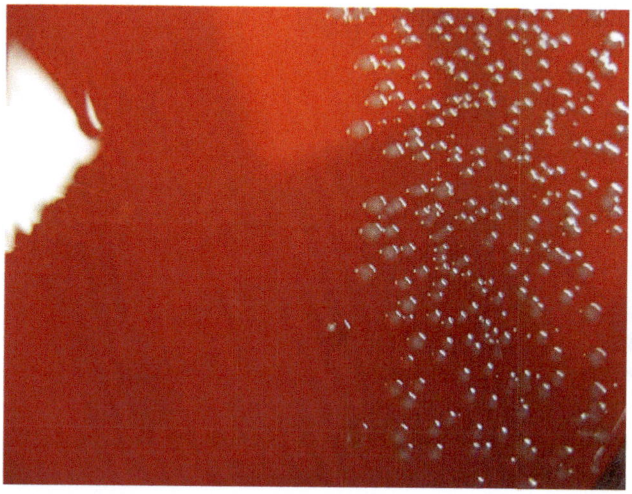

Fig. 27.25 Colonies on blood agar.

had recently (10 days ago) returned from a 3-month trip to South East Asia, India, and South America. He had received no medical advice nor interventions before his trip.

Construct a differential diagnosis, and plan a diagnostic strategy to evaluate him for these possibilities.

Answer

The clinical features (fever and evidence of liver disease) prompt a broad differential diagnosis (considering that he has had exposure to multiple pathogens, he might have more than one disease).

- Viral infections: hepatitis A, B, C, E, yellow fever
- Bacterial infection: typhoid fever, leptospirosis

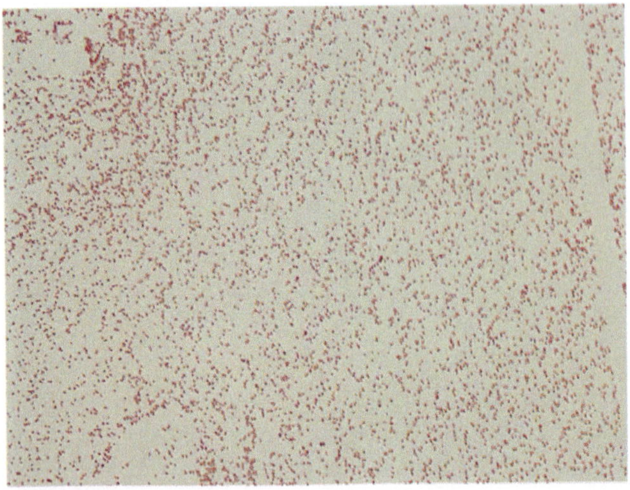

Fig. 27.26 Small Gram-negative coccobacilli.

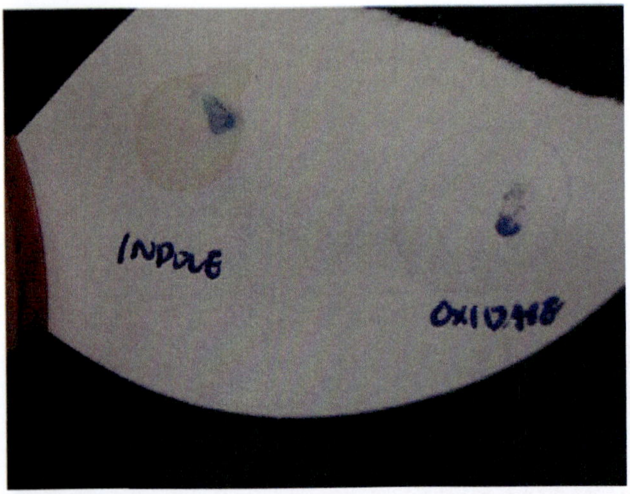

Fig. 27.27 A positive oxidase test (*right*) and a positive indole test (*left*).

- Protozoal infection: malaria, amebiasis
- Worm (fluke) infection: liver or bile duct

The tests used for diagnosing each of these infections include the following.

- Hepatitis A (HAV): anti-HAV IgM
- Hepatitis B (HBV): HBV surface antigen, HBV anti-core IgM
- Hepatitis C (HCV): HCV RNA
- Hepatitis E (HEV): anti-HEV IgM
- Yellow fever: yellow fever IgM, blood PCR, culture of blood for virus
- Typhoid fever: blood culture (routine), culture of stool and urine, serology
- Leptospirosis: serology
- Malaria: blood smear, thin and thick
- Amebiasis: liver ultrasound; serology
- Liver fluke: eggs in stool

Case 29

A 19-year-old man presented with a history of fever and sore throat for 1 week. He had not traveled nor been exposed to animals, but he had multiple sexual partners. On examination, he was mildly ill-appearing and febrile. He had marked cervical lymphadenopathy and enlarged, red tonsils with overlying exudate. He had mild hepatomegaly and splenomegaly.

1 What is your differential diagnosis?
2 How would you test for these possibilities?

Differential diagnosis

- Acute HIV infection
- Infectious mononucleosis (EBV infection)
- Secondary syphilis

Diagnostic tests

- For acute HIV, a serologic test might be negative because it is too early in the infection for the virus to be detected. This is the "window" period. With newer serologic tests, this window period is becoming shorter. A test detecting virus, an RNA PCR test, should therefore be used, in addition to a serologic test.
- Tests for EBV infection are serologic tests for IgG and IgM antibodies to viral capsid antigen (VCA), antibodies to early antigen (EA), and to nuclear antigen (NA), which appear only after several weeks. The presence of antibodies to NA would indicate a past infection and exclude EBV as a cause of his current illness.
- An RPR (non-treponemal test) is an excellent first test for syphilis. A positive should be confirmed with a treponemal test, such as an ELISA or fluorescent treponemal antibody test.

Case 30

A 16-year-old boy with cystic fibrosis (CF) has increasing cough and progressive deterioration in his lung function. Sputum repeatedly grows large mucoid colonies of Gram-negative rods, and non-mucoid Gram-negative rods.

How should they be identified, and what significance should be placed on the identification?

Answer

The microbiology of cystic fibrosis is very complex and the role of the laboratory is to identify "key" pathogens, to decrease deterioration in lung function. The progressive colonization/infectious etiologies are *Staphylococcus aureus*, occasionally the respiratory pathogen *Haemophilus influenzae*, and, of great importance, *Pseudomonas aeruginosa* and organisms grouped as *Burkholderia cepacia* complex. Many other species of Gram-negative rods may play a role in the pathogenesis of disease progression, depending on the individual patient. In this case, the specimen should be plated on selective and differential media to select for *B. cepacia* and other non-fermentative rods (lactose-negative rods as clear, colorless colonies on MacConkey agar) for further work-up by both conventional phenotypic and genotypic methods. Oxidase-positive, Gram-negative rods with characteristic "azure blue" color and grape-like odor would indicate *Pseudomonas aeruginosa,* the most commonly seen mucoid and non-mucoid organism isolated from CF cases. Unique among these organisms is the requirement to perform susceptibility testing using non-broth methods like Kirby–Bauer disk diffusion or E-test strips for accurate results.

Case 31

A 21-year-old sexually active female college student presents with a history of yellow vaginal discharge. Examination reveals a profuse vaginal discharge and erythematous vaginal and cervical mucosa. The rest of her examination is normal.

What tests, both in the clinic (point-of-care) and in the laboratory, should be performed to demonstrate the cause of her problem?

Differential diagnosis

- Bacterial vaginosis
- Trichomoniasis
- Gonorrhea
- *Chlamydia trachomatis* infection
- Candidiasis

In the clinic

Make a suspension of vaginal fluid in saline and a smear.

- Addition of KOH: whiff test (fishy odor) for bacterial vaginosis.
- Wet preparation: clue cells (see Gardnerella in Chapter 12) for bacterial vaginosis, *Trichomonas vaginalis.*
- Gram stain: Gram-negative diplococci (*Neisseria gonorrhoeae*).
- A paucity of Gram-positive rods resembling lactobacilli, and a predominance of Gram-positive rods resembling Gardnerella, as well as of Gram-negative rods and curved Gram-variable rods indicates bacterial vaginosis.
- Yeasts can also be seen on Gram stain.
- Rapid screen for HIV infection.

In the laboratory

- Culture of vaginal swab on Thayer–Martin agar for *N. gonorrhoeae.*
- Urine NAAT for *Chlamydia trachomatis* and *N. gonorrhoeae, and Trichomonas vaginalis.*
- RPR (for syphilis).

Further reading

Centers for Disease Control and Prevention (2010) Sexually transmitted disease treatment guidelines. MMWR 59(RR12): 56–7.

Gilligan PH (2014) Infections in patients with cystic fibrosis. Diagnostic microbiology update. Clin Lab Med 34: 197–217.

SECTION VII

Appendices

APPENDIX 1

Taxonomy of infectious agents infecting humans and lists of infectious agents according to their source

Largely as a result of advances in genetics and the consequent ability to better classify organisms, taxonomy and nomenclature are changing rapidly. The following websites offer the most up-to-date classifications and nomenclatures of microorganisms.

- Viruses: http://ictvonline.org
- Bacteria: www.bacterio.net/-alintro.html
- Fungi: Mycobank (www.mycobank.org) and Index Fungorum (www.indexfungorum.org/)
- Parasites: www.cdc.gov/dpdx

Human (close contact – other than sexual, droplet)

Viruses: all herpes viruses, pox viruses, parvovirus, adenovirus, HBV, respiratory viruses, enteroviruses, rhinoviruses, coronaviruses, diarrhea-causing viruses (rotavirus, caliciviruses, enteric adenoviruses, astroviruses), mumps, measles, papilloma viruses

Bacteria: *Chlamydia pneumoniae, Mycoplasma pneumoniae, Streptococcus pyogenes, Streptococcus pneumoniae, Streptococcus agalactiae, Staphylococcus aureus, Neisseria meningitidis, Haemophilus influenzae, Corynebacterium diphtheriae, Arcanobacterium haemolyticum, Mycobacterium tuberculosis, Mycobacterium leprae*, Shigellae, *Bordetella pertussis*

Fungi: dermatophytes

Ectoparasites: lice, *Sarcoptes scabiei*

Human (sexual contact)

Viruses: HSV, CMV, HIV, HBV, papilloma virus, HTLV

Bacteria: *Chlamydia trachomatis, Treponema pallidum, Klebsiella granulomatis, Haemophilus ducreyi, Neisseria gonorrhoeae, Mycoplasma genitalium*

Protozoon: *Trichomonas vaginalis*

Ectoparasite: Phthirus pubis

Practical Medical Microbiology for Clinicians, First Edition. Frank E. Berkowitz and Robert C. Jerris.

Table A1.1 Taxonomy of human pathogens, their usual sources, and the main clinical syndromes they cause.

Organism	Usual source	Clinical syndromes
Prions		Spongiform encephalopathy
Kuru	H	Cannibalism
Sporadic Creutzveld–Jacob disease	H	Surgical implantation
Variant Creutzveld–Jacob disease	A	Ingestion
Viruses		
DNA VIRUSES		
Alphaherpesvirinae		
Herpes simplex 1	H, Mat	Gingivostomatitis, fever blisters, encephalitis, keratitis, neonatal infection
Herpes simplex 2	H, Mat	Genital ulcers, neonatal infection
Varicella zoster virus	H	Chickenpox (varicella), zoster (shingles)
Betaherpesvirinae		
Cytomegalovirus	H, Mat	Infectious mononucleosis, hepatitis, OI – pneumonia, retinitis, enteritis
Human herpes virus 6	H	Fever, roseola, OI – hepatitis, encephalitis
Human herpes virus 7	H	Fever, roseola
Gammaherpesvirinae		
Epstein–Barr virus	H	Infectious mononucleosis, hepatitis, OI – posttransplant lymphoproliferative disease, Burkitt lymphoma, other cancers
Human herpes virus 8	H	Kaposi sarcoma
Adenoviridae		
Multiple types (52)	H	Conjunctivitis, respiratory infection, cystitis, OI, enteritis (types 40, 41)
Hepadnaviridae		
Hepatitis B virus	H, Mat	Hepatitis, chronic liver disease, liver cancer
Polyomaviridae		
JC virus	H	OI – progressive multifocal leukoencephalopathy
BK virus	H	Nephropathy following renal transplant
Merkel cell virus	H	Merkel cell cancer (skin)
SV 40	A	Monkey kidney cell contamination of polio vaccine
Papillomaviridae		
Papillomavirus	H	Warts – skin, genital, laryngeal papillomas
Multiple types		Cancer – cervical, oropharyngeal, anal, penile, vulvar
Parvoviridae		
Parvovirus B19	H	Erythema infectiosum, aplastic crisis, hydrops fetalis, chronic anemia, OI – hepatitis, encephalitis
Boca virus	H	Respiratory infection
Poxviridae		
Chordopoxvirinae		
Orthopoxvirus		

Table A1.1 Continued

Smallpox virus	H	Smallpox
Vaccinia virus	H, L	Vaccinia – local, disseminated, encephalitis, myopericarditis
Cowpox virus	A	
Monkeypox virus	A	Monkeypox
Molluscipox virus Molluscum contagiosum	H	Molluscum contagiosum
Parapoxvirus Orf virus	A	Orf
Yatapoxvirus		
Tanapoxvirus		
RNA VIRUSES		
Picornaviridae		
Enteroviruses, ABCD, including polio, Coxsackie, and Echoviruses	H	Poliomyelitis, encephalitis, fever, rash, meningitis
Rhinoviruses	H	Upper respiratory tract infection, sinusitis
Parechovirus	H	Myocarditis, neonatal infection
Hepatovirus Human hepatitis A virus	H	Acute hepatitis
Caliciviridae		
Noroviruses	H	Diarrhea
Sappoviruses		
Paramyxoviridae		
Measles virus	H	Measles
Mumps virus	H	Mumps
Respiratory syncytial virus	H	Bronchiolitis, pneumonia, URI
Parainfluenza viruses (1,2,3,4)	H	Respiratory tract infection
Human metapneumovirus	H	Respiratory tract infection
Hendra virus	B	Respiratory tract infection, encephalitis
Nipha virus	B	Encephalitis
Reoviridae		
Rotavirus	H	Diarrhea
Coltivirus: Colorado tick fever virus	Tick	Colorado tick fever
Coronaviridae		
SARS coronavirus	A→H→H	Severe acute respiratory syndrome (SARS)
MERS coronavirus	A→H→H	MERS-cov severe pneumonia
Other coronaviruses	H	Diarrhea, upper respiratory infection
Astroviridae		
Human astrovirus	H	Diarrhea
Orthomyxoviridae		
Influenza viruses A,B,C	H, A	Influenza
Retroviridae		
Lentiviruses		
Human immunodeficiency virus (HIV1 and HIV2)	H	HIV infection, AIDS

Continued

Table A1.1 Continued

Deltaretrovirus Human T-cell lymphotropic virus (HTLV)	H, Mat	T-cell leukemia, tropical spastic paraparesis
Togaviridae		
Rubivirus Rubella virus	H, Mat	Rubella,congenital rubella syndrome
Alphaviruses Many arthropod-borne viruses	Mos	Fever, myalgia, encephalitis Equine encephalitides, Chikungunya
Flaviviridae		
Hepacavirus Hepatitis C	H	Hepatitis, chronic liver disease, liver cancer
Many arthropod-borne viruses	Mos, Tick	Yellow fever, dengue, West Nile, Japanese B encephalitis, tick-borne encephalitis
Arenaviridae		
Lassa fever virus	A	Hemorrhagic fever
Lymphocytic choriomeningitis virus (LCMV)	A, Mat	Meningitis, congenital infection
Junin, Machupo, Sabia, Guanarito	A	South American hemorrhagic fevers
Rhabdoviridae Lyssaviruses Rabies virus	A, B	Rabies – encephalitis
Several others	B	
Filoviridae Ebola virus	A→H→H	Hemorrhagic fever
Marburg virus	A→H→H	Hemorrhagic fever
Bunyaviridae Orthobunyavirus California, La Crosse encephalitis Many arthropod-borne viruses	Mos	Encephalitis
Phlebovirus Sandfly fever virus	Fly	Sandfly fever
Rift Valley fever virus	Mos	Rift Valley fever
Hantavirus Hantaan virus	A	Korean hemorrhagic fever
Western hemisphere hantaviruses	A	Pulmonary syndrome
Nairovirus Congo Crimean hemorrhagic fever virus	Tick	Hemorrhagic fever
Hepatitis D virus	H	Chronic hepatitis associated with hepatitis B HBV
Hepeviridae Hepatitis E virus	H, A	Hepatitis E
BACTERIA		
Obligate intracellular replication, lack cell wall		
Chlamydiaceae		
Chlamydia trachomatis		
Serovars B, Da, Ia, D-K	H, Mat	Genital tract infection, neonatal conjunctivitis, pneumonia

Table A1.1 Continued

Serovars A, B, Ba, C	H, Fly	Trachoma
Serovars L1, L2, l2a, L3	H	Lymphogranuloma venereum
Chlamydophila psittaci	A	Pneumonia, systemic infection
Chlamydophila pneumoniae	H	Pneumonia
Rickettsiaceae		
R. prowazekii	Louse	Epidemic (louse-borne) typhus
R. mooseri(typhi)	Flea	Endemic (flea-borne) typhus
R. rickettsii	Tick	Rocky Mountain spotted fever
R. conori	Tick	Mediterranean spotted fever
R. africae	Tick	African tick typhus
Orientia tsutsugamushi	Mite	Scrub typhus
R. akari	Mite	Rickettsialpox
R. sibirica	Tick	Spotted fever
R. australis	Tick	Queensland tick typhus
Several others		
Anaplasmataceae	Tick	Ehrlichiosis
Ehrlichia chaffeensis		Human monocytic ehrlichiosis
Anaplasma phagocytophilum		Human granulocytic ehrlichiosis
Ehrlichia ewingii		Human ehrlichiosis
Mollicutes		
Mycoplasma pneumoniae	H	Pneumonia, rash, encephalitis
M. hominis	H, Mat	Puerperal infection, neonatal pneumonia
M. genitalium	H	Urethritis
Ureaplasma urealyticum	H, Mat	Puerperal infection, neonatal pneumonia
Spirochetes		
Treponema pallidum subsp. *pallidum*	H	Syphilis
Treponema pallidum subsp. *pertenue*	H	Yaws
Treponema pallidum subsp. *endemicum*	H	Bejel
Treponema carateum	H	Pinta
Leptospira interrogans	A, E	Leptospirosis
Multiple serogroups		
Borrelia		
B. burgdorferi	Tick	Lyme disease
B. recurrentis	Louse	Relapsing fever (louse-borne)
B. duttoni	Tick	Relapsing fever
B. hermsii, turicatae, parkeri	Tick	Relapsing fever (tick-borne)
Spirillum minus	A (rat)	Rat-bite fever
Gram-negative bacilli		
Enterobacteriaceae		
Ferment glucose, oxidase negative, reduce nitrate to nitrite, facultative anaerobes, many inhabit the intestine; cause infections arising from the intestine, such as abdominal abscesses, urinary tract infections, and healthcare-associated infections.		
Escherichia coli	A, H, End	Urinary infection, neonatal sepsis, sepsis, HCAI
Specific virulence factors associated with specific forms of enteric infection		
Shigella: four species: *dysenteriae, flexneri, boydii, sonnei*	H	Diarrhea, dysentery, encephalopathy

Continued

Table A1.1 Continued

Klebsiella		
pneumoniae	H, End	UTI, pneumonia, HCAI
oxytoca	H, End	
granulomatis	H	Granuloma inguinale
rhinoscleromatis		Nasal granuloma
Enterobacter	H, End	UTI, HCAI
cloacae		
aerogenes	H, End	UTI, HCAI
Pantoea agglomerans	End	HCAI
Cronobacter sakazakii		Infant formula, HCAI
Citrobacter	H,	HCAI
freundii		
koseri	H, End	HCAI
Serratia	E	HCAI
marcescens		
liquifaciens		
Proteus	H, End	UTI, HCAI
mirabilis		
vulgaris		
Morganella morganii	H, End	UTI, HCA
Providencia	H, End	UTI, HCAI
rettgeri		
stuartii		
Hafnia alvei	H	OI, diarrhea
Edwardsiella tarda	E	Wound
Salmonella	A	Diarrhea, dysentery, bacteremia
>2400 serotypes		
Serotype *typhi, paratyphi* A	H	Typhoid (enteric) fever
Yersinia	A, Flea	Plague
pestis		
enterocolitica	A	Diarrhea, dysentery, pseudoappendicitis
pseudotuberculosis		
Non-fermenters (many families)		
Pseudomonas aeruginosa	E	HCAI, pneumonia especially associated with cystic fibrosis
Burkholderia cepacia	E, H	Pneumonia associated with cystic fibrosis, OI in chronic granulomatous
Burkholderia pseudomallei	E	Melioidosis
Ralstonia pickettii	E	
Stenotrophomonas maltophilia	E	HCAI
Acinetobacter calcoaceticus-baumanni	E	HCAI, earthquakes
Acinetobacter lwoffii	E	HCAI
Elizabethkingae meningoseptica	E	Neonatal sepsis, meningitis
Achromobacter xylosoxidans	E	OI
Small Gram-negative bacilli (coccobacilli)		
Bordetella pertussis	H	Pertussis
Bordetella parapertussis	H	Pertussis-like illness
Bordetella holmesii	H	Pertussis-like illness
Bordetella bronchiseptica	A	Respiratory

Table A1.1 Continued

Brucella (several species, including *abortus*, *melitensis*, *suis*, *canis*)	A	Brucellosis
Capnocytophaga canimorsis	A	Dog-bite wound
Campylobacter (several species, including *jejuni*, *coli*, *upsaliensis*, *concisus*)	A	Diarrhea, dysentery
Campylobacter fetus	A	Sepsis (OI)
Helicobacter pylori	H	Gastritis
Francisella tularensis	A, arthropods	Tularemia
Haemophilus influenzae type b, other capsular types (a, c–f)	H	Bacteremia, meningitis, septic arthritis, epiglottitis
Non-typeable	H, End	Otitis media, sinusitis, pneumonia, conjunctivitis Brazilian hemorrhagic fever
Haemophilus ducreyi	H	Chancroid
Legionella pneumophila	E	Legionnaires disease (pneumonia)
Other *Legionella* species	E	Legionnaires disease (pneumonia)
Aggregatibacter *aphrophilus* *actinomycetemcomitans*	End	IE
Cardiobacterium hominis	End	IE
Eikenella corrodens	End	Mouth, bite-wound, skeletal
Kingella kingae	End	Skeletal infections
Streptobacillus moniliformis	A (rat)	Rat-bite fever
Bartonellae		
bacilliformis	Fly	Oroya fever
quintana	Louse	Trench fever, bacillary angiomatosis, IE
elizabethae	A	IE
Coxiella burnetii	A, tick	Q fever (hepatitis, pneumonia)
Non-enterobacteriaceae, fermenters		
Aeromonas	E	Water-associated infections
hydrophila *veronii* *caviae*	E	Soil-associated infections (OI)
Chromobacterium violaceum	E	Wounds, OI
Pasteurella multocida	A	Animal bite-wound infections
Plesiomonas shigelloides	E	Diarrhea
Vibrios		
Vibrio		
cholerae	E, H	Cholera
parahaemolyticus	E	Diarrhea
vulnificus	E	OI – sepsis, oyster-associated
Gram-negative cocci		
Neisseria		
meningitidis	H	Septicemia, meningitis
gonorrhoeae	H, Mat	Gonorrhea, neonatal conjunctivitis
Moraxella catarrhalis	End	Otitis media, sinusitis
Gram-positive cocci		
Staphylococci		
Coagulase-positive: *S. aureus*	H, End	Skin, wound, abscesses, bacteremia, pneumonia, HCAI, IE, FB

Continued

Table A1.1 Continued

Coagulase-negative: many species, including *S. epidermidis, haemolyticus*	H, End	HCAI, FB, IV catheters, ventricular shunts
S. saprophyticus	Cystitis	
S. lugdunensis	End	IE
Streptococci		
S. pneumoniae	H, End	Otitis media, sinusitis, conjunctivitis, pneumonia, bacteremia, meningitis
S. pyogenes (group A)	H	Tonsillitis, impetigo, wound, pneumonia, sepsis, toxic shock-like syndrome, acute glomerulonephritis, rheumatic fever
S. agalactiae (group B)	H, Mat	Neonatal sepsis, meningitis, OI – diabetics
Viridans streptococci, which contain six groups: *mutans, salivarius, sanguinis, mitis, bovis*	End	Dental caries, endocarditis, visceral abscesses, OI – sepsis
S. bovis (group D)	End	Endocarditis (colonic cancer associated)
S. dysgalactiae (group C)	H	Pharyngitis
S. canis (group G)	H	Pharyngitis
Enterococci		
E. faecalis	End	IE, HCAI, FB
E. faecium	End	IE, HCAI, FB
E. casseliflavus	E	Vancomycin resistant
Pediococcus spp.	E	OI – sepsis, HCAI (vancomycin resistant)
Abiotrophia spp.	End	Focal infections, IE
Granulicatella spp.	End	Focal infections, IE
Gemella spp.	End	Focal infections, IE
Leukonostoc spp.	E	HCAI, FB, OI (vancomycin resistant)
Rothia mucilaginosa	End	OI – sepsis
Gram-positive bacilli, non-spore-forming		
Actinomyces several species (microaerophilic), including:		
A.israeli, odontolyticus, meyeri, naeslundii	End	Actinomycosis – chronic infection mouth, gut, female genital
Nocardia, several species, including	E	OI – lung, brain
N. asteroides complex		
N. farcinica complex		
brasiliensis, transvalensis	E	Cutaneous infections
Corynebacteria	H	Diphtheria (throat, larynx, myocarditis, neuropathy
C. diphtheriae		
C. urealyticum	End	Chronic urinary tract infection
C. jeikeium	End	HCAI, FB
Many others	End	HCAI, FB
Arcanobacterium haemolyticum	H	Pharyngitis, scarlatiniform rash
Rhodococcus equi	E	OI – pneumonia
Gordonia	E	FB
Tsukamurella	E	FB
Listeria monocytogenes	E, Mat	Bacteremia, meningitis newborn, elderly, OI
Erysipelothrix rhusiopathiae	A, E	Cellulitis, IE (vancomycin resistant)
Rothia denticariosa	End	
Streptomyces somaliensis	E	Mycetoma

Table A1.1 Continued

Oerskovia spp.	E	IE
Tropheryma whipplei	Unknown	Whipple disease
Mycobacteria		
M. tuberculosis	H	Tuberculosis
M. bovis	A	Tuberculosis
M. africanum	H	Tuberculosis
M. leprae	H, Armadillo	Leprosy
M. avium complex	E	Lymphadenitis, OI
M. marinum	E	Water-associated granuloma
M. scrofulaceum	E	Lymphadenitis
M. ulcerans	E	Buruli ulcer
Many other species	E	Focal infections, OI
Gram-positive bacilli, spore-forming, aerobic		
Bacillus		
B. anthracis	A, E	Anthrax
B. cereus	E	Acute food-poisoning, eye
Several other species	E	FB, OI
Anaerobic bacteria		
Gram-positive rods, spore-forming (Clostridia)		
C. tetani	E	Tetanus
C. botulinum	E	Botulism – classic, wound, infant
C. difficile	E, H	Antibiotic-associated Colitis, HCAI
C. perfringens	E, End	Food-poisoning, myonecrosis (gas gangrene), hemolysis
Other histotoxic Clostridia	E	Myonecrosis, OI
Anaerobic Gram-positive rods, non-spore-forming		
Propionibacterium acnes	End	Acne, FB
Gardnerella vaginalis	End	Bacterial vaginosis
Lactobacillus spp.	End	OI – bacteremia, IE
Bifidobacterium spp.	End	
Eubacterium spp.	End	Dental plaque
Leptotrichia spp.		
Mobiluncus spp.	End	Bacterial vaginosis
Arachnia spp.		
Actinomyces spp. See aerobic Gram-positive rods		
Anaerobic Gram-negative bacilli		
Bacteroides	End	Infections arising from mucosae: mouth, gut, female genital tract
Prevotella spp.	End	As for *Bacteroides*
Porphyromonas spp.	End	As for *Bacteroides*
Fusobacterium spp.	End	Infections from mucosae, septic thrombophl
Bilophilia spp.		
Sutterella spp.		
Selenomonas spp.		
Anaerobic Gram-negative cocci		
Veillonella	End	Rare pathogen
Anaerobic Gram-positive cocci		
Peptococcus spp.	End	As for *Bacteroides*
Peptostreptococcus spp.	End	As for *Bacteroides*

Continued

Table A1.1 Continued

Anaerococcus spp.		
Finegoldia spp.		
Micromonas spp.		
Parvimonas spp.		
Peptoniphilus spp.		
FUNGI		
Yeasts		
Candida, many species, including *albicans, tropicalis, parapsilosis, krusei, glabrata, lusitaniae, guillermondii*	End, HCAI	Mucosal infections: mouth, vagina, HCAI, FB, fungemia, esophagitis, OI
Cryptococcus neoformans	E	OI – pneumonia, meningitis
C. gatti	E	Pneumonia, meningitis
Rhodotorula spp.	E	OI
Malassezia spp. *furfur, pachydermatis, globosa, restricta, sympodialis*	End	Pityriasis versicolor, HCAI, TPN
Trichosporon spp.	E	White piedra, OI
Saccharomyces cerevisiae (brewers' yeast)	E	OI
Dimorphic fungi		
Histoplasma	E	Histoplasmosis: pneumonia, OI
capsulatum		
capsulatum var. *duboisi*	E	African histoplasmosis
Coccidioides immitis	E	Coccidioidomycosis: fever, pneumonia, OI
Blastomyces dermatitidis	E	Blastomycosis: pneumonia, skin metastase
Paracoccidioides brasiliensis	E	South American blastomycosis
Sporothrix schenkii	E	Sporotrichosis (cutaneous, lymphangitis)
Penicillium marneffei	A	OI – pneumonia, systemic
Other fungi	H	OI – pneumonia
Pneumocystis jiroveci		
Filamentous fungi (molds)		
Aspergillus spp. including *fumigatus, niger, flavus, terreus, nidulans*	E	Aspergillosis – lung, OI, angioinvasion,allergic bronchopulmonary aspergillosis
Mucorales (formerly Zygomycetes)	E	Mucormycosis – OI, angioinvasion
Rhizopus spp.	E	OI, angioinvasion
Mucor spp.		
Lichtheimia spp.		
Rhizomucor spp.		
Cunninghamella		
Apophysomyces spp.		
Fusarium spp.		
Scedosporum spp.	E	OI
Pseudallescheria boydii	E	OI
Many other molds belonging to many families	E	Cutaneous, subcutaneous, deep tissue, and systemic infections as a result of injury or immunodeficiency in the host (see Chapter 21)
Dermatophytes		
Trichophyton spp.	H	Tinea (ringworm)
Epidermophyton spp.	H	

Table A1.1 Continued

Microsporum spp.	H, A	
***Microsporidium* spp:**		OI – diarrhea, cholangitis, sinusitis, keratitis (contact lens)
encephalitozoon		
bieneusi		
intestinalis		
CHROMISTA	H	Diarrhea
Blastocystis spp.		
PARASITES	H	Amebiasis: colitis, liver abscess
Intestinal protozoa		
Amebae		
Entamoeba histolytica		
Non-pathogenic amebae:		
Entamoeba dispar		
Entamoeba moshkovskii		
Entamoeba coli		
Entamoeba hartmanni		
Entamoeba polecki		
Iodamoeba buetschlii		
Endolimax nana		
Flagellate		
Giardia intestinalis	H, A	Diarrhea
Apicomplexa (Sporozoa)		
Cryptosporidium spp.	H, A	Diarrhea
hominis, parvum,		
meleagridis, felis, canis		
Cyclospora cayetanensis	H	Diarrhea
Cystoisospora belli	H	Diarrhea – OI
Sarcocystis spp.	A	Diarrhea,myositis
hominis		
suihominis		
Ciliate		
Balantidium coli	H, A	Colitis
Tissue and blood protozoa		
Amebae		
Naegleria fowleri	E	Meningoencephalitis (fresh water)
Acanthameba spp.	E	Meningoencephalitis, keratitis, OO
Balamuthia mandrillaris	E	Meningoencephalitis
Sappinia diploidea	E	Encephalitis
Flagellates		
Leishmania spp. (many):	Fly	Leishmaniasis
donovani, infantum		Visceral
mexicana, tropica		Cutaneous
braziliensis, panamenensis		Cutaneous and mucocutaneous
Trypanosomes		
cruzi	Reduviid bug	S. American trypanosomiasis (Chagas disease)
brucei rhodesiense	Fly	African trypanosomiasis (sleeping sickness)
brucei gambiense	Fly	African trypanosomiasis
Trichomonas vaginalis	H	Vaginitis
Apicomplexa (Sporozoa)		
Plasmodium falciparum	Mos	Falciparum malaria (malignant tertian)
Plasmodium vivax	Mos	Vivax malaria (benign tertian)
Plasmodium ovale	Mos	Ovale malaria

Continued

Table A1.1 Continued

Plasmodium malariae	Mos	Quartan malaria
Plasmodium knowlesi	Mos	Knowlesi malaria
Babesia microti and other species	Tick	Babesiosis
Toxoplasma gondii	A, Mat	Toxoplasmosis: mononucleosis, congenital, OI – encephalitis
Intestinal worms		
Roundworms (nematodes)		
Enterobius vermicularis (pinworm)	H	Perianal itching
Ascaris lumbricoides	H→E	Ascariasis, visceral larva migrans, intestinal obstruction
Toxocara canis and *cati*	A→E	Visceral larva migrans, pneumonitis, ocular granuloma
Baylisascaris procyonis	A→E	Encephalitis
Hookworms:	H→E	Anemia, hypoalbuminemia
Necator americanus		
Ancylostoma duodenale	H→E	Anemia, hypoalbuminemia
Strongyloides stercoralis	H→E	Diarrhea, OI – visceral disease
Trichuris trichiura	H→E	Chronic diarrhea, rectal prolapse
Anisakis spp.	A (raw fish)	Gastritis
Cestodes (tapeworms)		
Taenia solium	A (pig)	Tapeworm
Taenia solium larval stage	H	Cysticercosis
Taenia saginata	A (beef)	Tapeworm
Echinococcus multilocularis	A→E	Cysts
Multiceps multiceps	A→E	Cysts
Echinococcus granulosus	A→E	Hydatid disease (larval stage)
Diphyllobothrium latum (fish)	Fish	Vitamin B12 deficiency, intestinal obstruction
Spirometra mansonoides	Water, copepod	Sparganosis: eye, visceral disease
Dipylidium caninum	Flea	
Hymenolepis nana	H (+/- arthropod)	Anorexia
Trematodes (flukes)		
Fasciolopsis buski	Water plants	Intestinal disease
Heterophyes heterophyes	Fish	Intestinal disease
Metagonimus yokagawai	Fish	Intestinal disease
Tissue and blood worms		
Trematodes (flukes)		
Paragonimus spp.	Crustacean	Lung disease
Fasciola hepatica	Sheep, water plants	Liver disease
Clonorchis sinensis	Fish	Biliary tract
Opisthorchis viverrini	Fish	Biliary tract
Schistosomes:		
S. mansoni	H→W	Intestinal, liver fibrosis
S. japonicum	H, A→W	Intestinal, liver disease
S. haematobium	H→W	Urinary tract disease
S. mekongi	H, A→W	Intestinal, liver disease
S. intercalatum	H→W	Intestinal, liver disease
Nematodes		
Wuchereria bancrofti	Mos	Filariasis, elephantiasis
Brugia malayi	Mos	Filariasis
Onchocerca volvulus	Fly	River blindness, subcutaneous nodules

Table A1.1 Continued

Loa loa	Fly	Subcutaneous nodules, conjunctival disease
Dracunculus medinensis (Guinea worm)	W→copepod	Skin ulcer, cellulitis
Trichinella spiralis	A	Trichinosis (myositis)
Ancylostoma braziliense and *caninum*	A→E	Cutaneous larva migrans
Gnathostoma spinigerum	Fish	Subcutaneous, visceral swellings
Angiostrongylus cantonensis	Crustacean	Eosinophilic meningitis
Angiostrongylus costaricensis	Slug	Abdominal granulomas
Ectoparasites		
Mites	H	Scabies
Sarcoptes scabiei		
Eutrombicula alfreddugesi	(chigger) E	Dermatitis
Lice		
Pediculus humanus capitis	H	Head pediculosis
Pediculus humanus corporis	H	Body pediculosis
Phthirus pubis (crab louse)	H	Pubic pediculosis
Flies		Myiasis
Calliphoridae		
Cuteribridae		
Sarcophagidae		

A, animal, including birds; B, bat; CGD, chronic granulomatous disease; E, environment; End, endogenous; FB, infections on foreign bodies, such as catheters; H, human; HCAI, healthcare-associated infection; IE, infective endocarditis; L, laboratory; Mat, maternal; Mos, mosquito; OI, opportunistic infection; TPN, total parenteral nutrition; URI, upper respiratory tract infection; UTI, urinary tract infection.

Human (endogenous bacteria)

Staphylococci, streptococci (especially viridans group), enterococci, *Moraxella catarrhalis*, *Corynebacterium* spp., *Actinomyces* spp., many Enterobacteriaceae, e.g. *E. coli*, *Haemophilus* spp., many anaerobic bacteria, *Candida* spp.

Ingestion (human feces-contaminated food, water, or soil)

Viruses: HAV, enteroviruses, diarrhea-causing viruses (rotavirus, caliciviruses, enteric adenoviruses, astroviruses)

Bacteria: Salmonella, Shigella, *E. coli* (specific pathogenic types) *Campylobacter* spp.

Protozoa: *Entamoeba histolytica*, *Giardia intestinalis*, *Cryptosporidium* spp., *Cyclospora cayetanensis*

Worms: *Taenia solium* (cysticercosis), intestinal flukes, *Ascaris lumbricoides*, *Enterobius vermicularis*

Ingestion (animal excreta-contaminated food, water, or soil)

Viruses: henipa viruses

Bacteria: *Listeria monocytogenes, Brucella* spp., *E. coli* (specific pathogenic types), *Salmonella* spp., *Campylobacter* spp., *Yersinia enterocolitica, Mycobacterium bovis* (milk), *Leptospira interrogans*

Protozoa: *Giardia intestinalis, Cryptosporidium* spp., *Trypanosoma cruzi, Toxoplasma gondii*

Worms: *Toxocara* spp., *Baylisascaris procyonis, Echinococcus granulosus*

Ingestion (uncooked animal tissue)

Bacteria: *Bacillus anthracis, Listeria monocytogenes, Salmonella* spp., *E.coli, Campylobacter* spp., *Yersinia enterocolitica, Vibrio parahaemolyticus, Vibrio vulnificus*

Protozoon: *Toxoplasma gondii*

Worms: *Taenia solium* (pig), *Taenia saginata* (beef), *Hymenolepis diminuta* (rat fleas), *Diphyllobothrium latum* (fish), *Trichinella spiralis* (pig, carnivore), *Anisakis* spp. (fish), *Paragonimus* spp., liver flukes, *Gnathostoma spinigerum*

Animal (contact with animal, animal tissue, animal excreta, bite, scratch)

Viruses: Ebola virus, Congo-Crimean hemorrhagic fever, Rift Valley fever, lymphocytic choriomeningitis virus, Lassa fever, hantaviruses, monkeypox, cowpox, orf, rabies virus, herpes simiae

Bacteria: *Chlamydophila psittaci, Bacillus anthracis, Erysipelothrix rhusiopathiae, Streptobacillus moniliformis, Spirillum minus, Lepstospira interrogans, Pasteurella multocida, Yersina pestis, Francisella tularensis, Capnocytophaga canimorsus, Coxiella burnetii, Brucella* spp., *Bartonella henselae*

Fungi: *Microsporum canis*

Worm: Sparganosis (cestode)

Fresh water (non-enteral exposure)

Bacteria: *Leptospira interrogans, Erysipelothrix rhusiopathiae, Mycobacterium marinum, Legionella* spp., *Pseudomonas aeruginosa, Burkholderia pseudomallei, Aeromonas* spp., *Acinetobacter* spp.

Protozoa: *Naegleria fowleri, Acanthamoeba* spp.

Worm: *Schistosoma* spp.

Sea water (enteral)

Vibrio cholerae, Vibrio parahaemolyticus, Vibrio vulnificus

Sea water (non-enteral)

Erysipelothrix rhusiopathiae, Mycobacterium marinum, halophilic vibrios

Inanimate environment, non-enteral (soil, air)

Bacteria: *Bacillus cereus, Rhodococcus equi,* non-tuberculous mycobacteria, *Nocardia* spp., Enterobacteriaciae, *Aeromonas* spp., *Chromobacterium violaceum, Pseudomonas* spp., *Acinetobacter* spp., *Elizabethkingae meningosepticum, Legionella* spp., *Clostridium perfringens, Clostridium tetani*

Fungi: dimorphic fungi (*Histoplasma capsulatum, Blastomyces dermatitidis, Coccidioides* spp., *Paracoccidioides brasiliensis*), *Cryptococcus* spp., *Aspergillus* spp., many other filamentous fungi

Inanimate (human fecal-contaminated)

Viruses: enteroviruses, respiratory viruses

Bacteria: *Mycobacterium tuberculosis,* Enterobacteriaceae

Worms: hookworms, *Strongyloides stercoralis*

Mother

Rubella virus, CMV, HSV, HBV, HCV, HIV, HTLV, VZV, enteroviruses, lymphocytic choriomeningitis virus

Bacteria: *Treponema pallidum, Chlamydia trachomatis, Streptococcus agalactiae, Listeria monocytogenes,* Enterobacteriaceae, *Mycobacterium tuberculosis*

Protozoa: *Toxoplasma gondii, Plasmodium* spp., *Babesia* spp., *Trypanosoma cruzi, Candida* spp.

Hospital

Viruses: VZV, respiratory viruses, enteroviruses, diarrhea-causing viruses, measles, filoviruses

Bacteria: Enterobacteriaceae, *Pseudomonas aeruginosa, Stenotrophomonas maltophilia, Acinetobacter* spp., *Staphylococcus aureus,* coagulase-negative staphylococci, enterococci, *Clostridium difficile, Candida* spp., *Aspergillus* spp.

Blood

Viruses: HBV, HCV, HIV, CMV, EBV, West Nile virus, parvovirus B19

Bacteria: *Treponema pallidum, Yersinia enterocolitica, Listeria monocytogenes*

Protozoa: *Trypanosoma cruzi, Plasmodium* spp., *Babesia* spp., *Toxoplasma gondii*

Worm: Filaria

Tissue

Prions
Viruses: CMV, HIV, West Nile virus, rabies virus
Bacteria: any contaminating bacteria

Arthropod-transmitted infections

See Table A1.2.

Table A1.2 Arthropods as vectors of infectious agents.

	Viruses	Bacteria	Protozoa	Helminths
Mosquito	Alphaviruses Flaviviruses Bunyaviruses		*Plasmodium* spp.	*Wuchereria Brugia* Dirofilaria
Tick	Colorado tick Congo-Crimean tick-borne encephalitis	*Rickettsia* spp. *Coxiella burnetii Ehrlichia* spp. *Borrelia duttoni Borrelia burgdorferi Francisella tularensis*	*Babesia* spp.	
Flea		*Yersinia pestis Rickettsia typhi*		
Louse		*Borrelia recurrentis Rickettsia prowazekii Bartonella quintana*		
Fly	Sandfly fever	*Francisella tularensis*	*Leishmania Trypanosoma brucei*	*Onchocerca Loa loa*
Triatomid bug			*Trypanosoma cruzi*	
Mite		*Rickettsia akari Orientia tsutsugamushi*		

Acknowledgment

The lists in this appendix are adapted from *Case Studies in Pediatric Infectious Diseases*, Frank Berkowitz. Copyright © Frank E Berkowitz. Reprinted with the permission of Cambridge University Press.

APPENDIX 2

Clinical syndromes and their causative organisms

Table A2.1 Causative organisms of bloodstream infections according to the type of host.

Host/underlying condition	Organisms
Neonate	*S. agalactiae, E. coli* and other Gram-negative rods, *Listeria monocytogenes*
Premature neonate	*S. agalactiae, E. coli* and other Gram-negative rods, *Listeria monocytogenes*, coagulase-negative staphylococci, *Candida* spp.
Infant, young child	*S. pneumoniae, H. influenzae* type b*, *S. aureus, Neisseria meningitidis, Salmonella*
Older child, adult	*S. aureus, Neisseria meningitidis*
Elderly	*S. pneumoniae, S. aureus, Listeria monocytogenes*, Gram-negative rods
Neutropenia	*S. aureus*, streptococci, Gram-negative rods, *Candida* spp.
Sickle cell disease	*S. pneumoniae, H. influenzae* type b*, *Neisseria meningitidis*
Iron overload	*Yersinia enterocolitica*, other Gram-negative rods, *Mucorales* spp., *Vibrio vulnificus*
Nephrotic syndrome	*S. pneumoniae*, Gram-negative rods
Liver disease with ascites	*S. pneumoniae*, Gram-negative rods, *Vibrio vulnificus*
AIDS	*S. pneumoniae*, Gram-negative rods, esp. *Salmonella* spp., *Pseudomonas aeruginosa, Listeria monocytogenes, Mycobacterium* spp., Fungi: *Histoplasma, Penicillium marneffei*

*Where *H. influenzae* type b vaccination is not widespread.

Table A2.2 Respiratory tract infections and their causative organisms, including in special hosts.

Syndrome	Host	Organisms
Middle ear infection	Normal	*S. pneumoniae, H. influenzae, Moraxella catarrhalis*, respiratory viruses*
Sinusitis	Normal	*S. pneumoniae, H. influenzae, Moraxella catarrhalis*, respiratory viruses*
	Abnormal	Above + fungi
	Chronic	Above + *S. intermedius*, anaerobes

Continued

Table A2.2 Continued

Pharyngitis	Acute	Respiratory viruses*, HSV, *S. pyogenes*, *Corynebacterium diphtheriae*, *Arcanobacterium haemolyticum*, *N. gonorrhoeae*, Epstein-Barr virus
Epiglottitis	Acute	*H. influenzae* type b†, *S. pneumoniae*, *S. pyogenes*
Laryngitis	Acute	Respiratory viruses*, *Corynebacterium diphtheriae*
Bacterial tracheitis		*S. aureus*, *S. pyogenes*
Bronchitis	Acute	Respiratory viruses*, *Bordetella pertussis*
	Chronic	*S. pneumoniae*, *H. influenzae*
Bronchiolitis	Acute	Respiratory viruses*, especially RSV
Pneumonia	Neonate	*S. agalactiae*. *E. coli*, HSV, *Chlamydia trachomatis*
	Child	Respiratory viruses*, *S. pneumoniae*, *H. influenzae*, *S. aureus*, *Mycoplasma pneumoniae*, *Chlamydophila pneumoniae*
	Adult	*S. pneumoniae*, *M. pneumoniae*, *Chlamydophila pneumoniae*, *Legionella pneumophila*
	Healthcare-associated	*S. aureus*, Gram-negative rods, respiratory viruses*
	Immuno-compromised	Above + fungi, *Pseudomonas aeruginosa*, *Nocardia* spp., *Rhodococcus equi*, *Pneumocystis jiroveci*, cytomegalovirus
	Cystic fibrosis	*H. influenzae*, *S. aureus*, *Pseudomonas aeruginosa*, *Burkholderia cepacia*
	Chronic	*Mycobacterium tuberculosis*, fungi, anatomic abnormalities, e.g. foreign body
	Specific exposures	*M. tuberculosis*, *Histoplasma capsulatum*, *Blastomyces dermatitidis*, *Coccidioides immitis*, *Cryptococcus neoformans*, *Chlamydophila psittaci*, *Coxiella burnetii*, *Burkholderia pseudomallei*, *Legionella* spp.
Empyema		*S. aureus*, *S pneumoniae*, *S. pyogenes*, anaerobes
Lung abscess		*S. aureus*, *S pneumoniae*, *S. pyogenes*, anaerobes

H, *Haemophilus*; M, *Mycobacterium*; RSV, respiratory syncytial virus.
*Respiratory syncytial virus, parainfluenza viruses, influenza viruses, adenoviruses, human metapneumovirus.
†Where *H. influenzae type* b vaccination is not widespread.

Table A2.3 Eye infections and their causative organisms.

Syndrome	Host	Organisms
Keratitis	Normal	Adenovirus, HSV, VZV
	Trauma	Streptococci, staphylococci, Gram-negative rods, *Bacillus* spp., fungi
	Contact lens	Above + *Acanthamoeba* spp., fungi, e.g. *Fusarium* spp., *Microsporidium* spp.
	Newborn	HSV, *Neisseria gonorrhoeae*
Conjunctivitis	Newborn	*Neisseria gonorrhoeae*, staphylococci, *Chlamydia trachomatis* (oculogenital serovars), HSV
	Normal children and adults	Adenovirus, *S. pneumoniae*, *H. influenzae*, *S. aureus*
Trachoma		*Chlamydia trachomatis* (serovars A–C)
Anterior uveitis and endophthalmitis	Endogenous	Causes of bacteremia
	Exogenous (trauma)	*Bacillus* spp., staphylococci, Gram-negative rods, *Propionibacterium acnes*
Retinitis	Fetus	*Toxoplasma gondii*, cytomegalovirus, syphilis, HSV
	Healthcare associated	*Candida* spp., other causes of bloodstream infection
	Immunocompromised host	*Toxoplasma gondii*, cytomegalovirus, HSV, VZV
	Specific exposures	*Toxocara* spp., *Onchocerca volvulus*, *Loa loa*, *Mycobacterium tuberculosis*, *Brucella* spp.

Table A2.4 Infections of the nervous system and their causative agents.

Syndrome	Host	Organisms
Acute meningitis	Newborn	*S. agalactiae*, *E. coli*, other Gram-negative rods, *Listeria monocytogenes*
	Young child	Enteroviruses, *H. influenzae* type b*, *S. pneumoniae*, *N. meningitidis*
	Older child Adult	Enteroviruses, *S. pneumoniae*, *N. meningitidis*
	Elderly	*S. pneumoniae*, *Listeria monocytogenes*, Gram-negative rods
Chronic meningitis		*M. tuberculosis*, fungi, especially *Cryptococcus* spp., *Treponema pallidum*, malignancies, SLE, sarcoidosis
V/P shunt infection		Staphylococci, streptococci, *Corynebacterium* spp., Gram-negative rods, *Bacillus* spp.
Encephalitis	Newborn	HSV, enteroviruses
	Children, adults	HSV, enteroviruses, arboviruses, parvovirus, VZV, measles, EBV, CMV, HIV, rabies, *Bartonella henselae*, *Borrelia burgdorferi*, *Mycoplasma pneumoniae*, rickettsiae
	Immunocompromised	Above + *Toxoplasma gondii*, JC virus
Brain abscess		Streptococci, anaerobes, *S. aureus*, Gram-negative rods
Epi/subdural abscess		Streptococci, anaerobes, *S. aureus*, Gram-negative rods
Myelitis		Polio virus, West Nile virus, HTLV-1, EBV
Spinal abscess		*S. aureus*, many bacteria
Radiculopathy		CMV, VZV, *Schistosoma* spp.
Guillain–Barré syndrome		*Campylobacter jejuni*
Botulism		*Clostridium botulinum*
Tetanus		*Clostridium tetani*

*Where *Haemophilus influenzae* type b vaccination is not widespread.
SLE, systemic lupus erythematosus.

Table A2.5 Cardiovascular infections and their causes.

Syndrome	Host	Organisms
Endocarditis	Native valve	Viridans streptococci, enterococci, *S. aureus*, HACEK*, *Bartonella henselae*, *Coxiella burnetii*
	Prosthetic valve	Above + coagulase-negative staphylococci, *Candida* spp., *Corynebacteria* spp.
	Postoperative	Staphylococci, Gram-negative rods, *Candida* spp.
Myocarditis		Many organisms including enteroviruses, influenza viruses, *Corynebacterium diphtheriae*, *Trypanosoma cruzi*, *Toxoplasma gondii*
Pericarditis	Viral	Enteroviruses, influenza viruses
	Purulent	*S. aureus*, *S. pneumoniae*, *H. influenzae*†, *N. meningitidis*
	Chronic	*M. tuberculosis*
Septic thrombophlebitis	Sepsis associated	*S. aureus*
	Jugular	Anaerobes (*Fusobacterium necrophorum*), others
	Healthcareassociated	*S. aureus*, enterococci, *Candida* spp., Gram-negative rods

**Haemophilus* (now called *Aggregatibacter*) *aphrophilus*, *Actinobacillus* (now called *Aggregatibacter*) *actinomycetemcomitans*, *Cardiobacterium hominis*, *Eikenella corrodens*, *Kingella kingae*
†Where *Haemophilus influenzae* type b vaccination is not widespread.

Table A2.6 Intraabdominal infections and their causative organisms.

Syndrome	Host	Organism
Peritonitis	Bowel perforation	*E. coli*, other Gram-negative rods, anaerobes (*Bacteroides* spp. and others), enterococci
	Primary (bacteremic)	*S. pneumoniae*, Gram-negative rods
	Dialysis	Staphylococci, *Candida* spp., Gram-negative rods
Cholecystitis		Gram-negative rods, enterococci, anaerobes
Cholangitis		Gram-negative rods, enterococci, anaerobes
Abscesses	Bacterial	*E. coli*, anaerobes (*Bacteroides* spp. and others), enterococci
Hepatitis	Newborn	Rubella, CMV, *Toxoplasma gondii*, *Treponema pallidum*, bacteremia* , HSV, enteroviruses
	Children, adults	Hepatitis A, B, C, E viruses, EBV, CMV, adenovirus
	Granulomatous	*M. tuberculosis*, *Brucella* spp., fungi, *Toxocara* spp., CMV, *Bartonella henselae*
Hepatic abscess	Bacterial	Streptococci, enterococci, *S. aureus*, Gram-negative rods
	Amebic	*Entamoeba histolytica*

*See causes of neonatal bacteremia (Table 3).

Table A2.7 Infections of the gastrointestinal tract and their microbial causes.

Syndrome	Host	Organisms
Esophagitis		HSV, *Candida* spp., CMV
Gastritis		*Helicobacter pylori*
Acute gastroenteritis		Rotavirus*, norovirus, enteric adenovirus, astrovirus, *E. coli*, *Salmonella* spp., *Shigella* spp., *Campylobacter* spp., *Yersinia enterocolitica*, *Giardia intestinalis*, *Cryptosporidium hominis*, *Cyclospora cayetanensis*
	Immunocompromised	Above + *Cystoisospora belli*, *Microsporidium* spp.
Chronic diarrhea		*Giardia*, *Cryptosporidium*, *Cystoisospora*, *Microsporidium*, *Entamoeba histolytica*, *Balantidium coli*, ?*Blastocystis* spp., *Trichuris*
Colitis/dysentery		*E. coli*, *Salmonella* spp., *Shigella* spp., *Campylobacter* spp., *Yersinia enterocolitica*, *Clostridium difficile*, *Entamoeba histolytica*, *Balantidium coli*
	Immunocompromised	Above plus CMV
Colitis with perforation		*Clostridium difficile*, *Entamoeba histolytica*, *Balantidium coli*
Proctocolitis		See Table 12 for genital tract infections
Enteric fever		*Salmonella typhi*, other *Salmonella* spp., especially *S. paratyphi* A and B
Necrotizing enterocolitis	Neonate	Gram-negative rods, anaerobes

Table A2.8 Urinary tract infections and their causes.

Syndrome	Host	Organisms
Pyelonephritis		*E. coli*, other Gram-negative rods, enterococci
Cystitis		*E. coli*, other Gram-negative rods, enterococci, adenovirus, *S. saprophyticus*
	Renal transplant	Above + BK virus

Table A2.9 Skeletal, skin, and soft tissue infections and their causative organisms.

Syndrome	Host	Organisms
Osteomyelitis	Newborn	*S. agalactiae*, *S. aureus*, Gram-negative rods
	Child	*S. aureus*, *H. influenzae**, *Kingella kingae*, *S. pneumoniae*, *S. pyogenes*
	Adult	*S. aureus*, Gram-negative rods
	Sickle cell disease	*Salmonella* spp., *S. aureus*
	Chronic	*S. aureus*, *M. tuberculosis*, *Brucella* spp.
Septic arthritis	Newborn	*S. agalactiae*, *S. aureus*, Gram-negative rods
	Child	*S. aureus*, *H. influenzae**, *Kingella kingae*, *S. pneumoniae*, *S. pyogenes*
Impetigo/pyoderma		*S. pyogenes*, *S. aureus*
Furuncle		*S. aureus*
Cellulitis		*S. pyogenes*, *S. aureus*
Periorbital cellulitis		*S. pneumoniae*, *H. influenzae*, *S. aureus*
Fasciitis		*S. pyogenes*, *S. aureus*
Perineal/synergistic		Gram-negative rods, anaerobes
Myonecrosis/gas gangrene		Histotoxic *Clostridium* spp.
Omphalitis	Neonate	*S. agalactiae*, *S. aureus*, Gram-negative rods, anaerobes
Ecthyma gangrenosum	Neutropenia	*Pseudomonas aeruginosa*, other Gram-negative rods, *Aspergillus* spp., *Mucorales* spp.

**Haemophilus influenzae* type b, where immunization against it is not widespread.

Table A2.10 Genital tract and sexually transmitted infections and their causative organisms.

Syndrome	Organisms
Genital ulcer	HSV, *Treponema pallidum*, *Haemophilus ducreyi*, *Chlamydia trachomatis* vars L-1, 2, 2a, and 3, *Klebsiella granulomatis*
Inguinal lymphadenitis	*Treponema pallidum*, *Chlamydia trachomatis* serovars L-1, 2, 2a, and 3
Urethritis	*Neisseria gonorrhoeae*, *Chlamydia trachomatis* oculogenital serovars, *Mycoplasma genitalium*
Vaginitis	*Neisseria gonorrhoeae*, *Chlamydia trachomatis* oculogenital serovars, HSV, *Trichomonas hominis*, *Candida* spp.
Bacterial vaginosis	*Gardnerella vaginalis*, *Mobiluncus* spp., other anaerobes
Cervicitis	*Neisseria gonorrhoeae*, *Chlamydia trachomatis* oculogenital serovars, HPV, HSV
Pelvic inflammatory disease	*Neisseria gonorrhoeae*, *Chlamydia trachomatis* oculogenital serovars, anaerobes, Gram-negative rods
Acute epididymitis	*Neisseria gonorrhoeae*, *Chlamydia trachomatis* oculogenital serovars, Gram-negative rods
Proctocolitis	*Neisseria gonorrhoeae*, *Chlamydia trachomatis* oculogenital serovars, and serovars L-1, 2, 2a, and 3, *Treponema pallidum*, HSV

HPV, human papillomavirus; HSV, herpes simplex virus.

Table A2.11 Bites wounds: their causative organisms according to the inflicting animal.*,†

Animal	Organisms
Human	*S. aureus*, *S. pyogenes*, oral Gram-negative rods, anaerobes
Dogs, cats, other mammals	*S. aureus*, streptococci, *Pasteurella* spp., *Neisseria* spp., *Capnocytophaga* spp.
Sharks	Vibrios
Alligators, crocodiles	*Aeromonas* spp., *Pseudomonas aeruginosa*, other Gram-negative rods, anaerobes
Rat	*Streptobacillus moniliformis*, *Spirillum minus*
Monkey	Above (mammals)+ Herpes simiae

*Rabies and *Clostridium tetani* transmission should be considered in all animal bites.

†Risk of transmission of HIV, hepatitis B, and hepatitis C between biter and victim should be considered in all human bites.

Table A2.12 Viruses transmitted by rodent excreta ("rodex" viruses) and bat excreta ("batex" viruses), according to virus family, their geographic distributions, and animal sources.

Virus	Geographic distribution	Animal source	Clinical
Paramyxoviruses			
Hendra	Australia	Flying fox	Encephalitis, respiratory failure
Nipah	Malaysia, India, Bangladesh	Flying fox	Encephalitis
Bunyaviruses			
Hanta	Asia, Europe	Mouse	Hemorrhagic fever (HF), renal failure
Hanta	USA, S. America	Mouse	Pulmonary syndrome
Arenaviruses			
Lassa	West Africa	Mouse	HF
Lymphocytic choriomeningitis	Worldwide	Mouse, hamster	Meningoencephalitis, fetal infection
Bolivian HF	S. America		HF
Argentinian HF	S. America		HF
Venezuelan HF	S. America		HF
Lujo	Zambia		HF
Filiviruses			
Marburg	Africa	Bat	HF
Ebola	Africa	Unknown	HF
Coronaviruses			
SARS-cov	Asia	Bat	Respiratory syndrome
MERS-cov	Middle East	Camel/Bat	Respiratory syndrome

Table A2.13 Bacterial infections for which serologic tests are used for diagnosis.

Disease	Organism
Brucellosis*	*Brucella* spp.
Ehrlichiosis	*Ehrlichia* spp., *Anaplasma phagocytophilum*
Endemic treponematoses	*Treponema pallidum* subsp.
Leptospirosis	*Leptospira* spp.
Lyme disease	*Borrelia burgdorferi*
Mycoplasma infection	*Mycoplasma* spp.
Q fever	*Coxiella burnetii*
Relapsing fever	*Borrelia* spp.
Rickettsial infection (typhus, spotted fevers)	*Rickettsia* spp.
Syphilis	*Treponema pallidum*
Tularemia*	*Francisella tularensis*
Typhoid fever* (culture is optimal)	*Salmonella typhi*

*Culture for these organisms can be readily performed. *Brucella* spp. and *Francisella tularensis* can pose a significant hazard in the laboratory, and *Salmonella typhi* less so.

APPENDIX 3

General references and online resources

1 www.asmusa.org: this website provides updated information for the clinical microbiology laboratory.

2 www.cdc.gov: this is the website for the Centers for Disease Control and Prevention, US Public Health Service. This is a marvelous resource. Diseases or infectious syndromes can be searched alphabetically. Advice on diagnosis (and how to send specimens to the CDC, which provides reference laboratory services for some infectious agents, e.g. rabies) and prevention is provided.

Specific resources within the CDC, which have been used extensively for the preparation of this book, are:

- Guidelines for the Treatment of Sexually Transmitted Diseases, 2010 (updated 2015)
- Yellow Book: Health Information on International Travel, 2012 (updated 2014)
- Public Health Image Library (PHIL)
- Division of Parasitic Diseases (DPDx) – from which many pictures were taken. This laboratory provides identification services for parasitic diseases.

3 www.doctorfungus.org: this is a very extensive resource on fungi.

4 www.idsociety.org: this is the website of the Society of Infectious Diseases of America. It provides guidelines for the management of many infections, according to site and type of organism.

General references

Beck JW, Davies JE (1981) *Medical Parasitology*, 3rd edn. St Louis: CV Mosby.

Brooks GF, Morse SA, Carroll KC, Mietzner TA, Butel JS (2013) *Jawetz, Melnick, and Adelberg's Medical Microbiology*, 26th edn. New York: McGraw-Hill.

Feigin RD, Cherry JD, Demmler GJ, Kaplan SL (2004) *Textbook of Pediatric Infectious Diseases*, 5th edn. Philadelphia: WB Saunders.

Fisher RG, Boyce TE (2005) *Moffett's Pediatric Infectious Diseases. A Problem-Oriented Approach*, 4th edn. Philadelphia: Lippincott, Williams and Wilkins.

Jenson HB, Baltimore RS (2002) *Pediatric Infectious Diseases. Principles and Practice*, 2nd edn. Philadelphia: WB Saunders, 2002.

Katz M, Despommier DD, Gwadz RW (1988) *Parasitic Diseases*, 2nd edn. New York: Springer-Verlag.

Practical Medical Microbiology for Clinicians, First Edition. Frank E. Berkowitz and Robert C. Jerris.

Leventhal R, Cheadle RF (2002) *Medical Parasitology. A Self-Instructional Text*, 5th edn. Philadelphia: FA Davis.
Long SS, Pickering LK, Prober CG (eds) (2012) *Principles and Practice of Pediatric Infectious Diseases*, 4th edn. Philadelphia: Elsevier Saunders.
Mahon CR, Lehman DC, Manuselis G (2011) *Textbook of Diagnostic Microbiology*, 4th edn. Philadelphia: Elsevier Saunders.
Mandell GL, Bennett JE, Dolin R (eds) (2010) *Mandell, Douglas, and Bennett's Principles and Practice of Infectious Diseases*, 7th edn. Philadelphia: Churchill Livingston Elsevier.
Murray PR, Rosenthal KS, Pfaller MA (2009) *Medical Microbiology*, 6th edn. St Louis: Mosby Elsevier.
Pickering LK (ed) (2012) *Red Book: 2012 Report of the Committee on Infectious Diseases*, 29th edn. Elk Grove Village: American Academy of Pediatrics.
Visvesvara GS (2002) In vitro cultivation of Microsporidia of clinical importance. Clin Microbiol Rev 15: 401–13.
Winn W Jr, Allen S, Janda WM, et al. (2006) *Koneman's Color Atlas and Textbook of Diagnostic Microbiology*, 6th edn. Philadelphia: Lippincott Williams and Wilkins.

Specific references for Tables A2.7 and A2.10

DuPont HL (2014) Acute infectious diarrhea in immunocompetent adults. N Engl J Med 370: 1532–40.
Hodges K, Gill R (2010) Infectious diarrhea. Cellular and molecular mechanisms. Gut Microbes 1: 4–21.
Workowski K, Berman S (2010) Sexually transmitted diseases treatment guidelines, 2010. MMWR 59(RR12): 1–109.

Index

Note: Page numbers in *italics* refer to Figures; those in **bold** to Tables.

Practical Medical Microbiology for Clinicians, First Edition. Frank E. Berkowitz and Robert C. Jerris.

Printed and bound by CPI Group (UK) Ltd, Croydon, CR0 4YY
17/09/2026
14964291-0001